HOW SCIENCE WORKS

HOW SCIENCE WORKS

DK | Penguin Random House

Editorial Consultants
Robert Dinwiddie, Hilary Lamb, Professor Donald R. Franceschetti, Professor Mark Viney

Project Art Editors
Francis Wong, Clare Joyce, Duncan Turner, Steve Woosnam-Savage

Designer
Gregory McCarthy

Illustrators
Edwood Burn, Dominic Clifford, Mark Clifton, Phil Gamble, Gus Scott

Managing Art Editor
Michael Duffy

Jacket Designer
Suhita Dharamjit

Senior Jacket Designer
Mark Cavanagh

Sr. DTP Designer
Harish Aggarwal

Producer, Pre-production
David Almond

Senior Producer
Alex Bell

Producer
Anna Vallarino

Art Director
Karen Self

Contributors
Derek Harvey, Tom Jackson, Ginny Smith, Alison Sturgeon, John Woodward

Senior Editors
Peter Frances, Rob Houston

Project Editors
Lili Bryant, Martyn Page, Miezan van Zyl

Editors
Claire Gell, Nathan Joyce, Francesco Piscitelli

Managing Editor
Angeles Gavira Guerrero

US Editor
Kayla Dugger

Jacket Editor
Claire Gell

Jackets Design Development Manager
Sophia MTT

Jackets Editorial Coordinator
Priyanka Sharma

Managing Jackets Editor
Saloni Singh

Publisher
Liz Wheeler

Publishing Director
Jonathan Metcalf

First American Edition, 2018
Published in the United States by DK Publishing
345 Hudson Street, New York, New York 10014

18 19 20 21 22 10 9 8 7 6 5 4 3 2 1
001–299756–Mar/2018

Published in Great Britain by Dorling Kindersley Limited

A catalog record for this book is available from the Library of Congress.

ISBN 978-1-4654-6419-4

DK books are available at special discounts when purchased in bulk for sales promotions, premiums, fund-raising, or educational use. For details, contact: DK Publishing Special Markets, 345 Hudson Street, New York, New York 10014

SpecialSales@dk.com

Printed and bound in China

A WORLD OF IDEAS:
SEE ALL THERE IS TO KNOW

www.dk.com

CONTENTS

MATTER

ENERGY AND FORCES

LIFE

SPACE

EARTH

What makes science special?

Science isn't just a collection of facts—it is a systematic way of thinking based on evidence and logic. While it may not be perfect, it is the best way we have of understanding our Universe.

What is science?

Science is a way of finding out about and understanding the natural and social world, and applying the information gained. It is constantly updating information and changing our understanding of the world. Science is based on measurable evidence and must follow logical steps in generalizing that evidence and using it to make further predictions. The word "science" is also used to describe the body of knowledge we have accumulated using this process.

The scientific method

The scientific method varies from one discipline to another, but it generally involves: generating and testing a hypothesis; using data gathered through experiments to update and refine the hypothesis; and, hopefully, reaching a generalizable theory to explain why the hypothesis is true. To be confident in the data, it is important for experiments to be repeated, preferably in different labs. If different results are found the second time, the result might not be as reliable or generalizable as originally thought.

1 OBSERVATION

Science often starts with observations about the world—whether of unusual phenomena only seen under laboratory conditions or of everyday effects, such as noticing that peaches go bad faster in a fruit bowl than when refrigerated.

2 QUESTION

These observations are turned into questions—for example, a scientist might want to discover why a certain bacterium grows better in one medium than another or why peaches go bad faster in the fruit bowl.

3 RESEARCH

Researching the topic reveals whether others have asked (and answered) the same question. Related work may spark ideas—for example, maybe someone has studied the ripening of fruits other than peaches.

10 PEER REVIEW PUBLICATION

Papers scientists write about their findings are reviewed by other experts looking for problems in the experiment's method or the conclusions drawn from it. If accepted, the study is published, becoming available for others to read.

An ongoing process

Science is never finished. New data is constantly generated, and theories must be refined to include this information. Scientists understand that their work will probably be superseded by future experiments.

4 FORMULATE HYPOTHESIS

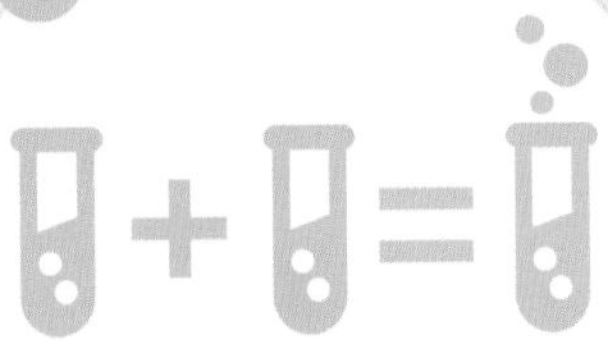

The next stage is to create a testable hypothesis—a prediction as to what is causing the occurrence. A hypothesis could be "the cooler temperature in the refrigerator stops peaches from spoiling."

5 DEVELOP TESTABLE PREDICTIONS

Predictions must follow logically from a hypothesis, be specific, and be testable experimentally. For example: "If temperature affects peach ripening, a peach kept at 72°F (22°C) will spoil faster than one kept at 46°F (8°C)."

6 GATHER EXPERIMENTAL DATA

Data is gathered to see if it is consistent with the hypothesis. Experiments must be designed carefully to attempt to ensure that there are no explanations for the outcome other than the one you are interested in.

7 ANALYZE DATA

The findings of an experiment must be analyzed statistically to ensure they aren't just the result of random fluctuation. To make this less likely, experiments should use as large a sample size as is feasible.

8 IS HYPOTHESIS SUPPORTED?

If the results agree with the predictions, confidence in the hypothesis increases. We can't ever prove a hypothesis, as future experiments could refute it, but the more experimental support we have, the more confident we can be.

9 REFINE, ALTER, OR REJECT

If the initial result of the experiment does not entirely agree with the predictions, there might be an indication as to why that is, and you can start the process again by refining your hypothesis, altering it, or rejecting it and formulating a new hypothesis.

IMPORTANT TERMS

HYPOTHESIS
A hypothesis is a potential explanation for an observation, based on current knowledge. To be scientific, it must be falsifiable.

THEORY
Theories are ways of explaining known facts. They are developed from many related hypotheses and supported by evidence.

LAW
A law doesn't explain anything; it simply describes something we have observed to be true every time it is tested.

FEATURES OF A HYPOTHESIS

SCOPE
Hypotheses with wide scopes explain a range of phenomena; hypotheses with narrow scopes may only explain one specific example.

TESTABLE
It must be possible to test a hypothesis. Unless it can be supported by evidence, a hypothesis should be rejected.

FALSIFIABLE
It must be possible to prove a hypothesis wrong. "Ghosts exist" is not scientific because no experiment can falsify it.

MATTER

What is matter?

Generally, matter is anything that takes up space and has mass. This means it is distinct from energy, light, or sound, which have neither property.

The structure of matter

At the most fundamental level, matter is made up of elementary particles, such as quarks and electrons. Combinations of elementary particles form atoms, which may sometimes be bonded together into molecules. The types of atom that make up the matter determine its properties. If the atoms or molecules form strong bonds with each other, the material is solid at room temperature. Weaker bonds lead to liquids or gases.

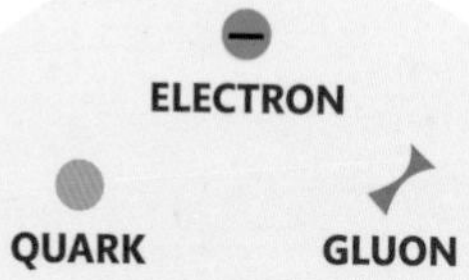

Elementary particles
Elementary particles known as quarks make up protons and neutrons in atoms. Gluons hold the quarks together in the nucleus. Along with electrons, quarks and gluons make up all known matter.

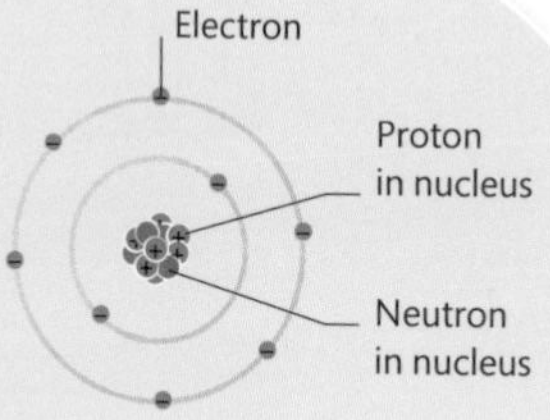

Atom
Atoms contain protons and neutrons in the nucleus with electrons orbiting around. Atoms of different elements have different numbers of protons in their nuclei.

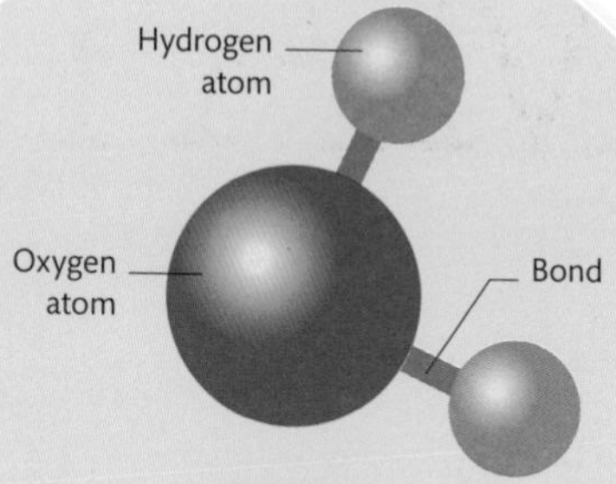

Molecule
Molecules can be made of different atoms, like water, which consists of two hydrogen atoms and one oxygen atom, or of identical atoms, like an oxygen molecule, which consists of two oxygen atoms.

States of matter

The main states of matter encountered in everyday life are solid, liquid, and gas. Other, more unusual states also exist, when matter becomes extremely cold or hot. Matter can change between the various states, depending on how much energy it has and the strength of the bonds between its constituent atoms or molecules. For example, aluminum has a lower melting point than copper because it has weaker bonds between its atoms.

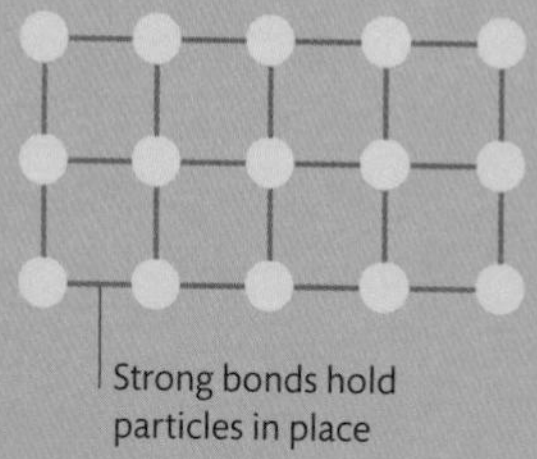

Solid
Atoms or molecules in solids are held in a rigid structure with strong bonds. The particles cannot move, so solids feel hard and keep their shape.

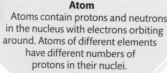

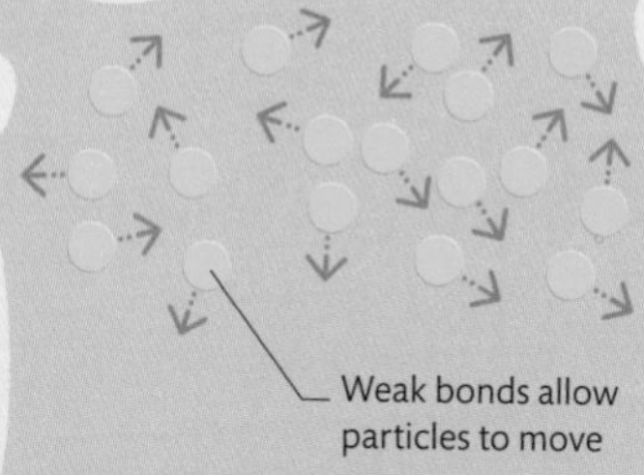

Liquid
Atoms or molecules in liquids have only weak bonds between them, so the particles can move around. This means liquids can flow, but tight packing of the particles prevents them from being compressed.

Mixtures and compounds

Atoms can combine in a huge variety of ways to produce different types of matter. When atoms are bonded together chemically, compounds are formed. Examples include water, a compound formed from oxygen and hydrogen. However, many atoms and molecules do not form bonds easily with others, so combining them does not change them chemically—we call what results a mixture. Mixtures include sand and salt, or air, which is a mixture of gases.

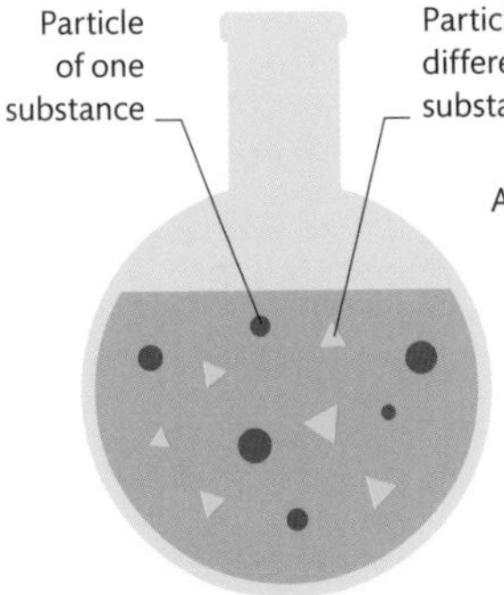

Mixture
In mixtures, the original chemicals are not changed, so they can be separated out again physically—for example, by sieving, filtering, or distillation.

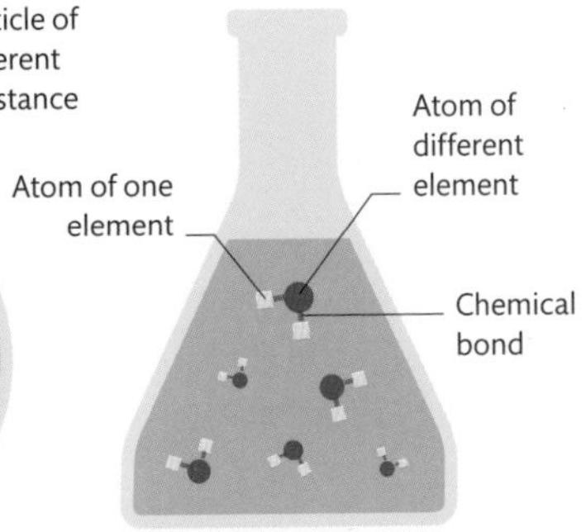

Compound
When atoms or molecules react, they form a new compound. They cannot be returned to their original forms physically; separating them requires breaking chemical bonds.

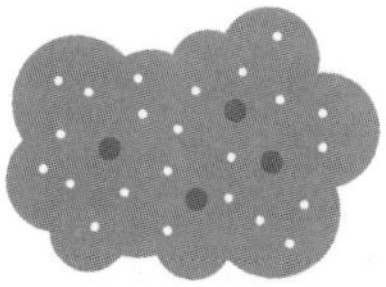

ABOUT **99 PERCENT OF ALL MATTER** IN THE UNIVERSE IS IN THE FORM OF **PLASMA**

THE CONSERVATION OF MASS

During most ordinary chemical reactions or physical changes (such as when a candle burns), the mass of the products equals the mass of the reactants. No matter is lost or gained. However, this "law" can be broken in certain extreme conditions, such as nuclear fusion reactions (see p.37), in which mass is converted into energy.

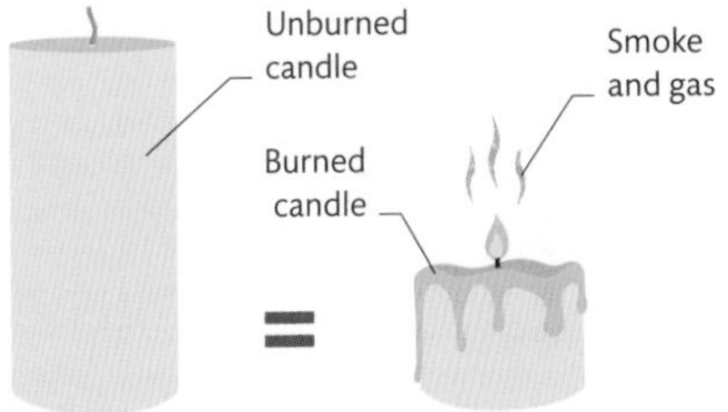

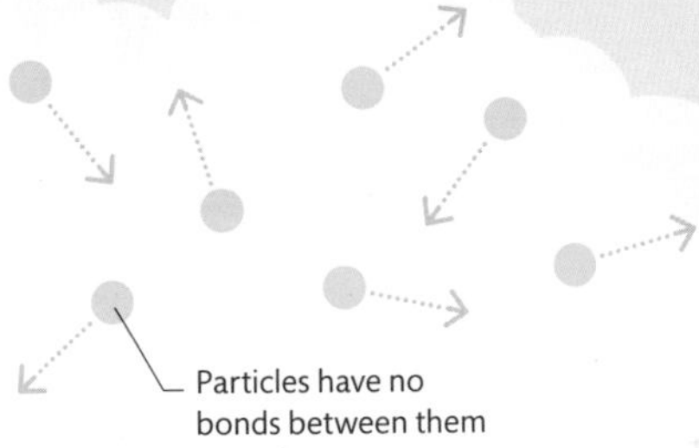

Gas
There are no bonds between the atoms or molecules in a gas, so they can spread out and fill their container. The particles are also far apart, so a gas can be compressed, although doing so increases the pressure.

HIGH AND LOW TEMPERATURE STATES

At very high temperatures, gas atoms split into ions (see p.40) and electrons, becoming plasma, which can conduct electricity. At low temperatures, Bose-Einstein condensates may form (see p.22), changing the properties of matter dramatically. In this state, the atoms start acting strangely, behaving like one single atom.

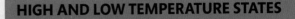

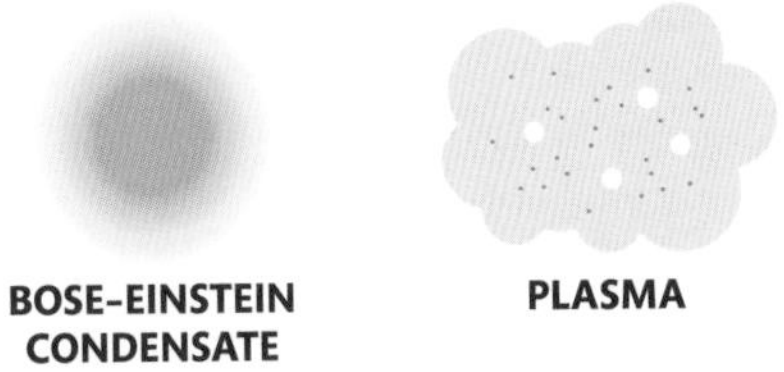

Solids

A solid is the most ordered state of matter. All of the atoms or molecules in a solid are connected together to form an object with a fixed shape and a fixed volume (although the shape can be altered by applying force). However, solids encompass a diverse group of materials, and other properties can vary greatly, depending on the exact solid involved.

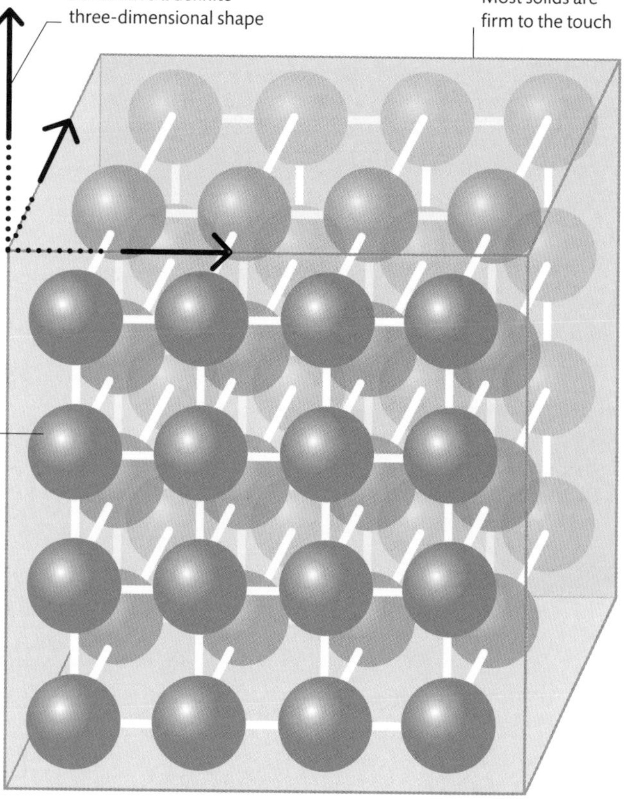

What is a solid?

Solids are firm to the touch and have a definite shape, rather than taking on the shape of their containers like liquids or gases. The atoms in solids are packed tightly together, so they cannot be compressed into a smaller volume. Some solids, like sponges, can be squashed, but that is because air is squeezed out of pockets in the material—the solid itself does not change size.

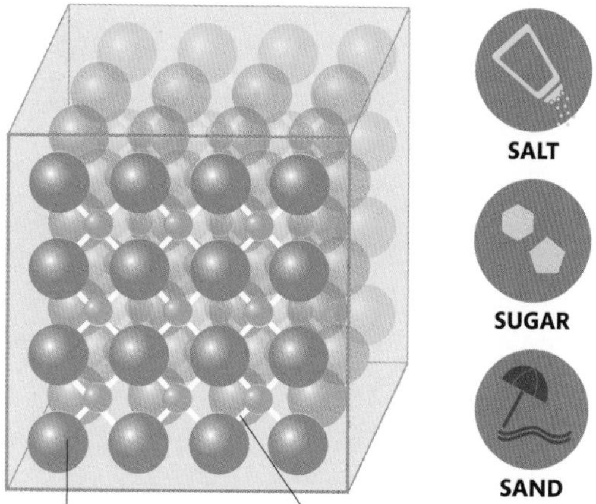

Crystalline solids
The atoms or molecules in crystalline solids are arranged in a regular pattern. Some substances, such as diamond (a crystalline form of carbon), form one large crystal. However, most are made up of lots of smaller crystals.

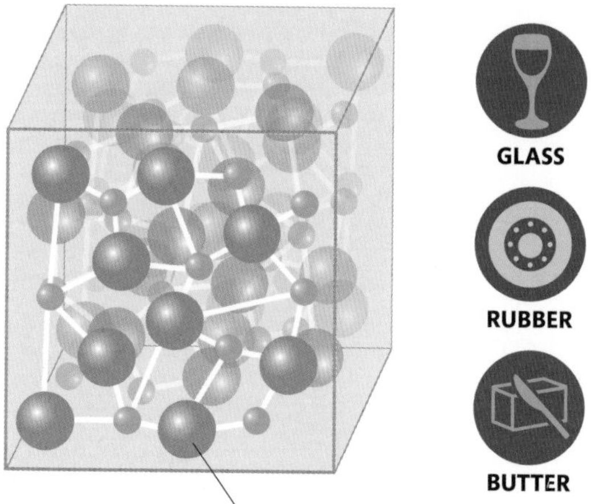

Amorphous solids
Unlike in crystalline solids, the atoms or molecules that make up amorphous solids are not arranged in a regular pattern. Instead, they are arranged more like those in a liquid, although they are unable to move around.

The properties of solids

Solids have a wide variety of properties; for example, they may be strong or weak, hard or relatively soft, and may return to their original shape after having been subjected to force or may be permanently deformed. A solid material's properties depend on the atoms or molecules that make it up, whether the solid is crystalline or amorphous, and whether or not there are defects in the material.

LONSDALEITE, A RARE FORM OF DIAMOND, IS THE HARDEST SOLID KNOWN, ALMOST 60 PERCENT HARDER THAN NORMAL DIAMONDS

Brittle
When under stress, brittle solids such as ceramics snap without changing shape much. Cracks pass easily through these materials because the atoms cannot move to absorb the stress. If the material can deform, it will be less brittle but also less hard.

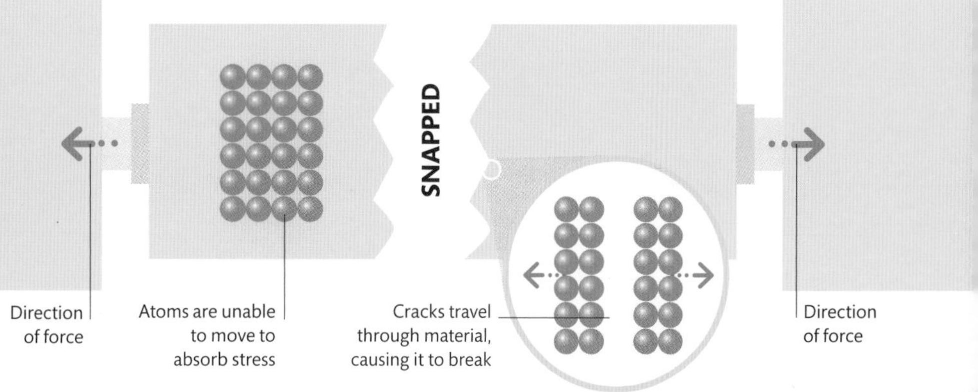
SNAPPED
Direction of force
Atoms are unable to move to absorb stress
Cracks travel through material, causing it to break
Direction of force

Ductile
Ductile materials change shape when stretched, so they can be drawn out into long wires. This type of deformation, in which a material changes shape permanently, is known as plastic deformation. Many metals are ductile because the bonds between the atoms allow the atoms to slide over each other.

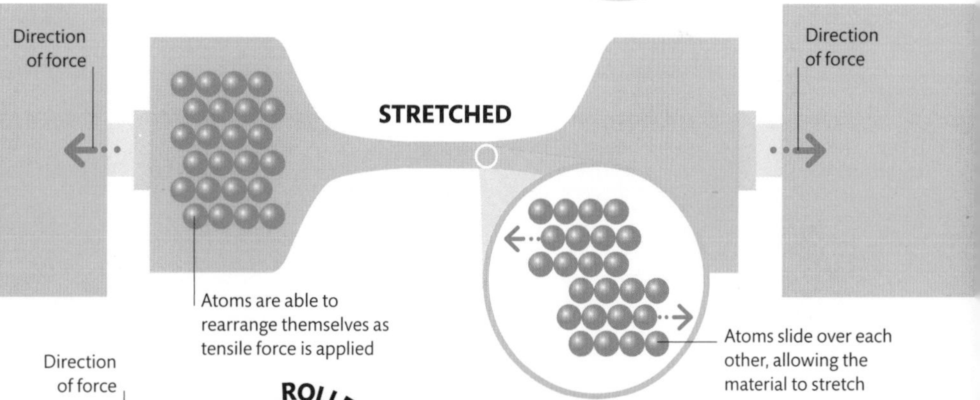
Direction of force
STRETCHED
Direction of force
Atoms are able to rearrange themselves as tensile force is applied
Atoms slide over each other, allowing the material to stretch

Malleable
Malleable solids can deform plastically when compressed. As a result, they can be flattened into sheets by rolling or hammering. Many malleable materials are also ductile, although the two properties do not always occur together; for example, lead is highly malleable but has low ductility.

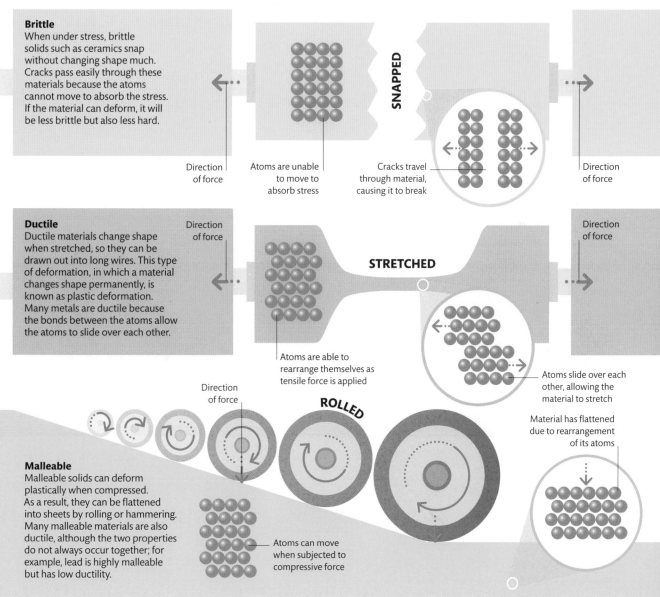

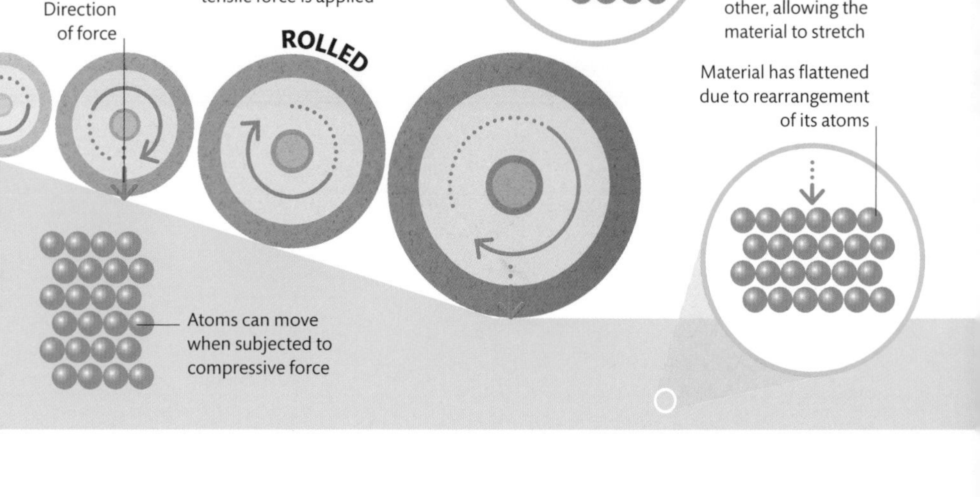
Direction of force
ROLLED
Material has flattened due to rearrangement of its atoms
Atoms can move when subjected to compressive force

Liquids

In liquids, the atoms or molecules are closely packed together. The bonds between them are stronger than in gases but weaker than in solids, allowing the particles to move freely.

Wetting

Wetting is the degree to which a liquid keeps contact with a solid surface. Whether a liquid wets a surface depends on the strength of the attractive forces within the liquid relative to the forces between the liquid and the surface.

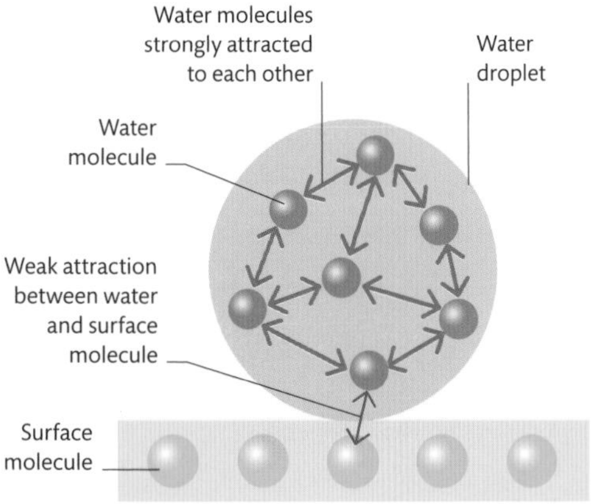

Water molecules strongly attracted to each other
Water droplet
Water molecule
Weak attraction between water and surface molecule
Surface molecule

No wetting
On waterproof surfaces, water forms droplets because the water molecules are less strongly attracted to the surface molecules than to one another.

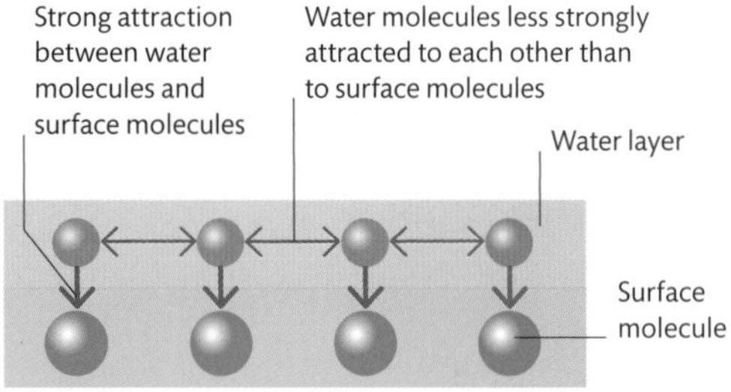

Strong attraction between water molecules and surface molecules
Water molecules less strongly attracted to each other than to surface molecules
Water layer
Surface molecule

Wetting
Water wets a surface—forms a layer on it—when the water molecules are more strongly attracted to the surface molecules than to other water molecules.

WHAT IS THE MOST VISCOUS LIQUID?

Pitch, used for road surfaces, is the most viscous liquid known. It is about 20 billion times more viscous than water at the same temperature.

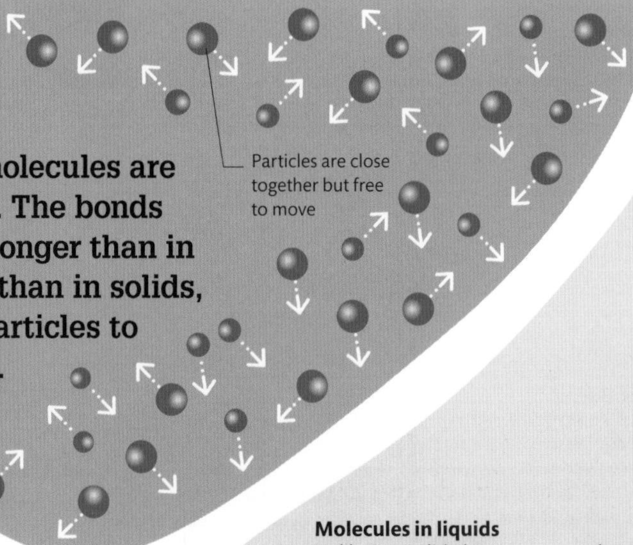

Particles are close together but free to move

Free flow

Liquids flow and take the shape of their containers. The atoms or molecules are close together, which means that liquids cannot be compressed. The density of liquids is higher than that of gases and is typically similar to, or slightly lower than, solids, except in the case of water (see pp.56–57).

Molecules in liquids
Unlike in a solid, the atoms or molecules in liquids are arranged randomly. There are bonds between the particles, but they are weak and continually break and reform as the particles move past each other.

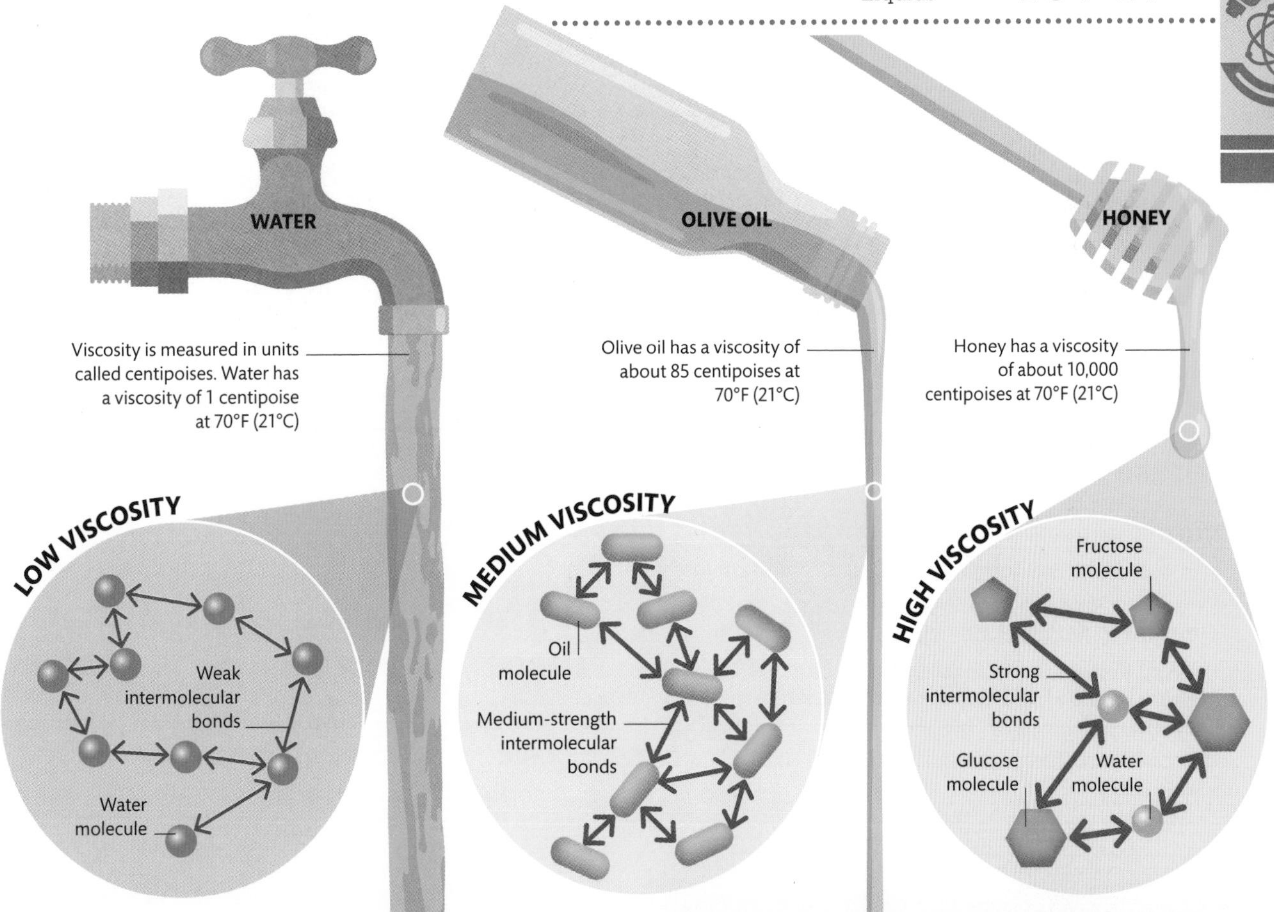
WATER
OLIVE OIL
HONEY
Viscosity is measured in units called centipoises. Water has a viscosity of 1 centipoise at 70°F (21°C)
Olive oil has a viscosity of about 85 centipoises at 70°F (21°C)
Honey has a viscosity of about 10,000 centipoises at 70°F (21°C)
LOW VISCOSITY
Weak intermolecular bonds
Water molecule
MEDIUM VISCOSITY
Oil molecule
Medium-strength intermolecular bonds
HIGH VISCOSITY
Fructose molecule
Strong intermolecular bonds
Glucose molecule
Water molecule

Liquid flow
Liquids with low viscosity, such as water, flow easily, because the bonds between the molecules are weak. In contrast, honey flows much less readily at the same temperature, due to the strength of its intermolecular bonds.

Viscosity

Viscosity is a measure of how easily a liquid flows. A liquid with low viscosity flows easily and is commonly said to be "thin," while a "thick," high-viscosity liquid flows less readily. Viscosity is determined by bonds between the liquid's molecules—the stronger the bonds, the more viscous the liquid. Increasing the temperature of a liquid decreases its viscosity, because the molecules have more energy to overcome the intermolecular bonds.

NON-NEWTONIAN LIQUIDS

Unlike Newtonian liquids, such as water, the viscosity of non-Newtonian fluids varies, depending on the force applied. For example, a cornflour and water mixture becomes thicker when a large force is applied, so a ball dropped into it from a great height will bounce off the surface while one dropped from a low height will sink.

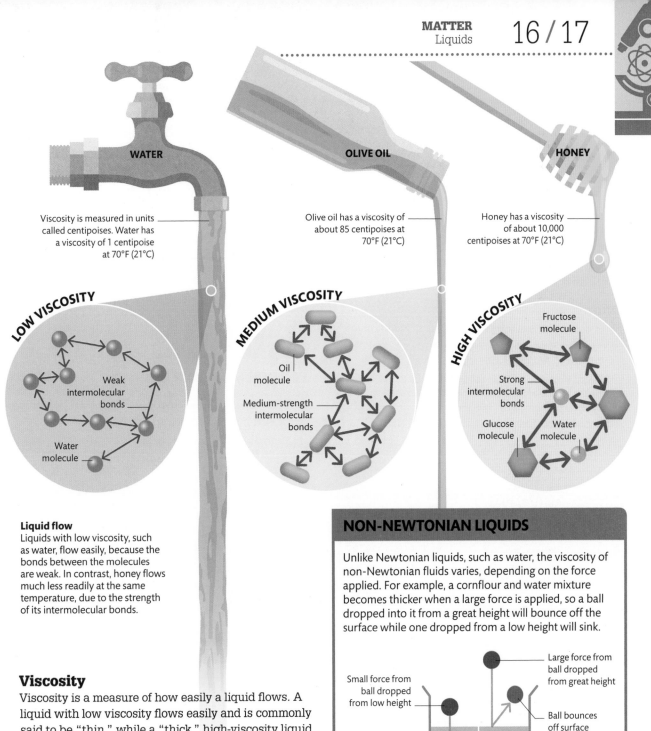

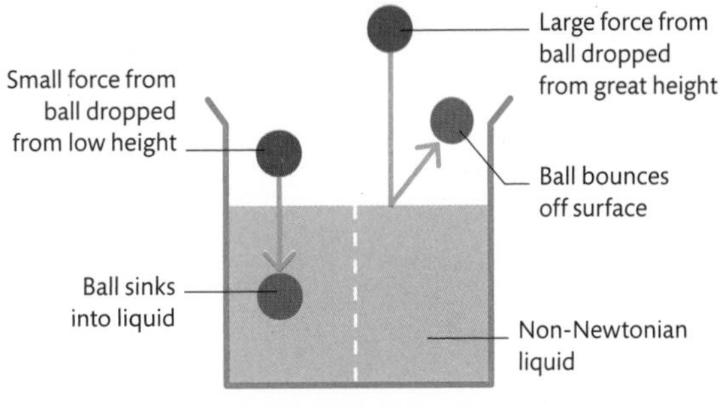
Large force from ball dropped from great height
Small force from ball dropped from low height
Ball bounces off surface
Ball sinks into liquid
Non-Newtonian liquid

NON-NEWTONIAN LIQUID

Gases

Gases are all around us but most of the time we don't give them much thought. However, along with solids and liquids, gases are one of the main states of matter and the way they behave is vital for life on Earth. For example, when we breathe in, we increase the volume of our lungs, which reduces the pressure inside and causes air to rush in.

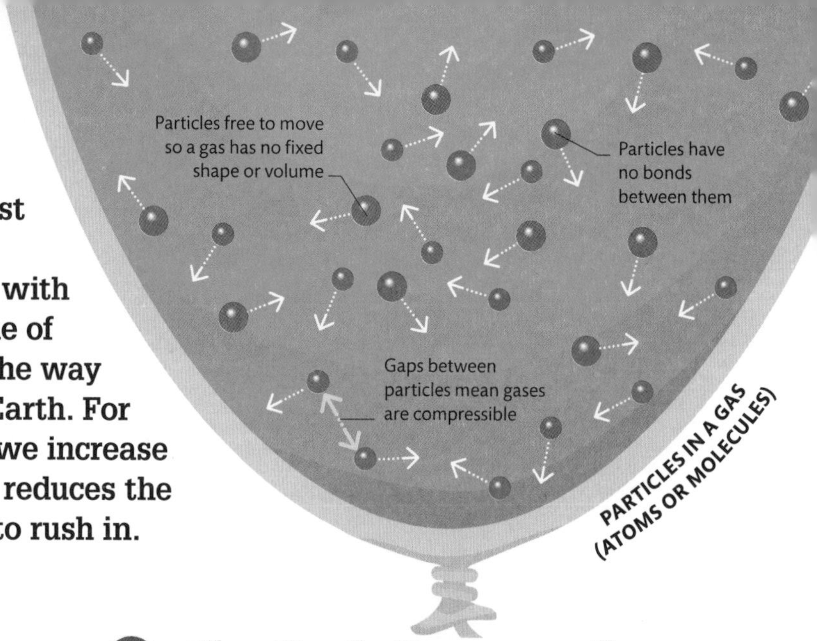

Particles free to move so a gas has no fixed shape or volume
Particles have no bonds between them
Gaps between particles mean gases are compressible
PARTICLES IN A GAS (ATOMS OR MOLECULES)

What is a gas?

Gases can be composed of individual atoms or of molecules of two or more atoms. These particles are very energetic and move rapidly, filling their container and taking its shape. There is plenty of space between the particles, so gases can be compressed.

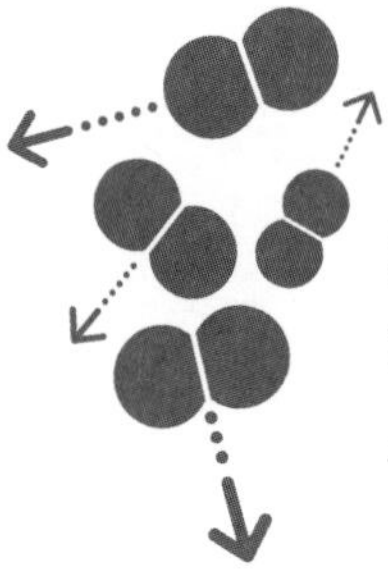

1,060mph

(1,700KPH)—THE SPEED AT WHICH OXYGEN MOLECULES MOVE AROUND AT ROOM TEMPERATURE

How gases behave

The way gases behave is described by a set of three gas laws. They relate a gas's volume, pressure, and temperature and show how each measure changes when the others change. The laws assume that all gases behave like an "ideal" gas. In such a gas, there are no interactions between individual gas particles, the particles move randomly, and they take up no space. Even though no real gas has these features, the gas laws do show how most gases behave at normal temperatures and pressures.

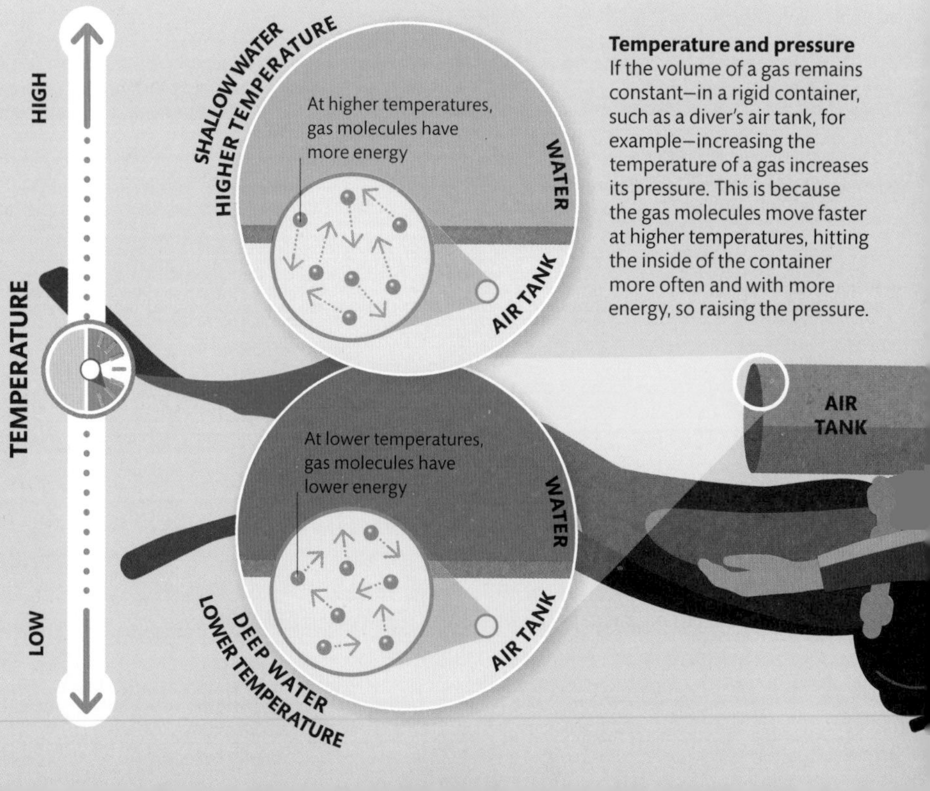

HIGH
TEMPERATURE
LOW
SHALLOW WATER HIGHER TEMPERATURE
At higher temperatures, gas molecules have more energy
WATER
AIR TANK
At lower temperatures, gas molecules have lower energy
WATER
AIR TANK
DEEP WATER LOWER TEMPERATURE
AIR TANK

Temperature and pressure

If the volume of a gas remains constant—in a rigid container, such as a diver's air tank, for example—increasing the temperature of a gas increases its pressure. This is because the gas molecules move faster at higher temperatures, hitting the inside of the container more often and with more energy, so raising the pressure.

AVOGADRO'S LAW

Avogadro's law states that, at the same temperature and pressure, equal volumes of all gases contain the same number of molecules. For example, even though molecules of chlorine gas have a mass about twice that of oxygen, there will be the same number of molecules of each in containers of the same size and at the same temperature and pressure.

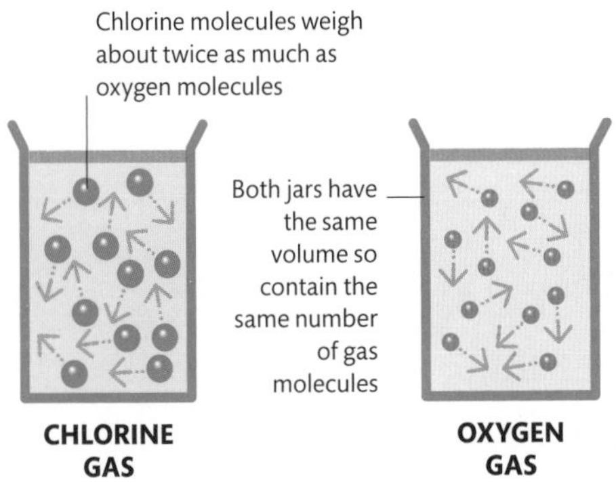
Chlorine molecules weigh about twice as much as oxygen molecules
Both jars have the same volume so contain the same number of gas molecules
CHLORINE GAS
OXYGEN GAS

Temperature and volume
If the volume of a gas is not restricted (by a rigid container, for example), a gas will expand as it is heated and the gas molecules gain more energy. The higher the temperature of the gas, the greater its volume. For example, if the air in an inflated dinghy is heated by the Sun, the air will expand and inflate the dinghy more.

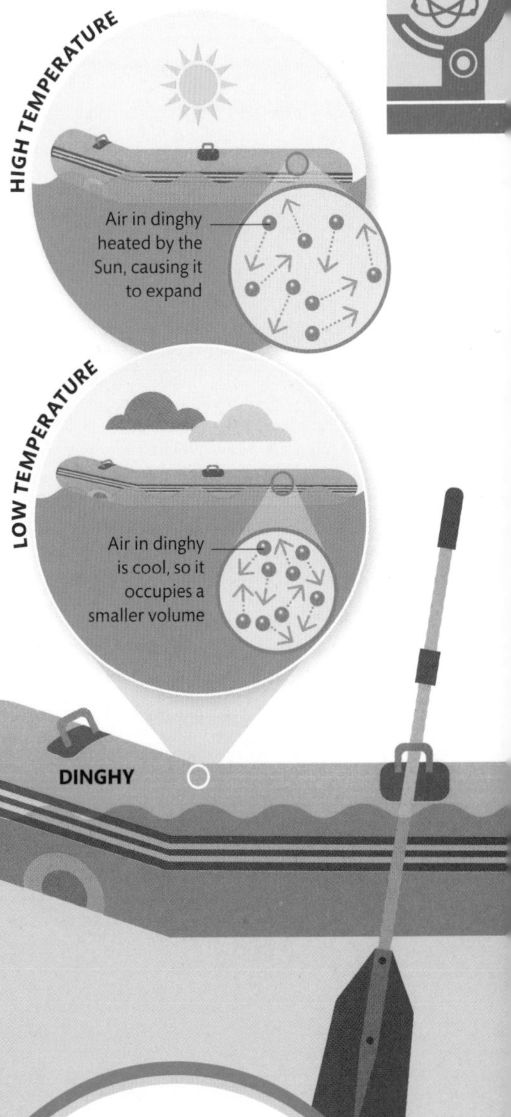
HIGH TEMPERATURE
Air in dinghy heated by the Sun, causing it to expand
LOW TEMPERATURE
Air in dinghy is cool, so it occupies a smaller volume
DINGHY

Pressure and volume
If the temperature of a gas remains constant, increasing the pressure on the gas reduces its volume. Conversely, reducing the pressure on a gas increases its volume. That is why bubbles expand as they rise to the surface of a liquid.

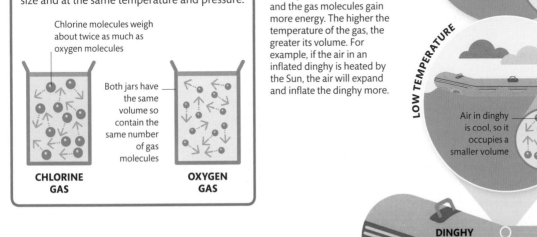

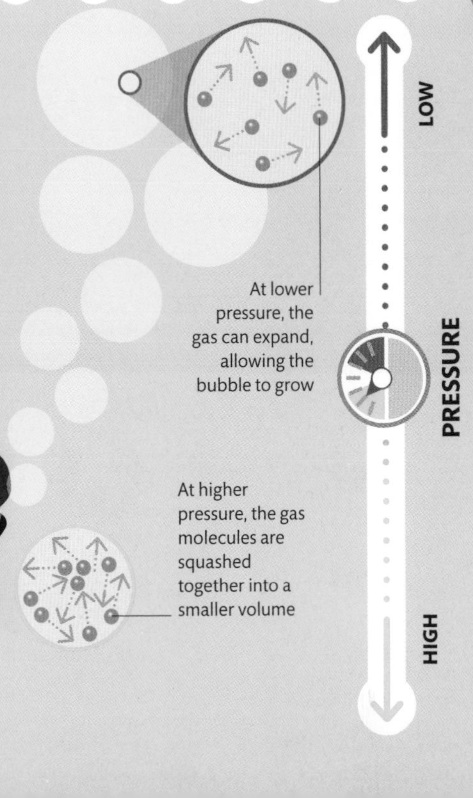
LOW
At lower pressure, the gas can expand, allowing the bubble to grow
PRESSURE
At higher pressure, the gas molecules are squashed together into a smaller volume
HIGH

WHY CAN'T WE SEE AIR?

Something is visible only if it affects light, by reflecting it, for example. Air affects light only slightly, so is usually invisible. But large amounts of air scatter blue light noticeably, which is why the sky looks blue.

Strange states

Solids, liquids, and gases are the most familiar states of matter, but they are not the only ones that exist. Superheated gases can become plasmas, consisting of high-energy charged particles that conduct electricity. At extremely low temperatures, some substances can become superconductors or superfluids, which have strange properties, such as zero electrical resistance or viscosity.

Where to find plasma

Plasma is common in the Sun. Natural plasma is rare on Earth, although it does occur in lightning and the aurorae, or northern and southern lights. Plasma can be created artificially by passing electricity through a gas, as occurs in arc welding and neon lights, for example.

Stars
In stars like the Sun, it is so hot that hydrogen and helium, which form most of the star's mass, ionize and take the form of plasma.

Aurorae
When plasma from the Sun reaches Earth, it interacts with the atmosphere, creating light shows in polar regions.

Lightning
Lightning bolts are visible trails of plasma left by the passage of electric charge from a thundercloud to the ground.

Neon lights
Electricity heats neon inside the light, causing it to form a plasma. Excited by the electric current, the plasma emits light.

Plasma arc welding
Electricity is used to create a jet of plasma that can reach around 50,000°F (28,000°C), high enough to melt metal.

Plasma

At normal temperatures and pressures, gases exist as atoms (made up of a nucleus of protons and neutrons orbited by electrons) or molecules. Plasmas are created by breaking up the atoms or molecules into their negatively charged electrons and positively charged nuclei or ions (see p.40). This can be achieved by heating a gas to a very high temperature or by passing an electric current through it.

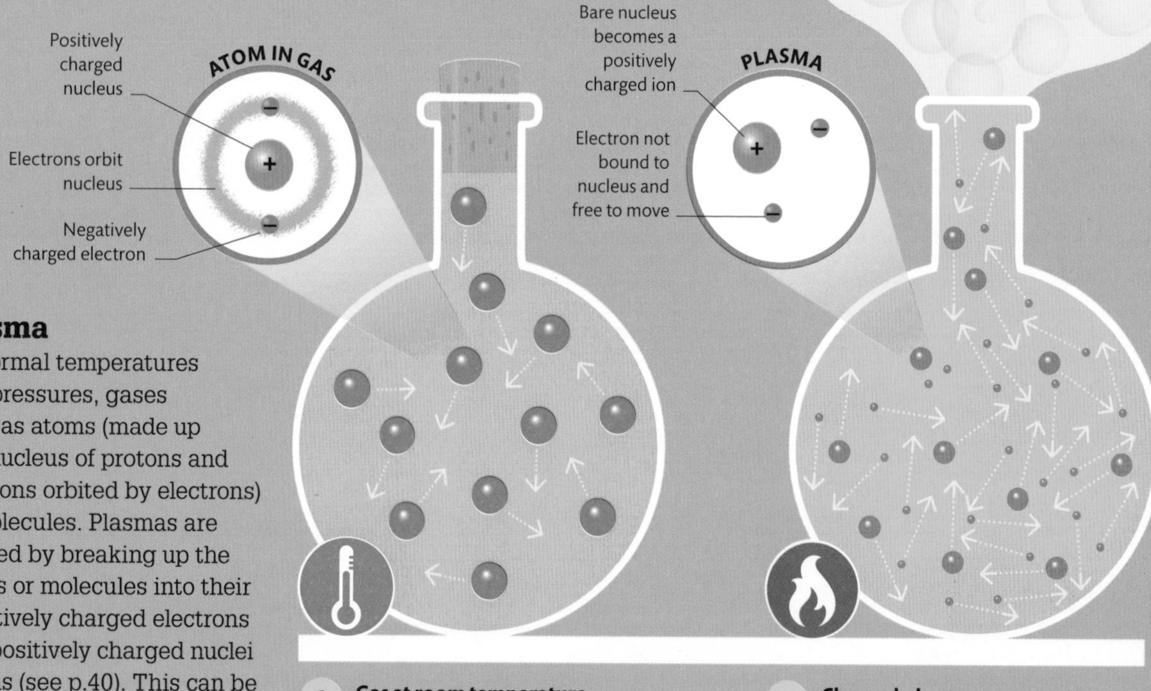

1 Gas at room temperature
In a gas at normal room temperature, negatively charged electrons orbit the nucleus of each atom and balance out the positive charge of the protons in each atomic nucleus. As a result, the atoms are neutral.

2 Charged plasma
In plasma, electrons have been ripped away from atoms, leaving negatively charged electrons and positively charged nuclei (ions). These electrons and ions can move freely, so plasma can conduct electricity.

Superconductors and superfluids

At temperatures below about 130 K (-226°F/-143°C), some materials become superconductors—they allow electricity to flow through them with no resistance. At even lower temperatures, the most common isotope (see p.34) of helium, helium-4, becomes a superfluid. Its viscosity drops to zero, and it flows with no resistance. At temperatures close to absolute zero (0 K/-459.67°F/-273.15°C), some substances form a strange state known as a Bose–Einstein condensate (see p.22). Normally, each atom in a substance behaves as an individual, but in a Bose–Einstein condensate, all the atoms act as one giant atom.

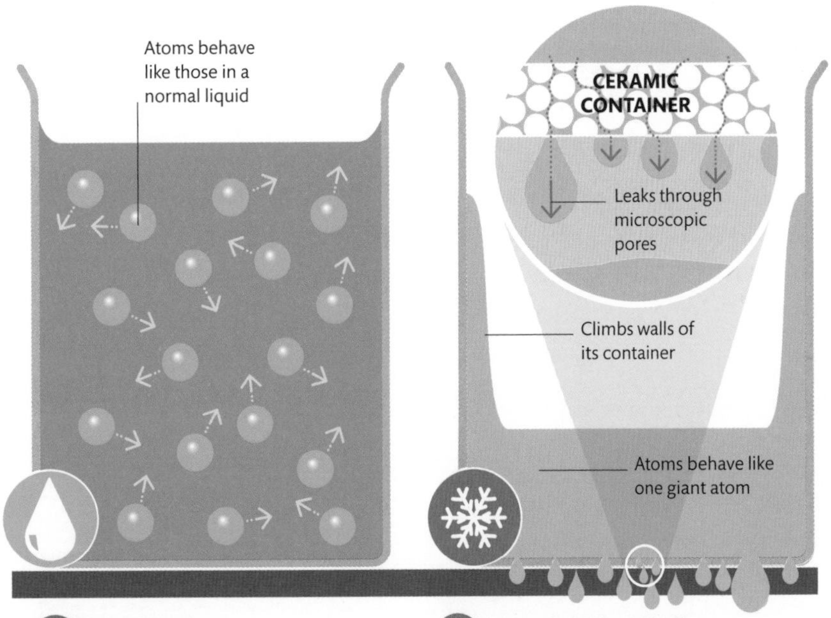
Atoms behave like those in a normal liquid
CERAMIC CONTAINER
Leaks through microscopic pores
Climbs walls of its container
Atoms behave like one giant atom

1 Liquid helium
At normal atmospheric pressure, helium-4 liquefies at about 4 K (-452°F/-269°C). At this temperature, it behaves like any other liquid, flowing to fill a container and staying in the container.

2 Superfluid liquid helium
At about 2 K (-456°F/-271°C), helium-4 becomes a superfluid. At this temperature, it exhibits strange behavior, such as flowing through microscopic pores in solid objects and climbing the walls of its container.

Uses of superconductors

Superconductors are mainly used to make extremely powerful electromagnets, which are vital for applications such as magnetic resonance imaging (MRI) scanners, maglev trains, and particle accelerators for investigating the structure of matter.

MRI scanner
Superconducting magnets are used in MRI scanners to help produce detailed images of body tissues, such as the brain.

Particle accelerators
Some particle accelerators rely on the great power of superconducting magnets to guide particles around the accelerators.

E-bombs
Superconductors are used in E-bombs to produce a powerful electromagnetic pulse that disables nearby electronic equipment.

Maglev trains
High-speed maglev trains use superconducting electromagnets to levitate the trains and also provide forward propulsion.

SUPERFLUID HELIUM WOULD SPIN FOREVER IF STIRRED

MEISSNER EFFECT

Superconductors do not allow magnetic fields to pass through them. In fact, they expel magnetic fields, a phenomenon known as the Meissner effect. If a magnet is placed above a superconducting material cooled to its critical temperature (the temperature at which the material becomes superconducting), the superconductor will repel the magnet, causing it to levitate.

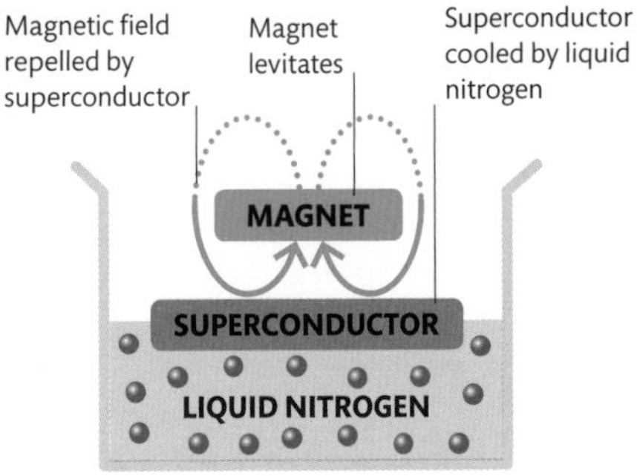
Magnetic field repelled by superconductor
Magnet levitates
Superconductor cooled by liquid nitrogen
MAGNET
SUPERCONDUCTOR
LIQUID NITROGEN

Transforming matter

Solid, liquid, gas, and plasma are the best-known states of matter, but there is also another, weird state known as the Bose–Einstein condensate. Transforming matter from one state to another involves adding or removing energy.

Gaining energy

When a substance gains energy, its particles (atoms or molecules) can vibrate or move more freely. If enough energy is added, bonds between particles in solids or liquids can be broken, changing the state of the substance. In a gas, the energy may separate the electrons from the particles, forming a plasma.

32.02°F (0.01°C) IS THE **TRIPLE POINT OF WATER,** AT WHICH IT CAN BE **SOLID, LIQUID, AND GAS** AT THE SAME TIME

SUBLIMATION

Some solids, such as frozen carbon dioxide ("dry ice"), go straight from the solid phase to a gas. Any substance can sublime under the right conditions of temperature and pressure, but sublimation is relatively rare under normal conditions.

MELTING

As the energy of a solid substance increases, the bonds holding the particles together vibrate more. Eventually the bonds break, and the substance becomes liquid. Its particles are still attracted to each other but can now move more freely.

LIQUID

In a liquid, the atoms or molecules are less tightly bonded together than in a solid and are free to flow.

ENERGY LEVEL

LOW

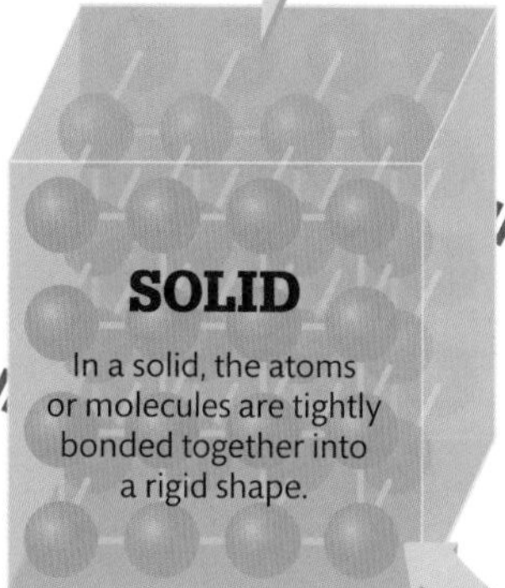

SOLID

In a solid, the atoms or molecules are tightly bonded together into a rigid shape.

BOSE–EINSTEIN CONDENSATE

A strange state of matter in which the atoms have so little energy they act as if all of them were everywhere at once, like one single atom. Most substances do not form Bose–Einstein condensates.

FREEZING

As a liquid loses energy, its atoms or molecules slow down and attractive forces between the particles pull them closer together. The particles may become arranged in an ordered way, forming a crystal, or more randomly, forming an amorphous solid.

SUPERCOOLING

Cooling the gaseous forms of some substances to within a few millionths of a degree above absolute zero (0K/-459.67°F/ -273.15°C) reduces the energy of the atoms so much that they are almost motionless and they all clump together.

IONIZATION

At high energy, electrons separate from their atoms or molecules, producing a plasma. Consisting of negatively charged electrons and positively charged ions (the atoms or molecules that have lost electrons), plasma is found in stars, neon lights, and plasma displays.

EVAPORATION

Even at low temperatures, some surface particles in liquids have enough energy to escape from the liquid as vapor. The more energy, the more evaporation. At a substance's boiling point, even nonsurface particles have enough energy to escape as vapor.

PLASMA

Sometimes called the fourth state of matter, plasma consists of a cloud of free electrons and positively charged ions.

HIGH

GAS

In a gas, the atoms or molecules move freely because there are no bonds between them.

RECOMBINATION

Recombination is when plasma changes back to gas. As the energy level of the plasma falls, the positive ions recapture the free electrons and the substance turns back into a gas—as happens, for example, when the power to a neon light is turned off.

CONDENSATION

The opposite process to evaporation, condensation occurs when the temperature falls and gas atoms or molecules lose energy to their surroundings. The gas particles move more slowly, and the gas condenses into liquid.

Losing energy

As a substance loses energy, its atoms or molecules move more slowly. With a large energy loss, the substance may change state, generally from plasma to gas, liquid, then solid. However, in some conditions, certain substances may skip states as they change—for example, during deposition of water vapor as frost.

DEPOSITION

The opposite process to sublimation, deposition is when a gas turns directly into a solid without first becoming a liquid. Frost is a common example—it occurs when water vapor in the air solidifies on surfaces in very cold conditions.

LATENT HEAT

Latent heat is the energy released or absorbed by a substance when changing phase. Sweating cools us down because the evaporation of sweat absorbs heat from the skin.

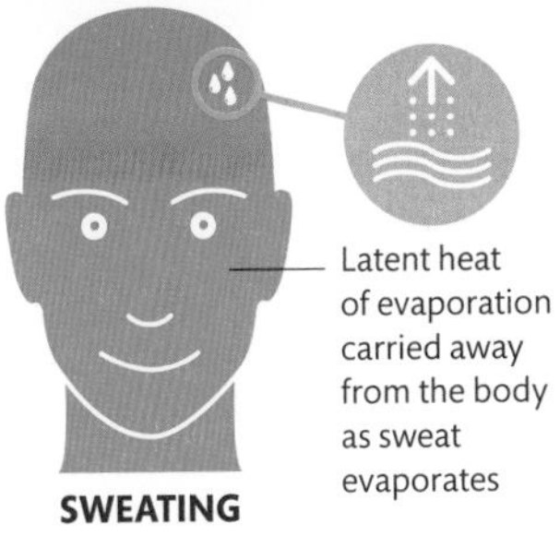
Latent heat of evaporation carried away from the body as sweat evaporates
SWEATING

Inside an atom

For a long time, atoms were thought to be indivisible, but we now know that they are made up of protons, neutrons, and electrons. The number of each of these particles determines what an atom is and its chemical and physical properties.

Structure of an atom

An atom consists of a central nucleus surrounded by one or more electrons. The nucleus contains positively charged protons and, except in the case of hydrogen, neutral neutrons. Most of the mass of an atom is found in the nucleus. Around the nucleus, tiny, negatively charged electrons orbit, held in place by attraction to the positively charged protons. An atom always has the same number of protons and electrons, so the positive and negative charges cancel out, making atoms electrically neutral.

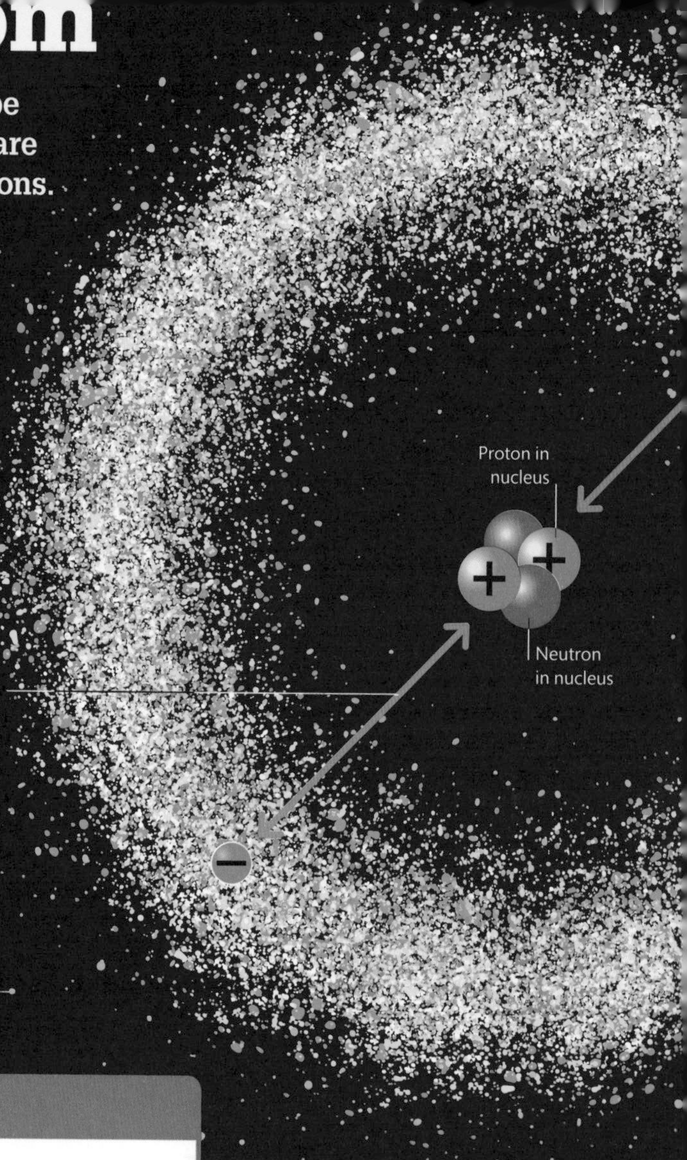

Structure of a helium atom
Each helium atom has two protons and two neutrons in its central nucleus, with two electrons orbiting around.

ATOMIC SIZES

The element with the smallest atom is hydrogen, which has only one proton and one electron. Its diameter is about 106 picometers (trillionths of a meter). Caesium is one of the largest atoms. It has 55 electrons orbiting its nucleus and is about six times wider than hydrogen, at approximately 596 picometers in diameter.

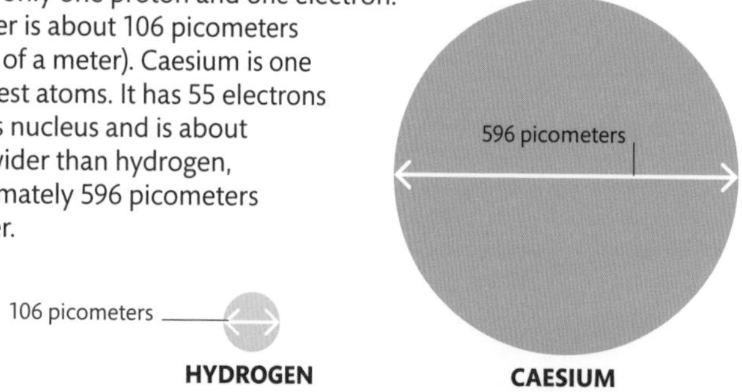

99 PERCENT OF A HYDROGEN ATOM IS EMPTY SPACE

Electron
Region where electrons most likely to be found

Electron orbitals

Electrons do not orbit the nucleus like planets orbiting the Sun. Because of quantum effects (see p.30), it is impossible to pinpoint their exact locations. Instead, they exist in regions called orbitals. These are areas of space around the nucleus where electrons are most likely to be found. There are four main types of orbital: s-orbitals, which are spherical; p-orbitals, which are dumbbell-shaped; and d- and f-orbitals, which have more complex shapes. Each orbital can hold up to two electrons, and the orbitals fill up in order, starting with the one closest to the nucleus.

Fluorine's orbitals
Fluorine's atoms have nine protons and nine orbiting electrons. The first four electrons fill the two s-orbitals, with two electrons in each orbital. The remaining five electrons are split between three p-orbitals.

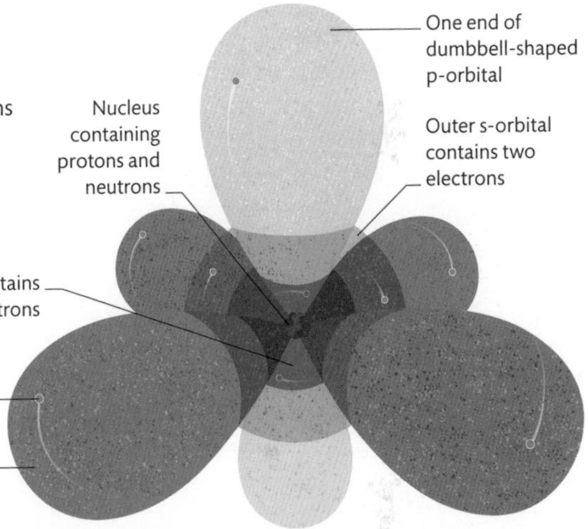
One end of dumbbell-shaped p-orbital
Nucleus containing protons and neutrons
Outer s-orbital contains two electrons
Inner s-orbital contains two electrons
Electron
Orbital is a region where there is a high probability of finding an electron

Atomic number and atomic mass

Scientists use several numbers and measurements to quantify the properties of atoms. These include the atomic number and various measurements of an atom's mass.

Quantity	Definition
Atomic number	The number of protons in an atom. An element is defined by its atomic number because all atoms of an element have the same number of protons. For example, all atoms with eight protons are oxygen atoms.
Atomic mass	The combined mass of an atom's protons, neutrons, and electrons. The number of neutrons in atoms of a particular element can vary, giving different isotopes of that element (see p.34). This means that different isotopes have different atomic masses. The unit used to measure atomic mass is called the atomic mass unit (amu)—one amu is one-twelfth of the mass of an atom of carbon-12, a common isotope of carbon.
Relative atomic mass	The average mass of an element's isotopes.
Mass number	The total number of protons and neutrons in an atom.

WHAT IS THE MASS OF AN ELECTRON?

An electron is extremely light, only about one two-thousandth the mass of a proton.

The subatomic world

Atoms are made up of smaller units called subatomic particles. They come in two types: those that form matter, and those that carry forces. Subatomic particles combine to form other particles and forces, including some with exotic properties.

IS THERE A GRAVITY PARTICLE?

Scientists think that the force of gravity may be carried by a particle called a graviton. The existence of gravitons has not yet been confirmed experimentally.

Subatomic structure

Electrons in an atom cannot be divided further, but protons and neutrons can. Each is made of three quarks—subatomic particles in a family called fermions. Fermions are matter particles, and all matter is made of quarks (in combinations of "flavors," or types) along with leptons (another class of fermions that includes electrons). Each fermion has a corresponding antiparticle with the same mass but opposite charge—for example, the electron's antiparticles are positrons. Combinations of antiparticles form antimatter.

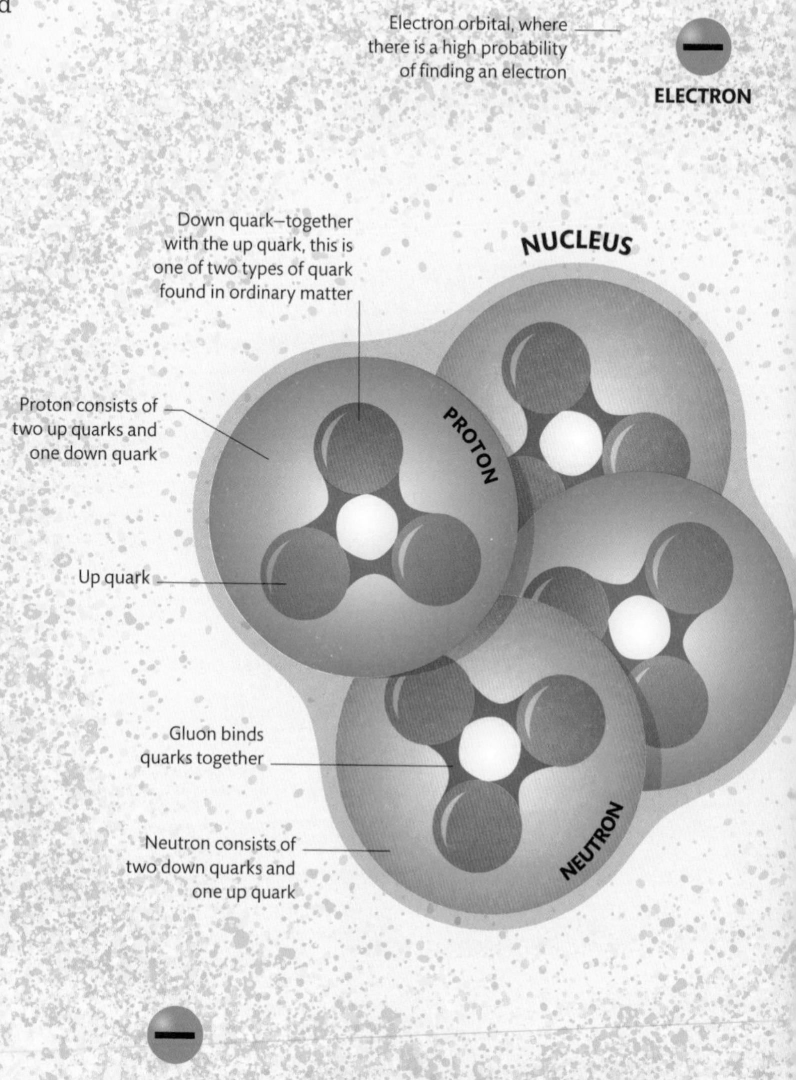

Electron orbital, where there is a high probability of finding an electron
ELECTRON
Down quark—together with the up quark, this is one of two types of quark found in ordinary matter
NUCLEUS
Proton consists of two up quarks and one down quark
PROTON
Up quark
Gluon binds quarks together
Neutron consists of two down quarks and one up quark
NEUTRON

Elementary particles
For a long time, scientists thought that protons and neutrons were elementary particles, which couldn't be divided, but we now know they are made of quarks. However, electrons and the quarks themselves do seem to be elementary.

THE TERM **"QUARK"** COMES **FROM** JAMES JOYCE'S NOVEL ***FINNEGAN'S WAKE***

SUBATOMIC PARTICLES

FERMIONS are matter particles. They make up the matter components of atoms, such as protons, neutrons, and electrons.

BOSONS are force-carrying particles. They act as messengers carrying forces between other particles.

ELEMENTARY FERMIONS are matter particles that are not made up of other particles.

HADRONS are composite particles made up of a number of quarks.

ELEMENTARY BOSONS are force-carrying particles that are not made up of other particles.

Quarks

- **Up**
- **Down**
- **Charm**
- **Strange**
- **Top**
- **Bottom**

Leptons

- **Electron**
- **Electron neutrino**
- **Muon**
- **Muon neutrino**
- **Tau particle**
- **Tau neutrino**

Baryons are composite fermions consisting of three quarks.

- **Proton** Two up quarks + one down quark + three gluons
- **Neutron** Two down quarks + one up quark + three gluons
- **Lambda particle** One down quark + one up quark + one strange quark + three gluons
- **Numerous others**

Mesons are composite bosons containing a quark and an antiquark.

- **Positive pion** One up quark + one down antiquark
- **Negative kaon** One strange quark + one up antiquark
- **Numerous others**

Elementary bosons

- **Photon**
- **Gluon**
- **W- boson**
- **W+ boson**
- **Z boson**
- **Higgs boson**

Fundamental forces

Rather than simple pushes and pulls, forces in the subatomic world are carried by particles. Think of two skaters throwing a ball on an ice rink; the ball carries energy from the first skater, exerting a force on the second, so the second skater moves when catching the ball.

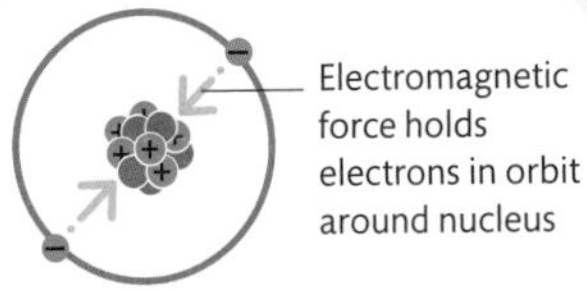
Electromagnetic force holds electrons in orbit around nucleus

Electromagnetic force
Interactions between charged particles are carried by photons, which are massless particles that move at the speed of light.

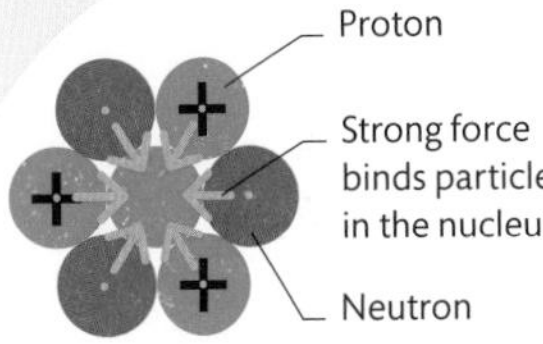
Proton
Strong force binds particles in the nucleus
Neutron

Strong force
The strong force binds quarks together, opposing electromagnetic repulsion within protons and neutrons. It acts over short ranges and is carried by gluons.

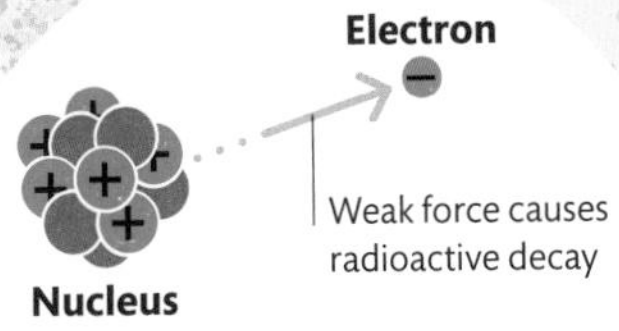
Electron
Weak force causes radioactive decay
Nucleus

Weak force
During radioactive decay, particles are pushed out of the nucleus as quarks change type—this is made possible by W and Z bosons, which carry the weak force.

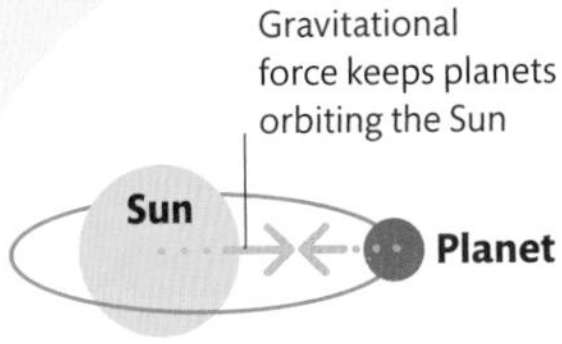
Gravitational force keeps planets orbiting the Sun
Sun
Planet

Gravity
Gravity is an attractive force that acts over an infinite range, so its yet-to-be-discovered particle must move at the speed of light.

Waves and particles

Waves and particles seem to be completely different: light is a wave and atoms are particles. However, sometimes waves (such as light) act as particles, and particles (such as electrons) act as waves. This is called wave-particle duality.

DO ALL PARTICLES ACT AS WAVES?

It seems it's not just small particles such as electrons that can act as waves. Some large molecules with more than 800 atoms behave like waves in double-slit experiments—although it's not known if all large molecules behave in this way.

Light as waves

The double-slit experiment is a simple way of showing that light can act as a wave. Light is shone through two screens, the first with one slit to produce a narrow beam of light and the second screen with two slits to split the light into two. After being split, the light hits the viewing screen, where it produces a series of alternating light and dark bands. If light acted as particles, the result would be very different.

Light particles
If light acted as simple particles, like grains of sand, some would pass through one slit and some through the other, producing only two distinct bands of light on the viewing screen. However, what actually happens when light is passed through two slits is different (see below).

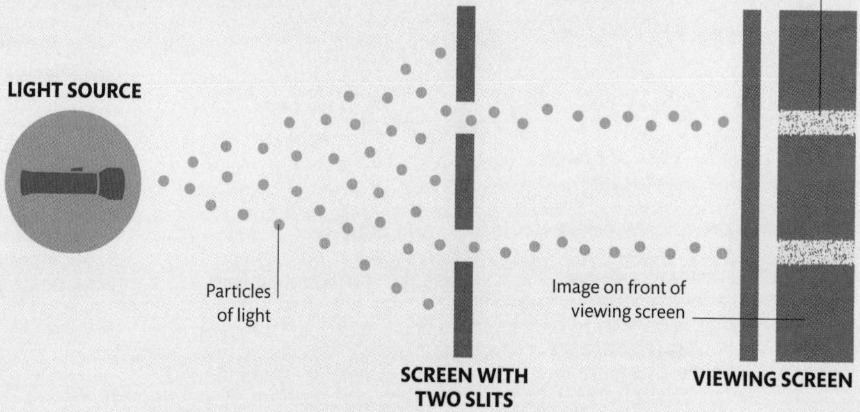

Light waves
After passing through the slits, waves form ripple patterns, like a stone thrown in a pond. The ripples interact, producing a series of several light and dark bands—an interference pattern—on the screen.

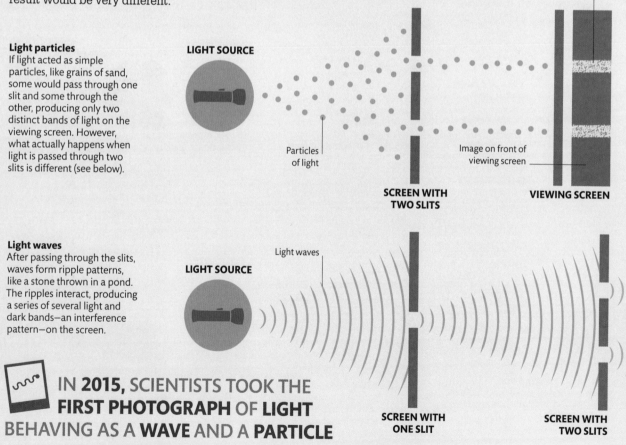

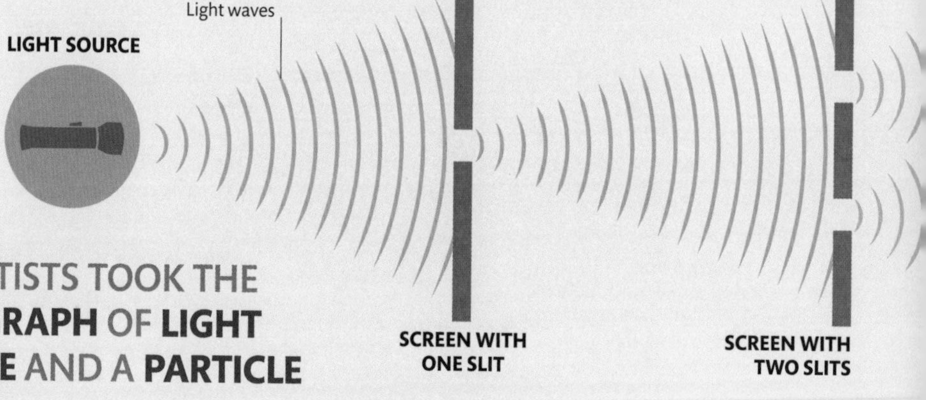

IN **2015,** SCIENTISTS TOOK THE **FIRST PHOTOGRAPH** OF **LIGHT** BEHAVING AS A **WAVE** AND A **PARTICLE**

Light as particles

When illuminated, metals can emit electrons, but only if the light is the right wavelength (color). This effect—called the photoelectric effect—occurs because light is acting as particles. The photons (particles) of long-wavelength red light have less energy than shorter-wavelength photons (such as those of green and ultraviolet light), and not enough to enable the metal's electrons to escape.

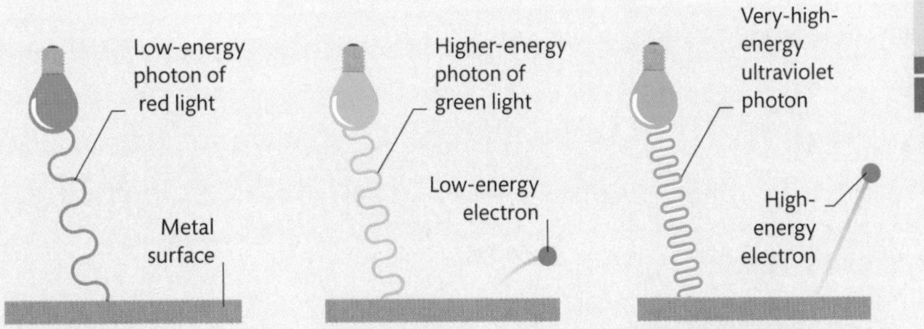

Red light
Photons of red light have too little energy to make most metals emit electrons from the surface, however bright the light.

Green light
Green-light photons have more energy than red—enough energy to enable electrons to escape from the metal's surface.

Ultraviolet light
Ultraviolet photons have very high energy, so they stimulate the release of high-energy electrons from the metal's surface.

Wave-particle duality

When the double-slit experiment is carried out with particles, such as electrons or atoms, interference patterns of light and dark bands are produced, just as happens with waves. The particles are therefore behaving like waves—this is wave-particle duality. If electrons are fired one by one, the same interference pattern results, because the particles' wavelike properties cause them to interfere with themselves.

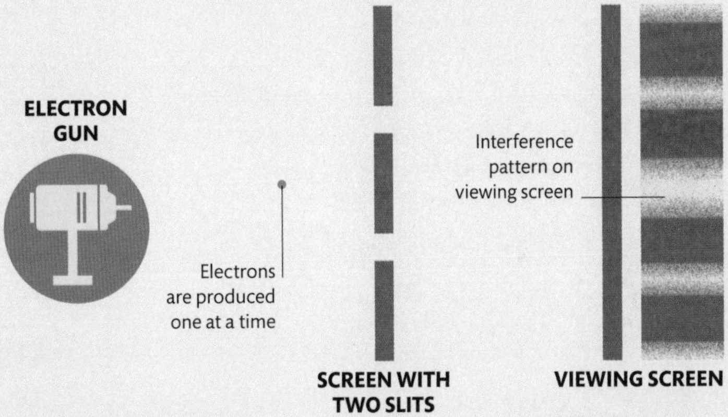

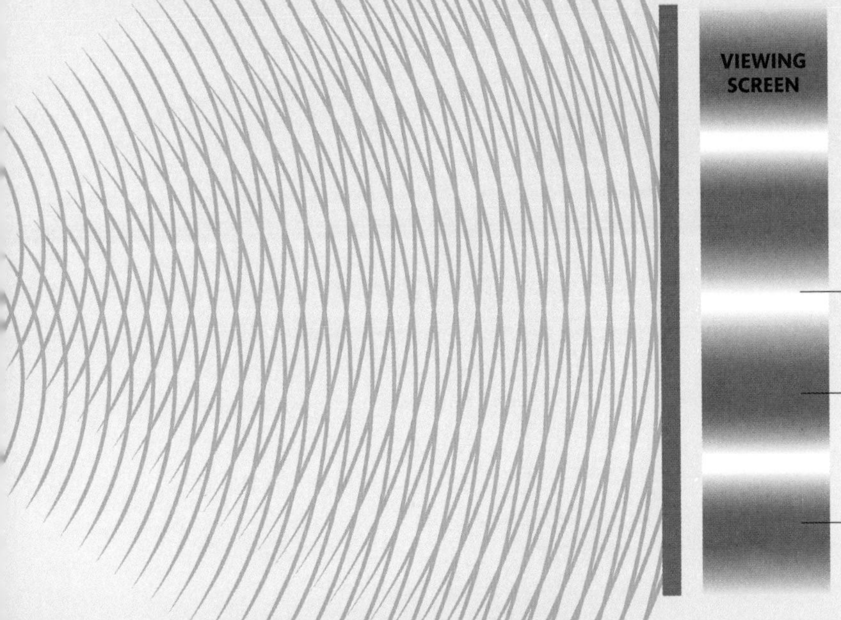

Interference
Where two waves are in phase, meeting at the same part of their cycle (peaks coincide with peaks, and troughs with troughs), they add together. Where they are out of phase (a peak meets a trough), they cancel each other out.

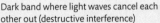

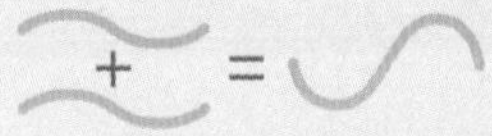

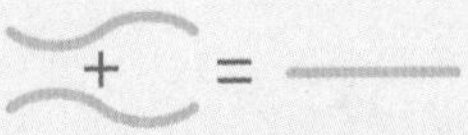

The quantum world

At the level of subatomic particles, things no longer behave the way we are used to in everyday life. Particles can act like waves as well as like particles, energy changes occur in jumps—called quantum leaps—and particles can be in an indeterminate state until observed.

Packets of energy

A quantum is the smallest possible amount of any physical property, such as energy or matter. For example, the smallest amount of electromagnetic radiation, such as light, is a photon. Quanta are indivisible—they can exist only as whole-number multiples of a single quantum.

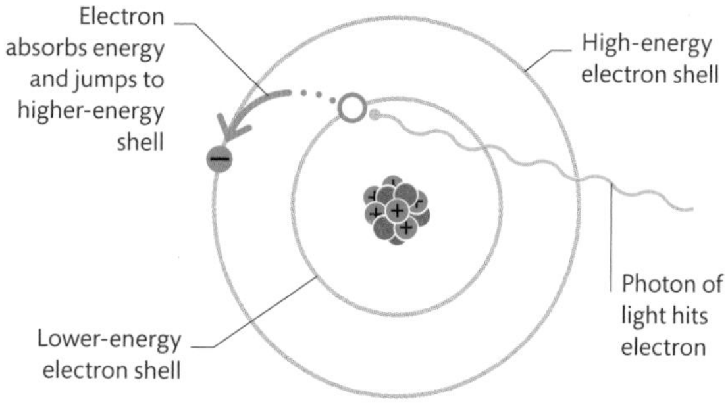

Quantum leap
Electrons in an atom can only jump directly from one energy level, or shell, to another—a "quantum leap"; they cannot occupy an intermediate energy level. When they move between levels, electrons absorb or emit energy.

The uncertainty principle

In the quantum world, it is impossible to know both the exact position and the exact velocity of a subatomic particle, such as an electron or photon. This effect, known as the uncertainty principle, occurs because measuring one property disturbs the particle, making other measurements inaccurate.

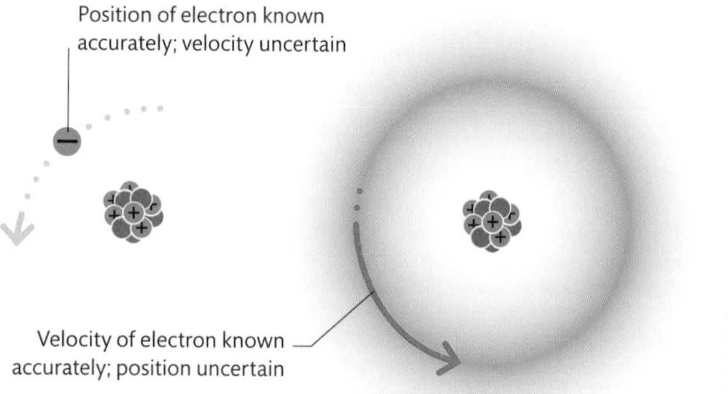

Position or velocity?
The position and velocity of an electron cannot both be known accurately. The more accurately its position is known, the more uncertain is its velocity, and vice versa.

QUANTUM ENTANGLEMENT

Quantum entanglement is a strange effect whereby a pair of subatomic particles, such as electrons, are linked, or entangled, and remain connected even when physically separated by an enormous distance (for example, in different galaxies). As a result, manipulating one particle instantaneously alters its partner. Similarly, measuring the properties of one particle immediately gives information about the properties of the other.

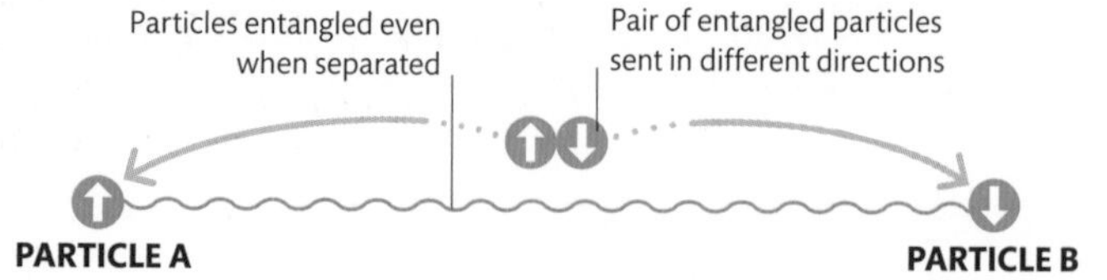

IS TELEPORTATION POSSIBLE?

Using quantum entanglement, researchers have teleported information over a distance of about 750 miles (1,200km). However, teleportation of physical objects is still science fiction.

Quantum limbo

In the quantum world, particles exist in a kind of limbo until they are observed. For example, a radioactive atom can be in an indeterminate state in which it has both decayed, releasing radiation, and not decayed. This in-between state is known as superposition. Only when a particle is observed or measured does it "decide" which option to adopt; in more technical terms, its superposition collapses. Superposition implies that subatomic events are never decided on until they are observed—an idea that led physicist Erwin Schrödinger to invent a famous thought experiment referred to as Schrödinger's cat.

Schrödinger's cat

A cat is shut in a box with a bottle of poison and some radioactive material. If the radioactive material decays and emits radiation, the radiation is detected by a Geiger counter, which triggers a hammer to break the bottle of poison, killing the cat. However, radioactive decay is random, so it is impossible to determine if the cat is alive or dead without looking in the box—in effect, the cat is both alive and dead until the box is opened.

ERWIN **SCHRÖDINGER** HAS A **MOON CRATER** NAMED AFTER HIM

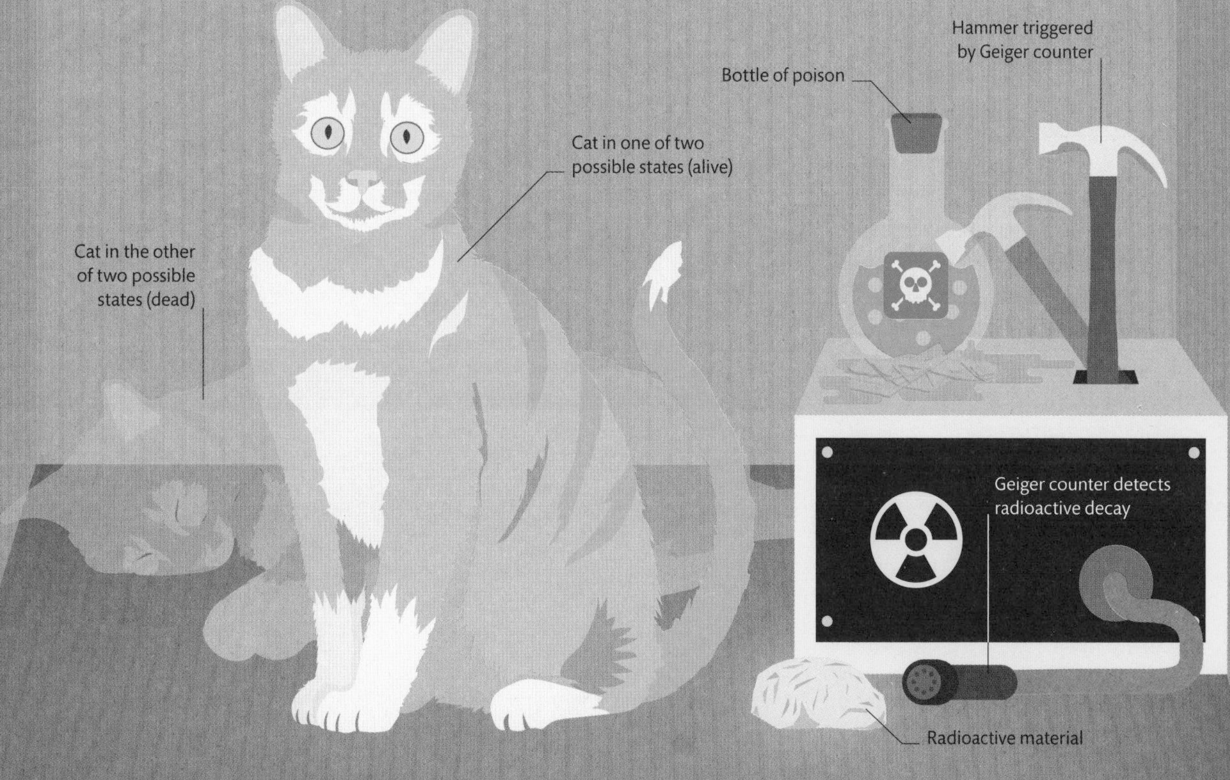

Particle accelerators

Particle accelerators are devices that propel subatomic particles at close to light speed to investigate fundamental questions about matter, energy, and the Universe.

How accelerators work

Particle accelerators use electric fields generated by high voltages and powerful magnetic fields to produce a beam of high-energy subatomic particles, such as protons or electrons, which are crashed together or fired at a metal target. Many particle accelerators are circular, so the particles can make many circuits, increasing in energy each time, before finally colliding.

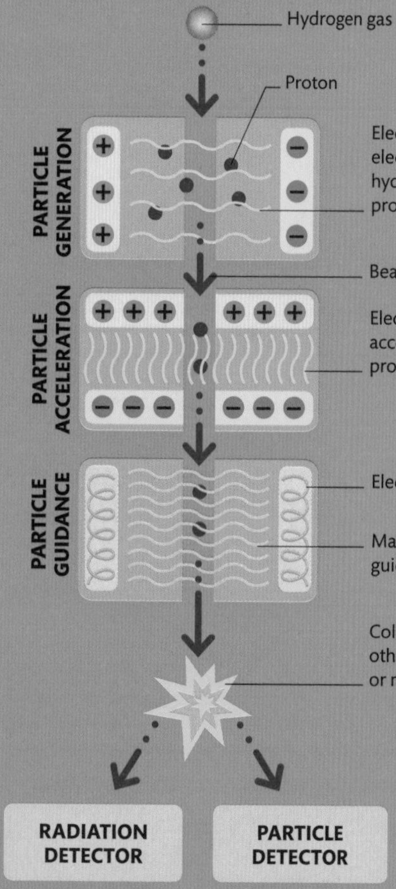

Hydrogen gas in
Proton
PARTICLE GENERATION
Electric field strips electrons from hydrogen to produce protons
Beam of protons
PARTICLE ACCELERATION
Electric field accelerates protons
PARTICLE GUIDANCE
Electromagnet
Magnetic field guides protons
Collision with other particles or metal target
RADIATION DETECTOR
PARTICLE DETECTOR

Subatomic collisions

Fast-moving protons are produced by passing hydrogen gas through electric fields. Guided by magnetic fields, the protons collide with other subatomic particles or atoms in a piece of metal. Detectors capture the radiation or particles from the collision.

Studying the subatomic world

Particle accelerators are used primarily to study matter and energy at the subatomic level, but they have also been used to investigate dark matter (see p.206) and conditions just after the Big Bang (see p.202). As well as being used to discover the Higgs boson, accelerators have also detected other exotic subatomic particles, such as pentaquarks, composite particles consisting of four quarks and one antiquark that might exist in supernovas.

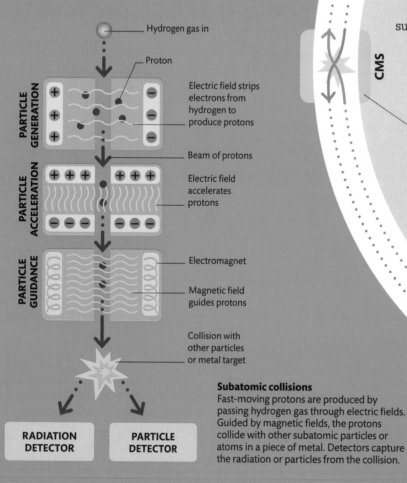

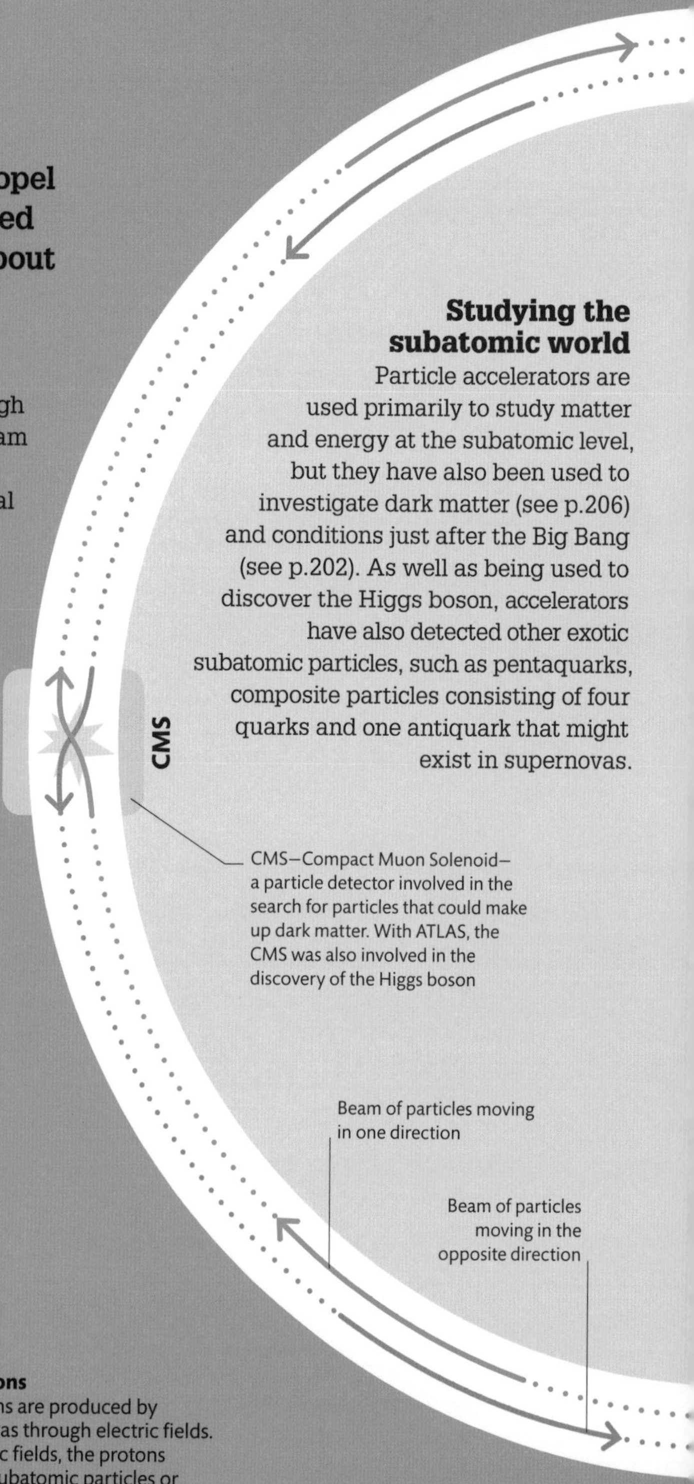

CMS
CMS—Compact Muon Solenoid—a particle detector involved in the search for particles that could make up dark matter. With ATLAS, the CMS was also involved in the discovery of the Higgs boson
Beam of particles moving in one direction
Beam of particles moving in the opposite direction

PARTICLES TRAVEL ROUND THE 17-MILE (27-KM) RING OF THE LHC MORE THAN 11,000 TIMES A SECOND

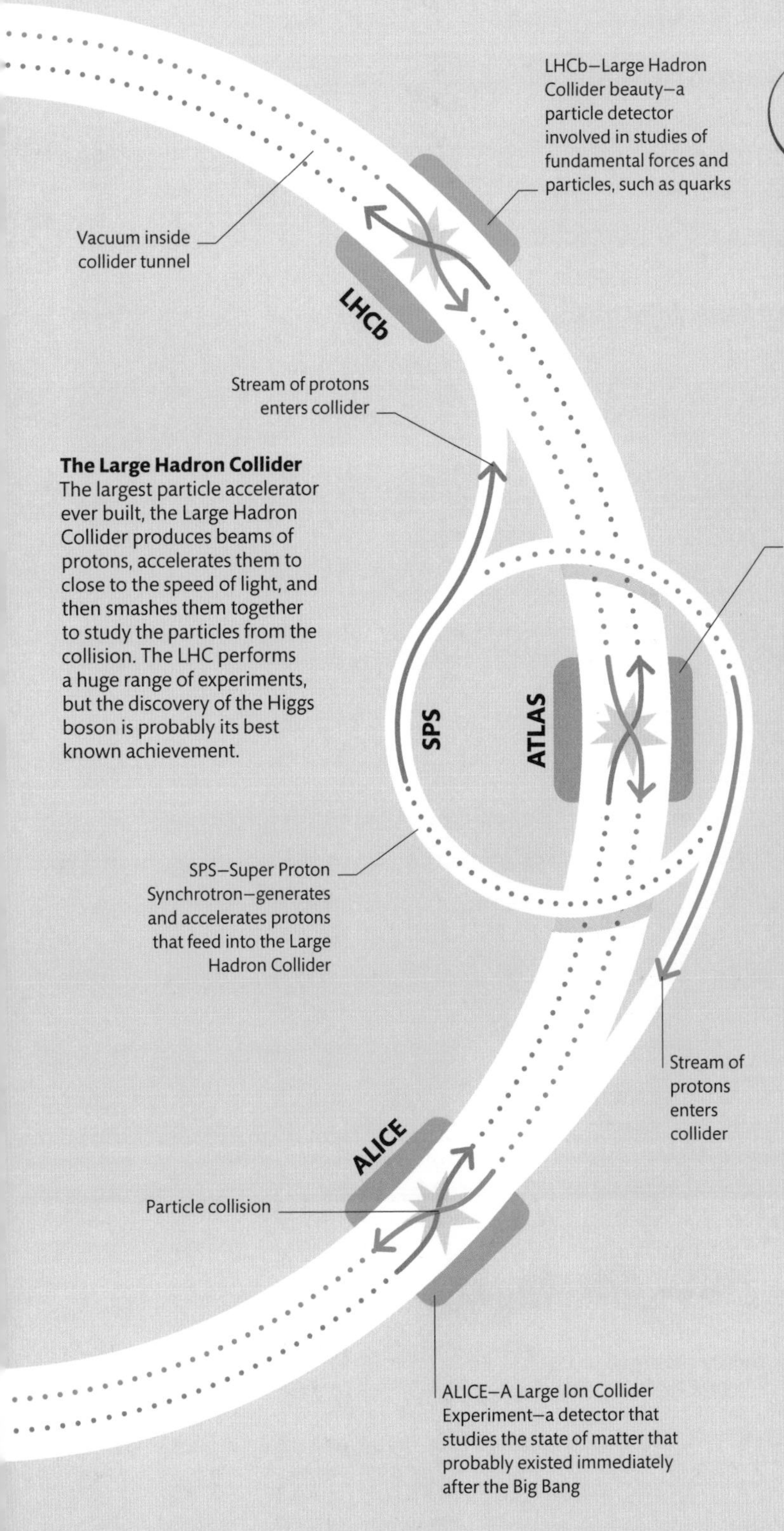
LHCb—Large Hadron Collider beauty—a particle detector involved in studies of fundamental forces and particles, such as quarks
Vacuum inside collider tunnel
LHCb
Stream of protons enters collider
ATLAS—A Toroidal LHC Apparatus—a high-energy particle detector that, with the CMS, was involved in the discovery of the Higgs boson
SPS
ATLAS
SPS—Super Proton Synchrotron—generates and accelerates protons that feed into the Large Hadron Collider
Stream of protons enters collider
ALICE
Particle collision
ALICE—A Large Ion Collider Experiment—a detector that studies the state of matter that probably existed immediately after the Big Bang

The Large Hadron Collider
The largest particle accelerator ever built, the Large Hadron Collider produces beams of protons, accelerates them to close to the speed of light, and then smashes them together to study the particles from the collision. The LHC performs a huge range of experiments, but the discovery of the Higgs boson is probably its best known achievement.

THE HIGGS BOSON

The Higgs boson is an aspect of a field—called the Higgs field—that produces mass through its interaction with other particles, such as photons and electrons. The Higgs boson can be thought of as being like a snowflake in a snowfield. The snowfield—the Higgs field—interacts differently with different objects: an object that interacts strongly with the field (sinks deeply into the snow) has a large mass; one that interacts weakly (sits on the surface of the snow) has small mass; and one that does not interact with the field at all has no mass.

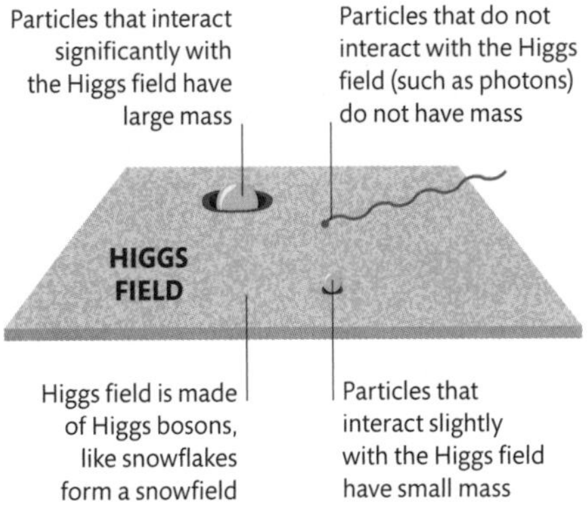
Particles that interact significantly with the Higgs field have large mass
Particles that do not interact with the Higgs field (such as photons) do not have mass
HIGGS FIELD
Higgs field is made of Higgs bosons, like snowflakes form a snowfield
Particles that interact slightly with the Higgs field have small mass

The elements

Elements contain only one type of atom, so cannot be broken down chemically into smaller parts. Atoms differ from each other in the number of protons, neutrons, and electrons they contain, but it is the protons that define an element. The periodic table is a way of organizing the elements according to the number of protons in the nucleus of an atom.

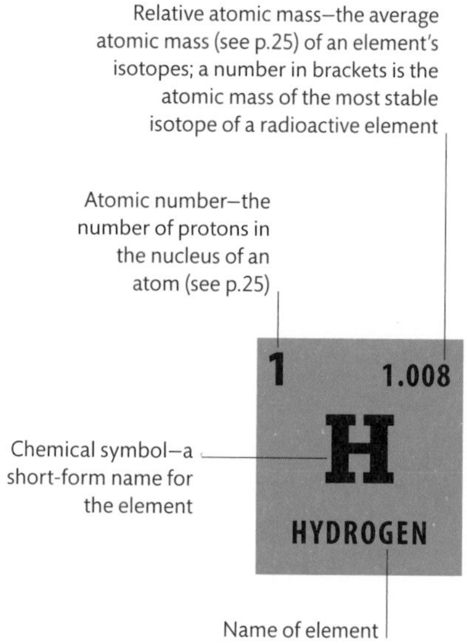

The periodic table

Elements are arranged in the periodic table by atomic number—their number of protons. In the table, the atomic number increases from left to right along a row. An element's position in the periodic table also tells you more about it; for example, elements in the same column react in similar ways.

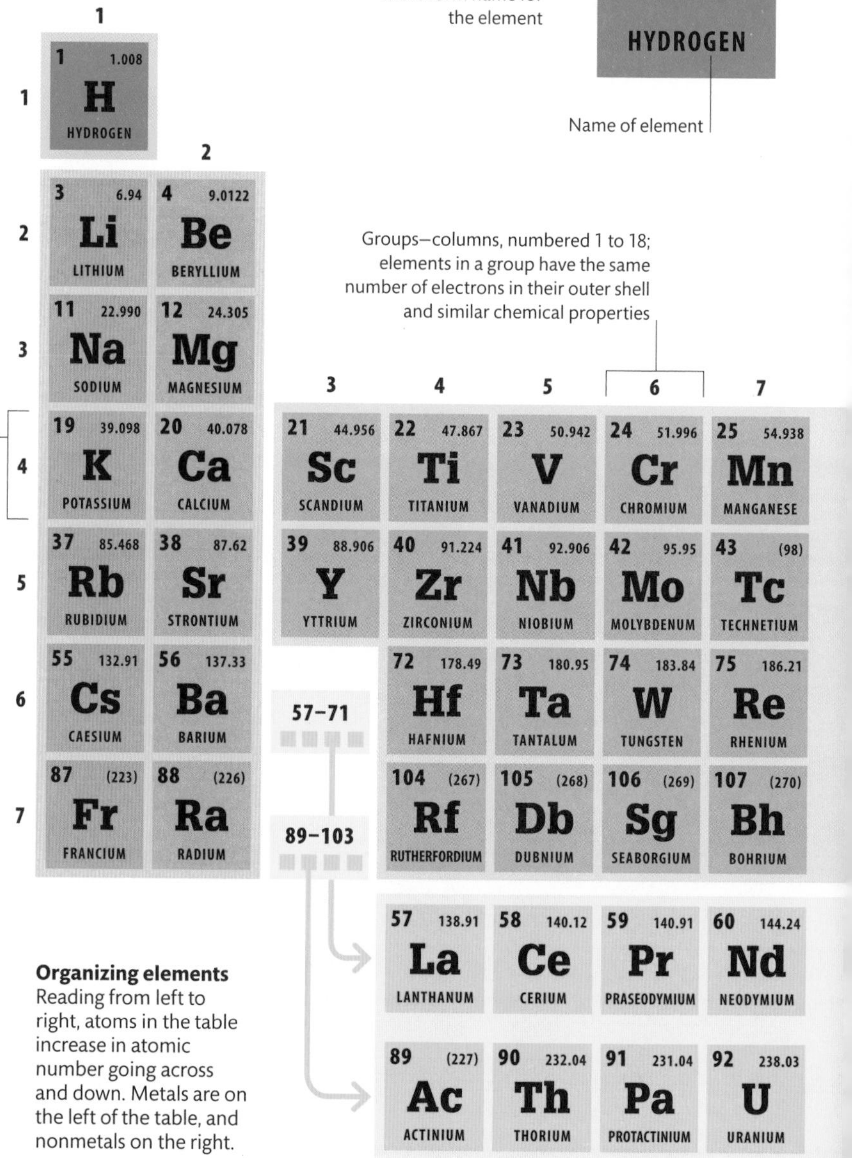

Organizing elements
Reading from left to right, atoms in the table increase in atomic number going across and down. Metals are on the left of the table, and nonmetals on the right.

ISOTOPES

Isotopes of an element have the same numbers of protons but different numbers of neutrons, so they differ in their atomic mass. For example, carbon isotopes exist naturally with 6, 7, or 8 neutrons. Isotopes react chemically in the same way but behave differently in other ways—some, for instance, are radioactive.

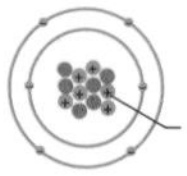

CARBON-12
6 neutrons + 6 protons = 12

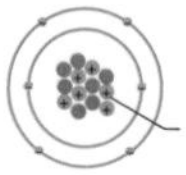

CARBON-13
7 neutrons + 6 protons = 13

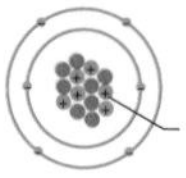

CARBON-14
8 neutrons + 6 protons = 14

KEY

- **Hydrogen**—a reactive gas

REACTIVE METALS

- **Alkali metals**—soft, very reactive metals
- **Alkaline earth metals**—moderately reactive metals

TRANSITION ELEMENTS

- **Transition metals**—a varied group of metals, many with valuable properties

MAINLY NONMETALS

- **Metalloids**—elements with properties between those of metals and nonmetals
- **Other metals**—mostly relatively soft metals with low melting points
- Carbon and other nonmetals
- **Halogens**—very reactive nonmetals
- **Noble gases**—colorless, very unreactive gases

RARE EARTH METALS

- Also called lanthanoids and actinoids, these are reactive metals—some are rare or synthetic

Periods, groups, and blocks

All the elements in a row, or period, have the same number of electron orbitals (see p.25). The columns in the periodic table, known as groups, contain elements that have the same number of electrons in their outer shells, and therefore react in similar ways. Four main "blocks" (see left) group together elements with similar properties, such as the transition elements, which are mostly hard and shiny metals. Hydrogen has a distinct set of properties, so it is in a group by itself.

8	9	10	11	12	13	14	15	16	17	18
										2 4.0026 He HELIUM
					5 10.81 B BORON	6 12.011 C CARBON	7 14.007 N NITROGEN	8 15.999 O OXYGEN	9 18.998 F FLUORINE	10 20.180 Ne NEON
					13 26.982 Al ALUMINUM	14 28.085 Si SILICON	15 30.974 P PHOSPHORUS	16 32.06 S SULFUR	17 35.45 Cl CHLORINE	18 39.948 Ar ARGON
26 55.845 Fe IRON	27 58.933 Co COBALT	28 58.693 Ni NICKEL	29 63.546 Cu COPPER	30 65.38 Zn ZINC	31 69.723 Ga GALLIUM	32 72.63 Ge GERMANIUM	33 74.922 As ARSENIC	34 78.97 Se SELENIUM	35 79.904 Br BROMINE	36 83.80 Kr KRYPTON
44 101.07 Ru RUTHENIUM	45 102.91 Rh RHODIUM	46 106.42 Pd PALLADIUM	47 107.87 Ag SILVER	48 112.41 Cd CADMIUM	49 114.82 In INDIUM	50 118.71 Sn TIN	51 121.76 Sb ANTIMONY	52 127.60 Te TELLURIUM	53 126.90 I IODINE	54 131.29 Xe XENON
76 190.23 Os OSMIUM	77 192.22 Ir IRIDIUM	78 195.08 Pt PLATINUM	79 196.97 Au GOLD	80 200.59 Hg MERCURY	81 204.38 Tl THALLIUM	82 207.2 Pb LEAD	83 208.98 Bi BISMUTH	84 (209) Po POLONIUM	85 (210) At ASTATINE	86 (222) Rn RADON
108 (277) Hs HASSIUM	109 (278) Mt MEITNERIUM	110 (281) Ds DARMSTADTIUM	111 (282) Rg ROENTGENIUM	112 (285) Cn COPERNICIUM	113 (286) Nh NIHONIUM	114 (289) Fl FLEROVIUM	115 (289) Mc MOSCOVIUM	116 (293) Lv LIVERMORIUM	117 (294) Ts TENNESSINE	118 (294) Og OGANESSON

61 (145) Pm PROMETHIUM	62 150.36 Sm SAMARIUM	63 151.96 Eu EUROPIUM	64 157.25 Gd GADOLINIUM	65 158.93 Tb TERBIUM	66 162.50 Dy DYSPROSIUM	67 164.93 Ho HOLMIUM	68 167.26 Er ERBIUM	69 168.93 Tm THULIUM	70 173.05 Yb YTTERBIUM	71 174.97 Lu LUTETIUM
93 (237) Np NEPTUNIUM	94 (244) Pu PLUTONIUM	95 (243) Am AMERICIUM	96 (247) Cm CURIUM	97 (247) Bk BERKELIUM	98 (251) Cf CALIFORNIUM	99 (252) Es EINSTEINIUM	100 (257) Fm FERMIUM	101 (258) Md MENDELEVIUM	102 (259) No NOBELIUM	103 (262) Lr LAWRENCIUM

Radioactivity

Radioactive materials have unstable nuclei that release energy, or radiation. Radioactivity is often thought of as dangerous, and it can be if handled incorrectly. However, it could also reduce our reliance on polluting fossil fuels.

What is radiation?

Radiation consists of streams of energetic waves or particles that can knock electrons off other atoms. In large amounts, radiation can damage the DNA in cells. In addition, it can create reactive free radicals in the body, which may also damage cells.

Types of radiation
An alpha particle consists of two neutrons and two protons (a helium nucleus). A beta particle is an electron or positron. Gamma rays are high-energy electromagnetic waves.

RADIOACTIVE ATOM

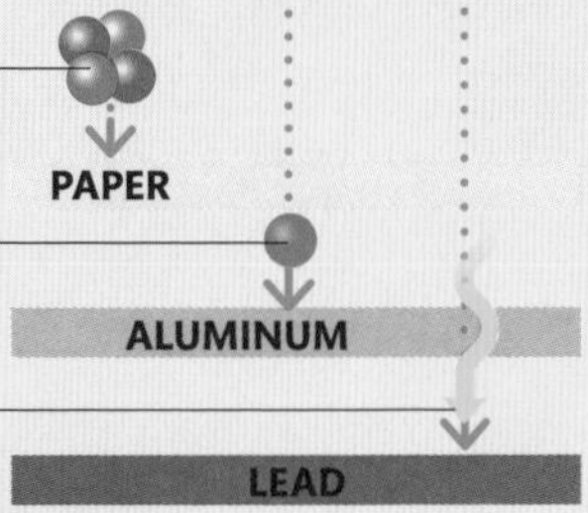
ALPHA
BETA
GAMMA
Alpha particles can be blocked by a sheet of paper
PAPER
Beta particles can be stopped by a thin sheet of aluminum
ALUMINUM
Gamma rays penetrate furthest but can be stopped by a thick layer of lead
LEAD

Nuclear energy

When atoms split apart or fuse together, energy is released—nuclear energy. This energy is in the form of heat, which can be used to boil water to power a turbine, just as in fossil fuel–powered electricity generators (see p.84).

Fission reactions
In fission reactions, atomic nuclei are split to release energy. In nuclear power stations, this process is carefully controlled to prevent a runaway chain reaction.

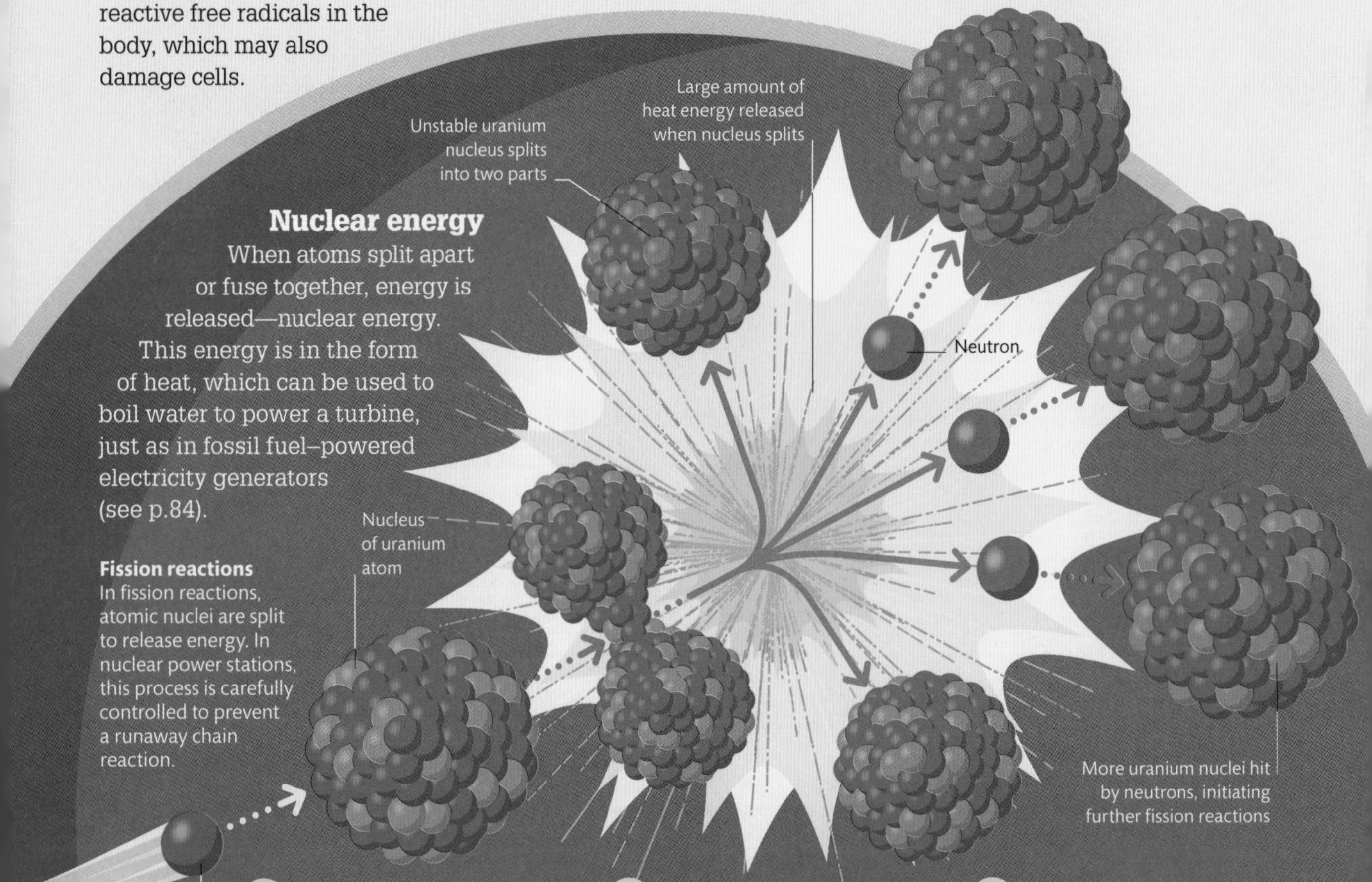
Large amount of heat energy released when nucleus splits
Unstable uranium nucleus splits into two parts
Neutron
Nucleus of uranium atom
More uranium nuclei hit by neutrons, initiating further fission reactions
High-energy neutron fired at nuclear material

1 Neutron hits atomic nucleus
Radioactive material (most commonly uranium) is bombarded with neutrons, some of which hit the nucleus of an atom and destabilize it.

2 Nucleus splits
The unstable nucleus splits into two. This fission releases a large amount of energy and also more neutrons.

3 Chain reaction
The additional neutrons released hit other atoms, which may in turn split and release even more neutrons, initiating a chain reaction.

HALF-LIVES AND DECAY

The half-life of a radioactive substance is the time taken for half the amount of original material to decay. Some substances decay very quickly but others take millions of years. For example, uranium-235, used in fission reactors, has a half-life of about 704 million years, which makes it problematic when disposing of nuclear waste.

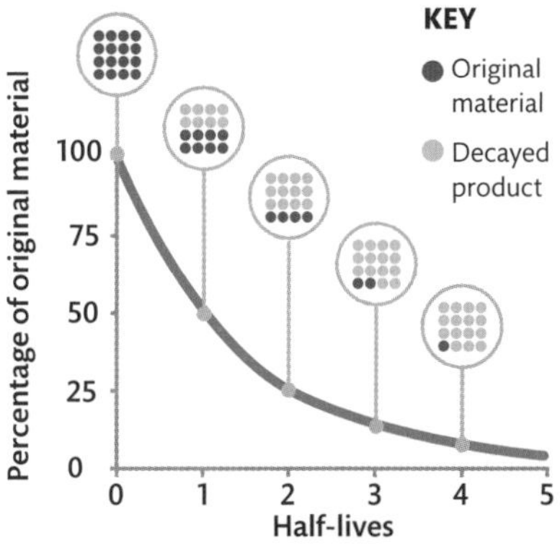

IS NUCLEAR FUSION SAFE?

There is no risk of a meltdown with a fusion reactor (unlike with a fission reactor) because a malfunction would cool the plasma and stop the reaction.

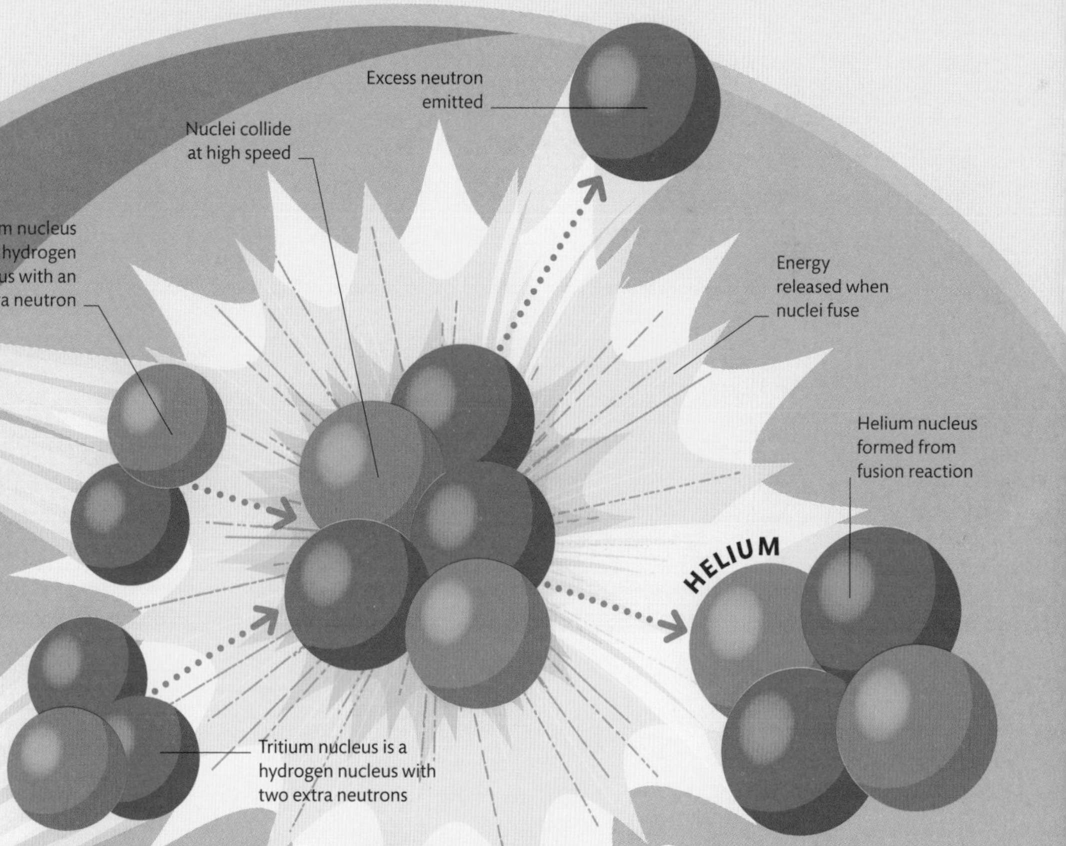

Fusion reactions
In fusion reactions, separate atomic nuclei are fused together into a single, larger nucleus. This process releases a huge amount of energy, and it is how the Sun and other stars produce light and heat (see p.193).

1 Nuclei heated
Tritium and deuterium nuclei are heated to high temperature to form a plasma and give them enough energy to overcome their natural repulsion.

2 Nuclei fuse
The high-energy deuterium and tritium nuclei collide. The collision makes the two nuclei fuse together.

3 Energy released
The merging of the two nuclei produces a helium nucleus and releases a vast amount of energy. An excess neutron is also emitted.

Mixtures and compounds

When different substances are mixed, one of two things can happen. They may react to form a new substance—a compound—or they may remain as individual substances but mixed together.

Mixtures

When mixed, many substances do not react but stay chemically the same, such as a mixture of sand and salt. The substances may be individual atoms, molecules of an element, or molecules with more than one element (compounds).

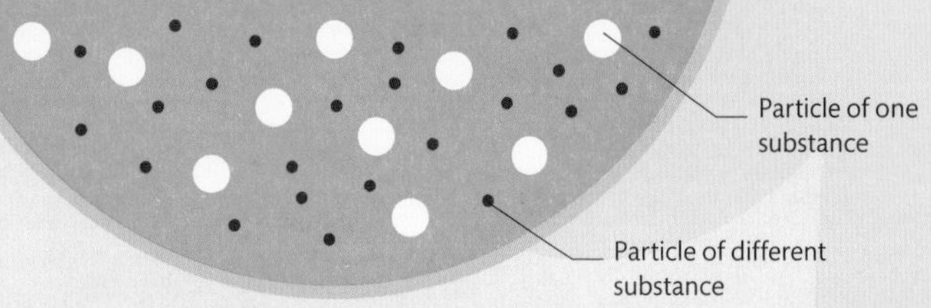

Compounds

Compounds contain atoms of two or more elements bonded together chemically. A compound's properties can differ greatly from those of its constituent elements; for example, hydrogen and oxygen are both gases but combine to form liquid water.

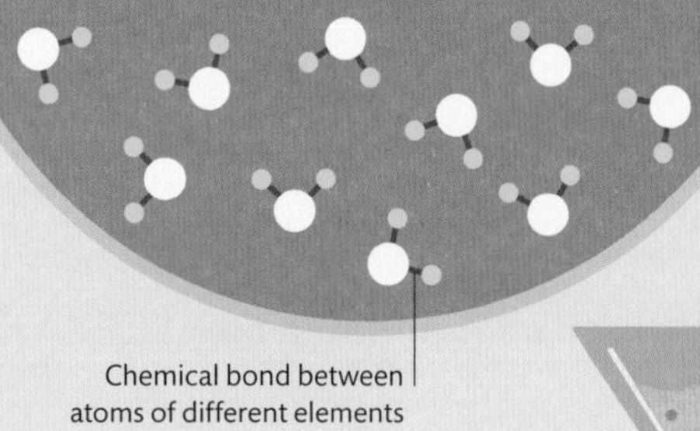

Separating mixtures

Mixtures can be separated by physical methods because their constituents are not chemically bonded. Suitable separation methods depend on the type of mixture. For example, mixtures in which only one component dissolves can be separated out by filtration. Other types of mixture require more complex methods, such as chromatography, distillation, or centrifuging.

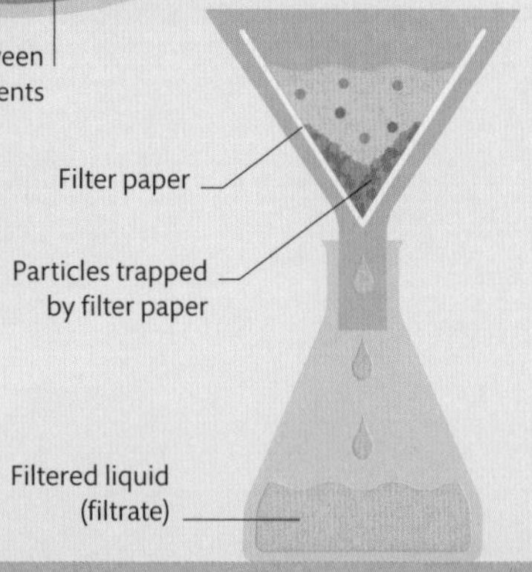

Filtration

Filters allow very small or soluble particles to pass through but trap bigger or insoluble particles. A salt solution, for example, will pass through a filter while any sand in the mixture will be trapped by the filter.

Types of mixture

There are different types of mixture, varying according to the solubility of their individual constituents and the size of the particles. Solutions form when a substance dissolves, such as sugar dissolving in water (see pp.62–63). In colloids and suspensions, the particles of the components do not dissolve but disperse into one another.

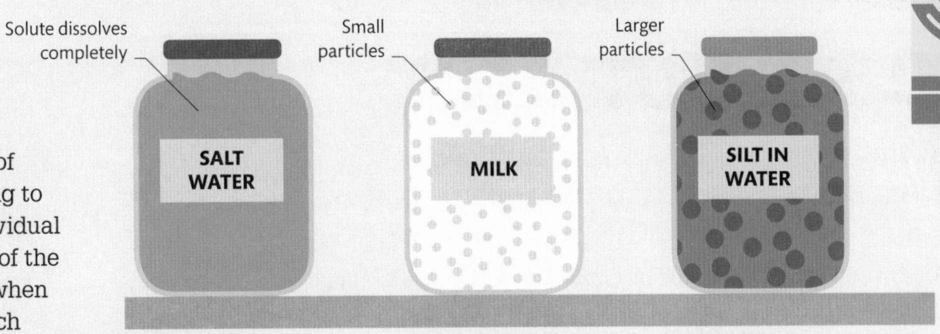

True solutions
In true solutions, such as salt dissolved in water, all the constituents are in the same state of matter—liquid in the example here.

Colloids
A colloid has tiny particles distributed evenly through the mixture. The particles are invisibly small and do not settle out.

Suspensions
Suspensions contain dispersed particles about the size of dust specks. They are visible to the naked eye, and can settle out.

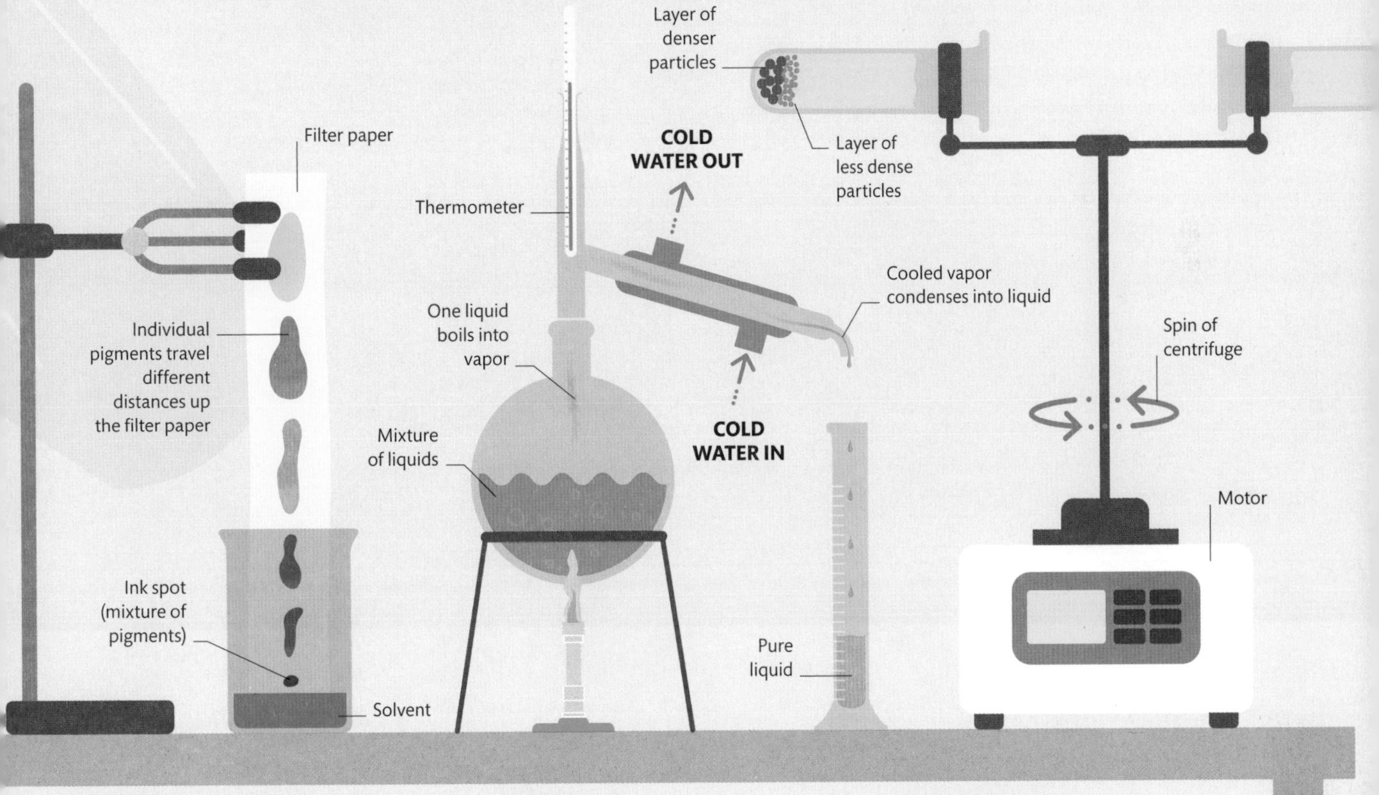

Chromatography
Components of a mixture can often be separated out with chromatography. The individual components are carried different distances by the solvent as it moves up the strip of filter paper.

Distillation
Mixtures of liquids with different boiling points can be separated using distillation. As the mixture is heated, the constituents boil off one at a time. As it boils off, each constituent is condensed back into liquid.

Centrifuging
A mixture of particles with different densities or particles suspended in a liquid can be separated out by spinning in a centrifuge. The denser or suspended particles form the lower layers.

Molecules and ions

A molecule consists of two or more atoms bonded together. The atoms can be of the same element or of different elements. They are bound by forces between their charged particles—the attraction being created by the transfer or the sharing of electrons.

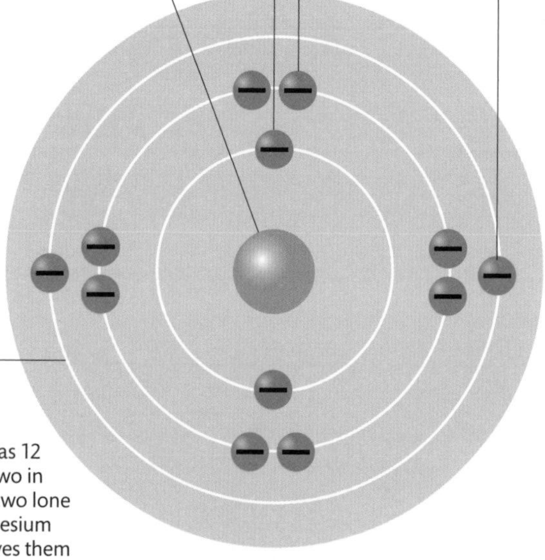

Nucleus contains 12 protons, balancing the electrons' charges and making the atom neutral
First shell contains two electrons
Second shell contains eight electrons
Third shell contains two electrons
Shells are pictured as round for simplicity, but their actual shapes are more complex

Magnesium's electron shells
A magnesium atom has 12 electrons, with only two in its outer shell. These two lone electrons make magnesium reactive—it readily gives them up to become more stable.

MAGNESIUM ATOM: Mg

Electron shells

Electrons orbit nuclei at distinct energy levels, or shells. Each shell can hold a fixed maximum number of electrons: the first shell can contain up to two electrons, the second and third up to eight. Atoms seek the most energetically stable arrangement of electrons, which often means having full outer shells.

What is an ion?

Atoms are electrically neutral—the positive charge of protons in their nucleus is balanced by the negative charge of their electrons. Atoms often acquire an overall electric charge in an effort to achieve stable electron arrangements—a charged atom (or a charged molecule) being known as an ion. Some atoms ionize by gaining electrons to fill one or two gaps in their outer shell. For others—for example, group I (alkali) metals such as sodium (see p.34)—it is better to give up their few outer electrons. Doing either gives the atom charge, because they no longer have equal numbers of electrons and protons.

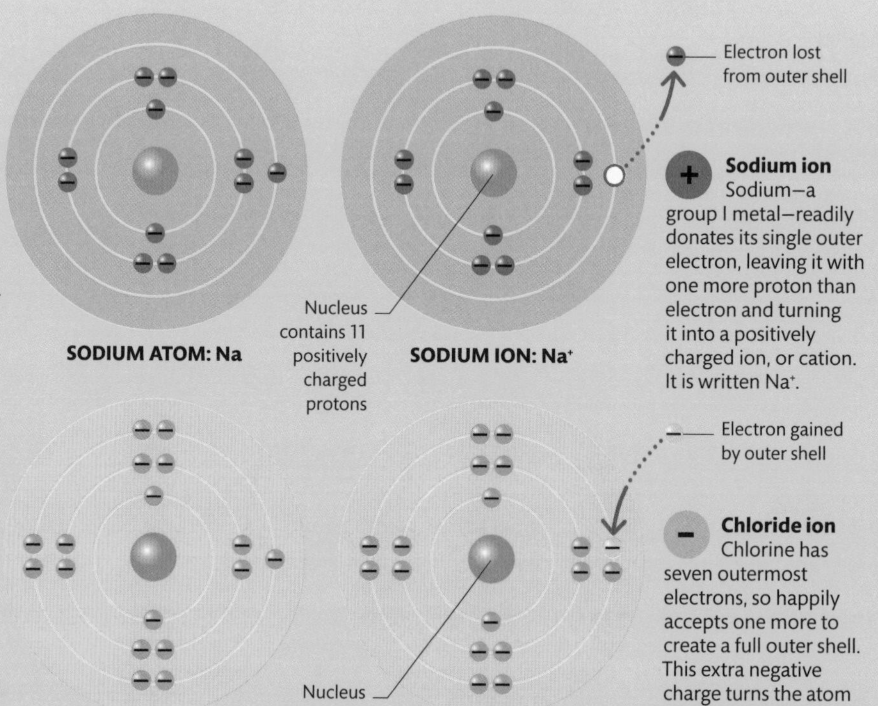

Electron lost from outer shell
Nucleus contains 11 positively charged protons
Electron gained by outer shell
Nucleus contains 17 protons

SODIUM ATOM: Na

SODIUM ION: Na^+

CHLORINE ATOM: Cl

CHLORIDE ION: Cl^-

Sodium ion
Sodium—a group I metal—readily donates its single outer electron, leaving it with one more proton than electron and turning it into a positively charged ion, or cation. It is written Na^+.

Chloride ion
Chlorine has seven outermost electrons, so happily accepts one more to create a full outer shell. This extra negative charge turns the atom into a negative ion, or anion, written Cl^-.

Sharing electrons

For some pairs of atoms, the easiest way to stabilize their electrons is to share them. Atoms that share electrons are tied together by forces known as covalent bonds. These bonds are common between two atoms of the same element or two different elements close to each other in the periodic table.

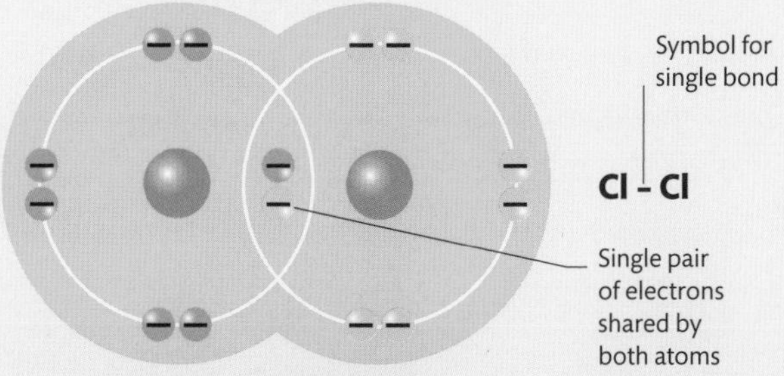
Symbol for single bond
Cl - Cl
Single pair of electrons shared by both atoms

CHLORINE MOLECULE: Cl_2

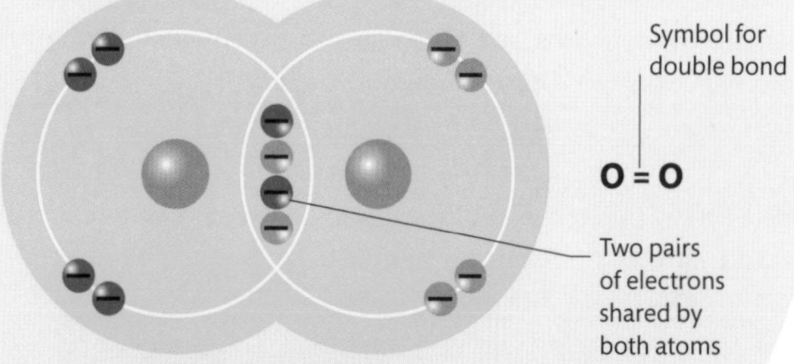
Symbol for double bond
O = O
Two pairs of electrons shared by both atoms

OXYGEN MOLECULE: O_2

Single bond
Chlorine has seven outer electrons, so pairs of atoms share one electron each to achieve full outer shells. These single bonds form Cl_2 molecules.

Double bond
Oxygen only has six outermost electrons, so it must share two pairs to become stable. This sharing of two electron pairs is known as a double bond.

Transferring electrons

When an atom with just one or a few outer electrons encounters an atom with gaps in its outer shell, it donates its outer electron (or electrons), forming positive and negative ions. Since unlike charges attract, these two ions are bound electrostatically, forming an ionic compound.

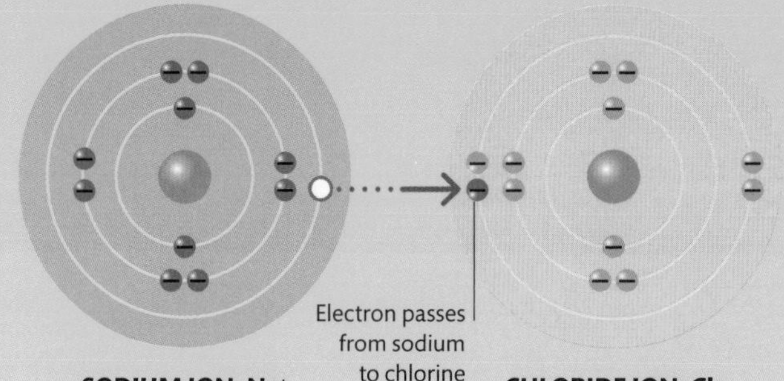
Electron passes from sodium to chlorine

SODIUM ION: Na^+

CHLORIDE ION: Cl^-

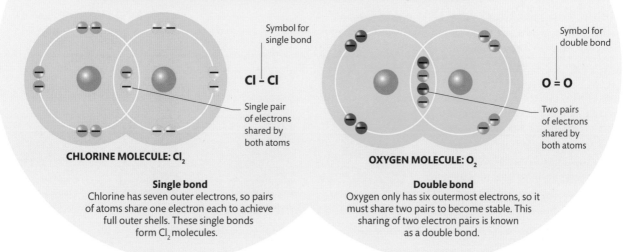

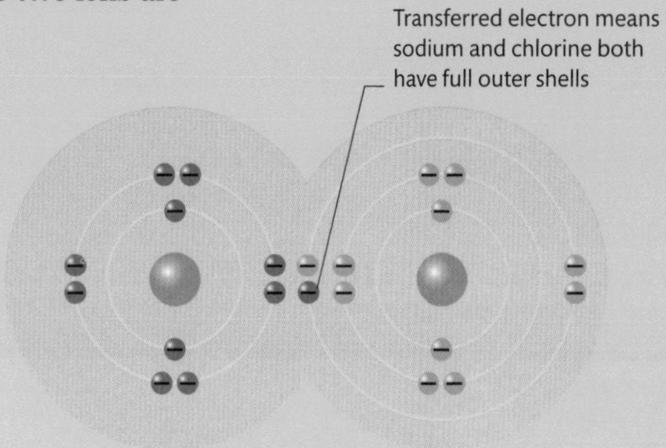
Transferred electron means sodium and chlorine both have full outer shells

SODIUM CHLORIDE COMPOUND: NaCl

1 Electron transfer
Sodium's outer electron transfers to chlorine, producing full outer shells on both atoms and ionizing them both, forming a sodium cation and a chloride anion. With other atom pairs, two, three, or more electrons may move.

2 Ionic bond formed
The cation and anion attract each other, forming a compound called sodium chloride (salt). The charges balance, so the compound is neutral overall. Ionic compounds tend to go on bonding to form giant lattices, often forming crystals (see p.60).

Getting a reaction

Chemical reactions are processes that change substances by breaking their atomic bonds and creating new ones. Many of these reactions take place within our bodies and are vital for our survival.

HOW IS COMMON SALT MADE?

Common salt can be created by mixing sodium and chlorine together. This causes a chemical reaction forming the compound sodium chloride (also known as common salt).

What is a reaction?

When chemicals react, their atoms are rearranged. These atoms are like Lego blocks—they fit together in different ways, but the number and types of blocks remains the same. Exactly how the atoms are rearranged depends on what they are reacting with. The substances that react together are called reactants, and the new substances they form are called products.

Irreversible reactions

Most reactions are irreversible, meaning they only occur in one direction, such as when hydrochloric acid (HCl) is mixed with sodium hydroxide (NaOH), creating sodium chloride (NaCl) and water (H_2O).

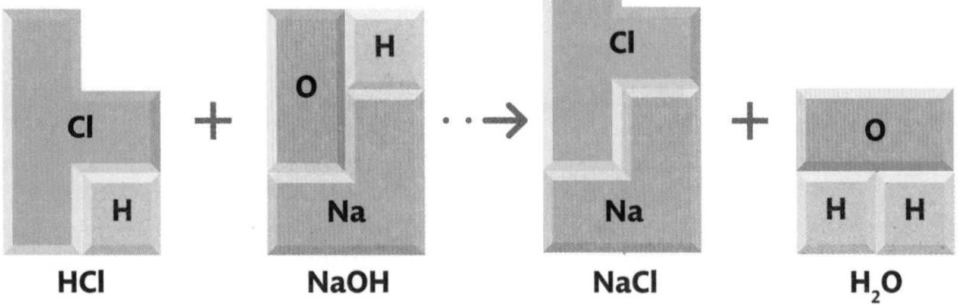

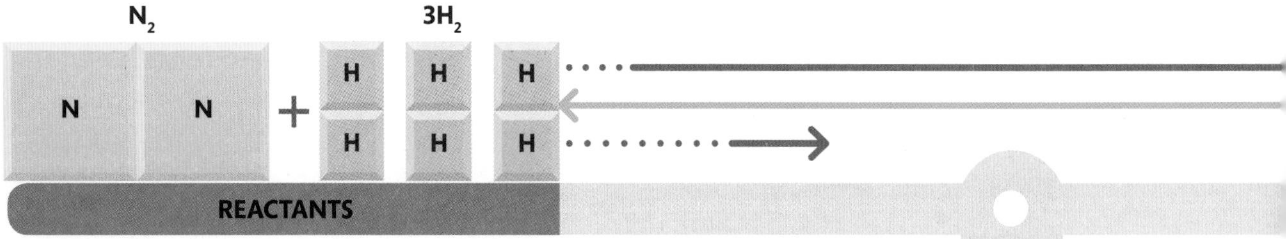

Dynamic equilibrium

In reversible reactions, the reaction starts when reactants are mixed, forming products (ammonia, in this example). But after a while, if nothing is added or removed, the amount of product stops increasing. At this point, reactions still occur in both directions, but they balance each other out. This is known as dynamic equilibrium.

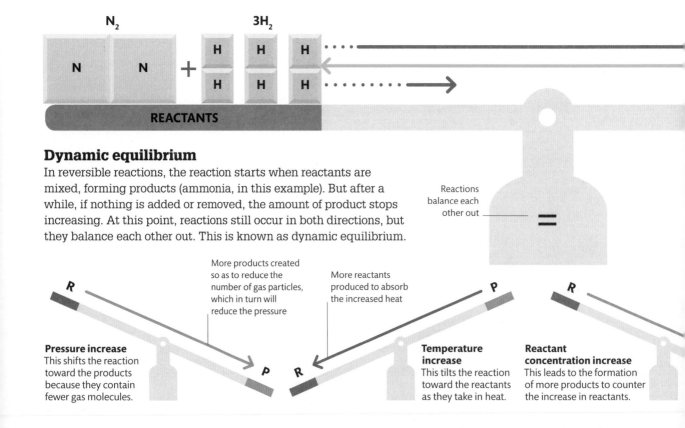

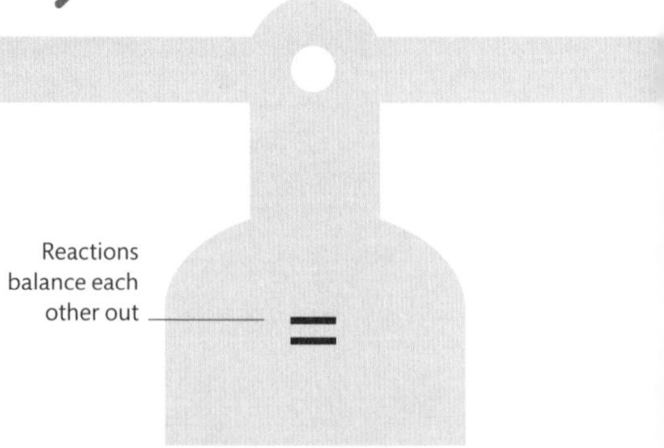

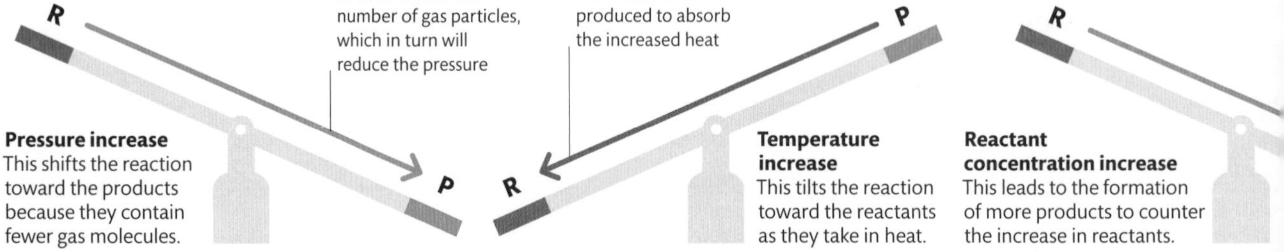

Pressure increase
This shifts the reaction toward the products because they contain fewer gas molecules.

Temperature increase
This tilts the reaction toward the reactants as they take in heat.

Reactant concentration increase
This leads to the formation of more products to counter the increase in reactants.

KEY

- Oxygen (O)
- Chlorine (Cl)
- Hydrogen (H)
- Sodium (Na)
- Nitrogen (N)

Reversible reactions
In some reactions, reactants can re-form from products, such as in the creation of ammonia (NH_3) from nitrogen (N_2) and hydrogen (H_2).

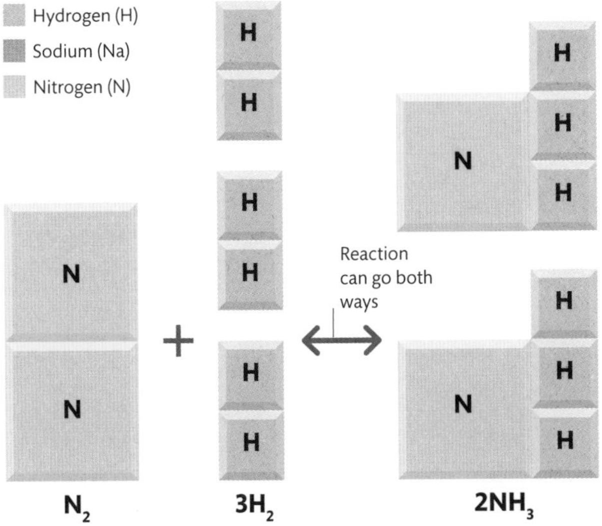

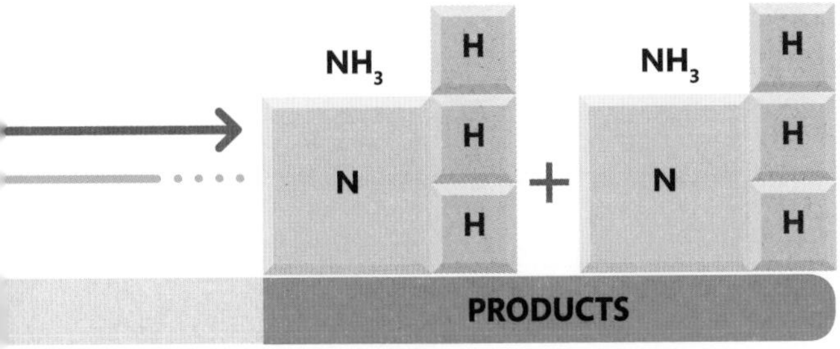

Tilting the scales
If you change something while the reactions are balanced, the equilibrium will shift to counteract that change. The four examples below show what happens when four different factors are changed during the creation of ammonia.

CHEMICAL REACTIONS ARE CONSTANTLY OCCURRING IN OUR 37.2 TRILLION CELLS

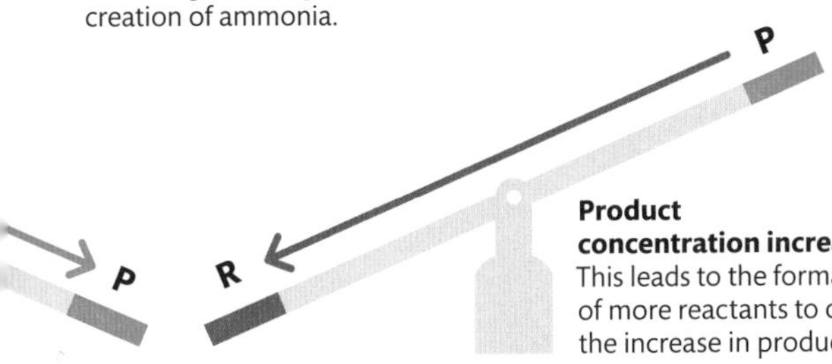

Product concentration increase
This leads to the formation of more reactants to counter the increase in products.

Common types of reactions

Chemical reactions can be grouped into several categories. Some involve molecules combining, while others split complex molecules into simpler ones. In other reactions, atoms swap positions, creating different molecules. Combustion (see pp.54–55) is a further type of reaction that occurs when oxygen reacts with another substance, creating enough heat and light to ignite.

Type of reaction	Definition	Equation
Synthesis	Two or more elements or compounds combine to form a more complex substance	A + B → AB
Decomposition	Compounds break down into simpler substances	AB → A + B
Single replacement	Occurs when one element replaces another in a compound	AB + C → AC + B
Double replacement	Occurs when different atoms in two different compounds exchange places	AB + CD → AC + BD

FIREWORKS

When fireworks are lit, a rapid chemical reaction takes place, releasing gas, which explodes outward into colored sparks. The colors depend on the type of metal used. Strontium carbonate, for example, produces red fireworks.

Reactions and energy

Reactions can only occur if the atoms involved have enough energy to start breaking and reforming their bonds. Very reactive substances need little extra energy to trigger reactions, but others have to be heated to high temperatures to cause them to react, due to the strength of their bonds.

CAN A REACTION GET OUT OF CONTROL?

If left unchecked, the rate of exothermic reactions can increase dangerously as the temperature rises. This can cause explosions, releasing toxic chemicals, as in Bhopal, India, in 1984.

Activation energy

To start a reaction, energy has to be put in, called the activation energy. The process is a bit like a hill that a snowboarder needs to climb in order to slide down the other side. Some reactions begin as soon as the reactants are combined. These reactions have a low activation energy, such as a reaction between a strong acid and an alkali.

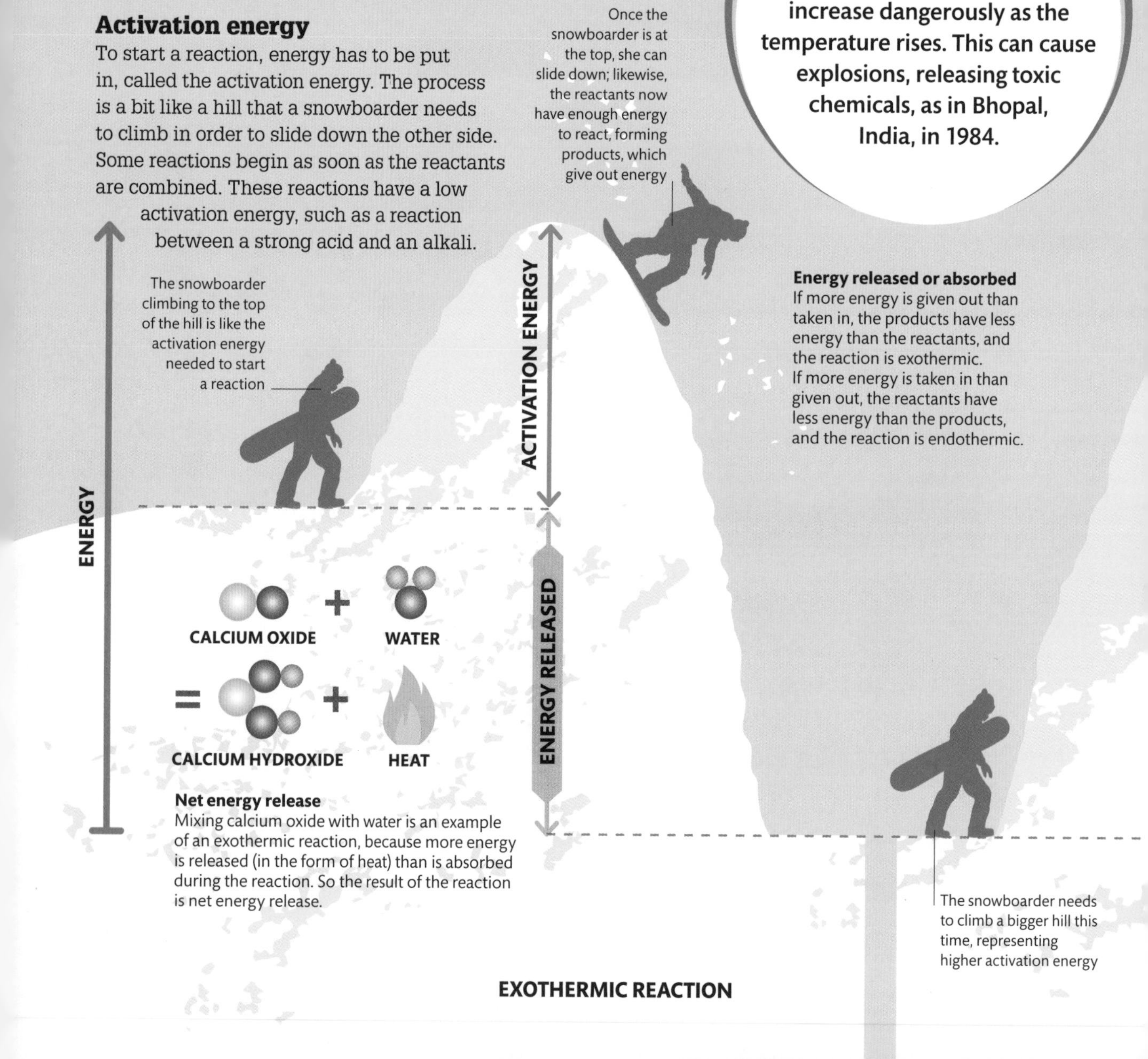

Once the snowboarder is at the top, she can slide down; likewise, the reactants now have enough energy to react, forming products, which give out energy
The snowboarder climbing to the top of the hill is like the activation energy needed to start a reaction
ENERGY
ACTIVATION ENERGY
ENERGY RELEASED
CALCIUM OXIDE
WATER
CALCIUM HYDROXIDE
HEAT
The snowboarder needs to climb a bigger hill this time, representing higher activation energy
EXOTHERMIC REACTION

Energy released or absorbed
If more energy is given out than taken in, the products have less energy than the reactants, and the reaction is exothermic.
If more energy is taken in than given out, the reactants have less energy than the products, and the reaction is endothermic.

Net energy release
Mixing calcium oxide with water is an example of an exothermic reaction, because more energy is released (in the form of heat) than is absorbed during the reaction. So the result of the reaction is net energy release.

SHERBET

When saliva comes into contact with the citric acid and sodium bicarbonate in sherbet, they dissolve and react, producing carbon dioxide bubbles, which makes sherbet fizzy. Because heat is absorbed by the reaction, the dissolved sherbet mixture will feel colder on the tongue.

The snowboarder travels down a shorter slope than the hill she climbed; likewise, less energy is released by the reaction than the activation energy that was put in at the start

CAESIUM IS SO REACTIVE IT **BURSTS INTO FLAME ON** CONTACT WITH AIR

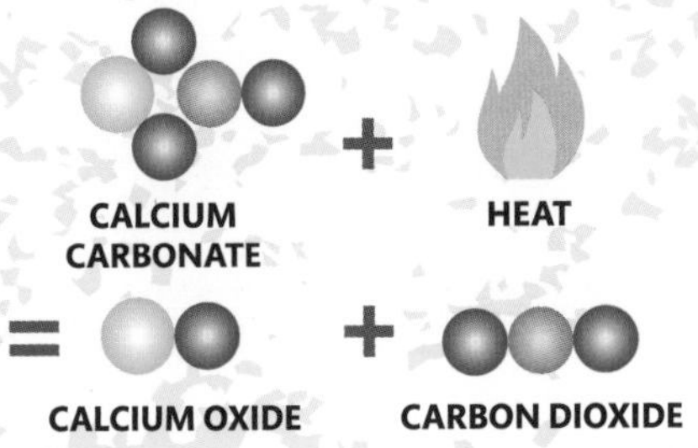

Net energy absorption
Heating calcium carbonate is an example of an endothermic reaction, because more energy is absorbed (in the form of heat) than is released during the reaction. So the result of the reaction is net energy absorption.

ENDOTHERMIC REACTION

Rates of reaction

Reactions can only occur when the atoms of the reactants collide with enough energy. Increasing the temperature, concentration, or surface area of the reactants or reducing the container volume will increase the number of collisions and speed up the rate of reaction.

Increasing concentration
More reactants lead to more collisions between atoms, so the rate of reaction rises.

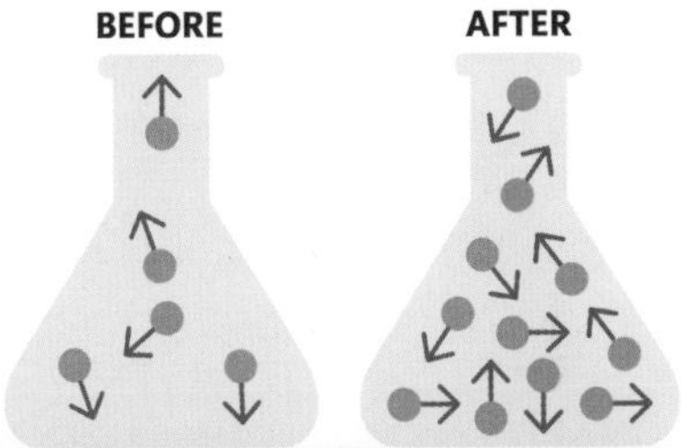

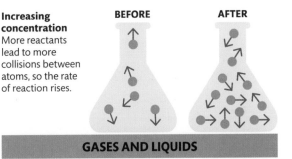

GASES AND LIQUIDS

Increasing temperature
This causes atoms to move more quickly, colliding more often and with more energy.

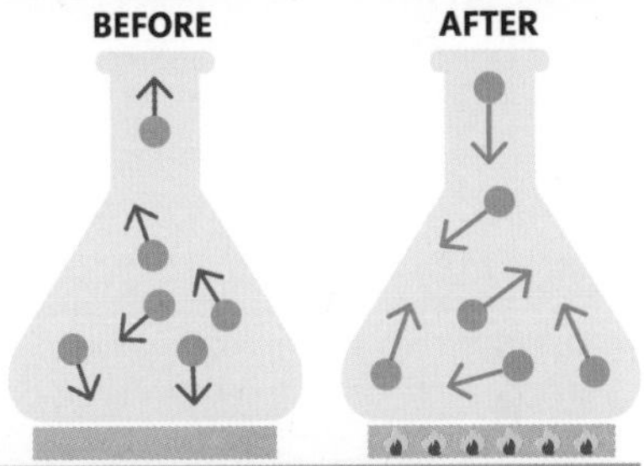

GASES, LIQUIDS, AND SOLIDS

Reducing volume
In a smaller container, atoms are squashed together, causing them to collide more often.

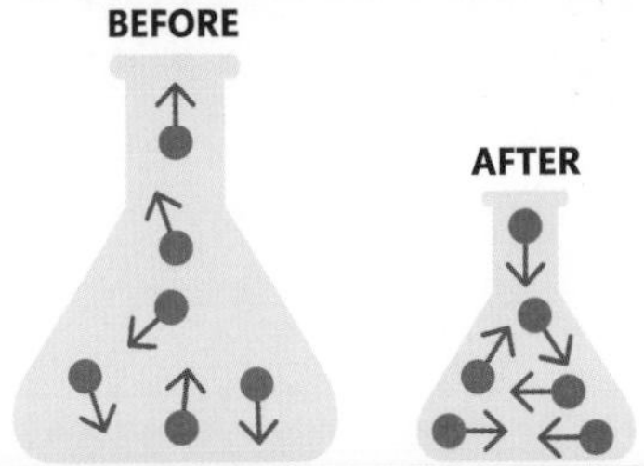

GASES ONLY

Increasing surface area of reactant
Collisions only occur on the surfaces of solids; increasing the surface area increases the reaction rate.

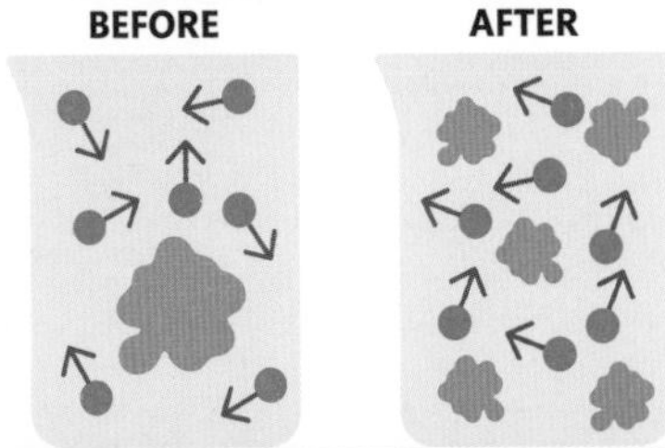

SOLIDS ONLY

The structure of metals
The outermost electrons in each atom are delocalized, so they are able to move between and around the ions. This is what gives metals their unique properties.

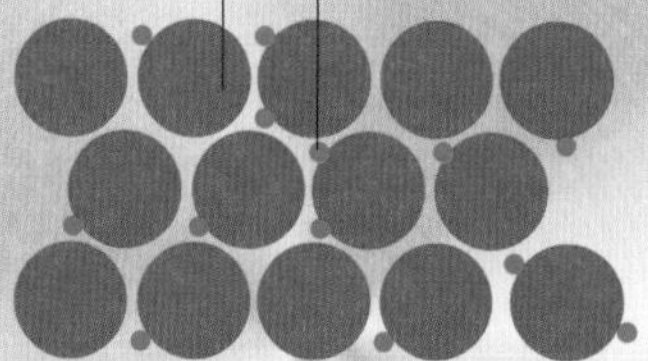

STRONG

Most metals are relatively strong yet tough. Their atoms compress a little after force is applied, before they return back to their previous positions.

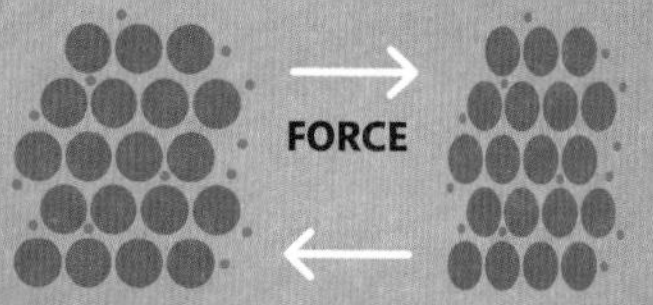

Examples
- Steel
- Tungsten
- Titanium

DENSE

The atoms in most metals are closely packed because of the strong attraction between the delocalized electrons and the positive ions. This makes them dense and heavy for their size.

Examples
- Osmium
- Platinum
- Plutonium

The atom can rearrar easily as the s of electrons mov with them, making metals ductile—for example, no bonds have to break for a metal to be stretched into a wire.

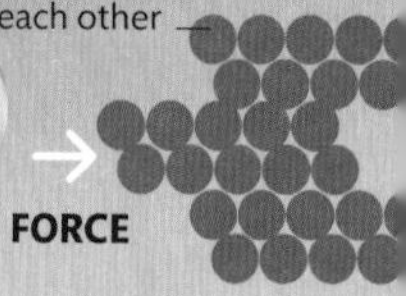

The electrons and ions in metals move without breaking thei bonds, so they can be hammered into shape

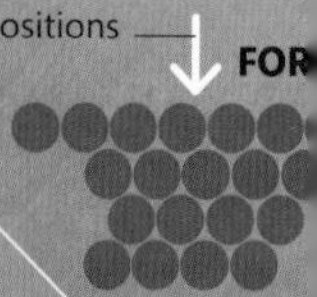

SHINY

There are many electrons orbiting each edge of a metal. When light encounters these, it is absorbed and then re-emitted, giving metals a lustrous glow.

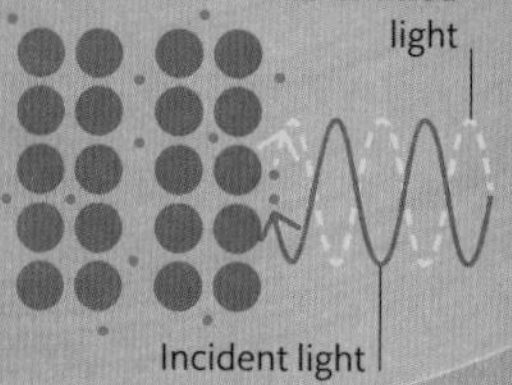

Examples
- Rhodium
- Aluminum
- Silver

HIGH MELTING AND BOILING POINTS

To melt a metal, the strong forces of attraction between the sea of negative electrons and the positive metal ions must be overcome. This takes a lot of energy in the form of heat.

Examples
- Tungsten
- Rhenium
- Osmium

ELECTRICAL CONDUCTOR

Electrons in a metal can move and carry charge, making metals good electrical conductors. However, some metals, like copper and silver, are better conductors than others.

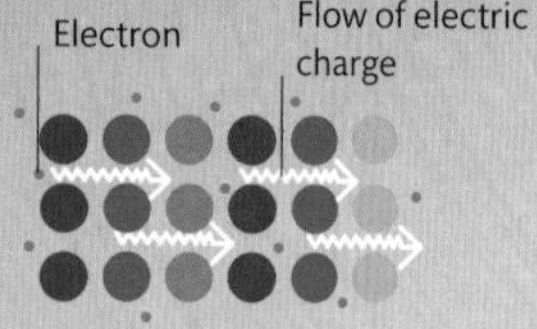

Examples
- Silver
- Copper
- Gold

HEAT CONDUCTOR

Electrons heated on one side of a metal are free to move to the other side, conducting the heat more rapidly than if it was passed from atom to atom, as in other materials.

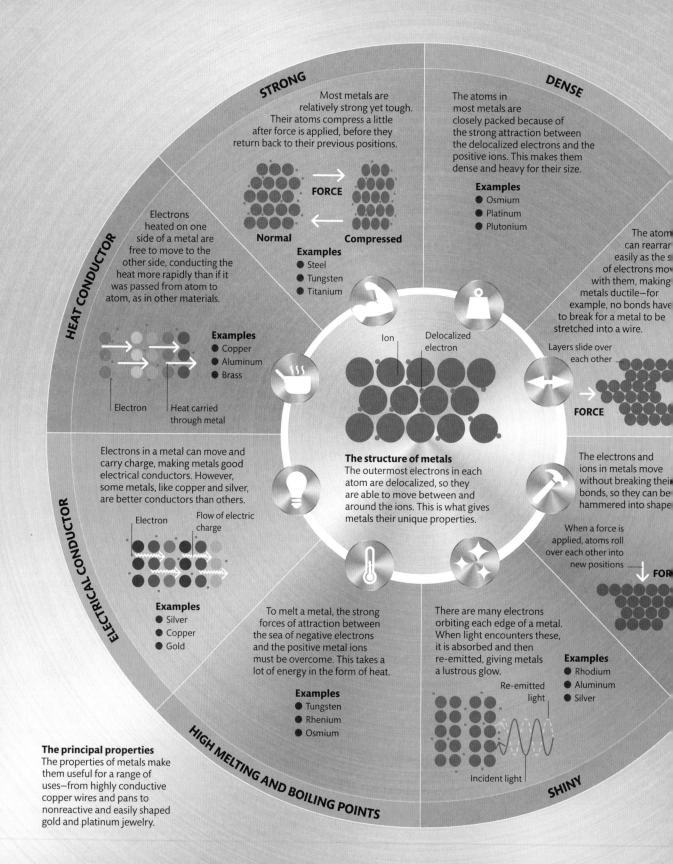

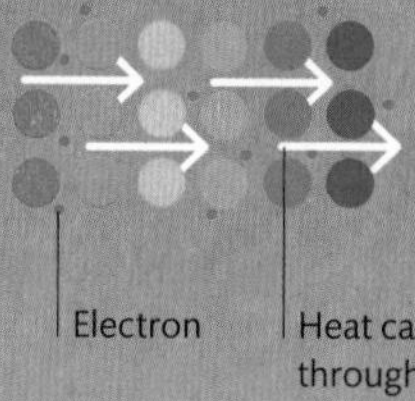

Examples
- Copper
- Aluminum
- Brass

The principal properties
The properties of metals make them useful for a range of uses—from highly conductive copper wires and pans to nonreactive and easily shaped gold and platinum jewelry.

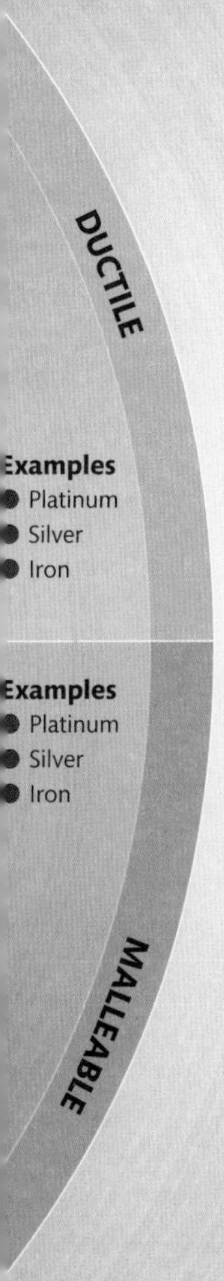

Metals

Metals make up over three-quarters of the elements found naturally on Earth, and vary dramatically in appearance and behavior. However, there are key properties that most metals share.

Properties of metals

Metals are crystalline substances, so they tend to be hard, shiny, and good conductors of electricity and heat. They are dense, with high melting and boiling points, but are easily shaped by a variety of methods. But some metals buck the trend. Mercury is liquid at room temperature because its outer electrons are very stable, so it does not tend to bond to other atoms.

RUST

Many metals are highly reactive, particularly the group 1 metals (see pp.34–35). Most metals form oxides when they combine with oxygen. For example, iron forms iron oxide, also known as rust, when exposed to the oxygen in the air or water.

Alloys

Most pure metals are too soft, brittle, or reactive for practical use. Combining metals or mixing metals with nonmetals forms alloys, often with improved properties. Varying the ratios and types of metals changes the properties of alloys. One common alloy is steel—a mixture of iron, carbon, and other elements. Adding more carbon makes harder steel, which is good for building. Adding chromium creates corrosion-resistant stainless steel. Other elements can also be included to increase heat resistance, durability, or toughness for use in items like car parts or drills.

IS AN OLYMPIC GOLD MEDAL ACTUALLY GOLD?

Only 92.5 percent of a gold medal is actually gold. The last solid gold Olympic gold medal was awarded in 1912.

Alloy composition
Copper forms two common alloys: bronze (tin is added to increase its hardness) and brass (zinc improves the alloy's malleability and durability). Stainless steel, another common alloy, varies in composition.

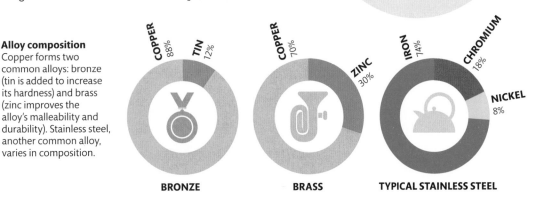

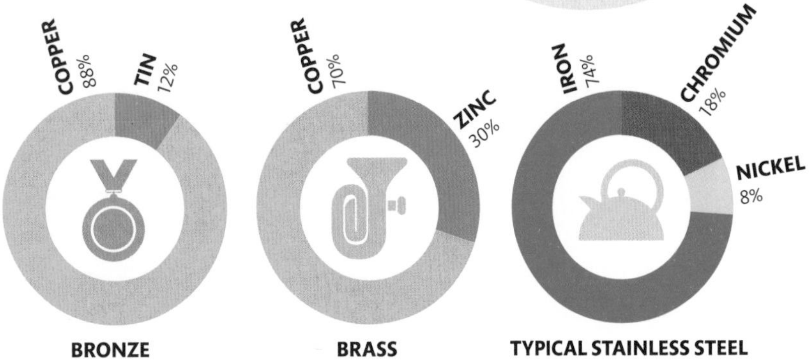

Hydrogen

The element hydrogen is thought to comprise 90 percent of the visible Universe. It is vital for life on Earth, principally because it is crucially involved in the formation of water and organic compounds called hydrocarbons. Hydrogen also has potential as a clean energy source for the future.

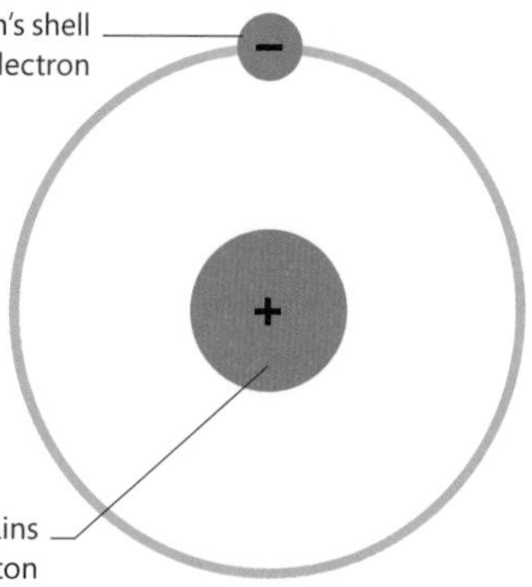

Hydrogen's shell has just one electron
Nucleus contains one proton

What is hydrogen?

Hydrogen is the main constituent of stars and the planets Jupiter, Saturn, Neptune, and Uranus. On Earth, at standard temperatures and pressures, it is a colorless, odorless, and tasteless gas. It is highly combustible and fairly reactive, so on Earth it mainly exists in molecular forms such as water, where it combines with oxygen. Hydrogen and carbon form millions of organic compounds called hydrocarbons, which form the basis of many living things.

The simplest element
Consisting of just one proton and one electron, hydrogen is the smallest, lightest, and simplest element in the Periodic Table (see pp.34–35). But it can react in complex ways, forming different types of atomic bonds, and allowing for interactions between acids and bases.

Inside a fuel cell
An electric current is generated by the movement of electrons to a cathode.

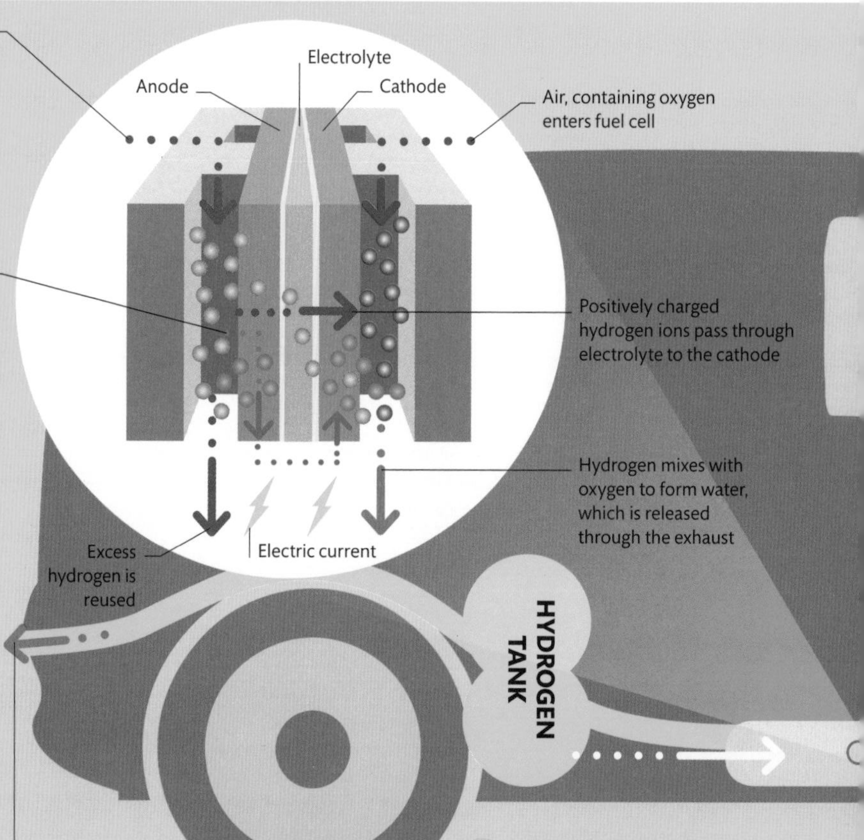

Hydrogen enters fuel cell and is split into positively charged hydrogen ions and negatively charged electrons
Negatively charged electrons travel along external circuit to cathode, creating a current to power the car's engine
Anode
Electrolyte
Cathode
Air, containing oxygen enters fuel cell
Positively charged hydrogen ions pass through electrolyte to the cathode
Hydrogen mixes with oxygen to form water, which is released through the exhaust
Excess hydrogen is reused
Electric current
HYDROGEN TANK
Water is released through exhaust

1 Supplying the fuel cell
Hydrogen is compressed in high-pressure tanks that can withstand extremes of heat and pressure. The gas is channeled from the tanks to the fuel cells.

REFUELING POINTS

A car's hydrogen tank can be refilled in about 5 minutes, and refueling points are becoming more widespread. However, it is currently a difficult fuel to transport because large pressurized tanks and pipes are required.

Harnessing hydrogen

Before hydrogen can be used as a fuel, it must be isolated. It can be extracted in a process that reacts steam with methane, but this produces greenhouse gases. A cleaner method called electrolysis uses electricity to split water into its constituent atoms. However, this is often inefficient and energy-intensive, so other methods are being developed to split water molecules using specialized catalysts.

How electrolysis works
Passing current through water causes hydrogen and oxygen atoms to gain and lose electrons respectively, turning the atoms into charged particles (ions). These travel to the anode and cathode, reunite with their electrons, and turn back into atoms of hydrogen and oxygen.

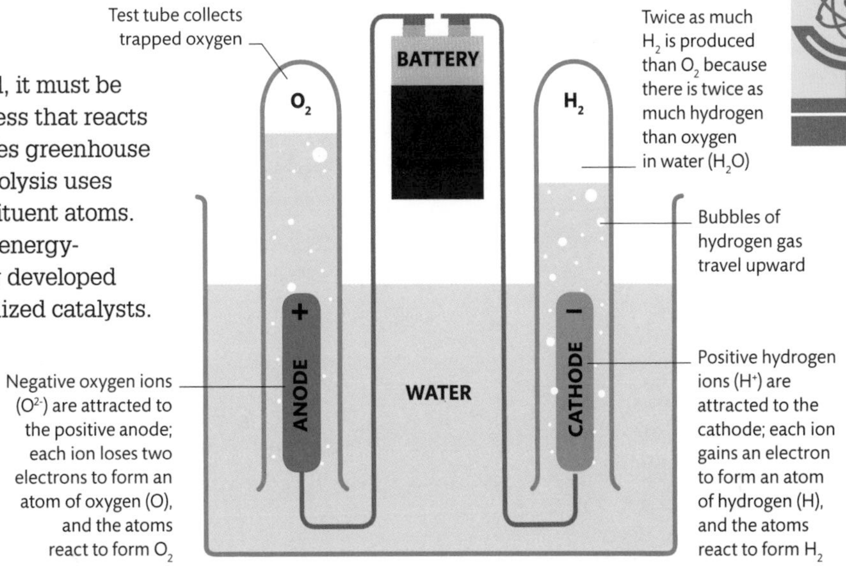
Test tube collects trapped oxygen
BATTERY
O_2
H_2
Twice as much H_2 is produced than O_2 because there is twice as much hydrogen than oxygen in water (H_2O)
Bubbles of hydrogen gas travel upward
+
ANODE
WATER
–
CATHODE
Negative oxygen ions (O^{2-}) are attracted to the positive anode; each ion loses two electrons to form an atom of oxygen (O), and the atoms react to form O_2
Positive hydrogen ions (H^+) are attracted to the cathode; each ion gains an electron to form an atom of hydrogen (H), and the atoms react to form H_2

Hydrogen-powered vehicles

Hydrogen's stored energy makes it a viable alternative fuel to petroleum. But because it is a gas, it contains less energy per unit volume than petroleum, so must be stored under pressure. This needs specialized equipment that requires energy, which in turn generates emissions. Scientists are developing improved storage and transportation methods, such as metal hydrides. These store hydrogen in a solid form, which then undergoes a reversible reaction (see pp.42–43) to release pure hydrogen when needed. This avoids some of the issues with storage, but introduces problems of its own, such as the weight of the compound.

Fuel of the future
Hydrogen-powered cars use tanks of compressed hydrogen, which supply hydrogen to fuel cells located in a stack. In the cells, hydrogen and oxygen undergo an electrochemical reaction, which generates electricity to supply the car's engine.

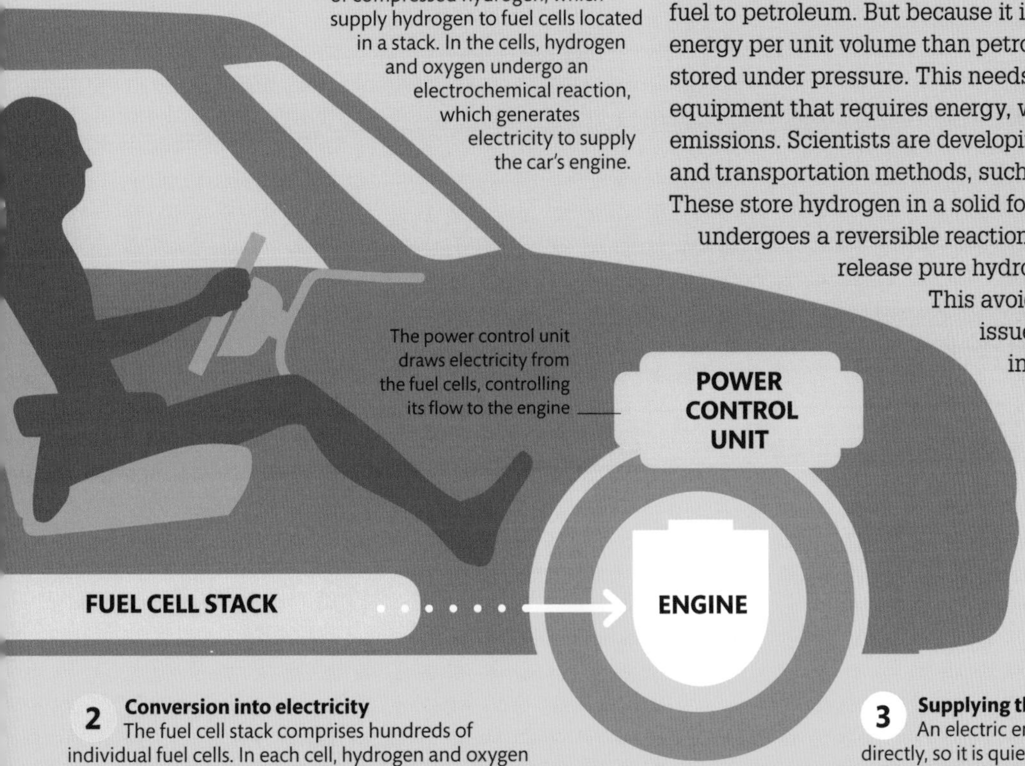
The power control unit draws electricity from the fuel cells, controlling its flow to the engine
POWER CONTROL UNIT
FUEL CELL STACK
ENGINE

2 Conversion into electricity
The fuel cell stack comprises hundreds of individual fuel cells. In each cell, hydrogen and oxygen are combined to generate electricity. This process is much more efficient than the combustion in a petroleum-driven car.

3 Supplying the engine
An electric engine drives the wheels directly, so it is quieter than internal combustion engines. Less energy is wasted, making the process more efficient.

Carbon

The element carbon accounts for 20 percent of all living things, and its atoms are the building blocks of the most complex molecules known to science. No other element has the structural versatility to behave in quite the same way.

WHAT DOES "ORGANIC" MEAN?

In the chemical sense, "organic" substances contain carbon. The term is usually restricted to compounds with carbon and hydrogen in combination, called hydrocarbons.

What makes carbon special?

Carbon atoms bond prolifically with others to form an incredible variety of molecular shapes. Each one has an outer ring of four electrons that can form four strong bonds. Most frequently, carbon atoms bond to smaller hydrogen atoms, or to one another—but other elements can be part of the mix, too. The results are molecules that contain an interconnected carbon "skeleton" and an outer "skin" of hydrogen: from simple methane, with just one carbon atom, to enormously long chains.

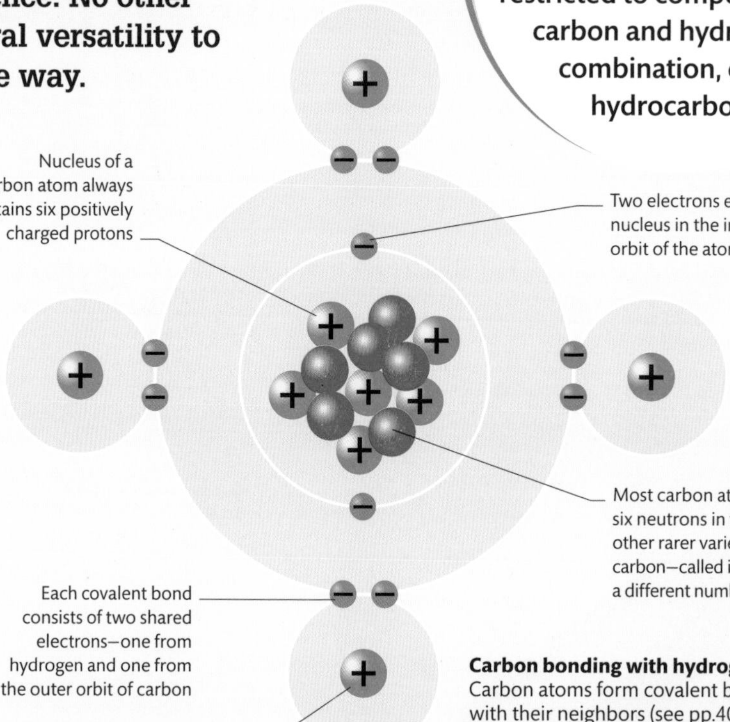

Nucleus of a carbon atom always contains six positively charged protons
Two electrons encircle the nucleus in the innermost orbit of the atom
Most carbon atoms contain six neutrons in their nucleus; other rarer varieties of carbon—called isotopes—have a different number of neutrons
Each covalent bond consists of two shared electrons—one from hydrogen and one from the outer orbit of carbon
Hydrogen nucleus consists of a single proton

Carbon bonding with hydrogen
Carbon atoms form covalent bonds with their neighbors (see pp.40–41). This means that electrons are shared in a strong connection. One carbon atom bonds with four hydrogen atoms to make a molecule of methane.

CHAINS AND RINGS

There are countless ways in which carbon and other kinds of atoms can bond together to form molecules. Each shape is a unique chemical compound with its own properties. The shortest chain is the two-carbon natural gas known as ethane (C_2H_6). When sufficiently long, the ends of a chain of carbon atoms can close together to form a ring—such as benzene (C_6H_6), a liquid component of crude oil.

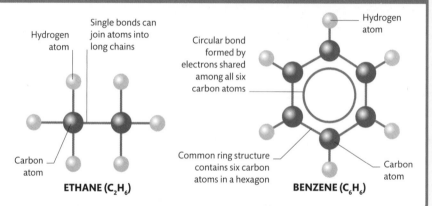

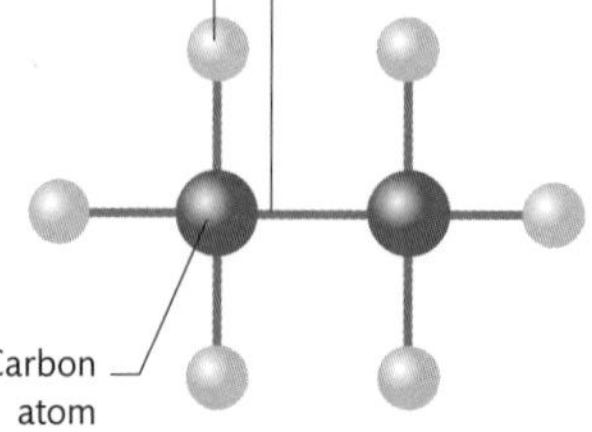

Hydrogen atom
Single bonds can join atoms into long chains
Carbon atom

ETHANE (C_2H_6)

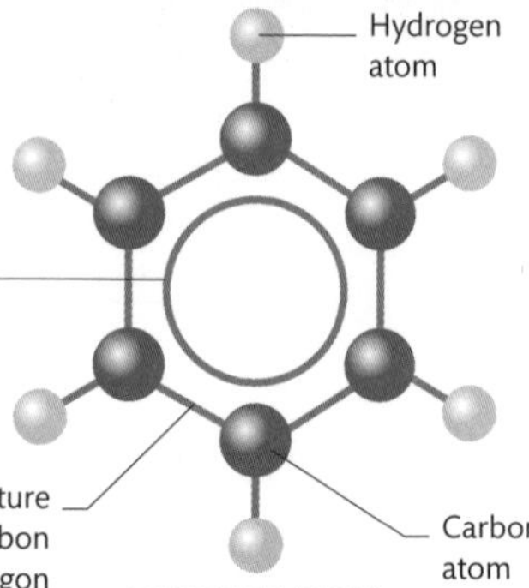

Hydrogen atom
Circular bond formed by electrons shared among all six carbon atoms
Common ring structure contains six carbon atoms in a hexagon
Carbon atom

BENZENE (C_6H_6)

Allotropes of carbon

The atoms of some elements in their purest forms can bond together in different ways to give varied physical states called allotropes. Solid carbon has three main allotropes—the layered, flaky structure of graphite, ultra-hard crystals of diamond, and the hollowed "cage" of fullerenes.

Graphite
Graphite is flaky because its carbon atoms are arranged in sheets, so they slide past one another. Each atom has three, rather than four, single bonds; the extra electron wanders through the sheet, which is what makes graphite electrically conductive.

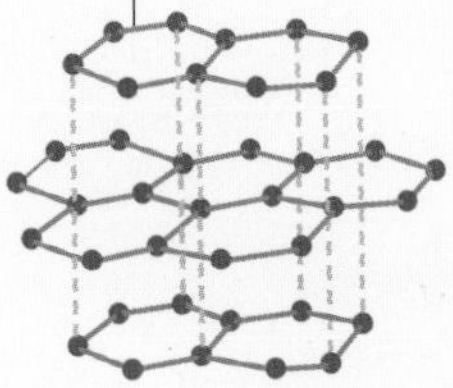

Diamond
In a diamond, the carbon atoms are arranged into a three-dimensional crystal, with each atom bonded to four others. This makes the entire structure strong and very hard. There are no free electrons, so unlike graphite, diamond does not conduct electricity.

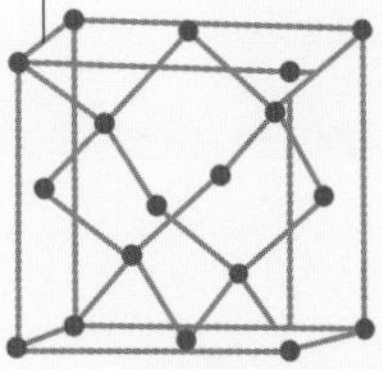

Fullerenes
Fullerenes have their atoms arranged into spherical or tubular "cages." Although hollow, their structures are rigid and strong, and the unique atomic arrangements have many applications, such as reinforcing graphite in tennis rackets.

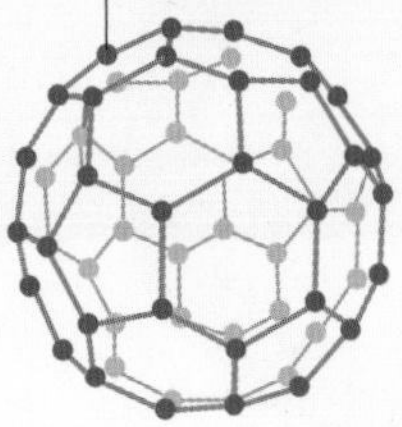

THE CULLINAN DIAMOND—THE WORLD'S LARGEST—WEIGHED 22OZ (621.35 GRAMS)

The building blocks of life

The most elaborate carbon-containing molecules of all are in the bodies of living things. Here, carbon atoms routinely incorporate oxygen, nitrogen, and a few other elements into their structures to form biochemicals—the molecules of life. Most of these fall into four main groups: proteins, carbohydrates, lipids, and nucleic acids. All are formed by complex sets of reactions known together as metabolism.

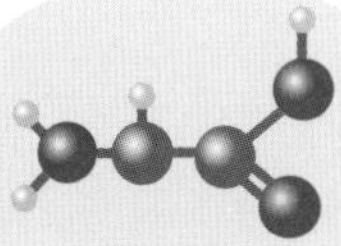

Proteins
Carbon-containing amino acids form chains called proteins that make up tissues such as muscles and also speed up reactions in cells.

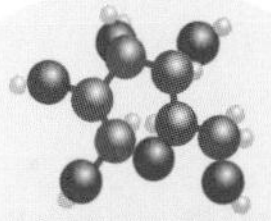

Carbohydrates
Carbon forms a crucial part of carbohydrates, the simplest of which are sugars, which are broken down to release energy.

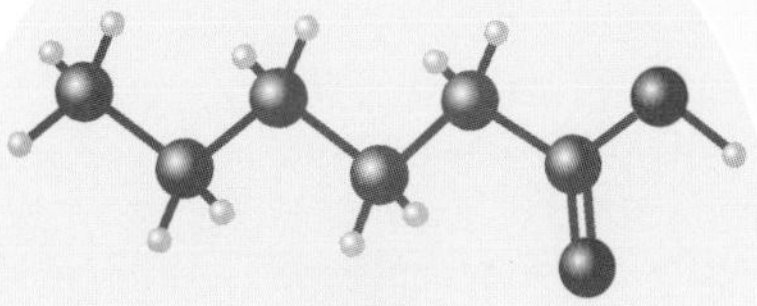

Lipids
Fats and oils—collectively called lipids—contain molecules known as fatty acids, formed of carbon, hydrogen, and oxygen. Many work as energy stores.

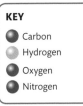

Nucleic acids
Nucleic acids—such as DNA—are complex molecules that carry genetic information; they comprise nitrogen, phosphorus, and carbon.

KEY
- Carbon
- Hydrogen
- Oxygen
- Nitrogen

Air

Air is the mixture of gases in the atmosphere. It is vital for survival, providing oxygen for respiration in animals and carbon dioxide for plants to use in photosynthesis. However, when air becomes polluted, it affects these processes and can damage our health.

The composition of air

Air is predominantly nitrogen, with around 20 percent oxygen; 1 percent argon; and smaller amounts of other gases, including carbon dioxide (CO_2). Water vapor content varies widely depending on location, so compositions usually exclude it, but it can make up as much as 5 percent of air in humid climates. Human behavior changes the composition of air, most notably by increasing the amount of CO_2.

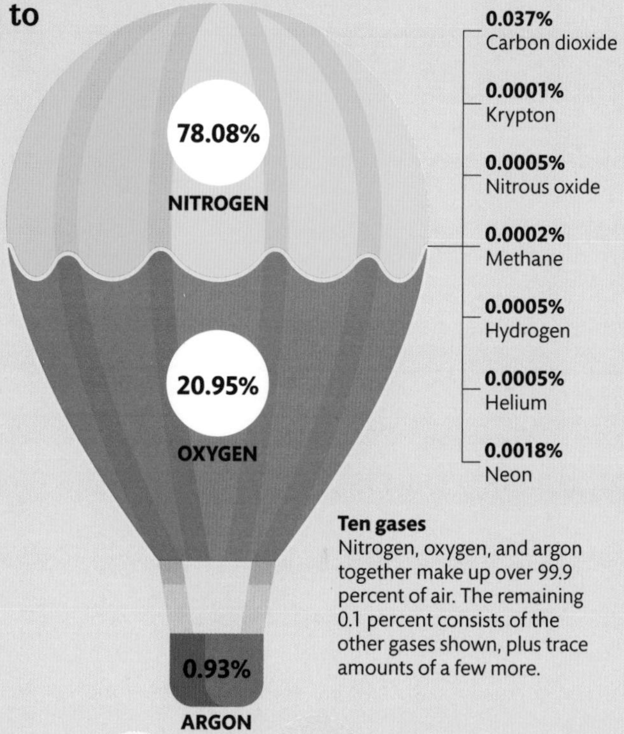

Ten gases
Nitrogen, oxygen, and argon together make up over 99.9 percent of air. The remaining 0.1 percent consists of the other gases shown, plus trace amounts of a few more.

92 PERCENT OF THE WORLD'S POPULATION BREATHES AIR THAT EXCEEDS W.H.O. SAFETY LIMITS

Air pollution

Air pollution is a huge problem—the World Health Organization (W.H.O.) found that more deaths are caused by poor air than tuberculosis, HIV/AIDS, and traffic accidents combined. In the developing world, the biggest air pollution source is burning wood or other fuels within the home. In cities, car exhausts and emissions from houses and industrial sites can lead to high pollution zones. These can exacerbate asthma and other respiratory illnesses. Particulate matter—a complex mixture of tiny airborne particles and liquid droplets—is especially damaging when fine enough to penetrate deep into the lungs.

Primary pollutants and their sources
There are six primary pollutants, which are released into the atmosphere directly, and six principal sources of primary pollutants. This color-coded graphic shows how much of each primary pollutant each of the sources contributes.

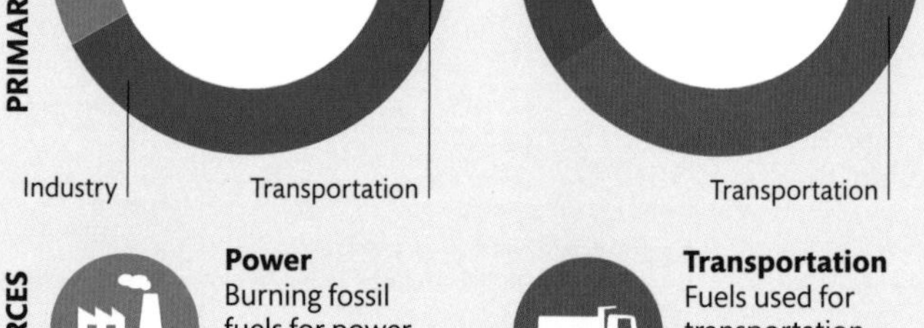

SOURCES

Power
Burning fossil fuels for power contributes a large proportion of the sulfur dioxide released into the atmosphere.

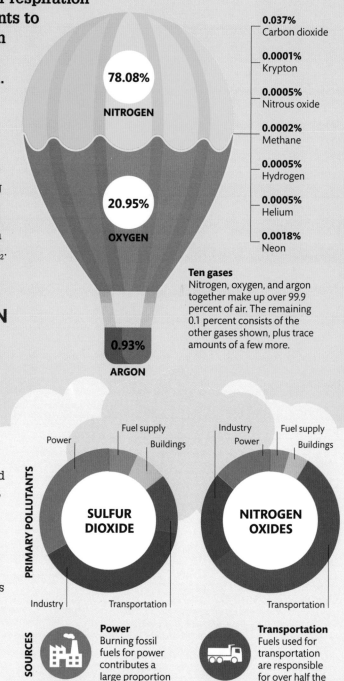

Transportation
Fuels used for transportation are responsible for over half the worldwide emissions of dangerous nitrogen oxides.

The changing color of the sky

The color of visible light depends on which light waves reach our eyes. Blue light, with its short wavelength, is scattered most by atmospheric particles. This creates the effect of a blue sky in the day (see p.107). Longer red and orange light is scattered the least, so they are not visible during the day, but they are at sunset, when the Sun is lower in the sky. Blood-red sunsets around cities occur largely because of suspended particles produced by internal combustion engines. These particles scatter out the violet and blue colors and enhance the red.

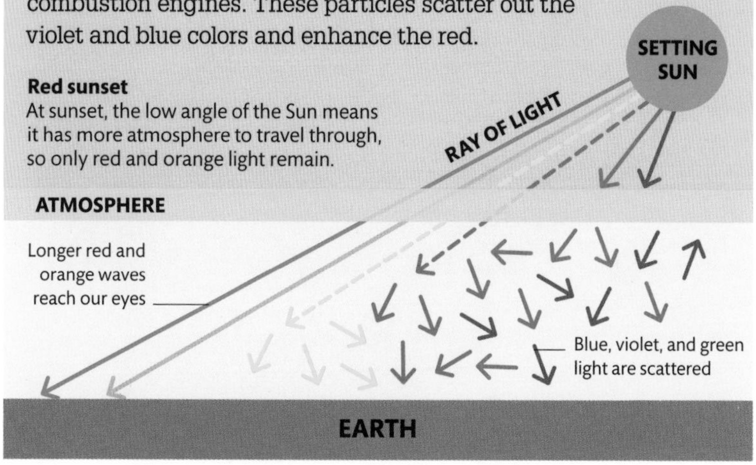

Red sunset
At sunset, the low angle of the Sun means it has more atmosphere to travel through, so only red and orange light remain.

POLLUTION IN THE HOME

The air in our homes can be hugely contaminated, too. Benzene from cigarette smoke, paint, and scented candles; nitrogen dioxide emitted from incomplete combustion in gas stoves; and formaldehyde from foam inside furniture are common in the home and all potentially dangerous to our health. Increasing the number of houseplants helps to absorb toxic chemicals, and air purifiers are increasingly effective at combating poor air quality.

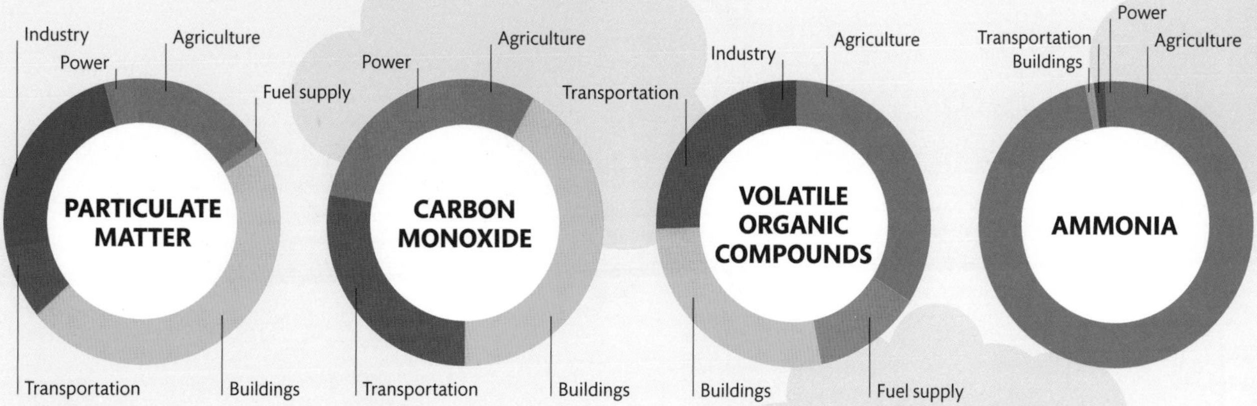

Industry
Factories are responsible for large quantities of sulfur dioxide, nitrogen oxides, and particulate matter emissions.

Buildings
The majority of carbon monoxide emissions are created by cooking and heating in homes, especially by solid fuel stoves.

Fuel supply
Extracting, transporting, and processing fuel produces pollution mostly as volatile organic compounds.

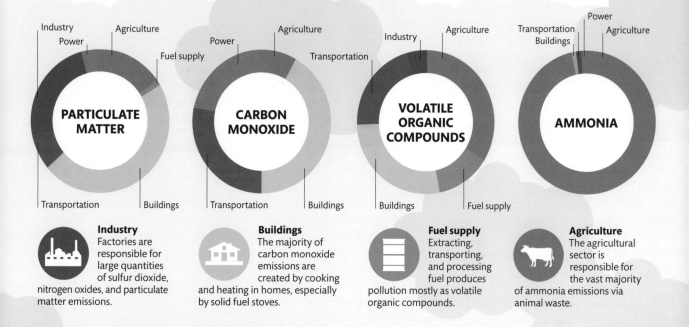

Agriculture
The agricultural sector is responsible for the vast majority of ammonia emissions via animal waste.

Burning and exploding

Taming fire has allowed humans to cook, ward off dangerous animals, generate electricity, and develop engines. But fire can cause great damage if it gets out of control, and simple combustion can become a devastating explosion, so it is vital to understand how fire works.

Combustion

Combustion, or burning, is a chemical reaction. A fuel, normally a hydrocarbon such as coal or methane, reacts with oxygen in the air, releasing energy as heat and light. In complete combustion, with plentiful oxygen, carbon dioxide, and water are produced. Once it has started, combustion continues unless the fire is extinguished or the fuel or oxygen runs out.

A FOREST FIRE MAY REACH TEMPERATURES OF 1,470°F (800°C) OR MORE

SPONTANEOUS COMBUSTION

Normally, an input of energy, such as a spark or flame, is needed to initiate combustion. However, some substances—such as hay, certain oils, or some reactive elements such as rubidium—may spontaneously ignite if they become hot enough.

HAY AND STRAW | **LINSEED OIL** | **RUBIDIUM**

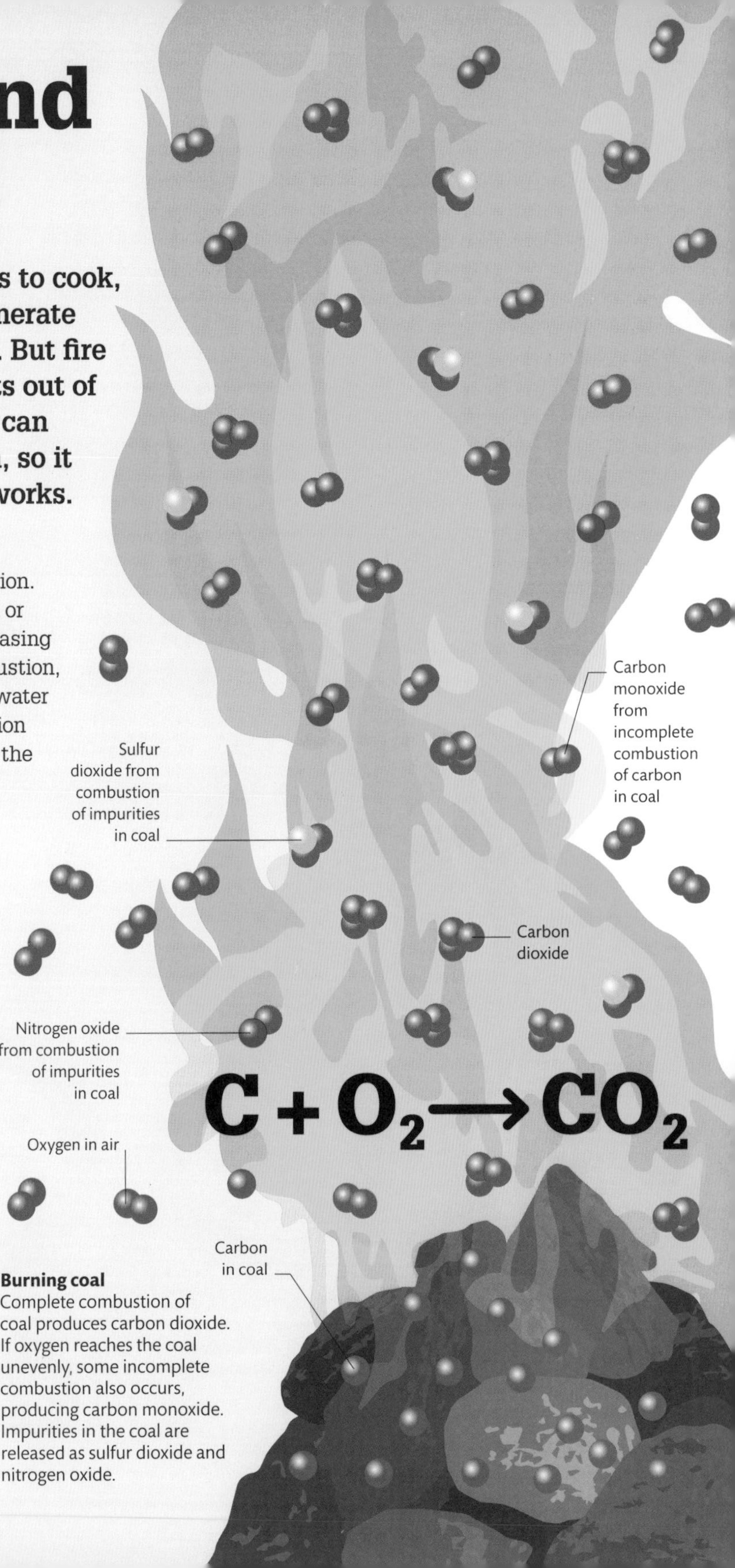

$$C + O_2 \longrightarrow CO_2$$

Burning coal
Complete combustion of coal produces carbon dioxide. If oxygen reaches the coal unevenly, some incomplete combustion also occurs, producing carbon monoxide. Impurities in the coal are released as sulfur dioxide and nitrogen oxide.

Extinguishing fires

Fire needs three things to burn: heat, fuel, and oxygen (often in the form of air). Removing any one of these can put out a fire. However, the best method for extinguishing a fire depends on the type. For example, using water to put out an electrical fire may lead to electrocution, and using water on an oil or grease fire may cause the burning oil or grease to spread.

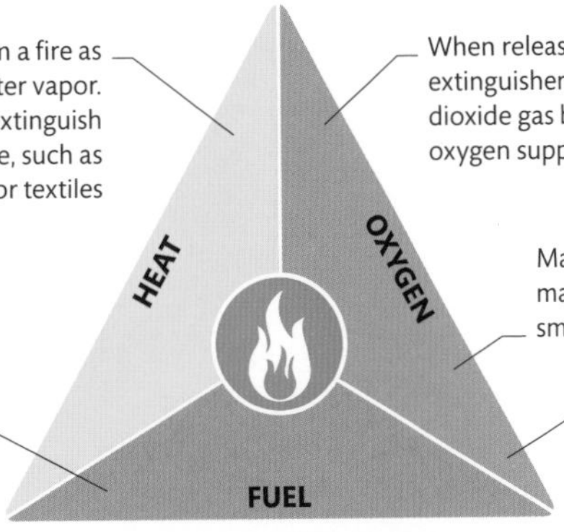
Water absorbs heat from a fire as it evaporates into water vapor. This can be enough to extinguish some types of fire, such as burning wood or textiles
When released from an extinguisher, carbon dioxide gas blocks a fire's oxygen supply
HEAT
OXYGEN
Made of fire-resistant material, fire blankets smother flames
In large forest fires, cutting down trees in the fire's path starves it of fuel, preventing it from spreading further
FUEL
Powder and foam extinguishers form a coating over the burning material, starving it of oxygen

THE FIRE TRIANGLE

Explosions

An explosion is a sudden release of heat, light, gas, and pressure. Explosions happen much more quickly than combustion. The heat from the explosion cannot dissipate, and the gases produced expand quickly, creating a shockwave that travels rapidly away from the explosion and may be powerful enough to cause injury and damage to property. Shrapnel forced outward by the blast causes further damage.

CAN YOU OUTRUN AN EXPLOSION?

No. In chemical explosions, the material given off by the explosion moves at more than 5 miles per second (8km per second), far faster than anybody can run.

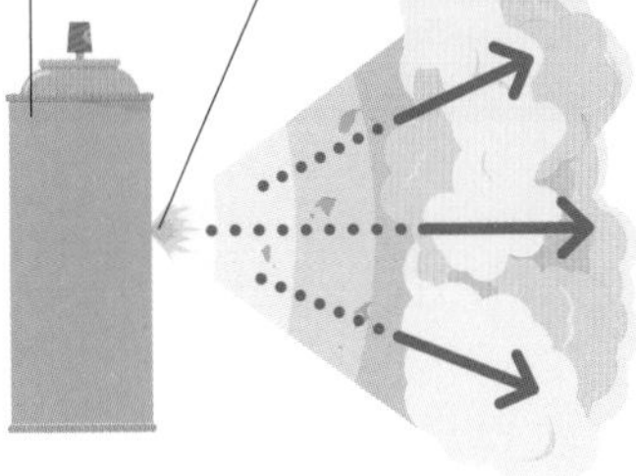
Liquid and gas under high pressure in container
Weak point in container ruptures, causing explosion

Physical explosion
A weak point in a pressurized container can rupture and allow its contents to escape suddenly. The pressure decrease causes the gases to expand very rapidly—resulting in an explosion.

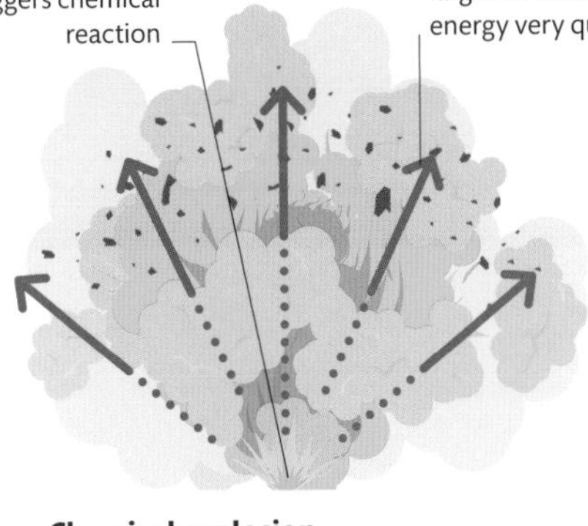
Applying energy, such as heat, triggers chemical reaction
Reaction releases large amount of energy very quickly

Chemical explosion
Chemical explosions are caused by rapid reactions that release large amounts of gas and heat. The reaction is often triggered by heat, as with gunpowder, or physical shock, as with nitroglycerine.

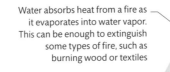

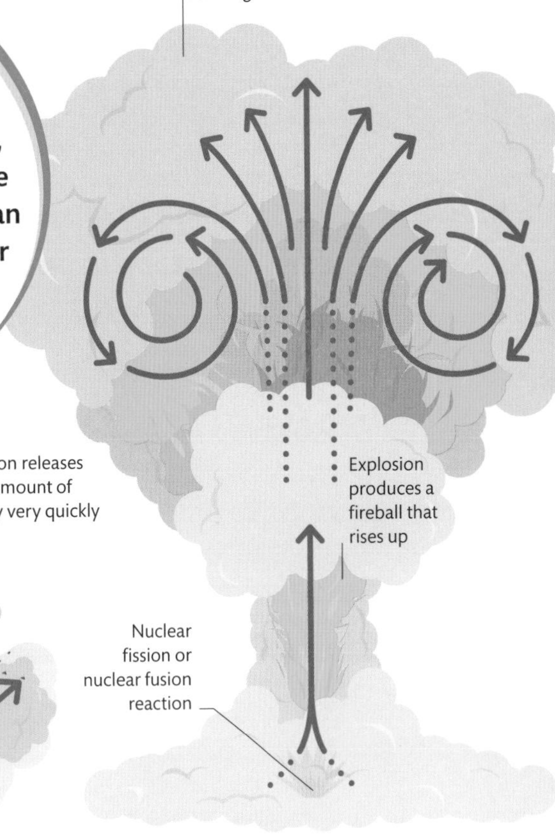
Fireball cools and condenses, forming mushroom cloud
Explosion produces a fireball that rises up
Nuclear fission or nuclear fusion reaction

Nuclear explosion
Nuclear explosions can be powered by fission (splitting) or fusion (joining together) of atomic nuclei. Either produces huge amounts of energy very rapidly, as well as radioactive fallout.

Ice
As water cools, its molecules slow down, allowing more hydrogen bonds to form. These bonds hold the molecules apart as water freezes and become locked in an open structure. This is why water expands as it freezes.

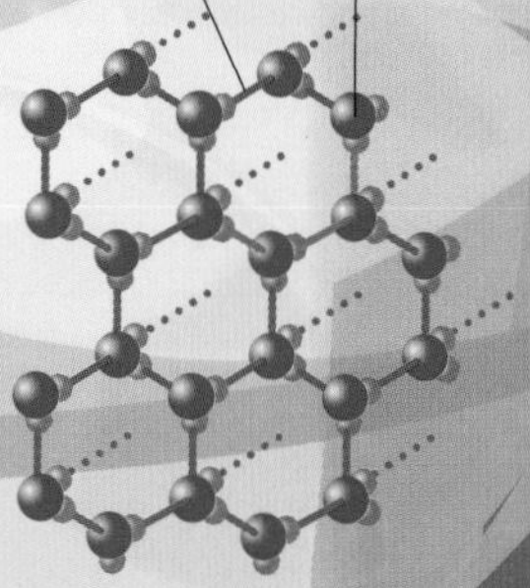

Water
When water is a liquid, its hydrogen bonds form and break repeatedly as the molecules move past each other. Without these bonds, water would be a gas at room temperature.

Water

Water may be an everyday substance, but it is extraordinary. It is the only substance that can exist as a solid, liquid, and gas at normal temperatures and pressures and the only one whose solid is less dense than its liquid.

Unique properties

Each water molecule consists of two hydrogen atoms bonded to one oxygen atom. One side of the molecule (where the oxygen is) has a weak negative charge, while the other side has a small positive charge. These different charges allow hydrogen bonds to form between molecules, which give water its unique properties.

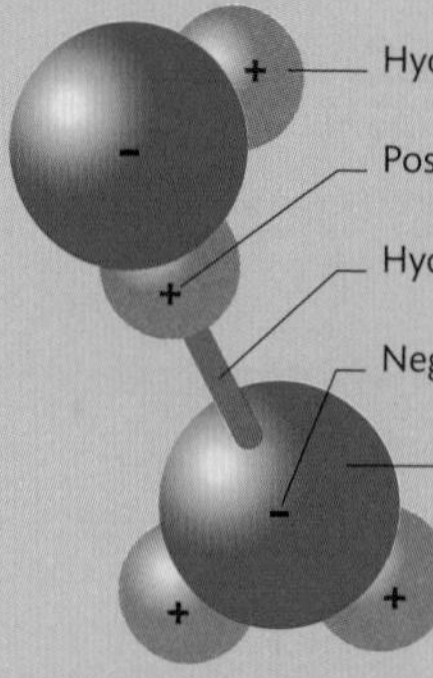

SURFACE TENSION

Water prefers to bond with itself rather than with air. As a result, water molecules at the surface form stronger bonds with their neighboring water molecules instead of bonding with the air molecules above. This, in effect, creates a layer at the surface that is strong enough to allow small insects to walk on it.

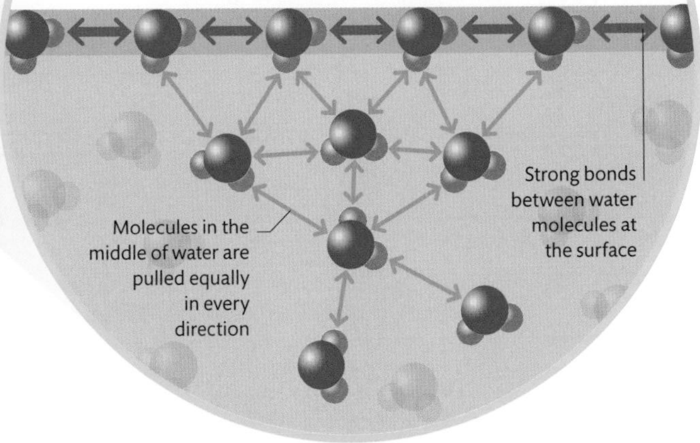
Strong bonds between water molecules at the surface
Molecules in the middle of water are pulled equally in every direction

CAPILLARY ACTION

Water molecules are attracted to some surfaces—how much depends on the material. In a thin glass tube, water creeps upward because attractive forces between the glass and water molecules are stronger than those between the water molecules themselves.

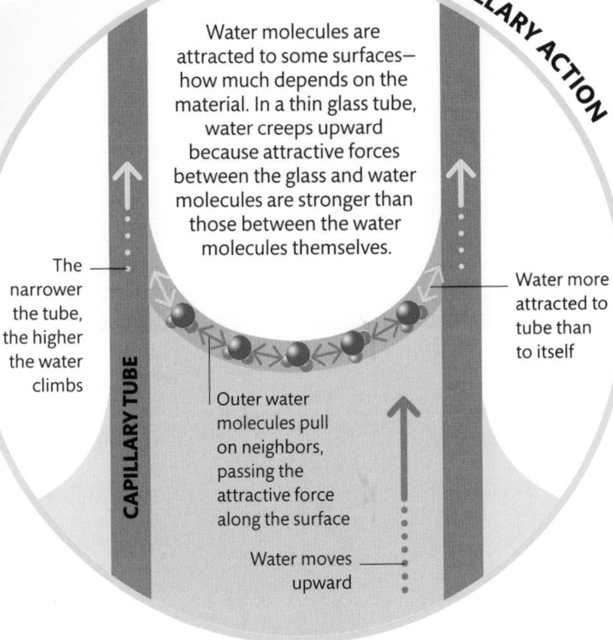
The narrower the tube, the higher the water climbs
CAPILLARY TUBE
Water more attracted to tube than to itself
Outer water molecules pull on neighbors, passing the attractive force along the surface
Water moves upward

WATER IN THE BODY

Water makes up about 60 percent of body weight in men and about 55 percent in women. The amount is lower in women because they have more body fat, which contains less water than lean tissue. On average, we need to drink 3 to 4 pints (1.5 to 2 liters) of water a day to replace that lost in urine, sweat, and the breath, although the exact amount depends on the climate and activity level.

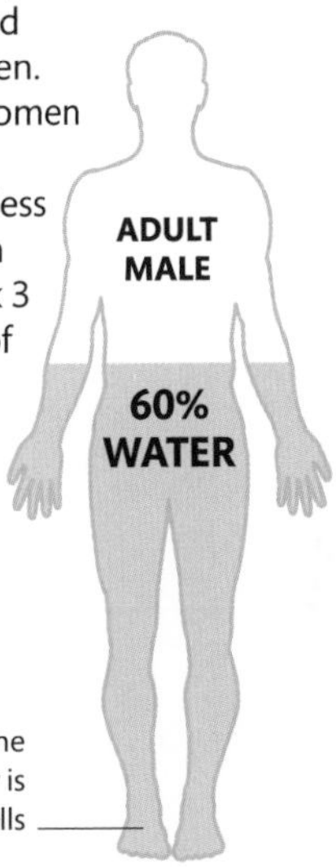
ADULT MALE
60% WATER
Most of the body's water is inside body cells

WHY DOES WATER SOMETIMES LOOK BLUE?

Water absorbs long wavelengths of light, at the red end of the spectrum, so the remaining light we see consists of the shorter wavelengths, at the blue end.

9 PERCENT

THE AMOUNT BY WHICH WATER EXPANDS WHEN IT FREEZES

Acids and bases

Even though they have opposite effects in chemical terms, both acids and bases are familiar to us as stinging or even dangerously corrosive substances. The strength of acids and bases varies over a wide range.

What is an acid?

Acids are substances with hydrogen atoms that are released as positively charged hydrogen ions when dissolved in water. The more of these ions it can release, the stronger the acid. For example, the gas hydrogen chloride behaves like this to form a solution called hydrochloric acid. This is one of the strongest of acids, with a hydrogen ion concentration a thousand times greater than the weaker acids found in some sour fruit.

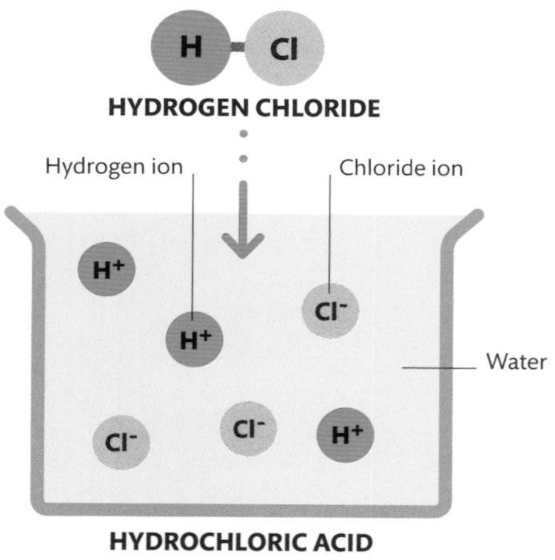

ACID RAIN

The corrosive effect of an acid is caused by its hydrogen ions, because these highly reactive chemical particles can break down other materials. Polluting sulfur dioxide gas, a by-product of industry, reacts with atmospheric water droplets to form sulfuric acid. When this falls as acid rain, it corrodes limestone buildings, as well as killing the foliage of trees and other plants.

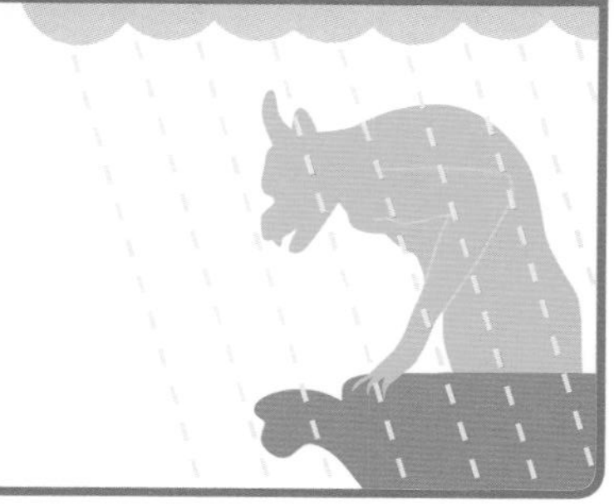

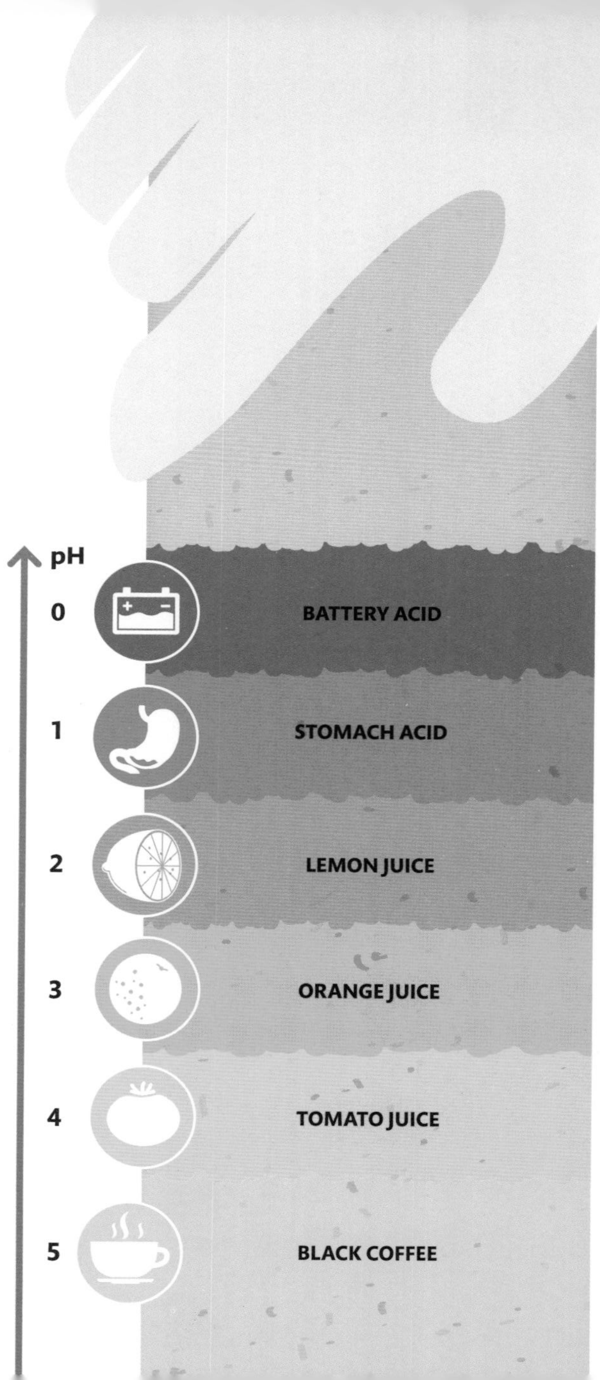

What is a base?

Bases are substances that are chemically antagonistic (opposed) to acids—but can be just as reactive. They counteract acids by neutralizing their hydrogen ions. Limestone or chalk is a basic rock because it reacts with acid in this way. The strongest bases, such as sodium hydroxide (caustic soda), dissolve in water and are called alkalis. In water, they release negatively charged particles called hydroxide ions.

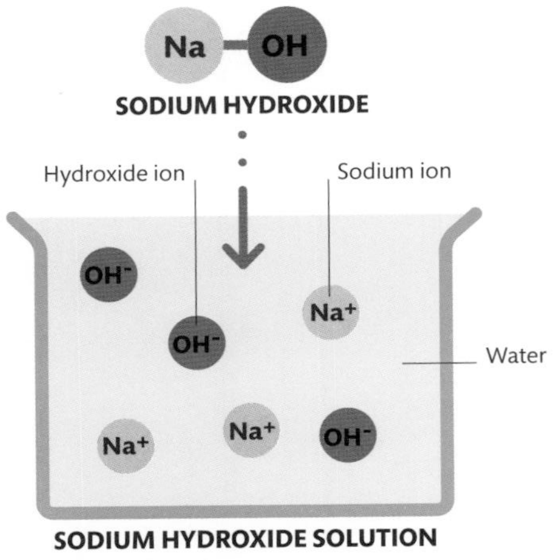

Acid-base reactions

The reaction between an acid and a base produces water and a different kind of substance called a salt. The type of salt formed depends on the types of acid and base involved. Hydrochloric acid and sodium hydroxide react to form sodium chloride (common table salt), with hydroxide and hydrogen ions coming together as water.

HOW DO ACIDS AND ALKALIS BURN YOU?

Acids and alkalis both damage protein in the skin, killing skin cells. Unlike acids, alkalis also liquefy the tissue, which can help them to penetrate deeper and cause more damage than acids.

Measuring acidity
The pH scale is a measure of the acidic or alkaline strength of a substance. It ranges from 0 for strong acids to 14 for strong alkalis. At each step up the scale, the hydrogen ion concentration is 10 times lower. A pigment called an indicator can be used to measure the pH of a substance. Reaction with the indicator produces colors from red for pH 0 to purple for pH 14, with green at pH 7 (neutral).

Crystals

From the hardest gemstone to a fleeting, delicate snowflake, the structure of a crystal can be a thing of beauty. This property comes from a precisely ordered microscopic arrangement of its atoms or other particles.

What is a crystal?

Crystalline solids (see p.14) are made up of neatly ordered particles: a repeated pattern of atoms, ions, or molecules runs right through their structure. This contrasts with amorphous (noncrystalline) materials, such as plastic or glass (see pp.70–71), in which particles are randomly jumbled together. Some solids, such as most metals, are only partly crystalline. They contain lots of tiny crystals called grains, but the individual grains are bonded in a random way.

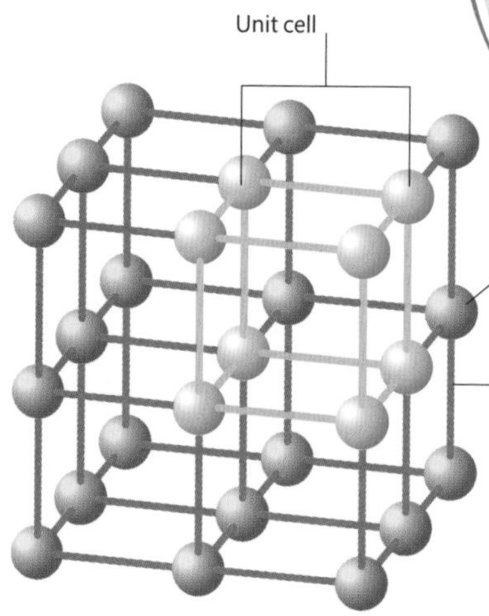
Unit cell
Atom
Bond between atoms

Crystal structure
A crystal contains a repeated arrangement of atoms, called a unit cell. The simplest unit cell—shown here—is a cube of eight particles. The planes of atoms run parallel, and crystals can be split along these planes.

WHY ARE SOME CRYSTALS COLORED?

Like any substance, crystals are colored if their atoms reflect or absorb certain wavelengths of light. Ruby is red, for example, due to its chromium atoms, which reflect red light.

Mineral crystals

Minerals—the chemical ingredients of rock—are crystallized from Earth's bedrock by geological processes. Crystals form when molten rock solidifies or when solid fragments recrystallize under heat and pressure. Crystals can also grow from solution, as water deposits dissolved minerals when they become too concentrated to carry. If such crystallization is stable over a long period (see right), crystals can grow to enormous size.

GIANT NATURAL **CRYSTALS** OF **GYPSUM** WEIGH UP TO **55 TONS (50 TONNES)**

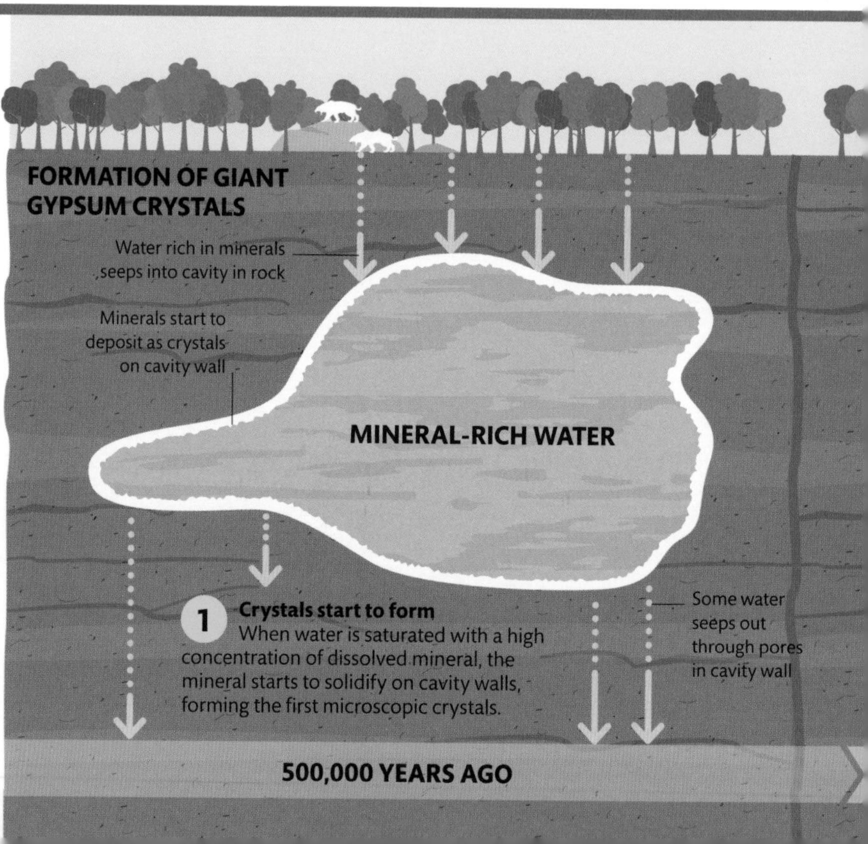
FORMATION OF GIANT GYPSUM CRYSTALS
Water rich in minerals seeps into cavity in rock
Minerals start to deposit as crystals on cavity wall
MINERAL-RICH WATER
Some water seeps out through pores in cavity wall
1 Crystals start to form
When water is saturated with a high concentration of dissolved mineral, the mineral starts to solidify on cavity walls, forming the first microscopic crystals.
500,000 YEARS AGO

Liquid crystals

Some materials flow but have crystalline properties. These liquid crystals exist in a state between liquid and solid. Their particles are neatly ordered, but they can also turn so they point in different directions. Like particles in solid crystals, they affect the way light is transmitted. The rotating molecules can "twist" polarized light (light that vibrates in one direction). This property forms the basis of a liquid crystal display, in which electricity controls the alignment of molecules to illuminate some pixels but not others.

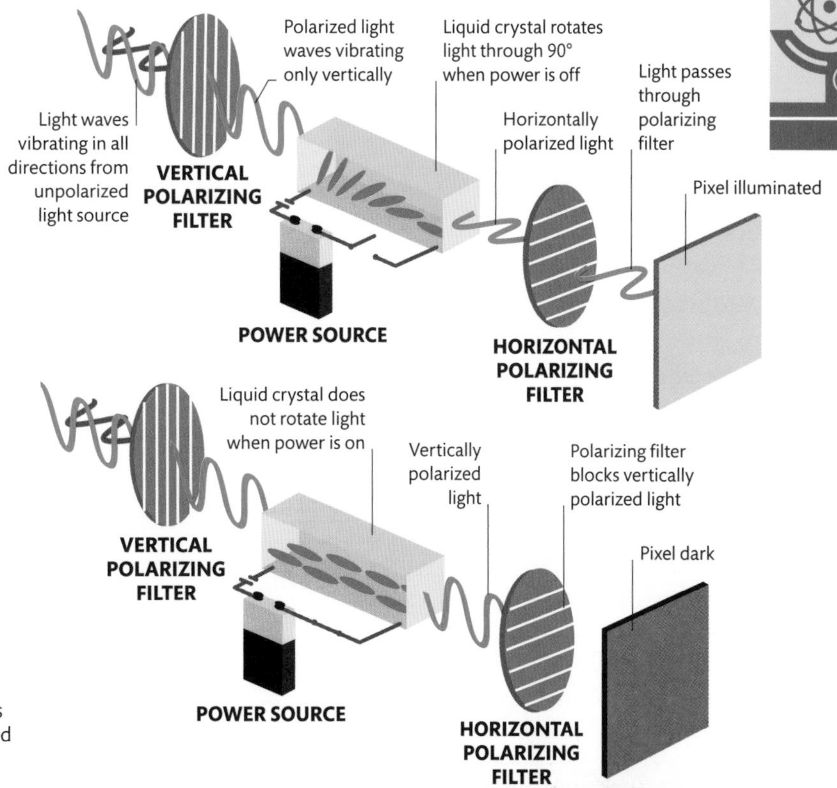

Liquid crystal display
In their "resting" state, liquid crystal molecules rotate polarized light to illuminate a pixel. But when aligned by an electrical current, light passes through untwisted—its vertical vibration is blocked by the horizontal filter, resulting in a dark pixel.

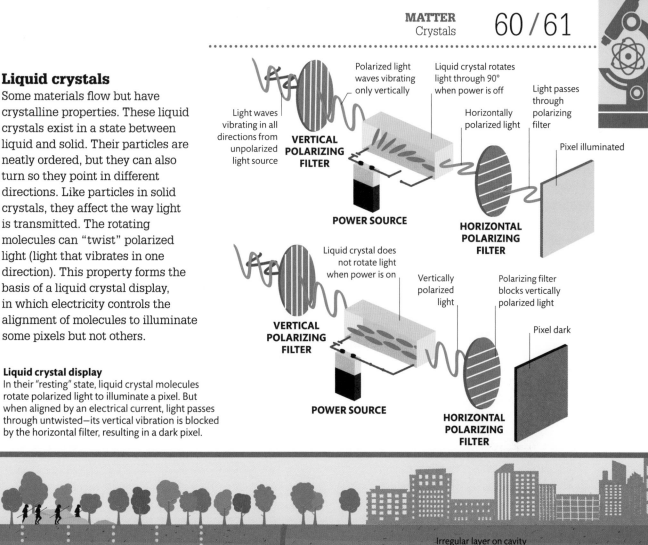

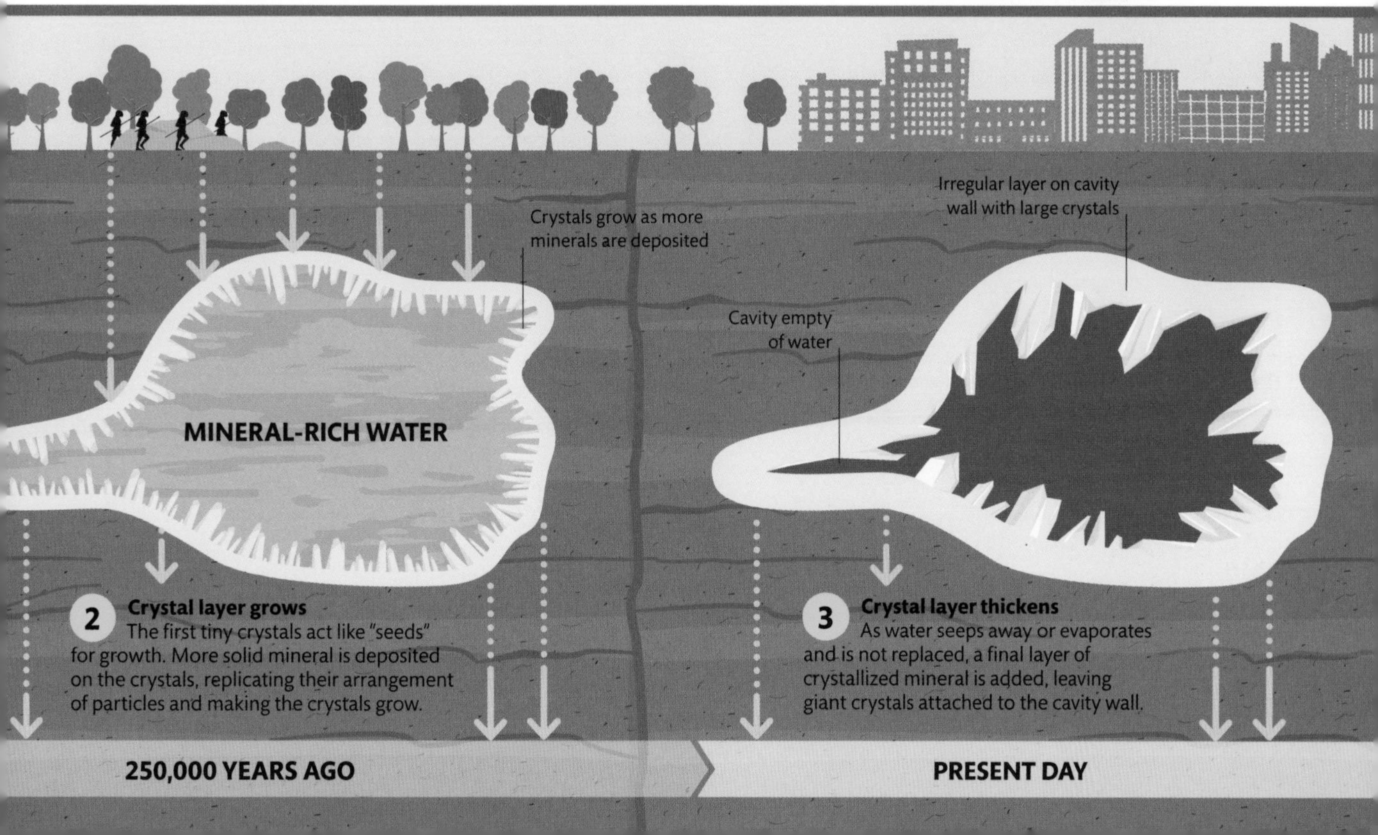

Solutions and solvents

Salt or sugar seem to disappear when they are added to water. But their tastes linger—proof that they have dissolved in the water and spread right through the solution.

CAN YOU DISPOSE OF A BODY USING ACID OR ALKALI?

A body placed in acid or alkali will eventually liquefy completely, although it may take several days, depending on the strength and temperature of the acid or alkali.

Types of solvents

When one substance dissolves in another, the substance that dissolves is known as the solute, and the substance that dissolves it is the solvent. There are two main types of solvent: polar and nonpolar. Polar solvents, such as water, have a small difference in electrical charge across the molecules, which interact with the opposite charges of polar solutes. Nonpolar solvents, such as pentane, lack these charges. They are good at dissolving noncharged atoms and molecules, such as oil and grease.

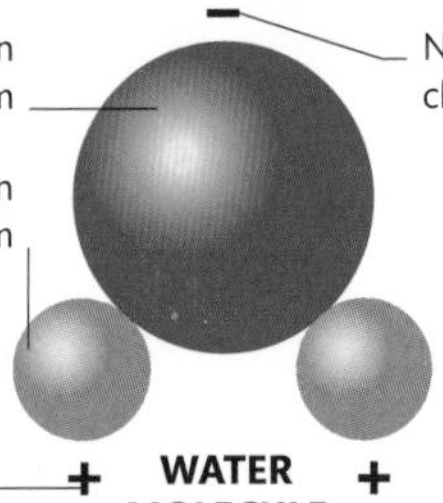

Polar solvent
In polar substances, such as water, one side of the molecule carries a negative charge and the other side a positive charge.

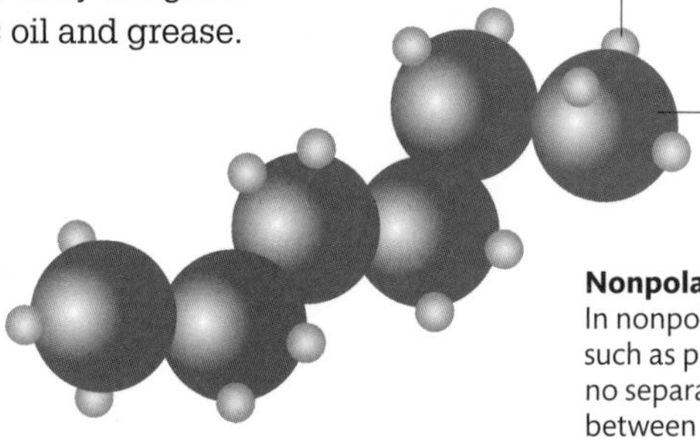

Nonpolar solvent
In nonpolar substances, such as pentane, there is no separation of charge between different parts of the molecule.

Types of solutions

When a solute dissolves in a solvent to form a solution, the two substances mix together so perfectly that their particles (atoms, molecules, or ions) completely intermingle. However, the particles do not react together, so they remain chemically unchanged. Solutions of solids in liquids are the most familiar type of solution, but there are also others, such as gases in liquids and solids in solids. When a solute dissolves, the resulting solution ends up with the same state (liquid, solid, or gas) as the solvent.

Solid in liquid
Sweetened coffee is a solution of a solid (sugar) dissolved in a liquid (coffee, consisting mostly of water with flavor molecules).

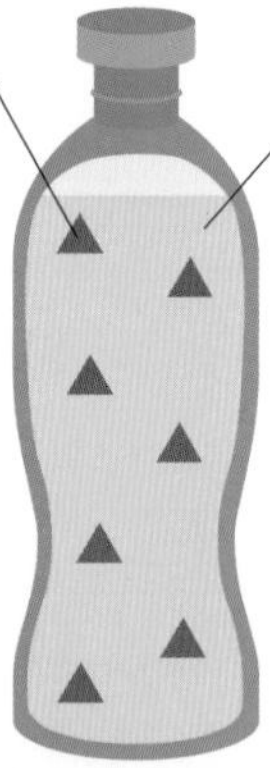

Gas in liquid
Ammonia is a gas and dissolves easily in water to form an alkali solution that is a component of some household cleaners.

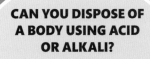

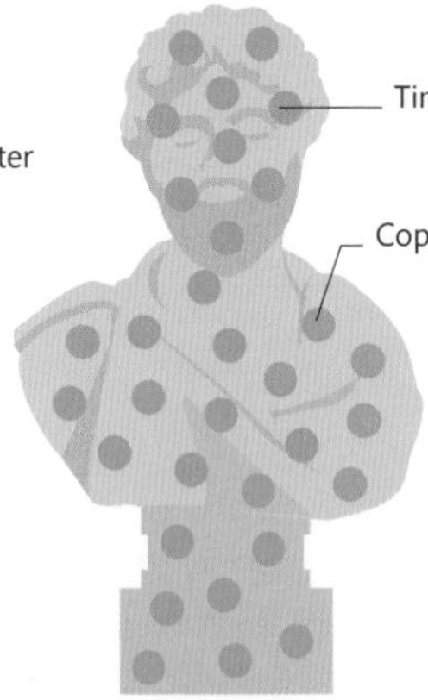

Solid in solid
Bronze is a solution of tin in copper. Copper is the solvent, because there is more of it than tin: about 88 percent compared to 12 percent.

Like dissolves like

Polar solvents dissolve polar solutes because their opposite charges attract each other to create weak bonds. Water is polar because its oxygen atoms are slightly negative and its hydrogen atoms are slightly positive. Nonpolar substances cannot intermix with polar ones, which is why oil and water do not mix. Only nonpolar particles can combine to form a solution.

WATER IS CALLED THE UNIVERSAL SOLVENT, BECAUSE IT DISSOLVES MORE SUBSTANCES THAN ANY OTHER LIQUID

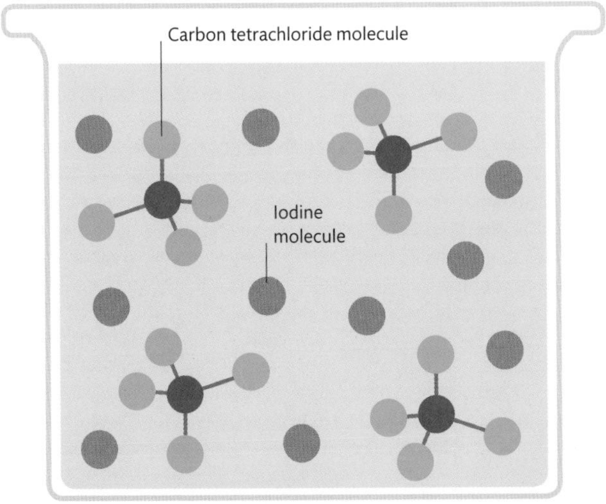

Nonpolar solute in nonpolar solvent
Nonpolar solvents, such as carbon tetrachloride, can dissolve nonpolar solutes, such as iodine, but not polar solutes.

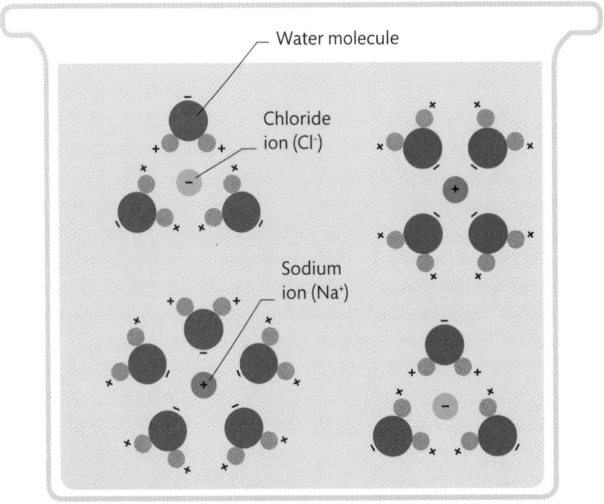

Polar solute in polar solvent
Polar solvents, such as water, can dissolve substances that have a charge, such as table salt (sodium chloride, NaCl) and sugar.

Solubility

Solubility is the degree to which a substance dissolves. It varies depending on temperature and, for gases, pressure. For example, more sugar will dissolve in hot than cold water, and more gas will dissolve in a liquid the higher the pressure of the gas. The maximum amount of solute that dissolves in a given amount of solvent at a specific temperature and pressure is called its saturation point.

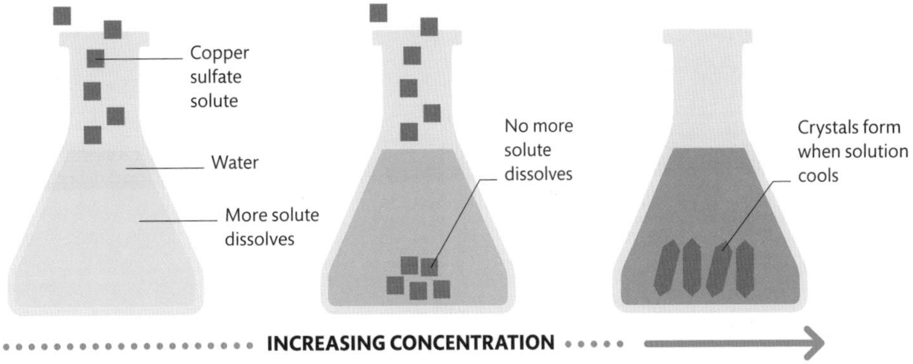

Unsaturated solution
In an unsaturated solution, more solute (here, copper sulfate crystals) will dissolve completely in the solvent.

Saturated solution
In a saturated solution, the maximum amount of solute has dissolved at that particular temperature.

Supersaturated solution
More solute dissolves with heating. Rapid cooling leaves the solution supersaturated before crystals solidify out.

Catalysts

Chemical reactions get faster at higher temperatures, when atoms and molecules have speedier collisions. Certain chemicals—called catalysts—can also boost reaction speeds. They are not themselves changed during reactions, so they can be reused.

How catalysts work

Particles need to have enough energy to react together. For some reactions, this activation energy (see p.44) is so large that the particles involved would not normally react at all. Catalysts work by lowering the activation energy, making it possible for the reaction to occur. Usually, only tiny amounts of catalyst are needed to do this.

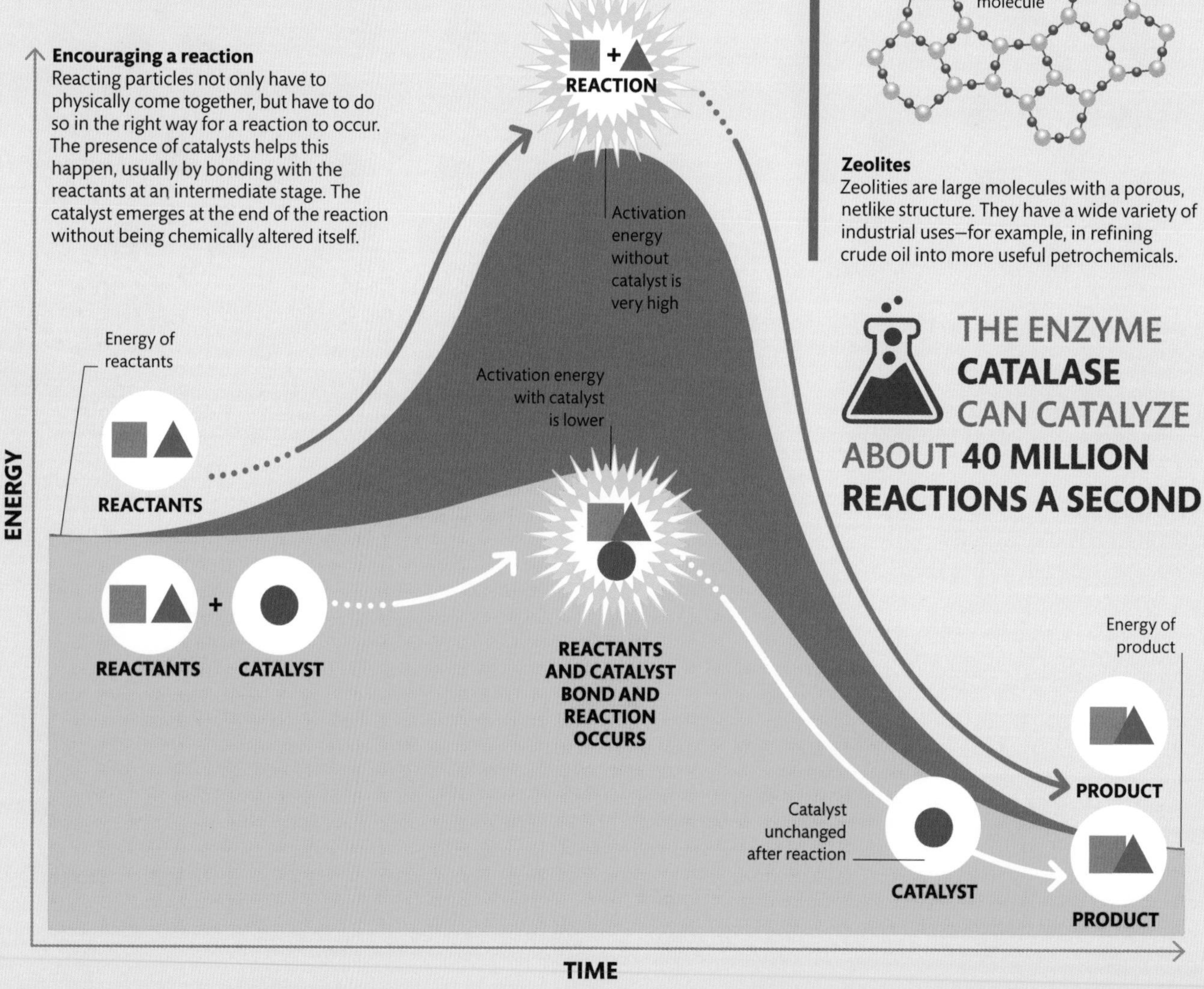

Encouraging a reaction
Reacting particles not only have to physically come together, but have to do so in the right way for a reaction to occur. The presence of catalysts helps this happen, usually by bonding with the reactants at an intermediate stage. The catalyst emerges at the end of the reaction without being chemically altered itself.
REACTION
Activation energy without catalyst is very high
Energy of reactants
REACTANTS
Activation energy with catalyst is lower
REACTANTS
CATALYST
REACTANTS AND CATALYST BOND AND REACTION OCCURS
Energy of product
PRODUCT
Catalyst unchanged after reaction
CATALYST
PRODUCT
ENERGY
TIME

Industrial catalysts

Various catalysts are used to make industrial chemical reactions more productive. Many are metals or metal oxides. Iron, for example, helps produce ammonia in the Haber process (see p.67). Most industrial catalysts are solids that can be easily separated for reuse.

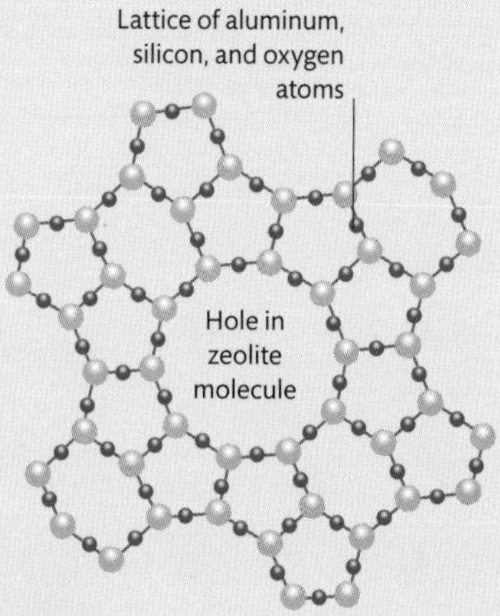

Lattice of aluminum, silicon, and oxygen atoms
Hole in zeolite molecule

Zeolites
Zeolities are large molecules with a porous, netlike structure. They have a wide variety of industrial uses—for example, in refining crude oil into more useful petrochemicals.

THE ENZYME CATALASE CAN CATALYZE ABOUT 40 MILLION REACTIONS A SECOND

Catalytic converters

The catalytic converters attached to modern cars contain ceramic "honeycombs" coated in platinum and rhodium catalysts. The arrangement provides a large surface area for the catalysts to work on the exhaust fumes, converting toxic gases into less harmful carbon dioxide, water, oxygen, and nitrogen. Heat from the car's engine makes the catalysts work at an effective speed.

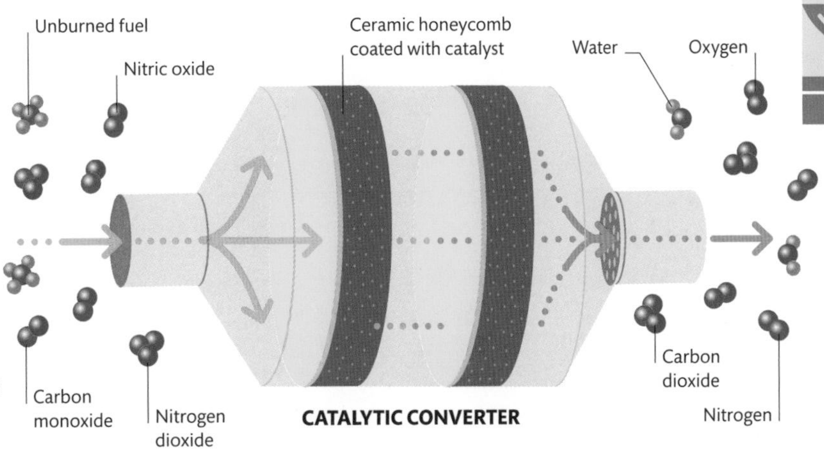

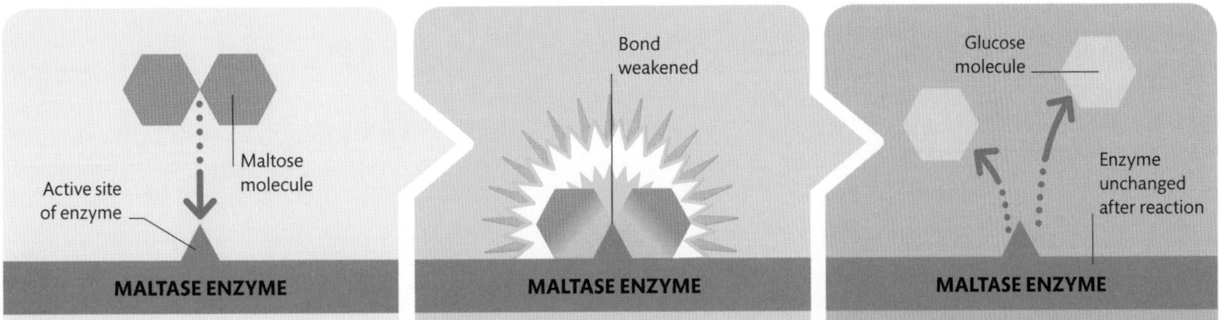

1 Maltose bonds to enzyme The reacting molecule—here, the sugar maltose—temporarily bonds to the active site (catalyzing part) of the enzyme maltase. Only maltose fits on maltase.

2 Maltose bond weakens The activation energy for splitting maltose is lowered when it binds to the enzyme's active site. This means maltose can be easily broken down by maltase.

3 Glucose splits off The chemical reaction at the active site rearranges chemical bonds, splitting the maltose into two glucose molecules. The enzyme is then ready to work again.

Biological catalysts

Most inorganic catalysts used in industry can catalyze a range of reactions, but in living bodies catalysts are more discriminating. Protein molecules called enzymes catalyze specific biological reactions, such as replicating DNA or digesting food. Each enzyme has a shape that locks on to particular kinds of reactants. Thousands of different enzymes are needed to drive the metabolism—the set of all chemical reactions needed to keep an organism alive.

BIOLOGICAL DETERGENTS

Like other catalysts, enzymes have useful applications. They are used wherever biological reactions are needed, such as cleaning stains from clothes. Biological detergents contain enzymes that digest fats in grease or protein in blood. And because enzymes work at body temperature—and are even destroyed if too hot—they work at lower water temperatures that are more energy efficient and less damaging to delicate fabrics.

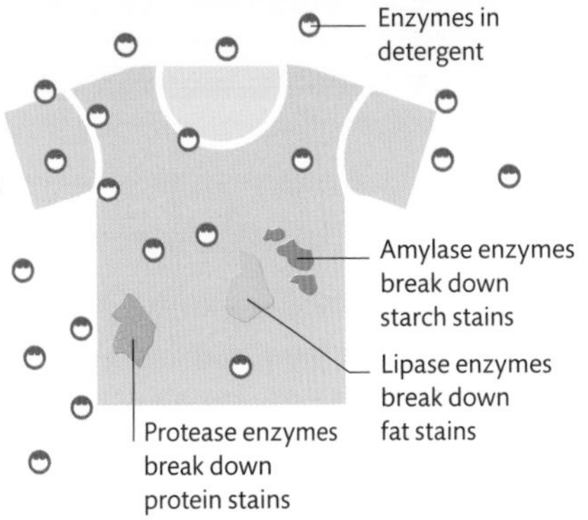

Making chemicals

Every day, we use man-made products, from plastics and fuels to medicines. Manufacturing many of these products requires basic chemicals, such as sulfuric acid, ammonia, nitrogen, chlorine, and sodium.

Sulfuric acid

Sulfuric acid is one of the most commonly used raw chemicals—it is used in drain cleaners and batteries, and to manufacture products from paper to fertilizers to tin cans. Various methods can be used to produce sulfuric acid, but the best known is the contact process.

The contact process
Liquid sulfur reacts with air to produce sulfur dioxide gas, which is cleaned, dried, and then converted to sulfur trioxide gas using a vanadium catalyst. Sulfuric acid is added to the gas to make disulfuric acid, which is then diluted with water to produce sulfuric acid.

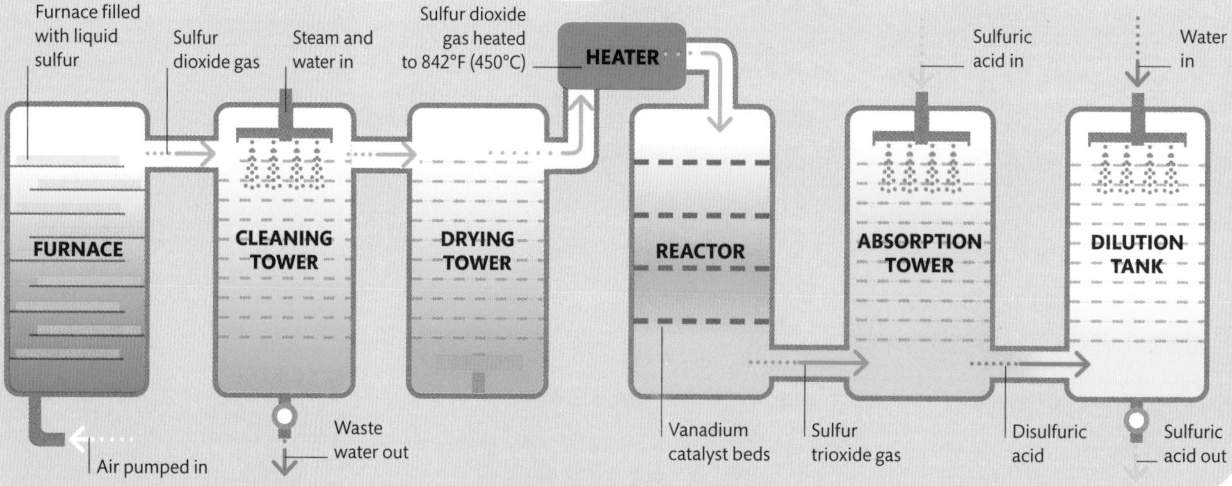

Chlorine and sodium

Chlorine and sodium are made from ordinary salt (sodium chloride) using a process called electrolysis, which is carried out on an industrial scale in a tank known as a Downs cell. The tank contains molten sodium chloride and iron and carbon electrodes. When an electric current is passed through the electrodes, the sodium and chloride ions move to the electrodes and turn into atoms of their elements, which are then collected.

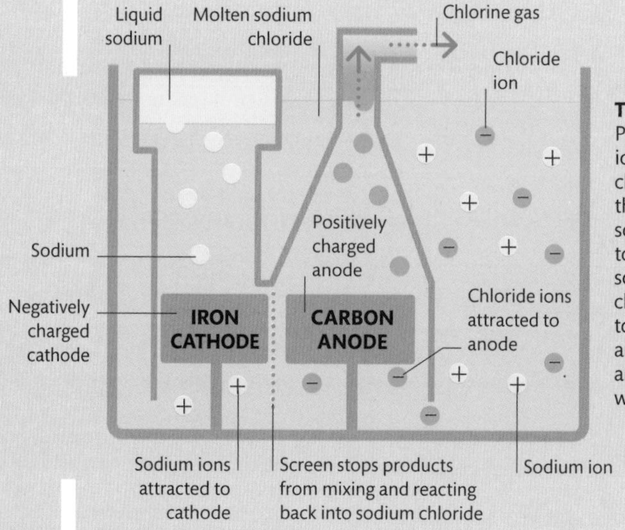

The Downs cell
Positively charged sodium ions move to the negatively charged cathode, where they gain an electron to form sodium metal. The metal floats to the surface of the molten sodium chloride. Negatively charged chloride ions move to the positively charged anode, where they lose an electron to form chlorine, which bubbles up as gas.

Nitrogen

Air consists of about 78 percent nitrogen and is the main source of pure nitrogen gas. The nitrogen is extracted from air by fractional distillation. The air is cooled to a liquid, then allowed to warm up. As it does so, the different components turn back to gas at different temperatures, corresponding to different heights up the distillation column. The oxygen remains as a liquid at the bottom.

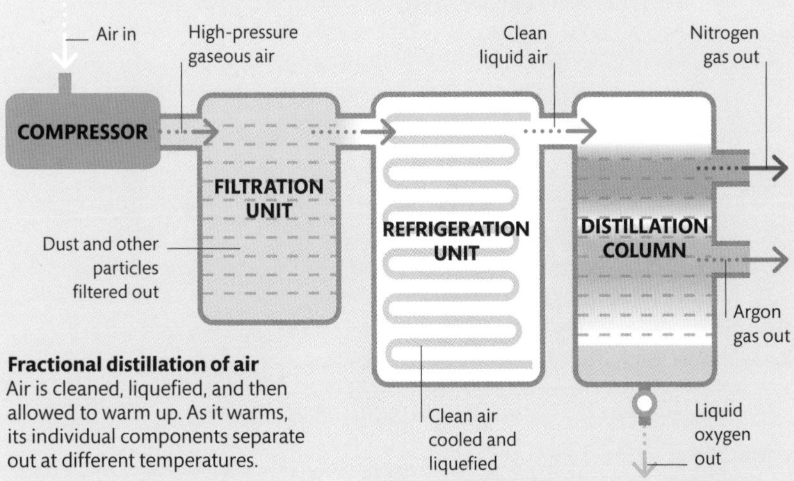

Fractional distillation of air
Air is cleaned, liquefied, and then allowed to warm up. As it warms, its individual components separate out at different temperatures.

PRODUCTS FROM OIL

Fractional distillation of crude oil produces a wide variety of useful products. Some are ready for use immediately—for example, natural gas, fuels such as petroleum and diesel, lubricant oils, and bitumen for road surfaces. Others are processed further into products such as plastics and solvents.

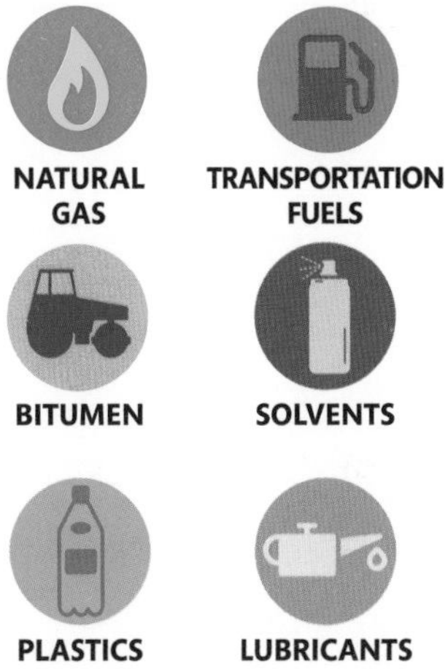

MORE THAN 250 MILLION TONS (230 MILLION TONNES) OF SULFURIC ACID ARE PRODUCED GLOBALLY EVERY YEAR

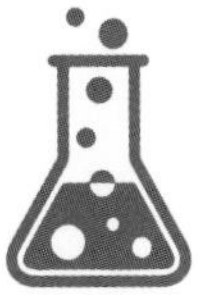

Ammonia

The Haber process makes ammonia from nitrogen and hydrogen gas. Ammonia is vital for manufacturing fertilizers, dyes, and explosives and is also used in cleaning products. Nitrogen is unreactive, so the Haber process uses an iron catalyst and high temperature and pressure in the reactor to improve the reaction speed and produce the highest yield of ammonia.

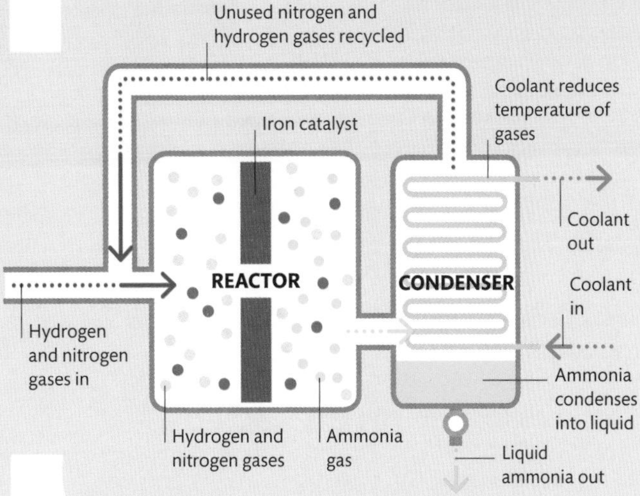

The Haber process
Hydrogen and nitrogen gases are mixed and passed over an iron catalyst, which encourages them to react and form ammonia. Cooling the mixture allows liquid ammonia to be tapped off. Unreacted nitrogen and hydrogen is recycled.

Plastics

Plastics are strong, light, and cheap, and they have transformed modern life. But because most plastics are made of fossil fuels and don't biodegrade, our growing plastic use also brings environmental problems.

Monomers and polymers

Plastics are a type of synthetic polymer—a long chain of molecules made of repeating units called monomers. Polymer chains can be hundreds of molecules long. Plastics made from different monomers have different properties and uses. Nylon, for example, becomes strong fibers for toothbrushes, while polyethylene is commonly used for lightweight bags.

Monomers
The monomers of many plastics have a carbon-carbon double bond (see p.41).

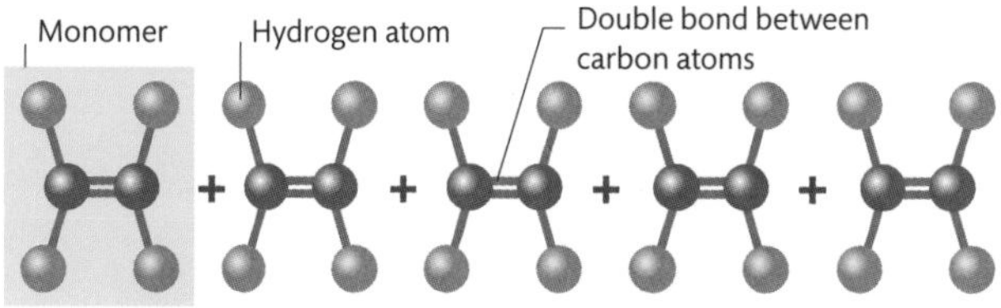

Polymers
To form a polymer, the double bond breaks, so each monomer can bond to its neighbor, creating long chains.

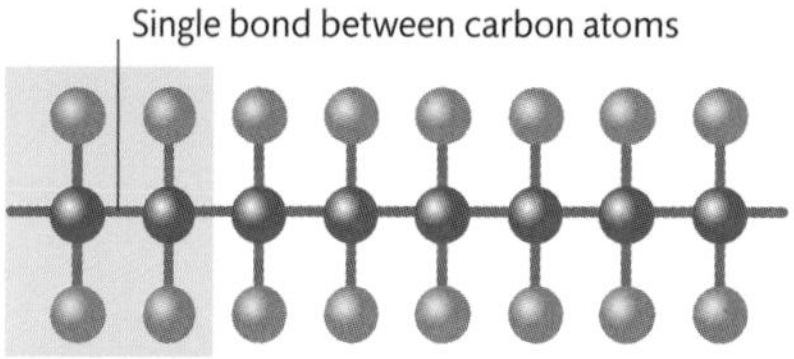

NATURAL POLYMERS

Polymers exist in nature, too—sugars, rubber, and DNA are all examples. DNA forms from monomers called nucleotides, which contain a sugar and a phosphate group (forming its backbone) and a nitrogen-containing base, which provides the code to make proteins.

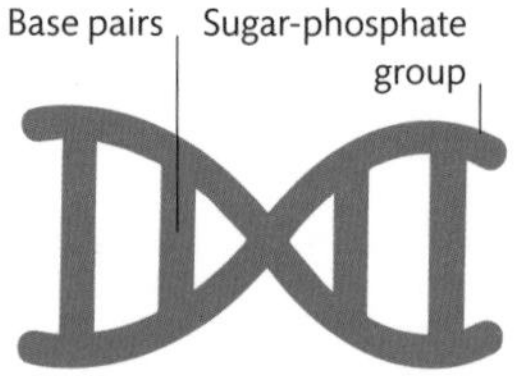

ENOUGH PLASTIC IS THROWN AWAY EACH YEAR TO CIRCLE THE EARTH FOUR TIMES

Manufacturing plastics

Most plastics are made from petrochemicals distilled from crude oil. Adding a catalyst while controlling temperature and pressure encourages these monomers to polymerize. Other chemicals can be added to change the plastic's properties. Once formed, plastic can be shaped into various products. Bioplastics, made from renewable sources, such as wood or bioethanol, exist, too, but are a small minority of the plastics made today. Plastics can be thermoset or thermoplastic. Thermosets can only be molded once, but thermoplastics can be melted and reshaped repeatedly.

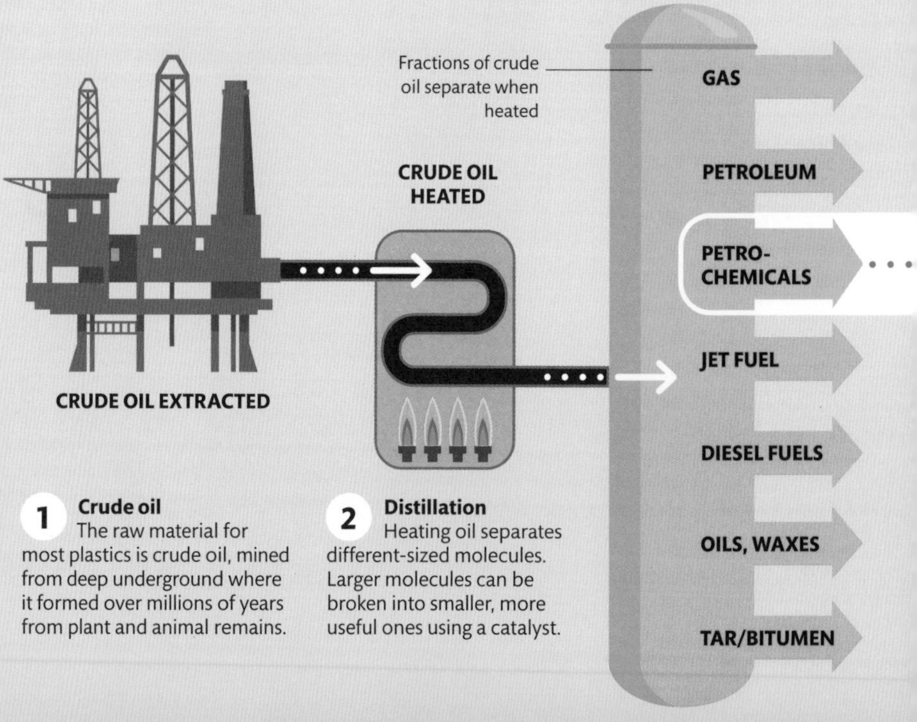

1 Crude oil The raw material for most plastics is crude oil, mined from deep underground where it formed over millions of years from plant and animal remains.

2 Distillation Heating oil separates different-sized molecules. Larger molecules can be broken into smaller, more useful ones using a catalyst.

Recycling

Some plastics can be recycled easily by chopping them up, melting them, and then reforming them. But for other types, alternative recycling methods are needed. One goal is to turn plastics into a liquid fuel or burn them to generate energy directly. Another idea is to create plastics that can be digested by bacteria, but these ideas cannot be implemented on a large scale yet.

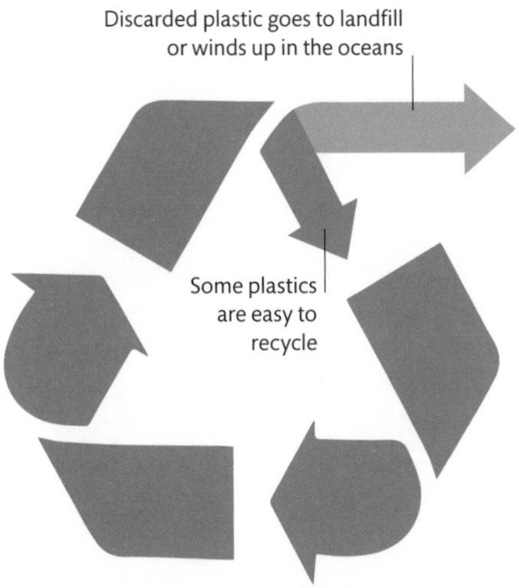

Waste
Most plastic waste will spend many thousands of years in landfill, leaching harmful chemicals into the soil. Some washes out to sea, breaking down into microplastics, which are harmful to wildlife.

BENEFITS AND DRAWBACKS OF PLASTICS

Benefits	Drawbacks
Plastics are cheap to make and do not rely on farming plants or animals or the resources that requires.	Plastics are mainly made from nonrenewable resources, and mining these resources is also damaging the environment.
Plastics are light and strong, so a small amount of material can make many useful products.	Plastics can break down into small pieces, working their way into our water systems and affecting wildlife and our food.
Plastics can be adapted to have a range of properties—hardness, flexibility, toughness, and more can all be controlled.	Plastics can experience fatigue and break after repeated use. UV from the Sun can also make them more brittle.
Synthetic fibers can be made stretchy and more resistant to creasing, water, and stains than natural fibers.	Synthetic clothes don't allow sweat to evaporate, so can be uncomfortable in hot weather. Static charges can also build up.
Some types of plastic can be recycled, making them more eco-friendly than their nonrecyclable counterparts.	Nonbiodegradable plastics contribute to global pollution, in seas and on land. They are also filling up landfill sites.

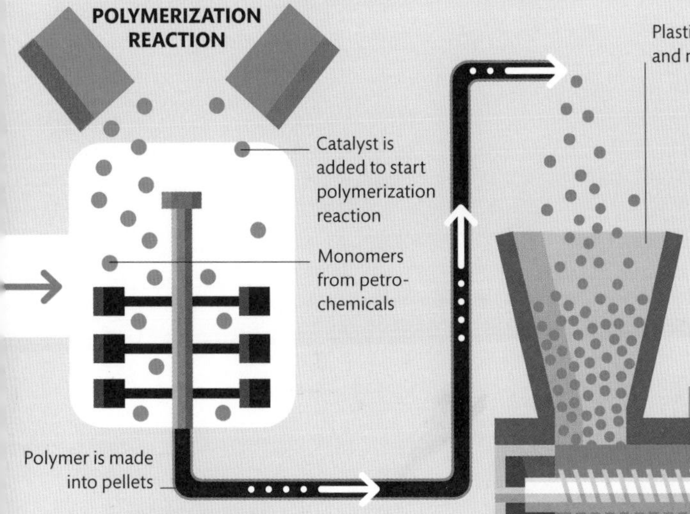

3 Polymerization
Adding a catalyst and controlling the temperature and pressure encourages monomers to react, becoming polymers. In some cases, small molecules, such as water, are produced as a by-product.

4 Molding plastics
Many plastics become flexible when heated, so they can be pressed or bent into shape, hardening again when cooled. Softened plastic can also be blown into a mold or pulled over one by a vacuum. If a plastic melts, it can be injection molded instead.

5 End product
Plastics are made into everything from drink bottles to TV remotes and fibers for clothing. Each product requires different properties, so the plastics used and the production methods differ.

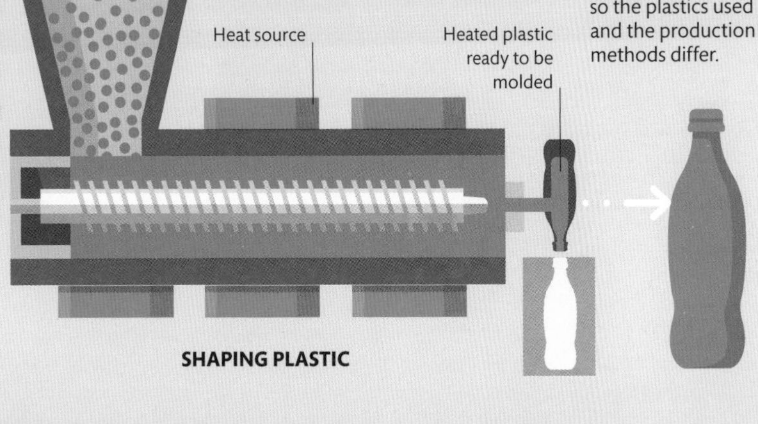

Glass and ceramics

Hard, corrosion-resistant, and often transparent, the glass we all know is mostly sand, or silicon dioxide. But the term "glass" is also used for a bigger group of materials, which are all types of ceramic.

The structure of glass

Glasses have amorphous structures, which means that there is little or no order to the arrangement of their molecules (or atoms). On an atomic scale, they look like immobile liquids (see pp.16–17). Glasses are, however, solid materials. Glasses are usually made by melting a substance and then cooling it so quickly that its atoms (or molecules) can't arrange themselves into their usual structure, whether that is crystalline or metallic. Instead, they are trapped in place, as disordered as when it was liquid.

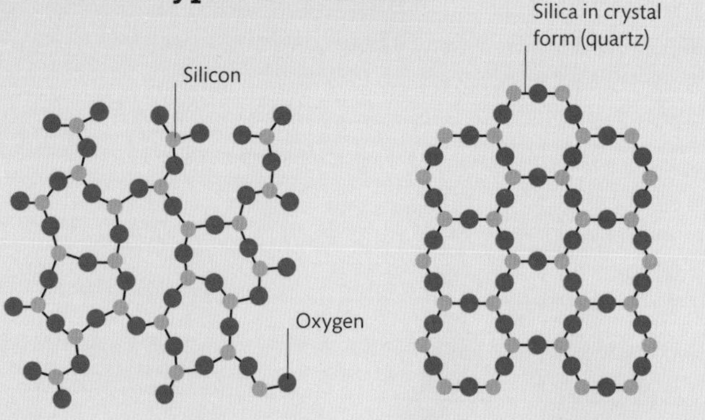
Silicon
Oxygen
Silica in crystal form (quartz)

AMORPHOUS STRUCTURE **CRYSTALLINE STRUCTURE**

Types of glass

We all know glass as the transparent, brittle material in windows. This is mainly silicon dioxide. But glasses can form from a range of materials—metals can be glassy, and some forms of polymer, or plastic, are technically glasses. Other chemicals are added to silicate glasses to change their properties. These chemicals might affect the color or clarity of the product, give it better heat resistance, as in borosilicate glass like Pyrex, or make them scratch-resistant, such as Gorilla Glass (used for many smartphone screens).

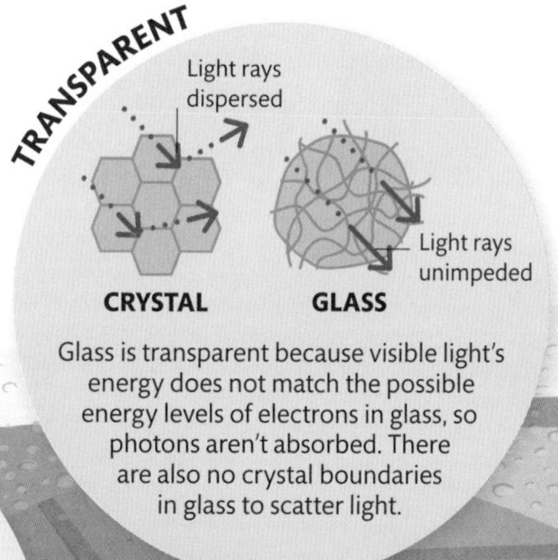
TRANSPARENT
Light rays dispersed
Light rays unimpeded
CRYSTAL
GLASS
Glass is transparent because visible light's energy does not match the possible energy levels of electrons in glass, so photons aren't absorbed. There are also no crystal boundaries in glass to scatter light.

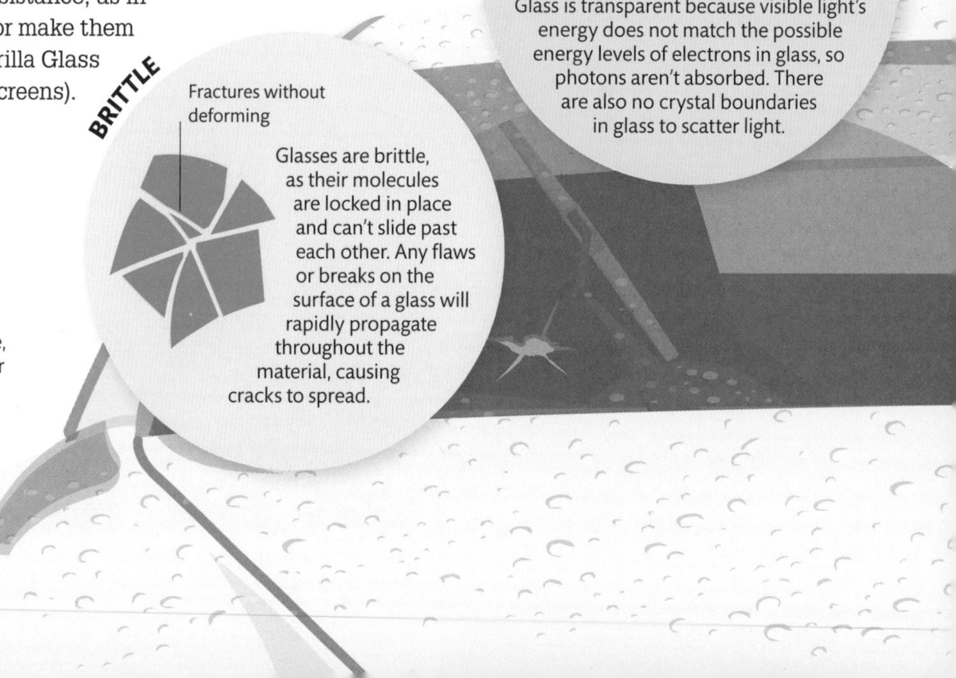
BRITTLE
Fractures without deforming
Glasses are brittle, as their molecules are locked in place and can't slide past each other. Any flaws or breaks on the surface of a glass will rapidly propagate throughout the material, causing cracks to spread.

Properties of glass
Glass's hardness, corrosion resistance, and low reactivity make it suitable for many products, but its most useful property is probably transparency, allowing widespread use for windows in buildings and cars.

Other ceramics

Glasses are a subset of materials called ceramics. The term "ceramic" traditionally describes clay-based products, but the scientific definition includes any nonmetallic solid that is shaped and then hardened by heating. Ceramics can have crystalline or amorphous structures and can be made of nearly any element. Like glass, they are usually hard but brittle and have high melting points. This makes ceramics ideal for thermal and electrical insulation, such as the ceramic titanium carbide, which is found on spacecraft heat shields.

DOES GLASS FLOW?

The description of glass as a slow-flowing liquid is false. Very old windows are thicker at the bottom because uneven panes were positioned that way around to ensure stability.

WATER-RESISTANT

Normal glass attracts water, so it forms a surface film. Water-repellent coatings make water bead up and run off the glass, improving visibility and cleaning the glass in the process.

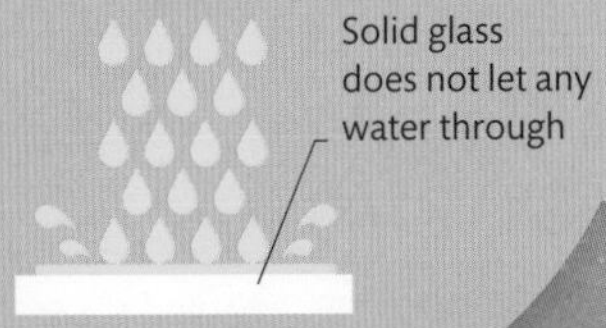

GLASS WAS FIRST MANUFACTURED IN EGYPT ABOUT **5,000 YEARS** AGO

TOUGHENED GLASS

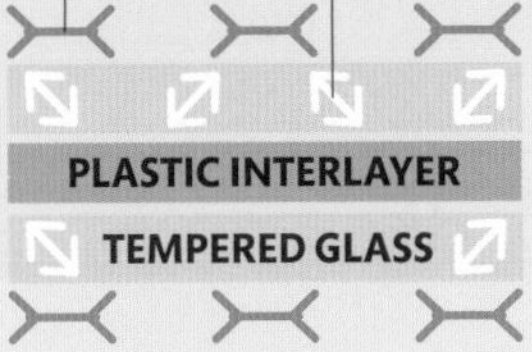

Toughened glass has a surface in compression and interior in tension to give it greater strength. If it does break, a plastic layer holds the pieces together.

TRANSPARENT ALUMINUM

Commonly known as transparent aluminum, aluminum oxynitride is a super-strong, transparent ceramic. The powdered mixture is compressed, heated to 3,630°F (2,000°C), and then cooled so that its molecules remain amorphous. It can withstand multiple impacts from armor-piercing bullets while remaining transparent. Its current high price means it is only used for specialty military applications, but it could be more widely used in future.

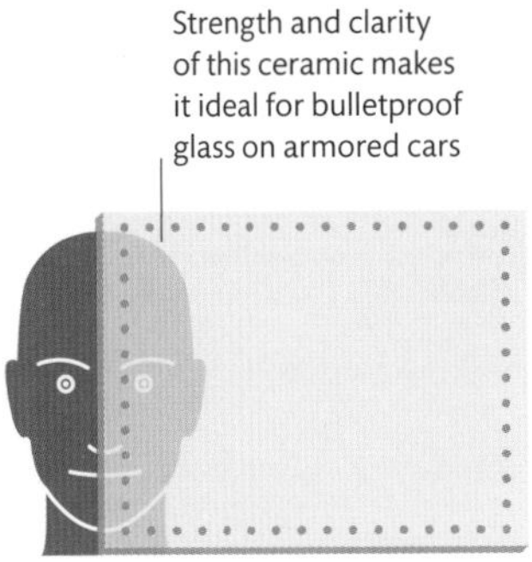

TRANSPARENT CERAMIC

Wonder materials

From super strength to incredible lightness, some of the materials we use have amazing properties. Many of these were invented, but others occur naturally. Some synthetic materials have been inspired by nature, a process called biomimicry.

ARE ALL COMPOSITES SYNTHETIC?

No; wood and bone are both examples of naturally occurring composites. Bone is composed of hard but brittle hydroxyapatite and soft, flexible collagen.

Composites

Sometimes no one material has the right balance of properties for a certain product. To get around this problem, two or more materials can be combined so that the finished product has the best properties of each. These materials are called composites. Concrete is the most common modern composite, but wattle and daub, used to cover walls 6,000 years ago, was an early example made from straw or branches and mud. New materials and techniques are used to created more advanced composites.

Relative strengths
Concrete is a composite consisting of stone aggregate in a cement matrix. Strong under compressive forces but weak under tension, concrete can't be used alone for buildings.

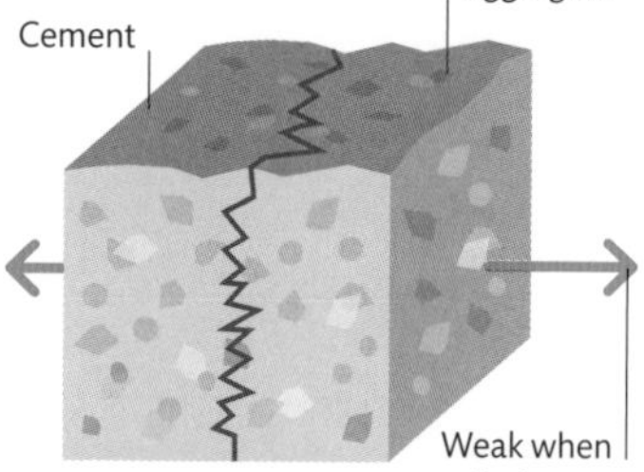

Adding tensile strength
In construction, concrete is reinforced using steel bars, which are strong in tension. Together these combine to create reinforced concrete—one of the most versatile materials in the modern world.

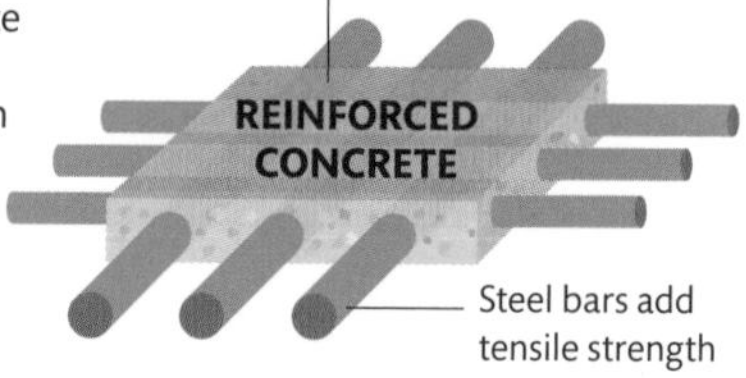

Advanced composites
High-tech composites include reinforced polymers, such as carbon fiber and fiberglass, in which fibers of carbon or glass are woven and sandwiched between layers of another polymer or mixed into resin while it is still liquid. Both are strong and light, although they are also expensive.

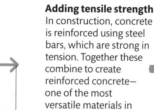

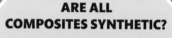

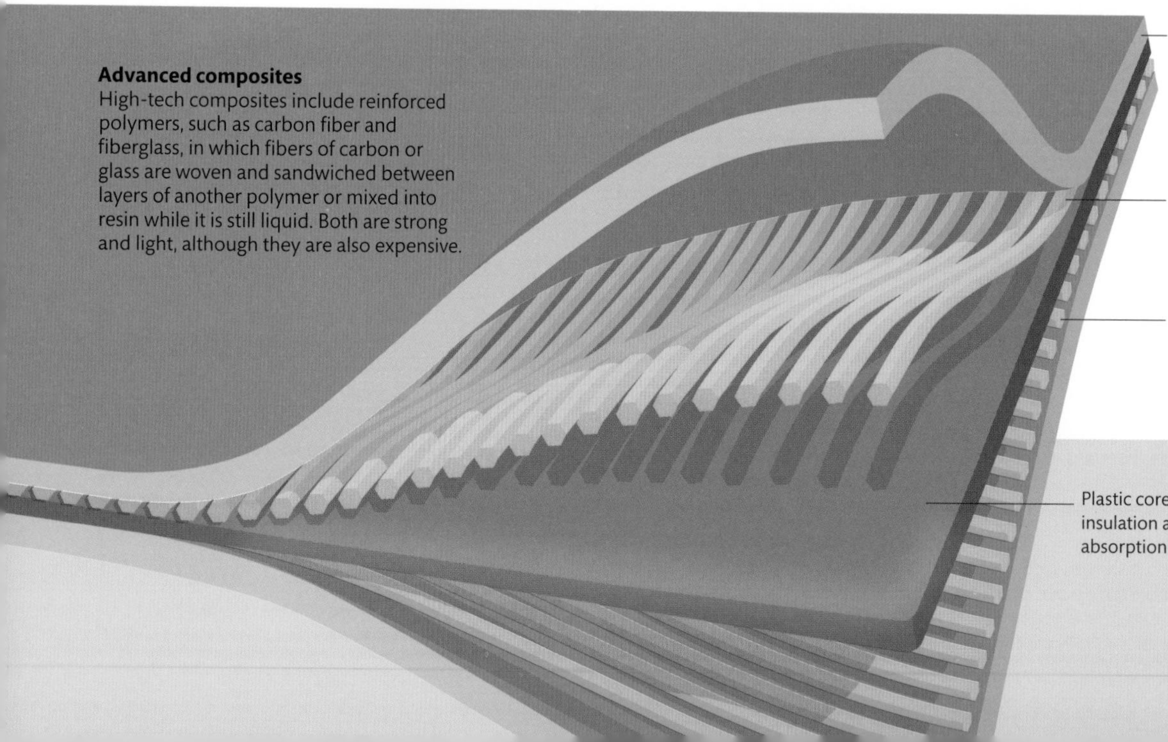

Spider silk
Large-scale production of spider silk could lead to new bulletproof materials. It is as strong as steel, but much lighter, and it is stretchable, so it resists breakage.

Aerogel
Replacing a gel's liquid with gas produces a super-lightweight solid. More than 98 percent air, aerogels are very good insulators.

Graphene
Made from layers of graphite one atom thick, graphene is stronger than steel, a good electrical conductor, transparent, flexible, and extremely light.

Amazing properties

Some materials, natural or man-made, have incredible properties. From flexible but bulletproof Kevlar to plastics that can repair themselves, these materials offer alternatives that could make our lives safer and easier. For example, new bone can grow through metal foam implants, integrating them into the body. And superhydrophobic window surfaces could remove the need for dangerous high-level cleaning.

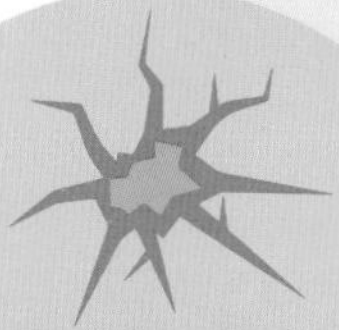

Self-healing plastic
Self-healing plastics may contain capsules that burst when damaged, allowing the liquids inside to react and solidify to fix any holes.

Metal foam
Forcing gas bubbles into molten metal can create foams. These are lightweight, while retaining many properties of the metal.

Kevlar
A super-strong plastic, Kevlar fibers can be woven into clothing or added to a polymer to form a composite.

Superhydrophobic material
Hydrophobic materials have tiny protrusions covering their surfaces, which hold water drops away so the material cannot get wet.

A SINGLE SHEET OF **GRAPHENE** CAN HOLD UP A CAT WEIGHING **9LB (4KG),** BUT WILL ITSELF **WEIGH LESS THAN** A **CAT'S WHISKER**

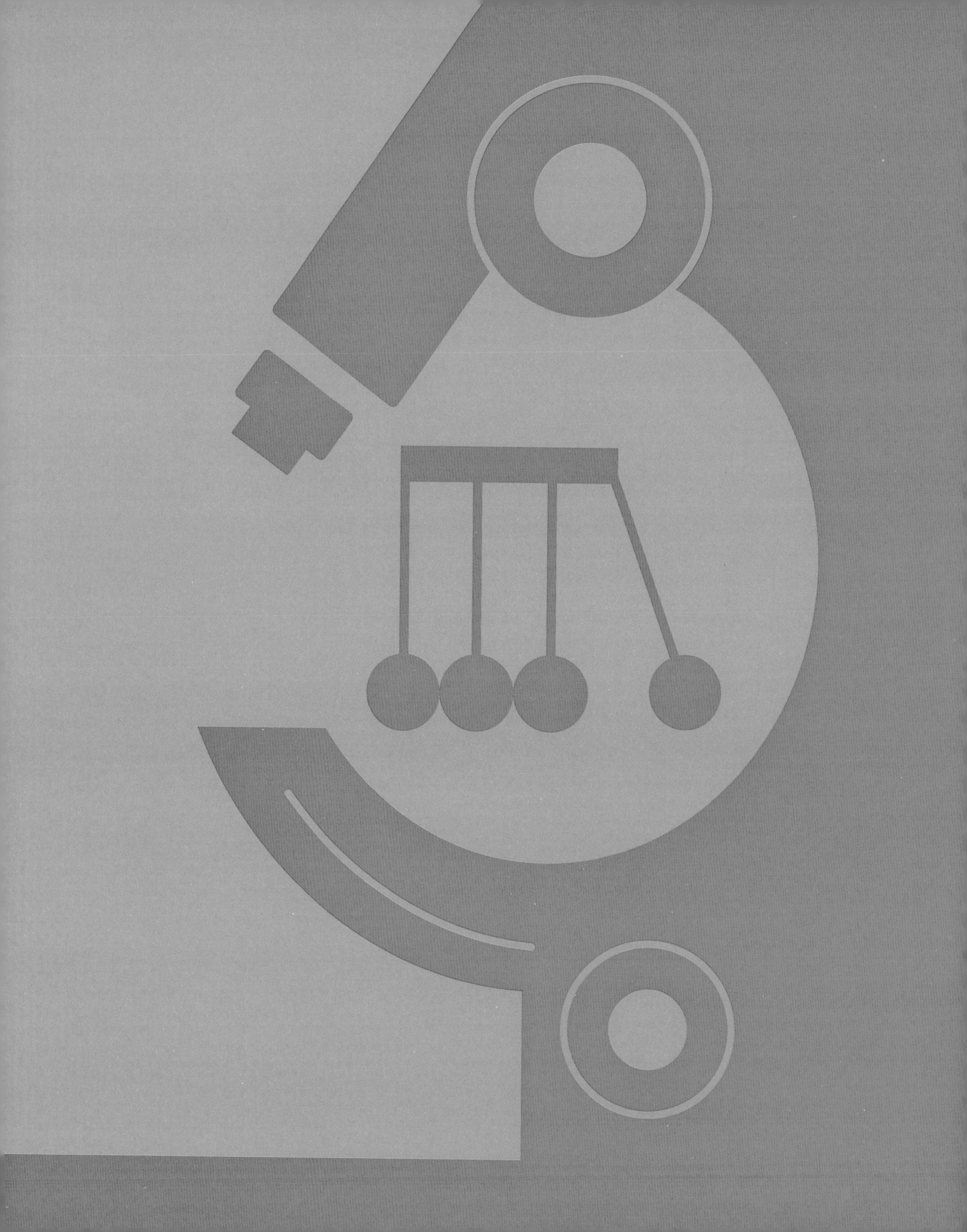

ENERGY AND FORCES

What is energy?

Physicists understand the Universe in terms of matter and energy in space and time. Many forms of energy exist and can be changed from one form to another. When a force is used to move an object, we say that work has been done on that object.

Types of energy

Energy is everywhere, it is indestructible, and it has existed since the beginning of time. However, to make things simpler to understand and measure, scientists categorize energy into different forms. Every natural phenomenon and every artificial process used by machines and technology occurs because one form of energy is making it work—and is converted into another form as it does so.

POTENTIAL ENERGY
This is stored energy that's not doing any work but which can be converted into some kind of useful energy.

Elastic potential
Stretched or squashed materials will release potential energy by bouncing back to their original shape.

Electrical potential
A battery is packed with potential energy that can be released as an electric current.

Gravitational potential
Objects lifted up high have the potential to fall down—releasing energy as motion.

Chemical energy
Burning and other chemical reactions are driven by the energy that holds atoms to one another.

Radiant energy
Light and other radiation is energy in the form of changing electrical and magnetic fields.

Acoustic energy
The energy carried in a sound wave squeezes and stretches the air (or other medium).

Nuclear energy
Radioactivity and nuclear explosions use the energy that holds an atom's nucleus together.

Electrical energy
An electrical current carries energy as a moving stream of charged particles, most often electrons.

Thermal energy
The motion of atoms, often as a vibration, is called thermal or heat energy. "Hot" atoms vibrate more.

Kinetic energy
Anything that moves, from electrons to galaxies, has kinetic energy, or the energy of motion.

Chemical release
If a man moves a heavy load, a chain of energy transformation takes place. At the start of the process, the body converts chemical energy stored in food into kinetic energy.

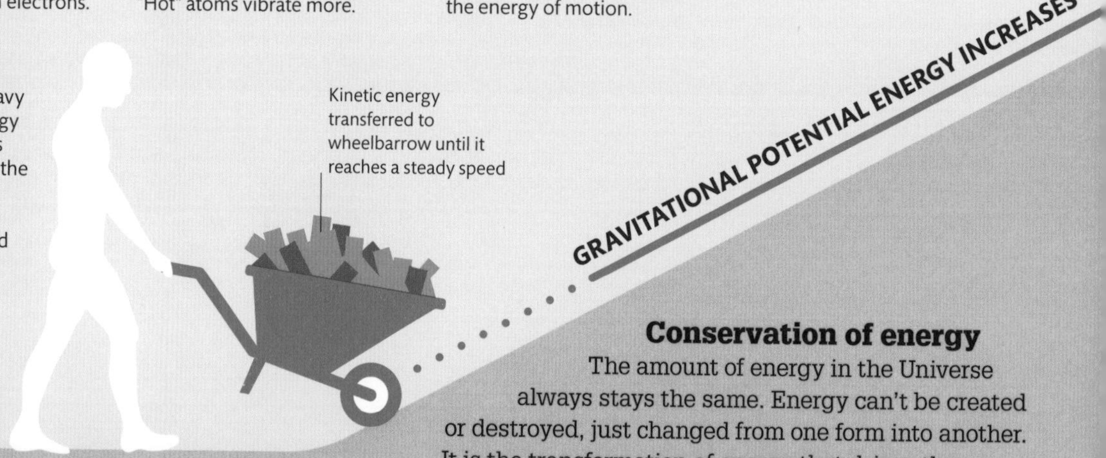

Conservation of energy

The amount of energy in the Universe always stays the same. Energy can't be created or destroyed, just changed from one form into another. It is the transformation of energy that drives the processes we see. Energy also spreads out or becomes more disordered and less "useful." So when considered by itself, every process always loses energy, most often ultimately as heat. So, a source of energy is needed to keep these processes going.

1 On the move
The body's kinetic energy is transferred to the wheelbarrow. The energy is used to overcome friction to make the wheelbarrow move. The body gets hotter as energy is converted to useless heat.

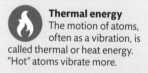

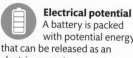

HOW MUCH ENERGY IS IN A CHOCOLATE BAR?

A 1¾oz (50g) bar of milk chocolate contains about 250 calories—the same amount of energy used by an average human body every 2.5 hours.

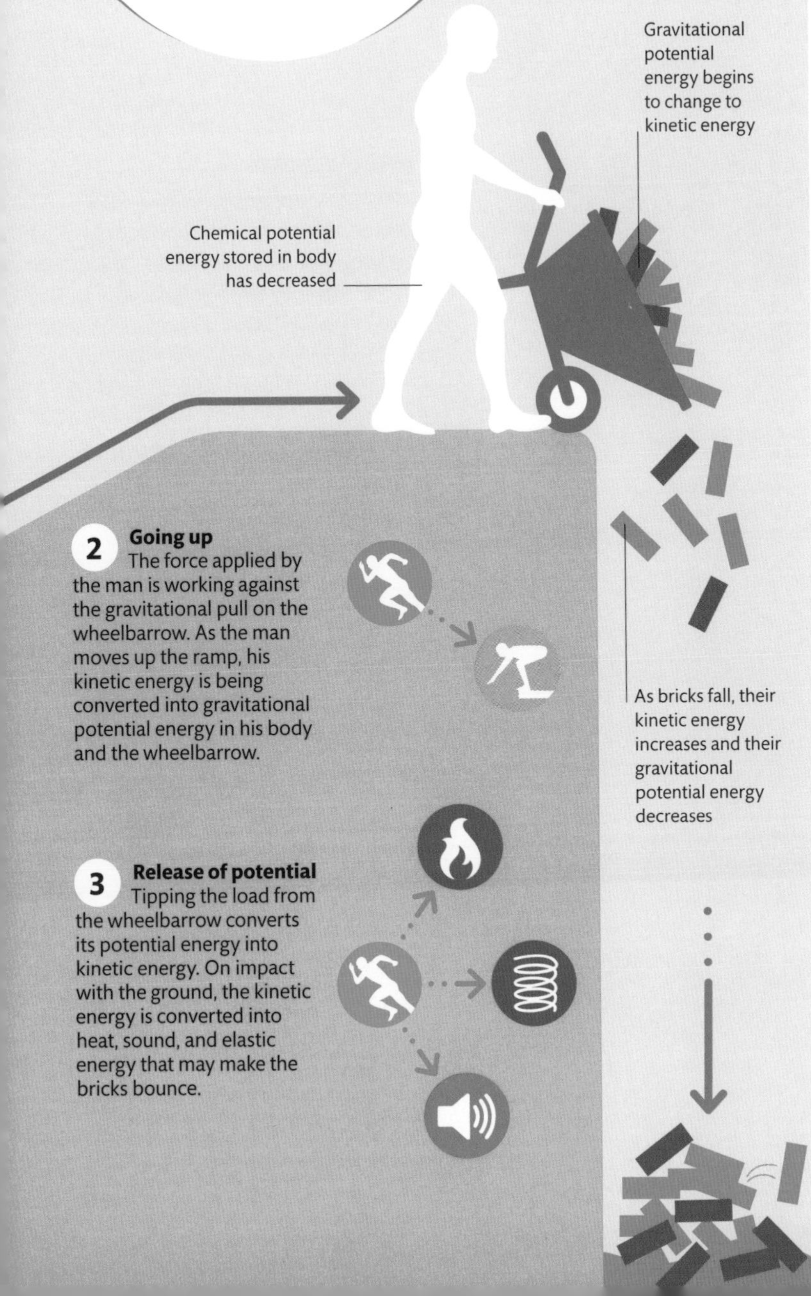

2 Going up
The force applied by the man is working against the gravitational pull on the wheelbarrow. As the man moves up the ramp, his kinetic energy is being converted into gravitational potential energy in his body and the wheelbarrow.

3 Release of potential
Tipping the load from the wheelbarrow converts its potential energy into kinetic energy. On impact with the ground, the kinetic energy is converted into heat, sound, and elastic energy that may make the bricks bounce.

Measuring energy

Energy is measured in units called joules (J). One joule is the energy needed to lift approximately 100g up by 1 meter. The energy in food is often measured in calories, which relates to how much heat the food produces when burned in a device called a calorimeter.

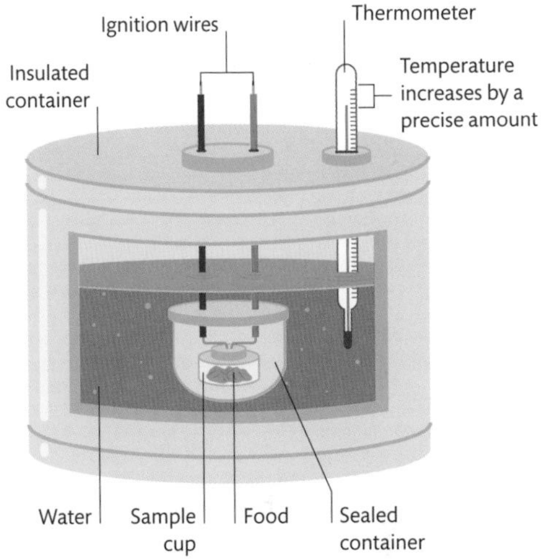

Measuring calories
As a sample of food burns, it raises the water temperature. The rise in temperature can be used to work out how many calories are in the food.

POWER

Rate of energy transformation is defined as power. Power is measured in watts (W); 1W equals 1J per second. A high-power process uses energy more quickly. A 100W light bulb has about the same rate of energy transformation, or power output, as an adult female human.

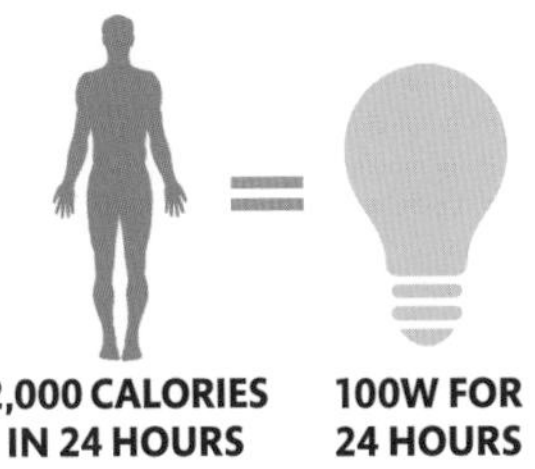

Static electricity

The most familiar form of electricity is the current supplied to the home, which is a largely artificial phenomenon. Most natural electrical effects, such as lightning, are due to static electricity.

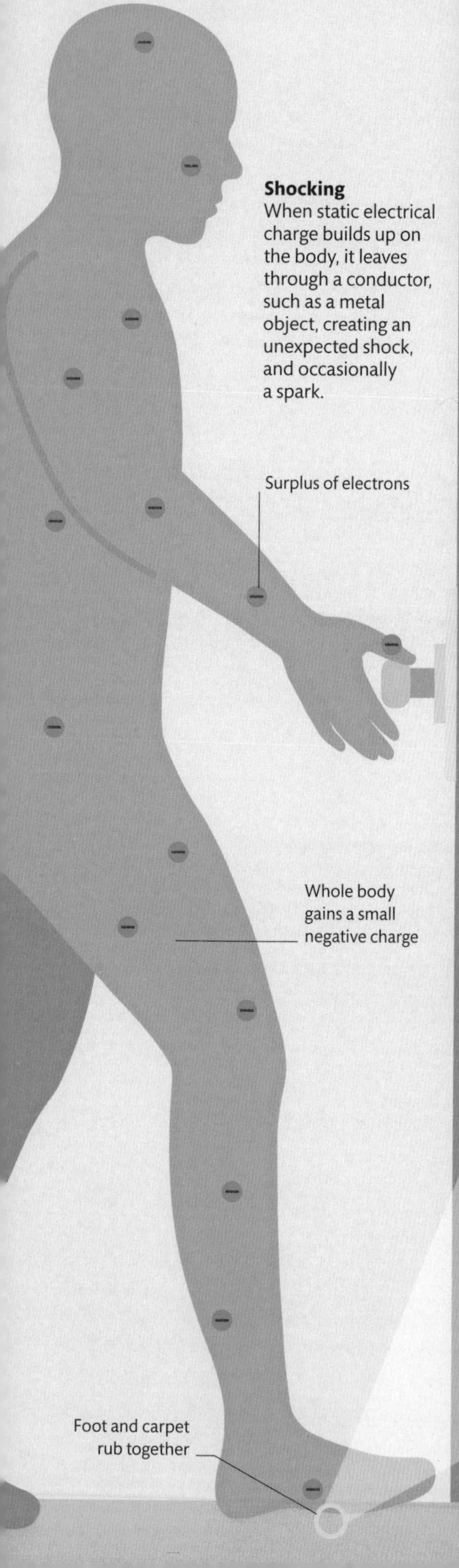
Surplus of electrons
Whole body gains a small negative charge
Foot and carpet rub together

Shocking
When static electrical charge builds up on the body, it leaves through a conductor, such as a metal object, creating an unexpected shock, and occasionally a spark.

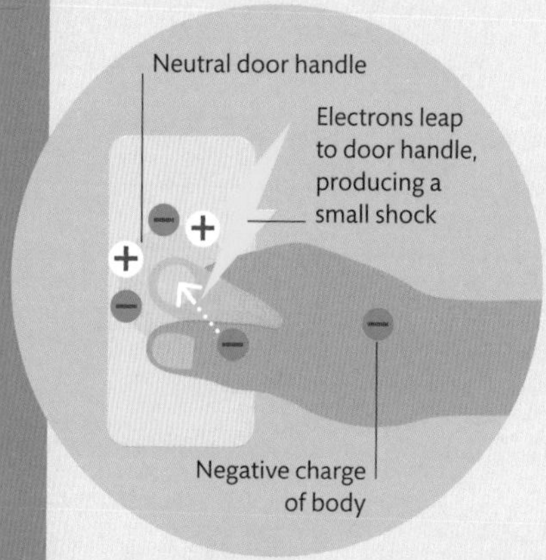
Neutral door handle
Electrons leap to door handle, producing a small shock
Negative charge of body

2 Discharge
The charge can escape through metal, often a door handle. When the hand touches it—or gets close—the excess of electrons in the body leap to the metal, giving a bit of a shock.

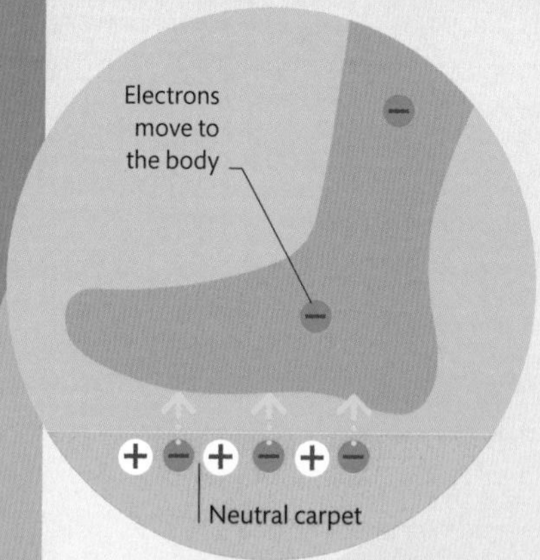
Electrons move to the body
Neutral carpet

1 Charge from friction
The foot rubbing on artificial fabrics in modern carpets makes electrons move from the ground to the body, giving it a small negative charge.

Electrostatics

Electricity is caused by a property of matter called charge. In every atom, protons are positively charged and locked in position, whereas the negatively charged electrons can be free to move to other objects. If an object acquires a surplus of electrons, it takes on a negative charge and will attract positively charged objects—those with a deficit of electrons. This force also makes electrons repel each other, and they eventually find a path to escape the charged object—creating a spark.

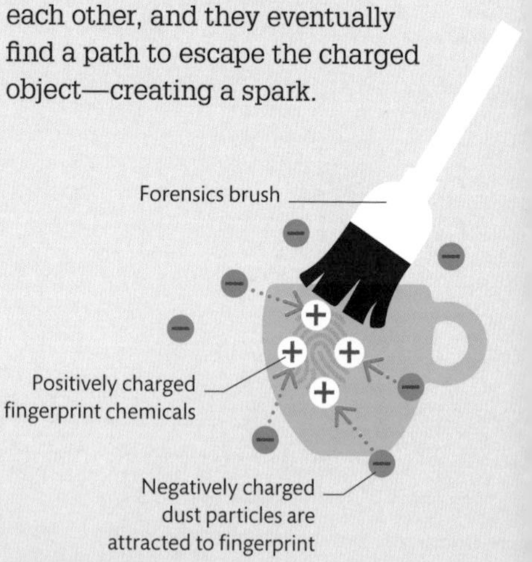
Forensics brush
Positively charged fingerprint chemicals
Negatively charged dust particles are attracted to fingerprint

Dusting for prints
Fingerprint investigators make use of static electricity. They brush negatively charged powder onto positively charged chemicals left behind in fingerprints at a crime scene.

Using static electricity

Static electrical charge is at work in many everyday situations. Generally the static charge is used to produce a small and easy-to-control force field, which attracts or repels other materials. Larger charges are dangerous but do have their uses, such as in defibrillators.

Conditioner
Shampoos make hairs electrically charged. The hairs repel each other; conditioner neutralizes the charge.

Defibrillator
A static generator builds up a large charge to be directed through a heart that has stopped beating properly.

Plastic wrap
Unrolling plastic wrap gives it a small electrical charge. This helps the film of the wrap cling to other objects.

Spray gun
Professional spray guns give paint a positive charge so that it is pulled onto a negatively charged object.

Ebook
The screen attracts or repels spheres containing oil particles with positive (white) or negative (black) charge.

Dust filters
Harmful particles in factory smoke are given a charge, then pulled out by highly charged collecting plates.

LIGHTNING STRIKES

Lightning is a massive discharge of static electricity. Air is a poor conductor of electricity, so charge in storm clouds cannot dissipate and builds up to enormous levels. Eventually a discharge zigzags through the air as it finds the easiest route to the ground.

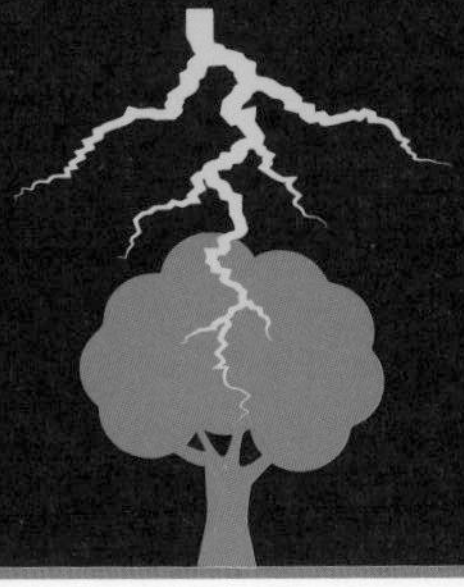

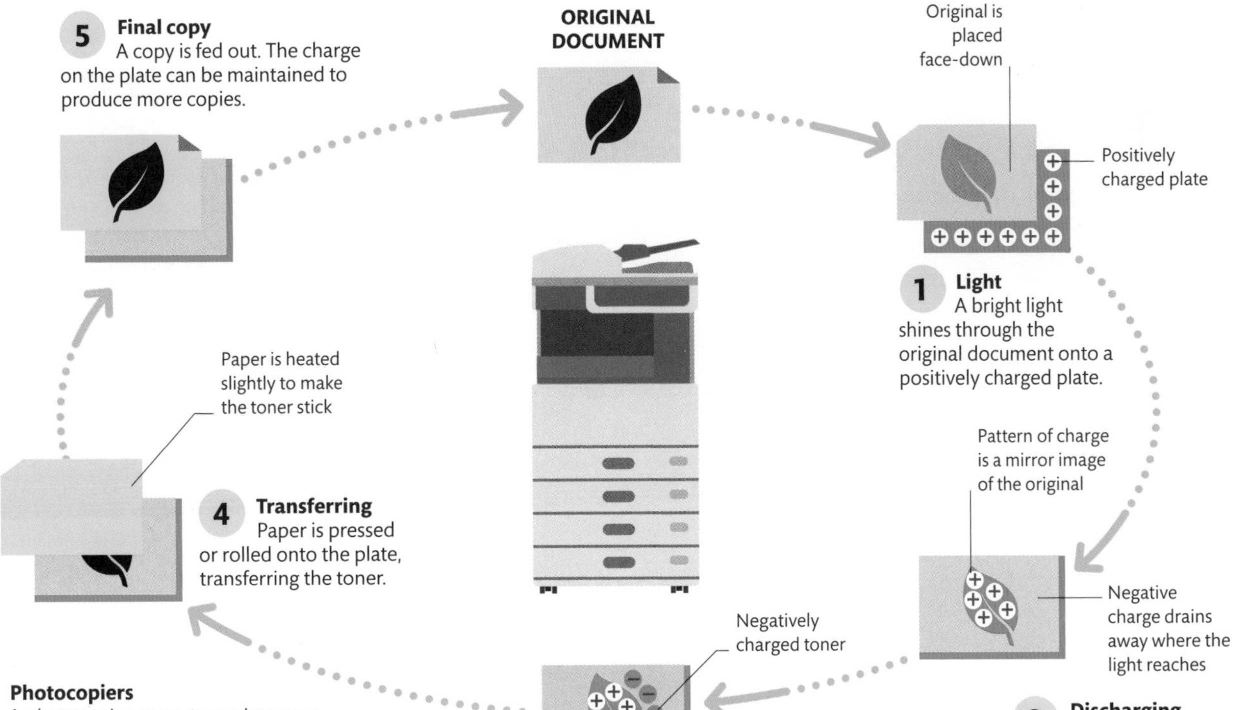

1 Light
A bright light shines through the original document onto a positively charged plate.

2 Discharging
The light discharges the plate apart from the areas in the shadow of the document.

3 Negative toner
The toner is negatively charged powder that clings to the positive regions of the plate.

4 Transferring
Paper is pressed or rolled onto the plate, transferring the toner.

5 Final copy
A copy is fed out. The charge on the plate can be maintained to produce more copies.

Photocopiers
A photocopier recreates an image or text as an invisible pattern of static charge. This pattern is then used to position the toner in the correct place to produce a very faithful copy.

Electric currents

An electric current is a flow of charge. In everyday examples, the charge is carried by the motion of electrons through metals, such as copper wires. Any material that is good at carrying a current is called a conductor. Insulators are not good at carrying a current.

Making a current

An electric current differs from a static charge, such as a spark or lightning bolt (see pp.78–79), in that the charge keeps moving. Charged particles move because they are being pulled toward an opposite charge. A spark moves because of a difference in charge between one place and another. The spark also removes the difference in charge that originally caused it. In a current, as produced by a battery, the difference in electrical charge keeps the current flowing.

QUANTITY	UNIT
Electric current is the flow of electric charge.	**A** **Ampere (amp)**
Voltage, or potential difference, is the force that pushes the current along.	**V** **Volt**
Resistance is the opposition to the movement of electrical current.	**Ω** **Ohm**

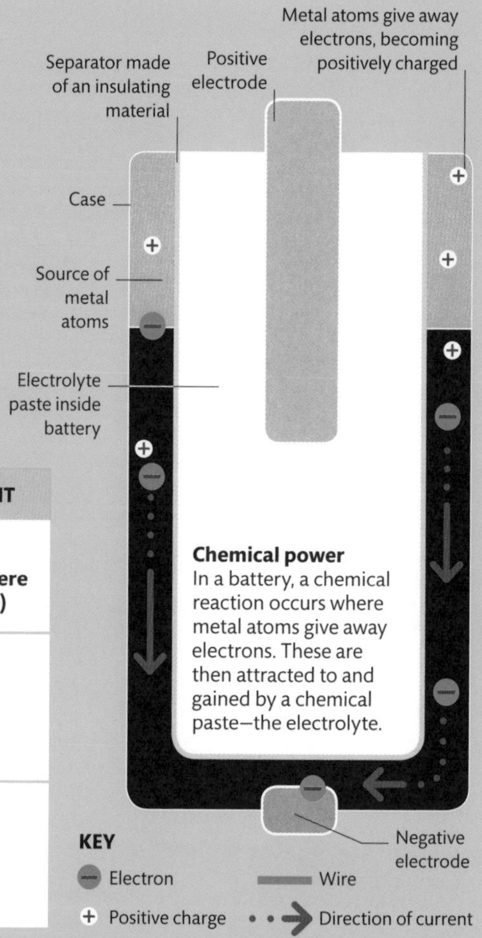

Chemical power
In a battery, a chemical reaction occurs where metal atoms give away electrons. These are then attracted to and gained by a chemical paste—the electrolyte.

Circuits

Electric currents carry energy that can be put to work. The flow of electrons is similar to water running downhill. The energy of the flowing water can be harnessed by a waterwheel to power a machine. Instead of a channel, electric current is sent around circuits, so its energy can be used by components such as lightbulbs, heaters, or motors. The way the energy is dispersed through the circuit depends on the circuit's design. There are two main types of circuit: series and parallel.

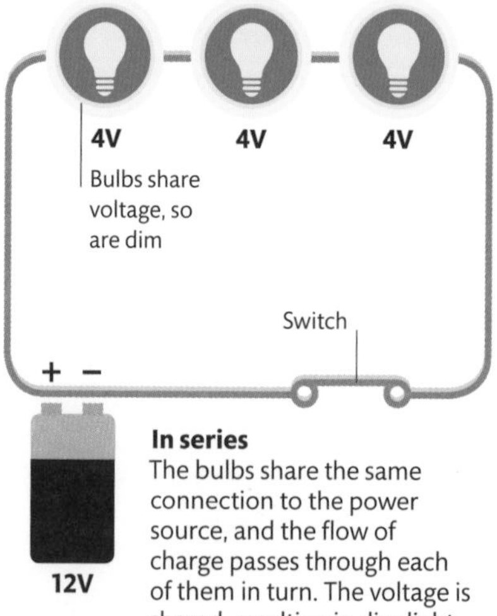

In series
The bulbs share the same connection to the power source, and the flow of charge passes through each of them in turn. The voltage is shared, resulting in dim light.

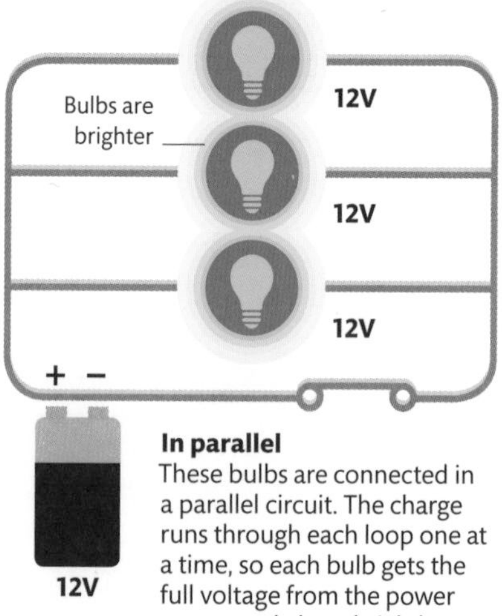

In parallel
These bulbs are connected in a parallel circuit. The charge runs through each loop one at a time, so each bulb gets the full voltage from the power source and glows brightly.

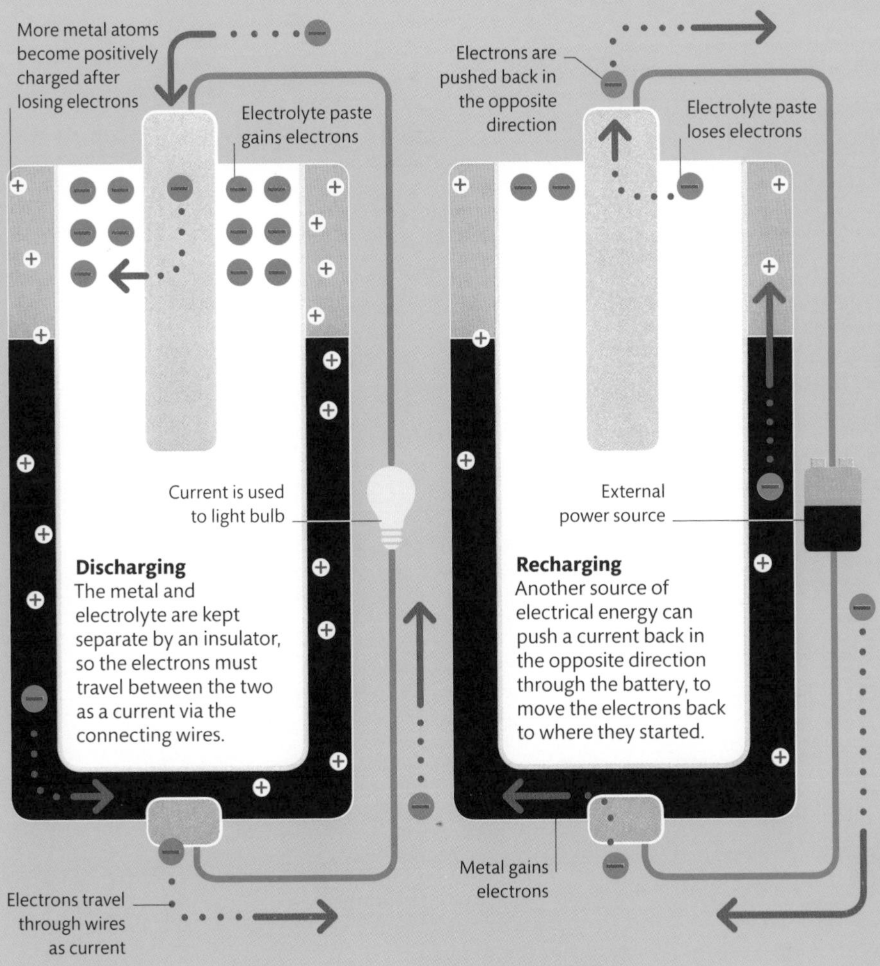

Discharging
The metal and electrolyte are kept separate by an insulator, so the electrons must travel between the two as a current via the connecting wires.

Recharging
Another source of electrical energy can push a current back in the opposite direction through the battery, to move the electrons back to where they started.

FREE ELECTRONS

Most metals, such as iron, are good conductors because the electrons on the shells in their atoms are free to move around to the shells of other atoms. If the electrons are given enough energy, then an electrical current can be formed. Atoms inside insulators—rubber, for example—hold the electrons tightly, so currents are harder to form.

Ohm's law

The link between voltage, current, and resistance is encapsulated by Ohm's law. Its formula (see right) can be used to calculate how much current passes through a component depending on the voltage of the power source and the resistance of items in the circuit.

$$\text{CURRENT} = \frac{\text{VOLTAGE}}{\text{RESISTANCE}}$$

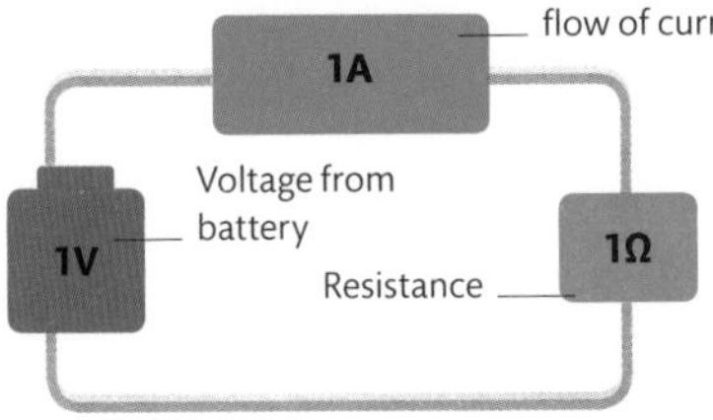

The ohm
Resistance is measured in units called ohms (Ω). A resistance of 1Ω allows a current of 1A to pass through when 1V is applied to it.

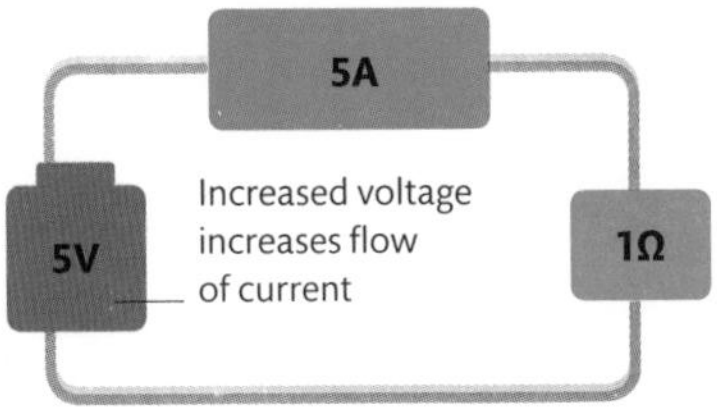

In proportion
The current is proportional to the voltage. If the voltage increases, so will the current, as long as the resistance stays the same.

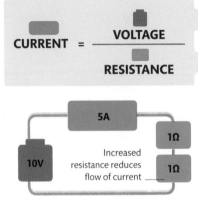

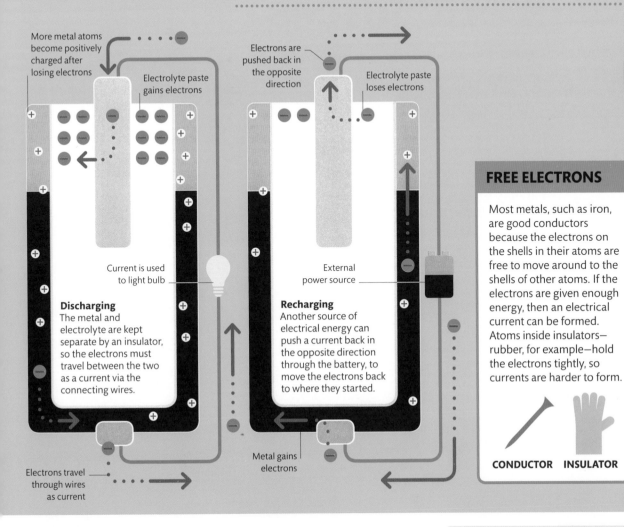

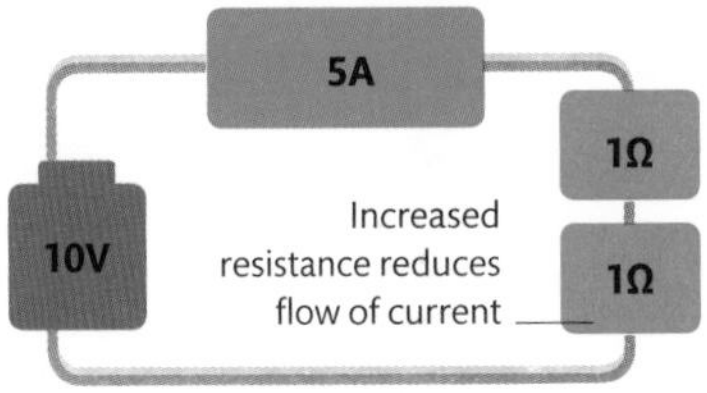

Increased resistance
Adding more resistance means the voltage is unable to push so much current through. Increasing voltage maintains the current.

Magnetic forces

Magnetic force between materials is a large-scale result of the behavior of particles inside the materials at the subatomic scale. Magnets have a range of uses and are integral to many devices.

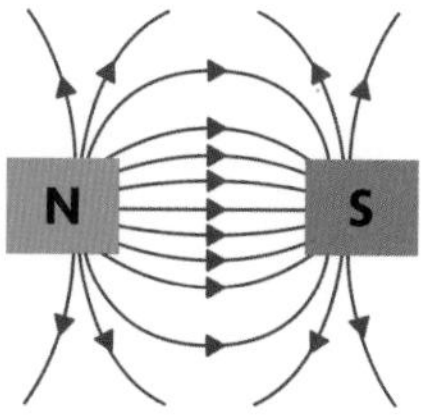

N
S

Opposites attract
The force of magnetism follows the rule of "opposites attract." The north pole of one magnet will attract the south pole of another. The attractive forces pull magnets together.

Magnetic fields

A magnet is surrounded by a force field that stretches out in all directions and reduces rapidly with distance. The force of magnetism has a direction, and the field emerges from a magnet at one point, called the north pole, and goes back in at the south pole. The magnetic field is densest at the poles, and so the effects of the force are strongest there.

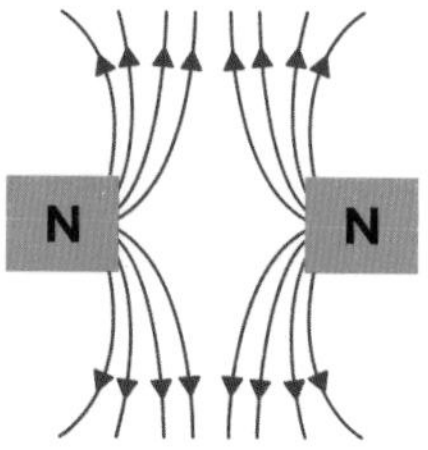

N
N

Likes repel
Two identical magnetic poles, a north pole and north pole, for example, will repel each other. The lines of force from both fields have the same direction and so are deflected away.

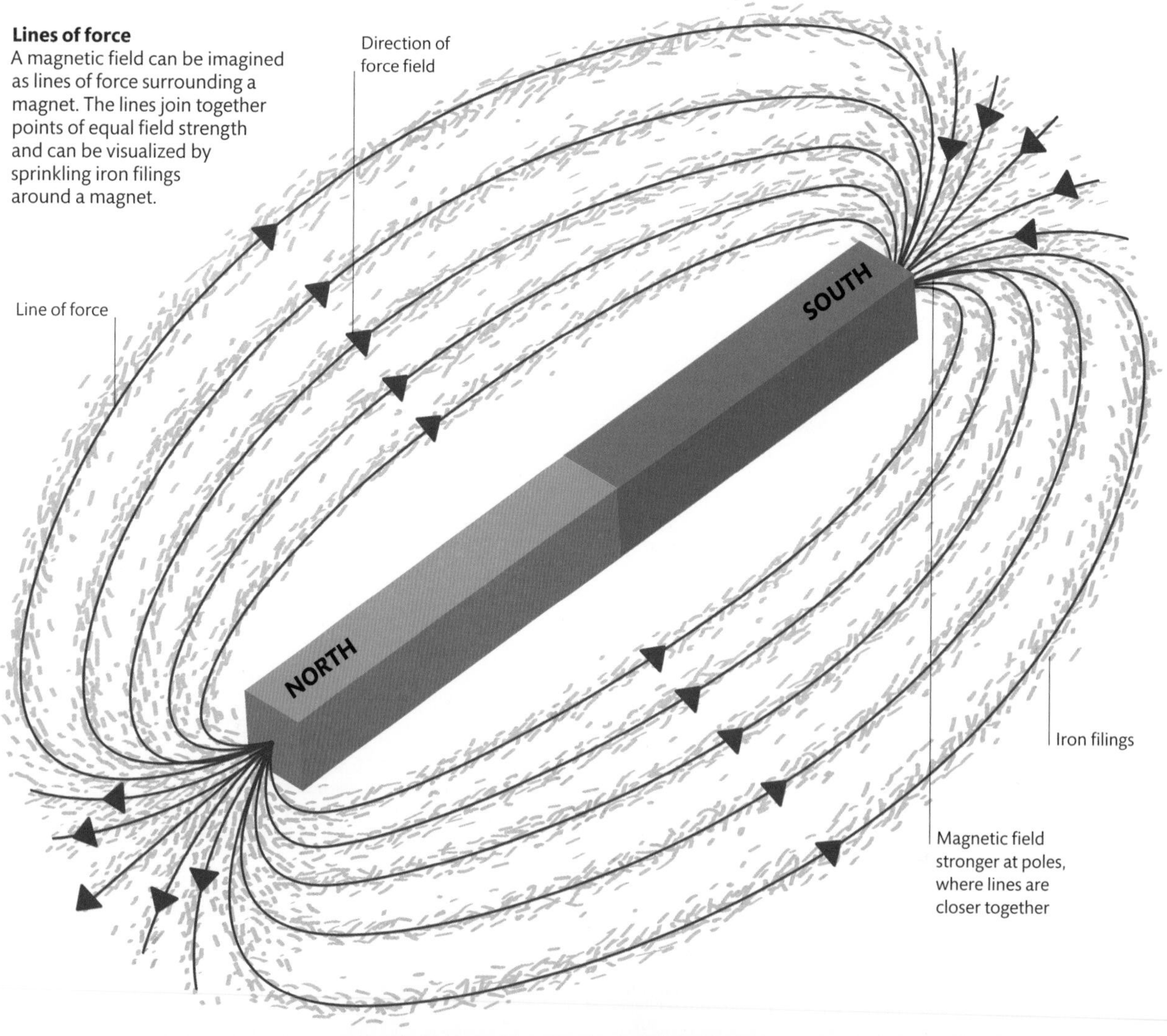

Lines of force
A magnetic field can be imagined as lines of force surrounding a magnet. The lines join together points of equal field strength and can be visualized by sprinkling iron filings around a magnet.
Direction of force field
Line of force
SOUTH
NORTH
Iron filings
Magnetic field stronger at poles, where lines are closer together

Types of magnetism

Every atom has its own tiny magnetic field, and normally these are all orientated in random directions, so they produce no overall effect. If the atoms in a substance are aligned by an exterior magnetic field, their tiny fields accumulate into a single, larger force field.

Diamagnetic materials
Materials including copper and carbon produce a magnetic field that will oppose an exterior field and repel magnets.

Paramagnetic materials
Most metals are paramagnetic. Their atoms align exactly with the exterior field and so are attracted to magnets.

Ferromagnetic materials
Atoms in iron and a few other metals remain aligned when the exterior field is removed and form permanent magnets.

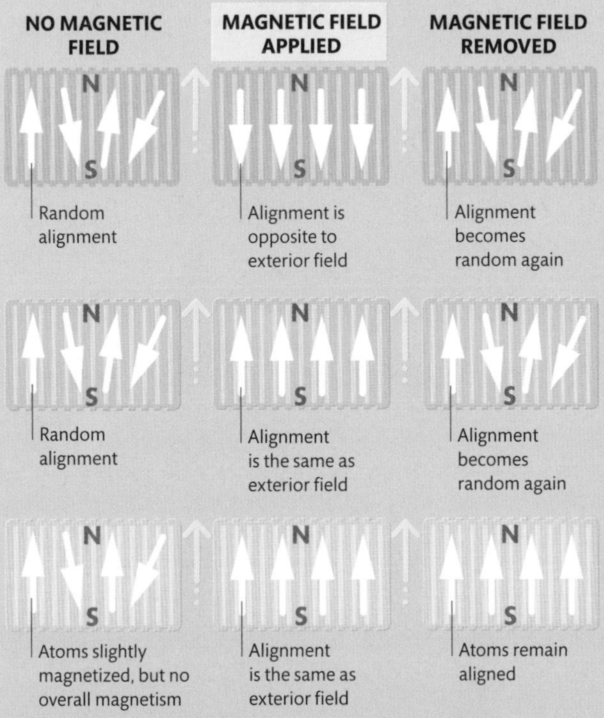

WHAT IS THE STRONGEST MAGNET?

Fast-spinning neutron stars called magnetars have magnetic fields that are 1,000 million million times stronger than Earth's.

MRI SCANNERS USE A **MAGNET COOLED TO –445°F (–265°C)** TO **MAGNETIZE THE WHOLE BODY** FOR A FRACTION OF A SECOND

Electromagnets

The magnetism of an electromagnet is created by an electric current running around an iron core. This means the magnetic field can be turned on and off. Electromagnets have a large number of uses in modern devices.

ELECTRIC MOTOR
An electric motor uses an electromagnet to create a force that pushes against a permanent magnet's poles, creating a spinning motion on demand.

COMPUTER HARD DRIVE
Data is stored on a hard drive as a coded pattern of magnetized and demagnetized zones. An electromagnet reads, writes, and erases this code.

LOUDSPEAKER
The force of an electromagnet is used to make a speaker cone wobble, which in turn wobbles the air in the correct pattern to make a sound wave.

INDUCTION STOVE
A powerful electromagnet is used to create a fluctuating magnetic field inside the metal structure of a pan, and this makes the pan get hot.

EARTH'S MAGNETISM

The liquid iron in Earth's outer core generates a strong magnetic field. Magnetic compass needles point north because they are aligning with the planet's magnetic field. The field reaches far out into space, forming a shield against the solar wind—a blast of hot, electrified gas produced by the Sun.

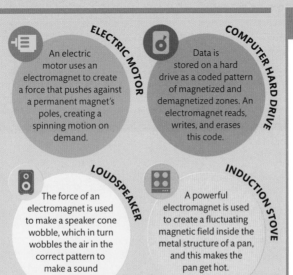

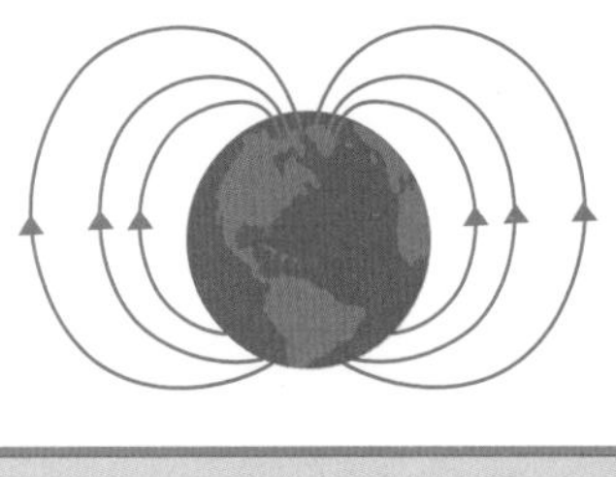

Generating electricity

Electricity is a very useful source of energy. It can be widely distributed for use far away from where it is produced, and it can provide power for all kinds of devices from computers to cars.

Inducing current

An electrical generator uses a process called induction to make electric current. When a wire is moved through a magnetic field, a voltage and current are produced within it. The wire's kinetic energy is converted into electrical energy, making a current run through it. A simple generator does this by spinning a loop of wire very quickly between the poles of a powerful magnet.

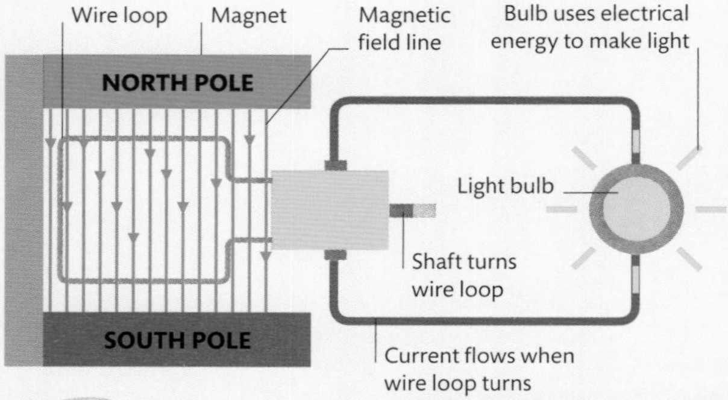

AC AND DC

Each time the wire loop passes through the magnetic field the direction of the current within it swaps over. This is an alternating current (AC). Power stations produce AC, because a constantly reversing current is needed in transformers (see below) to induce a current in the secondary coil. In direct current (DC), the wire loop's connection with the circuit is switched every turn so the charge only moves in one direction.

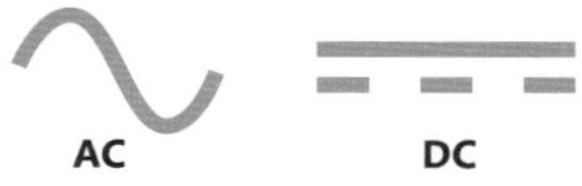

Thermal power plants

The role of a thermal power plant is to harness a source of heat energy that can be used to turn the rotor inside a generator. They make use of the heat released from burning a fuel and convert it into rotational energy using a steam turbine. Nuclear power plants use heat from the splitting of atoms.

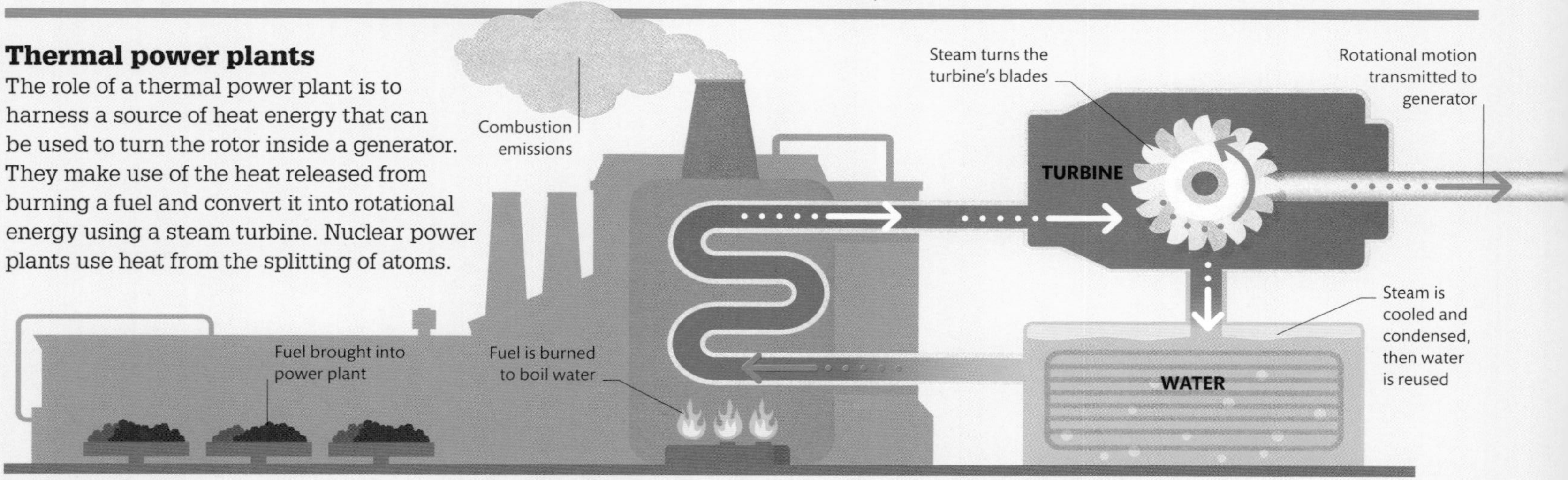

1 Fuel use
Fuels are substances that release large amounts of heat when burned. Common fuels are coal, natural gas, and oil. Power plants also burn wood, peat, and rubbish.

2 Furnace
Water flowing through tubes in the furnace is boiled by heat from the burning fuel. This produces high-pressure steam, which is directed to the turbine.

3 Turbine
The stream of steam flows through a turbine and turns the fan blades. The steam's pressure is converted into kinetic energy and transmitted to the generator.

4 Generation
The rotor in the generator is made to spin at about 3,600 times a minute and it produces an AC current and a voltage at around 25,000V. The voltage is then boosted by a transformer (see below) to 400,000V so it is more efficient to transmit over long distances.

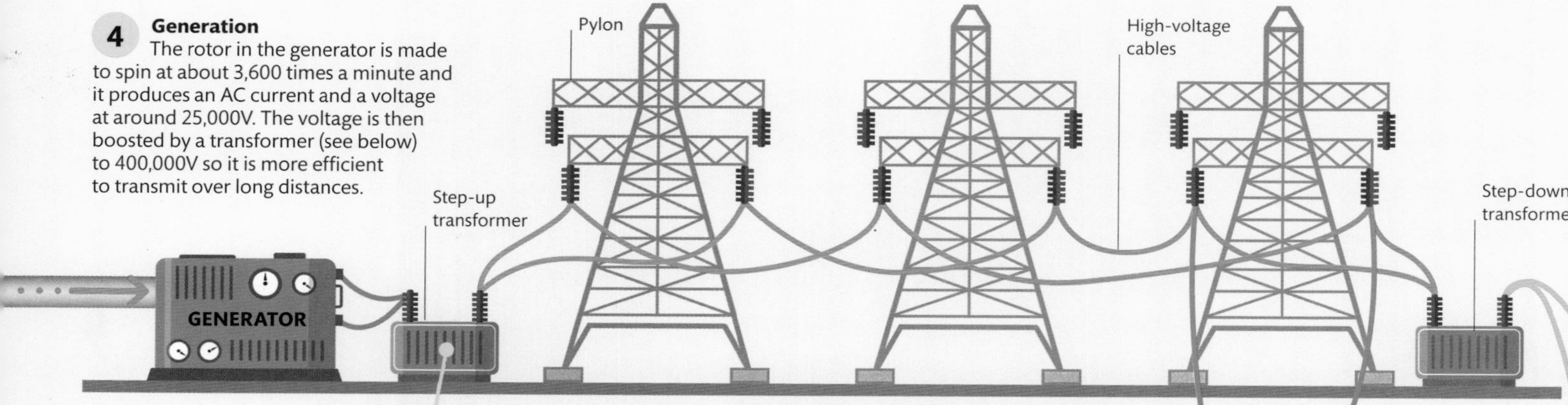

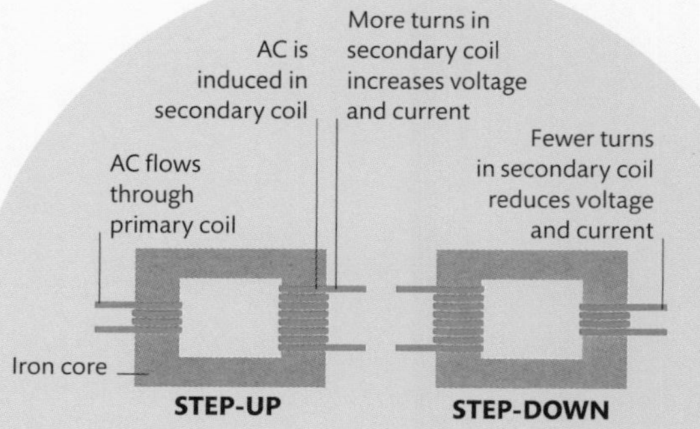

Transformers
A transformer is a device for modifying current and voltage. It is an iron ring with a coil of wire around each side. The system requires an AC supply, because AC has a constantly changing electrical field. A changing field in the primary coil induces an AC supply in the secondary coil.

A PERSON FROM QATAR USES 89 TIMES MORE ENERGY A YEAR THAN SOMEONE FROM SENEGAL

Factory
Factories use a range of voltages up to 33,000V. They may have their own substations, containing transformers.

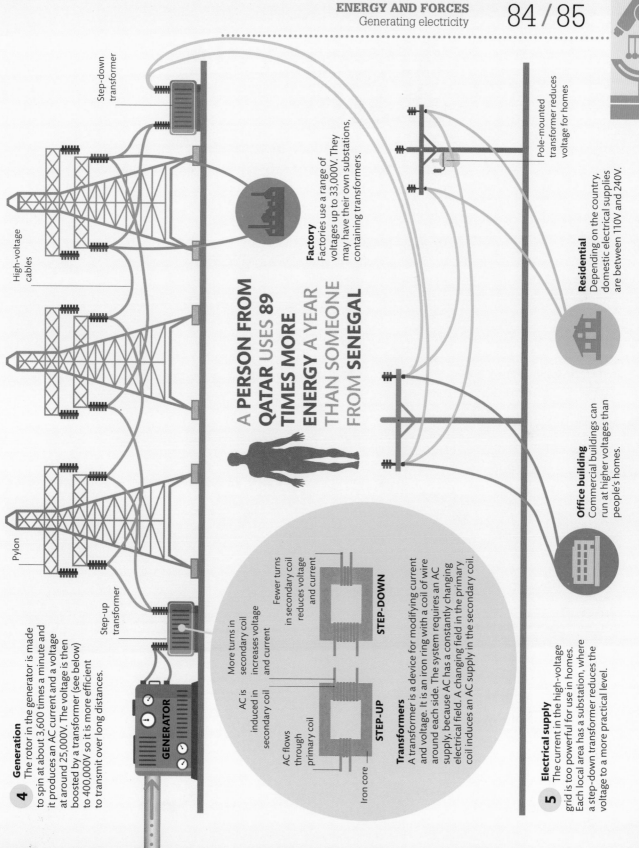

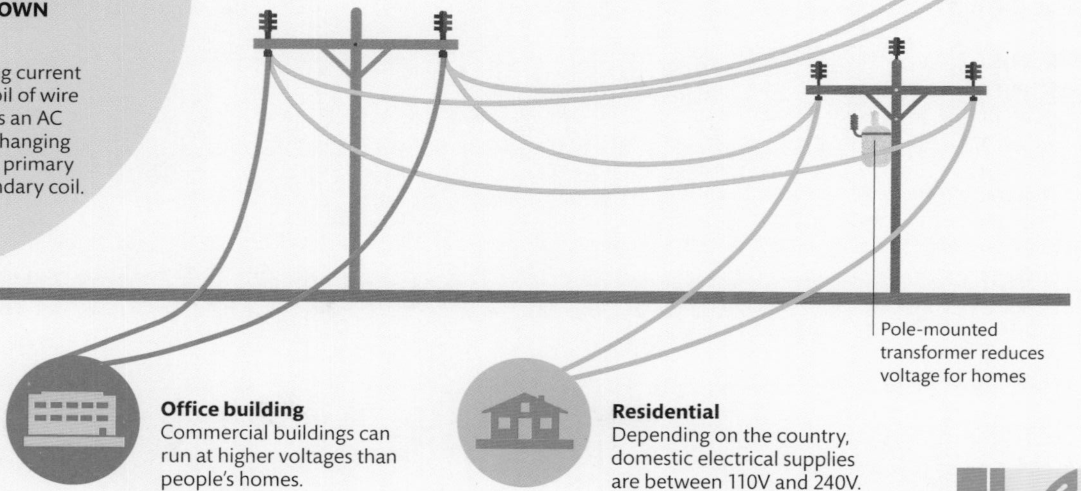

5 Electrical supply
The current in the high-voltage grid is too powerful for use in homes. Each local area has a substation, where a step-down transformer reduces the voltage to a more practical level.

Office building
Commercial buildings can run at higher voltages than people's homes.

Residential
Depending on the country, domestic electrical supplies are between 110V and 240V.

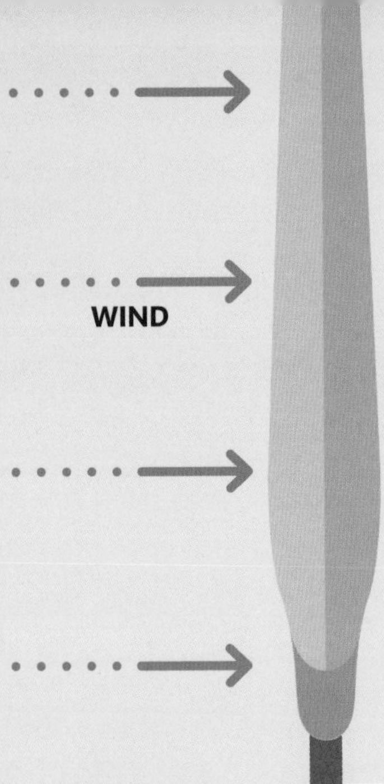

Alternative energy

Instead of fossil fuels, alternative power systems use another source of energy such as the natural motion of air or water, or heat from Earth or the Sun. This makes them less harmful to the environment.

Wind power

Wind is the motion of air from a region of high pressure to a region of low pressure. Such pressure difference is due to uneven heating of the atmosphere by the Sun. This air flow can be harnessed as a source of energy with wind turbines.

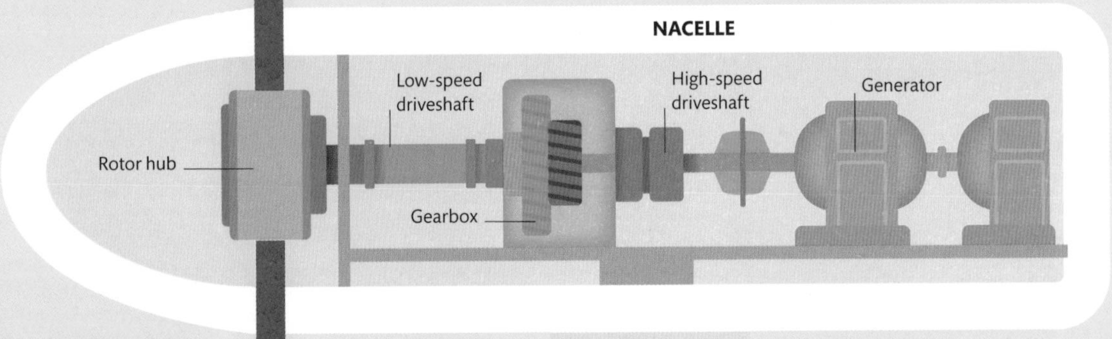

1 Turbine blades The curved blades work like a propeller in reverse. They are shaped precisely so they catch the air and convert its forward motion into rotational motion.

2 Gearing The blades turn about 15 times a minute, which is too slow to make a useful form of electricity. Gears boost the driveshaft rotation to around 1,800rpm.

3 Generator The driveshaft's rotational motion is converted into electricity by the generator. The generator can also be used as an electric starter motor—running current through it in the opposite direction gets the blades spinning after low winds.

CAN WE EVER STOP USING FOSSIL FUELS?

The supply of alternative energy is enough to meet our needs, but we must develop ways to store electricity in large amounts before we can dispense with all fossil fuels.

HYDROELECTRICITY

One of the problems with alternative power systems is finding a reliable supply of energy. Hydroelectric plants typically use dams that harness the flow of rivers. They produce two-thirds of all alternative power and nearly one-fifth of all electricity supplies. As water flows, its potential energy converts to kinetic energy, which is used to turn a water turbine inside a dam to generate electricity.

Natural heat

As well as the motion of air and water, natural sources of heat can be used to make electricity. Concentrated solar power plants have arrays of mirrors that are arranged to concentrate the Sun's light and use it to heat water, which boils and turns a turbine. Geothermal power plants are located in volcanic regions, where the heat from Earth's interior is especially near the surface and can be used as an energy source.

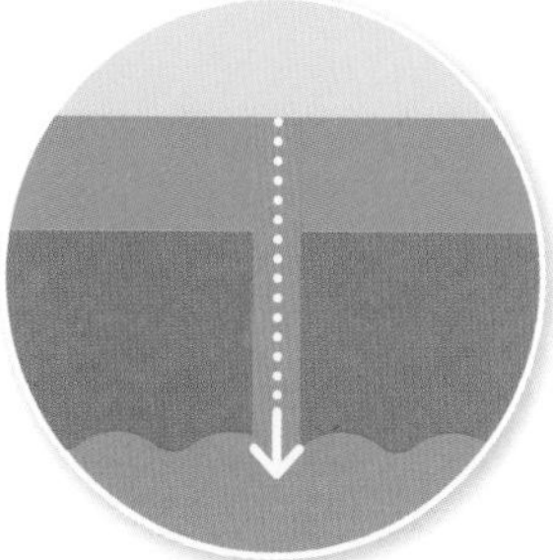

1 Water in
Cold water is pumped down a well, or borehole, at high pressure into a natural reservoir of groundwater deep under the ground—often 6,600ft (2,000m) or more below the surface.

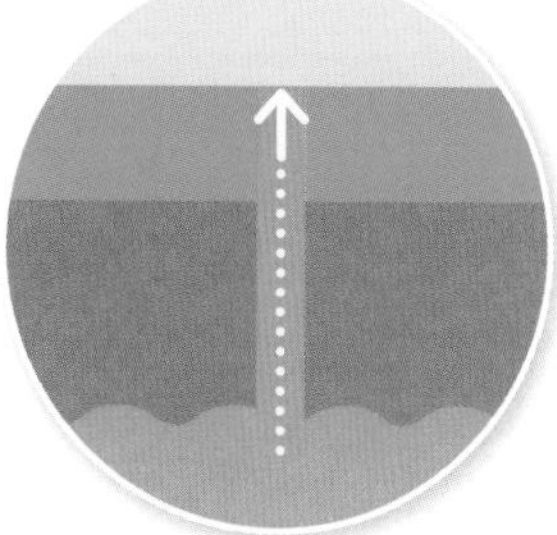

2 Water out
Volcanic heat warms the water to well above 212°F (100°C). Much of it stays as a liquid due to the high pressure conditions, so a mixture of hot water and steam is brought to the surface.

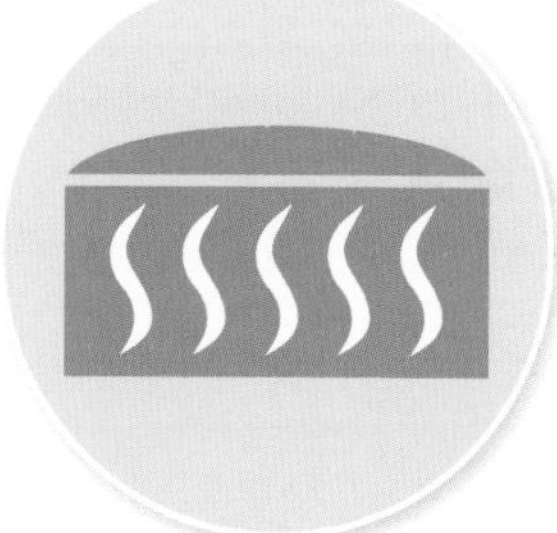

3 Making steam
Steam is separated from the water to create a high-pressure flow that is directed to the turbine. Any water reaching the surface is allowed to flow on into cooling towers.

4 Generator
The high-pressure steam spins turbine blades, as in a regular thermal power plant. That rotational kinetic energy is then transmitted to the generator to make electricity.

5 Cooling tower
The steam is left to cool and condense back into a liquid inside large cooling towers. Once cooled, the water is ready to be injected underground once more for the cycle to start again.

Biofuel

Biofuels are potentially less polluting alternatives to fossil fuels. They can be made by chemically altering raw materials that have been grown by living things. There are three main sources: grains, wood, and algae. Grains and wood are proving problematic for the environment but it is hoped that algae, although in its early stages of development, will eventually provide low-cost, low-pollution fuels.

Pretreatment
Processing starts by physically breaking up the source's raw materials into homogeneous materials and cleaning away unwanted contaminants.

Saccharification
Chemical treatments are used to break down the complex molecules in the starting materials into smaller, more useful molecules, such as sugars.

Fermentation
Similar to the production of alcoholic beverages, the sugars are converted into ethanol and other flammable substances that can be used as fuels.

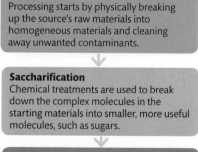

How electronics work

Electronics is the technology of electrical components and their use in circuits. They include transistors, which are used to control the flow of electricity, and most have no moving parts.

What makes a semiconductor?

Conductors have free electrons available to carry a current (see p.81), while insulators have a large energy barrier, or band gap, which stops electrons flowing and forming a current. Semiconductors, like silicon, have a small band gap, so can switch from being an insulator that blocks electricity to a conductor that carries it.

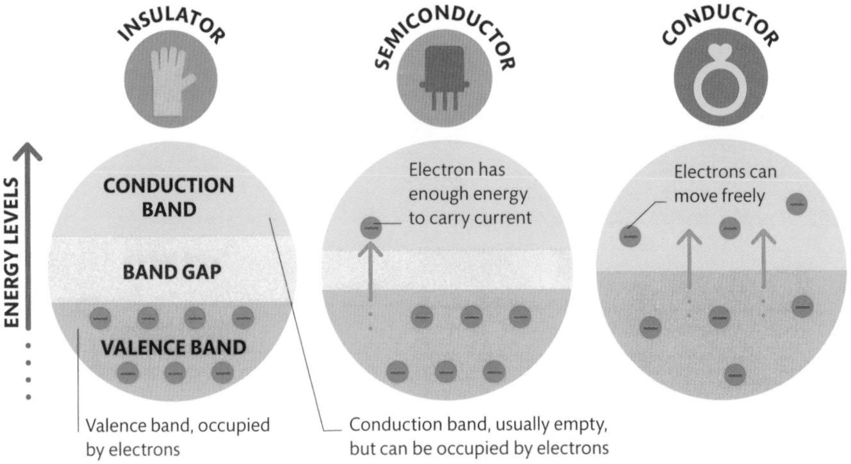

Inside a transistor

The brain of a computer is made up of electronic circuits on a chip. These circuits are told what to do by a set of instructions—the program. In the late 1940s, the transistor, a semiconductor device, was invented to replace early electronic devices that used vacuum tubes and were very unreliable. The transistor is made of crystals of silicon that has been "doped" or had other substances added to alter its electrical properties. The result is a device that can be made to control the flow of an electrical current very precisely.

THE EXPECTED **MINIMUM SIZE** FOR A TRANSISTOR IS EQUAL TO **2 SUGAR MOLECULES**

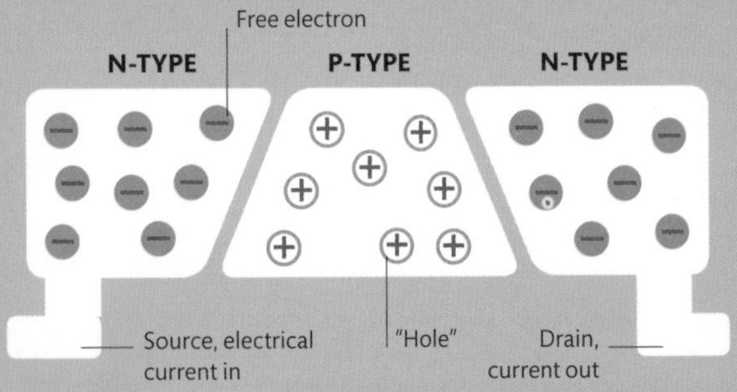

1 Basic structure
A transistor is made from a p-type semiconductor sandwiched between two n-type semiconductors. The n-type has a surplus of electrons and is negatively charged. The p-type contains "holes," which act as an excess of positive charges.

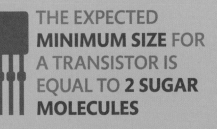

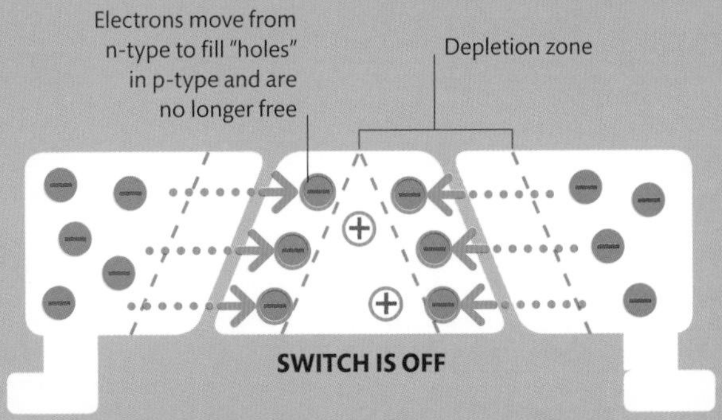

2 Depletion zone
Electrons from the n-type regions are pulled into the p-type by its positive charge. This creates depletion zones where there are no free electrons to carry a current. At this stage, no current can flow, and the transistor switch is "off."

MOORE'S LAW

In 1965, Gordon Moore, the cofounder of the Intel electronics company, predicted that transistors would halve in size every 2 years. So far, Moore's law has held broadly true. Today, standard transistors have a base length of 14 nanometers. This size can shrink further, but electronic technology will hit its limit in the next decade, where the base size becomes too small to form an effective barrier to a current.

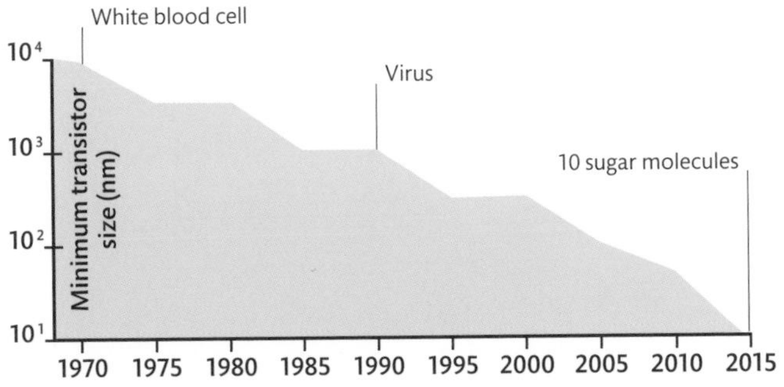

WHERE DOES SILICON COME FROM?

Silicon is the second most common element in Earth's crust. It is purified by burning sand, which contains silicon, mixed with molten iron.

Doping silicon
The purpose of doping silicon is to increase or reduce the number of electrons. Adding phosphorus atoms introduces an extra electron, while adding boron removes an electron, creating an empty space, or "hole," in the crystal.

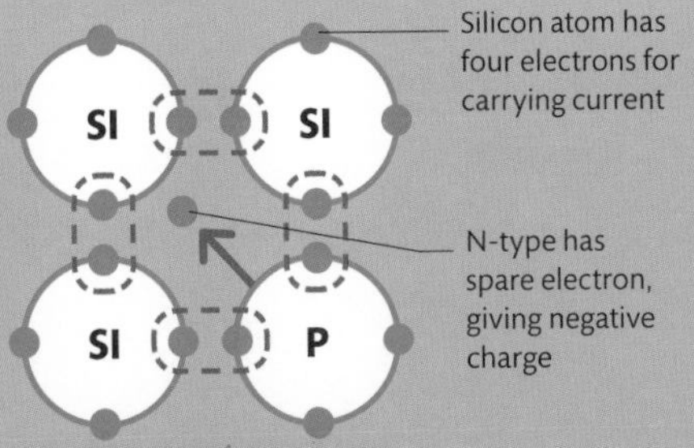

N-TYPE SILICON DOPED WITH PHOSPHORUS

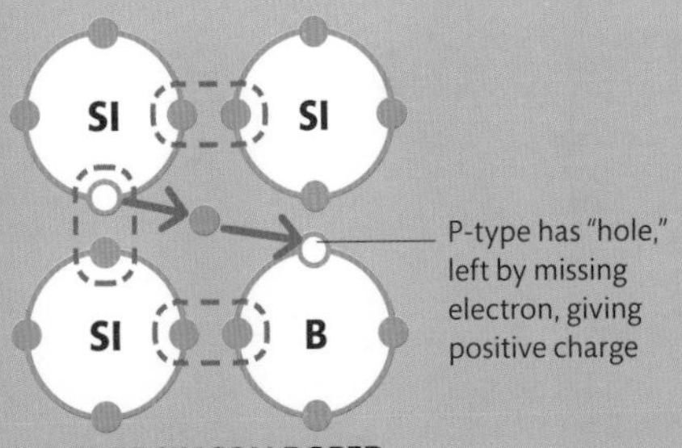

P-TYPE SILICON DOPED WITH BORON

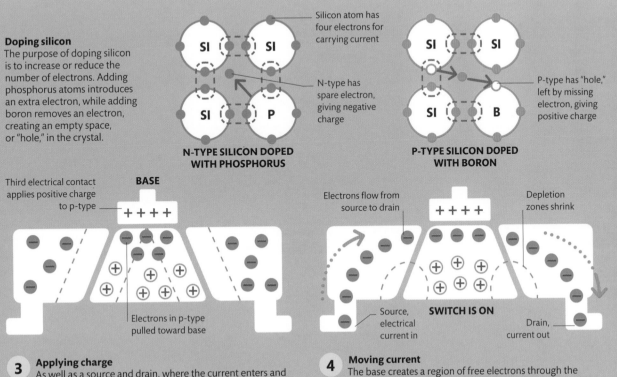

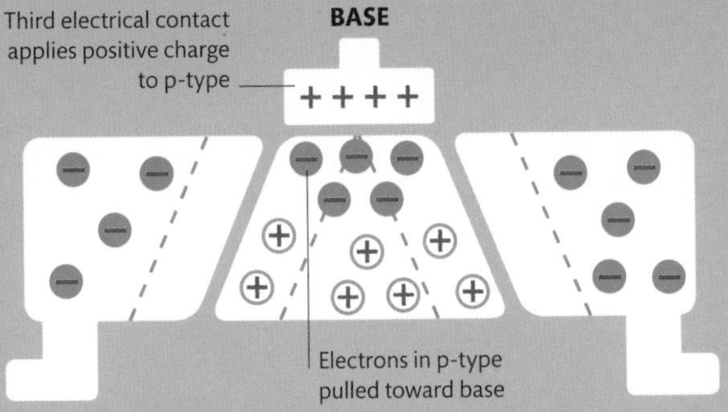

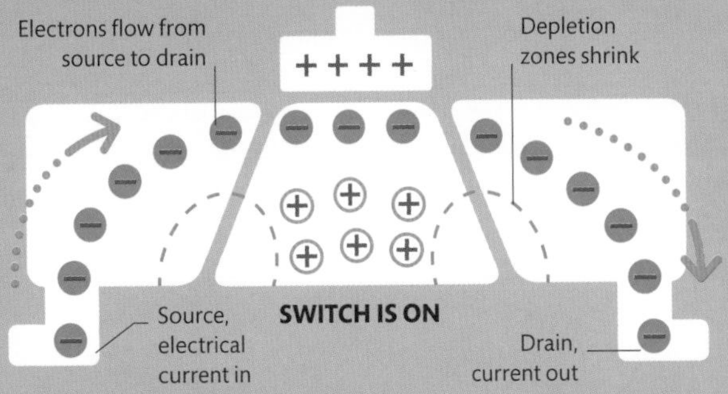

3 Applying charge
As well as a source and drain, where the current enters and leaves, a transistor has a third electrical contact called the base, which applies a positive charge to the p-type section. When turned on, the base pulls on the electrons in the depletion zones.

4 Moving current
The base creates a region of free electrons through the transistor, shrinking the depletion zones, so an electric current can move through it. In this state, the transistor is "on." When the base is off the electrons stop and the transistor switch is "off" again.

Microchips

The microchip is a piece of technology found in all kinds of everyday objects, from phones to toasters. Making a microchip involves incorporating tiny electronic components on a piece of pure silicon.

WHAT IS THE MICROCHIP IN A PET?

This type of microchip contains a tiny radio transmitter and is inserted under the animal's skin. When the reader is placed next to it, it gives out a unique code, which relates to the owner's details.

Making a microchip

A microchip is an integrated circuit, where all the components and the electrical connections between them are manufactured on a single piece of material. Microchip circuitry is etched into the surface of the silicon. Tiny wires are made from copper and other metals, while transistors and other electronics can be made by doping the silicon (see pp.88–89) and adding other semiconductors.

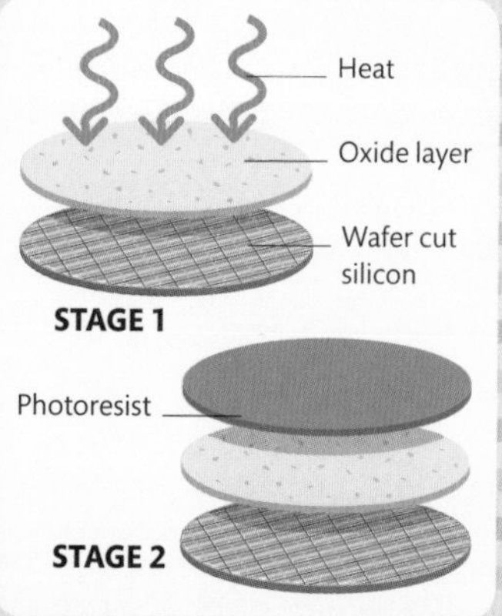

1 Coatings
A wafer of pure silicon is heated to create a fine layer of oxide on the surface. Then a light-sensitive coating called photoresist is added.

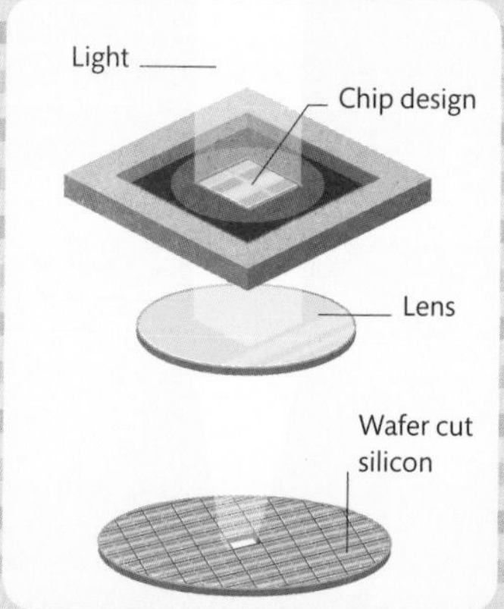

2 Exposure
A large negative of the chip's design is drawn on a transparent material. Lights focus the design onto the photoresist. Each wafer has room for many identical chips.

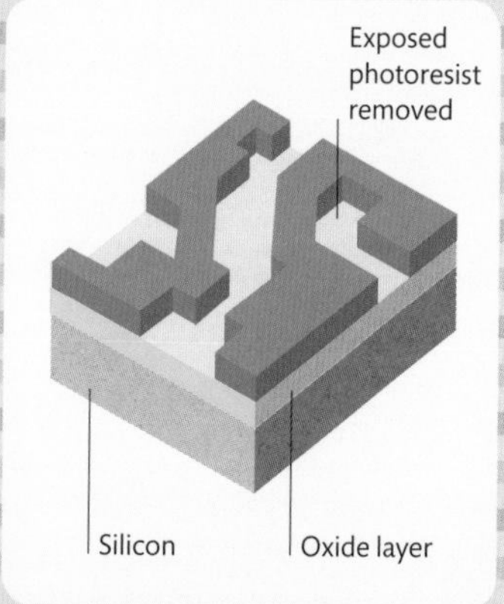

3 Development
The parts of the wafer exposed to light are washed away, exposing a pattern on the oxide layer underneath. Some features of the design are only a few dozen atoms wide.

Using logic

An integrated circuit makes decisions using combinations of transistors and diodes that form logic gates. Logic gates compare incoming electrical currents and send on a new current, based on the math of logic. Known as Boolean algebra, this type of logic has a set of operations where the answer is always true or false, represented by a 1 or a 0.

AND gate
This component has two inputs. It only switches on (outputs a 1) if both the inputs are 1.

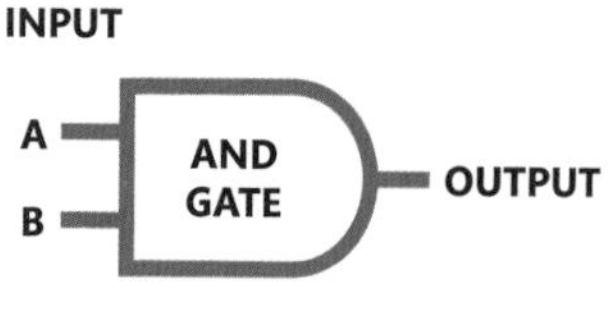

Input A	Input B	Output
0	0	0
0	1	0
1	0	0
1	1	1

OR gate
The opposite of an AND gate, the OR gate always outputs a 1 unless both inputs are 0.

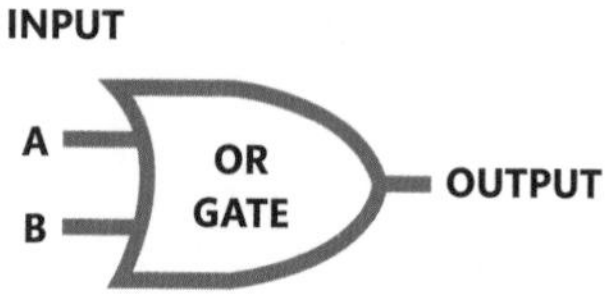

Input A	Input B	Output
0	0	0
0	1	1
1	0	1
1	1	1

ELECTRONIC COMPONENTS

Like other circuit components, electronics are represented by a set of symbols. Chip designers use these as they create new integrated circuits. Modern chips have many billions of components, so human designers will set out the high-level architecture of a chip, and a computer will then convert that into a circuit of logic gates. It takes more than a thousand people to create and test a new chip design.

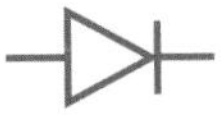

Diode
A one-way channel that lets current pass in one direction

Light-emitting diode
Uses a semiconductor to make electrons release a colored light

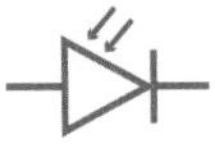

Photodiode
Generates current only when light is shining on it

NPN transistor
Switches on when a current is applied to the base

PNP transistor
Switches on when there is no current at the base

Capacitor
Stores charge, which can be released back into the circuit

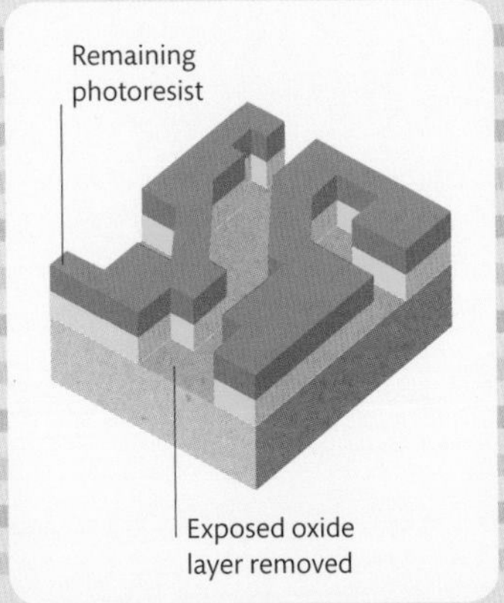

4 Etching
Chemicals are used to remove the exposed parts of the oxide layer, cutting precisely shaped channels into the surface of the silicon wafer.

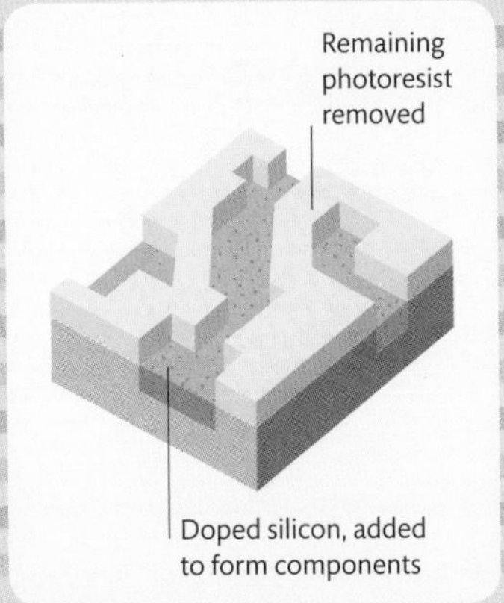

5 Doping
The silicon is doped to give it useful properties, and the channels are filled with precise mixtures of chemicals to create components.

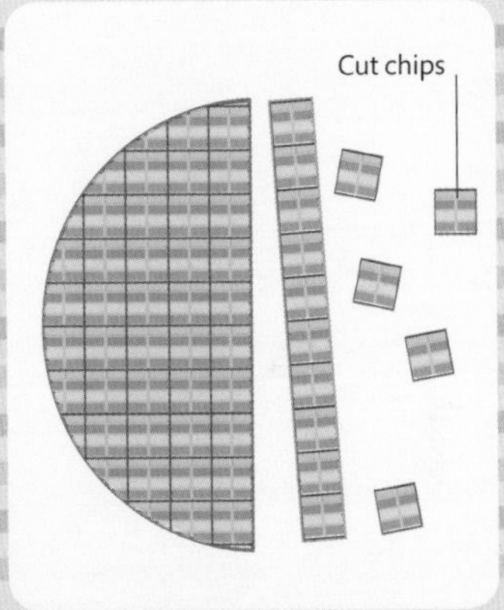

6 Cutting and mounting
The chips are cut from the wafer and given a protective coating of plastic or glass. When mounted on a circuit board, they will connect to other chips and a power source.

NOT gate
This logic gate switches the input and so always outputs whatever the input is not.

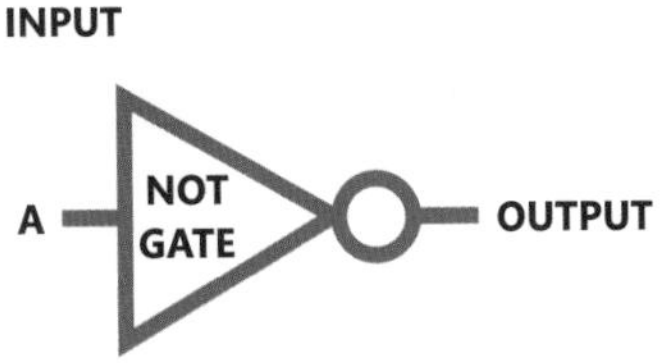

Input	Output
0	1
	0

XOR gate
The exclusive OR, or XOR, gate detects difference in the inputs and always outputs a 0 if the inputs are the same.

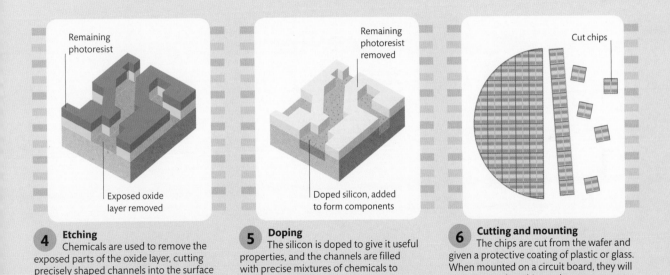

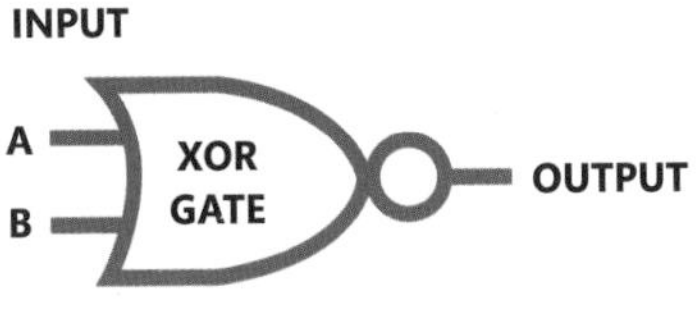

Input A	Input B	Output
0	0	0
0	1	1
1	0	1
1	1	0

HUNDREDS OF MILLIONS OF THE **LATEST TRANSISTORS** COULD FIT ON A **PINHEAD**

Computer basics

Common input devices include a mouse, keyboard, and microphone. These devices convert the user's activity into sequences of numbers that will usually be sent to the random access memory (RAM). The inputs are then called in order to the central processing unit (CPU), where calculations are performed on the input to produce an output. The output can be stored on the hard drive for later use, or it may be sent to an output device—for example, as a sound signal or as letters appearing on a screen as they are typed.

Internet

Data and instructions accessed on the internet can be used as an input for a computer. The computer can also output to the internet, and a user's data can be stored on the internet, or "in the cloud."

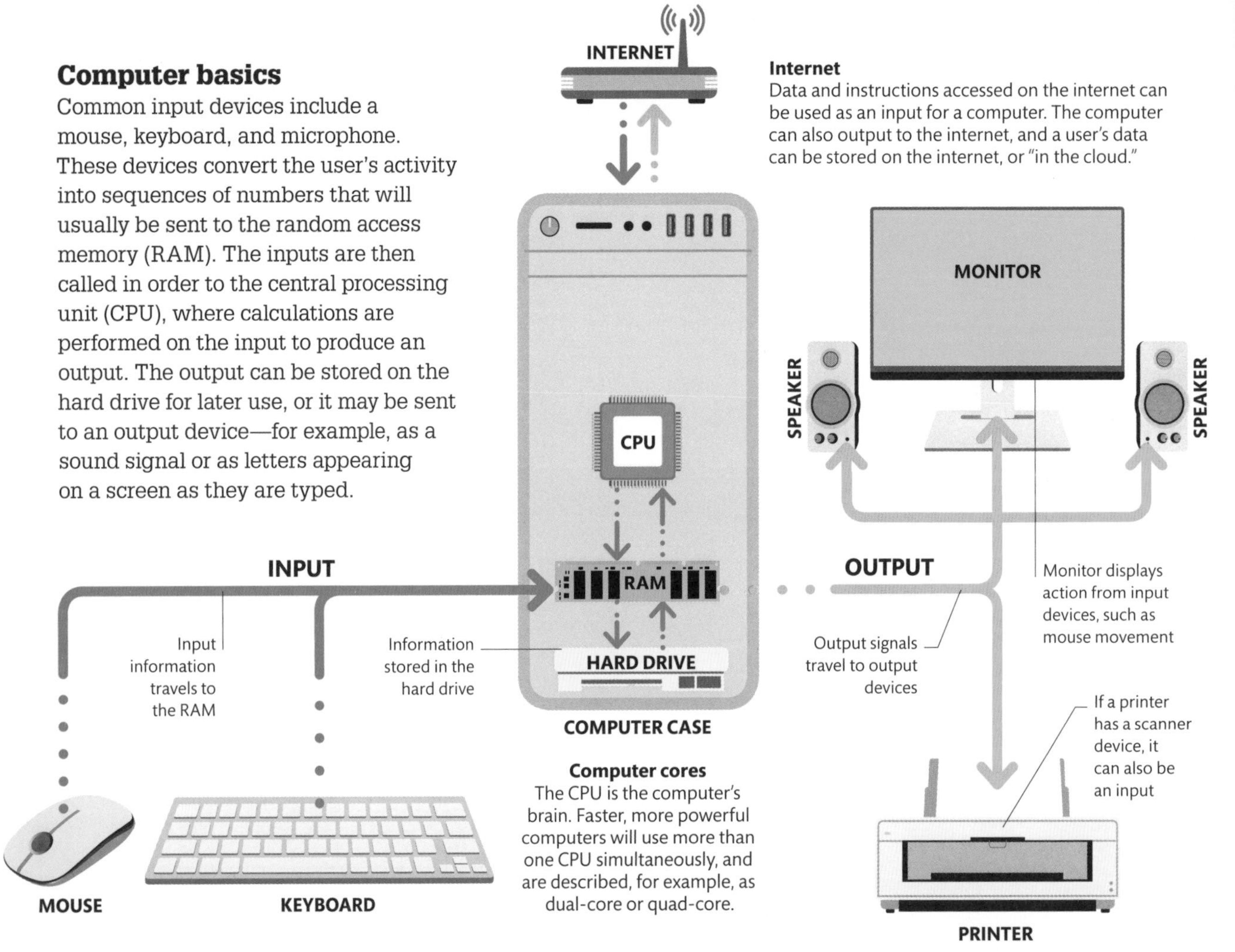

CPU

MONITOR

SPEAKER

SPEAKER

INPUT

OUTPUT

Monitor displays action from input devices, such as mouse movement

Input information travels to the RAM

Information stored in the hard drive

RAM

HARD DRIVE

Output signals travel to output devices

If a printer has a scanner device, it can also be an input

COMPUTER CASE

Computer cores

The CPU is the computer's brain. Faster, more powerful computers will use more than one CPU simultaneously, and are described, for example, as dual-core or quad-core.

How computers work

At its simplest, a computer is a device that takes an input signal and transforms it into an output signal according to a set of preprogrammed rules. The true power of such a system is that it can perform calculations far more quickly and accurately than a human can.

COMPUTER CODES

A CPU handles data using only 0s and 1s, in sequences of 8, 16, 32, or 64. Humans often simplify long binary code into hexadecimal, a counting system that uses 16 numerals: 0 to 9, then A to F represent 10 to 15.

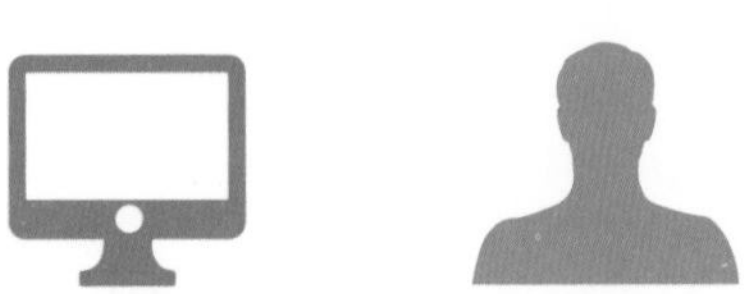

1111 BINARY = 15 = F HEXADECIMAL

How the internet works

In a computer network, computers are directly connected or communicate via other computers. The internet is a network without a central control point. Instead, data is sent from a source device to a recipient.

THE WORLD'S FASTEST SUPERCOMPUTER DOES **93 MILLION BILLION** CALCULATIONS A SECOND

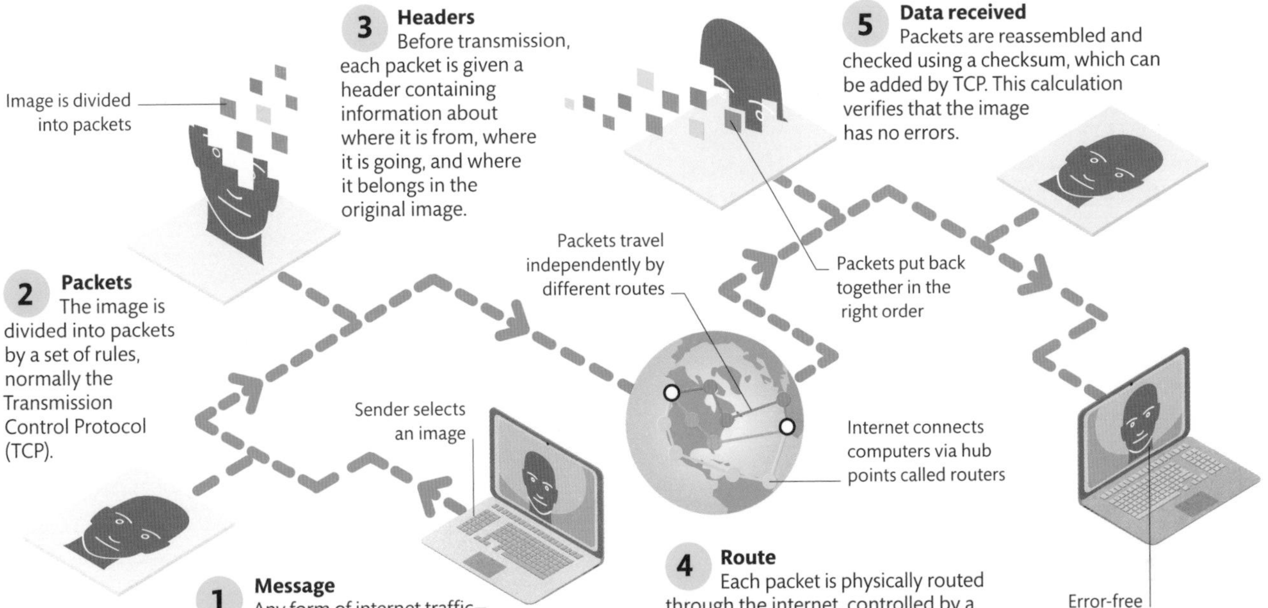
Image is divided into packets
Packets travel independently by different routes
Packets put back together in the right order
Sender selects an image
Internet connects computers via hub points called routers
Error-free image shown at receiver's end

1 Message
Any form of internet traffic—such as an image or email—may be split up, and is transmitted as a digital code of 1s and 0s.

2 Packets
The image is divided into packets by a set of rules, normally the Transmission Control Protocol (TCP).

3 Headers
Before transmission, each packet is given a header containing information about where it is from, where it is going, and where it belongs in the original image.

4 Route
Each packet is physically routed through the internet, controlled by a system called the Internet Protocol (IP), which sends each packet to the next nearest internet server.

5 Data received
Packets are reassembled and checked using a checksum, which can be added by TCP. This calculation verifies that the image has no errors.

Hard disk drives

Most desktop computers use a hard disk as their main storage. It records data as a physical pattern of magnetized and demagnetized zones. These patterns remain in place when the power supply is turned off. Each hard drive has several platters, each spinning thousands of times a minute. Some more recent computers, such as phones and slim laptops, use a solid-state flash memory instead of a hard disk, which stores data on interconnected memory chips.

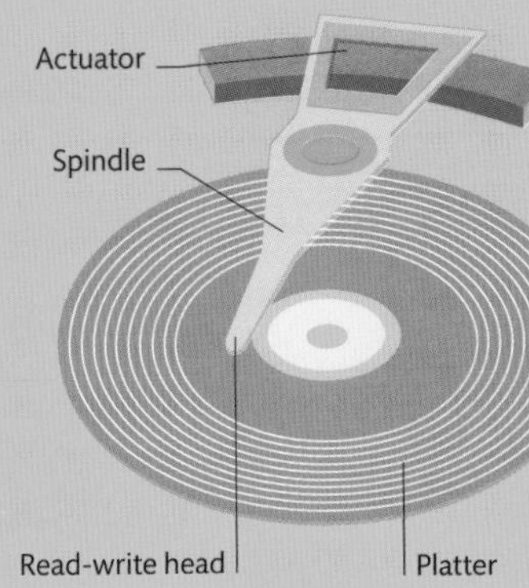
Actuator
Spindle
Read-write head
Platter

Read and write
Each platter is scanned by the read-write head. Its electromagnet detects patterns on the platter and can also write new patterns.

WHAT IS A BYTE?

One digit in a computer code is called a bit of data. These are often handled in sequences of eight bits, and that set of eight forms a byte of data. Four bits, or half a byte, is called a nibble.

Virtual reality

For many years, technology did not match our expectations of what virtual reality (VR) could achieve, and only now are VR applications becoming widespread. A VR headset has to do a great many things to convince a user that they are in another place.

Inside a VR headset

The term "virtual," when used in this context, refers to something that is not real but that can be viewed, manipulated, and interacted with as if it were real. A good example is a virtual image made in a mirror, where objects appear "behind" the glass. A VR headset uses a screen to fill a user's field of vision with part of a virtual scene. Moving the headset results in the view of the scene changing in response.

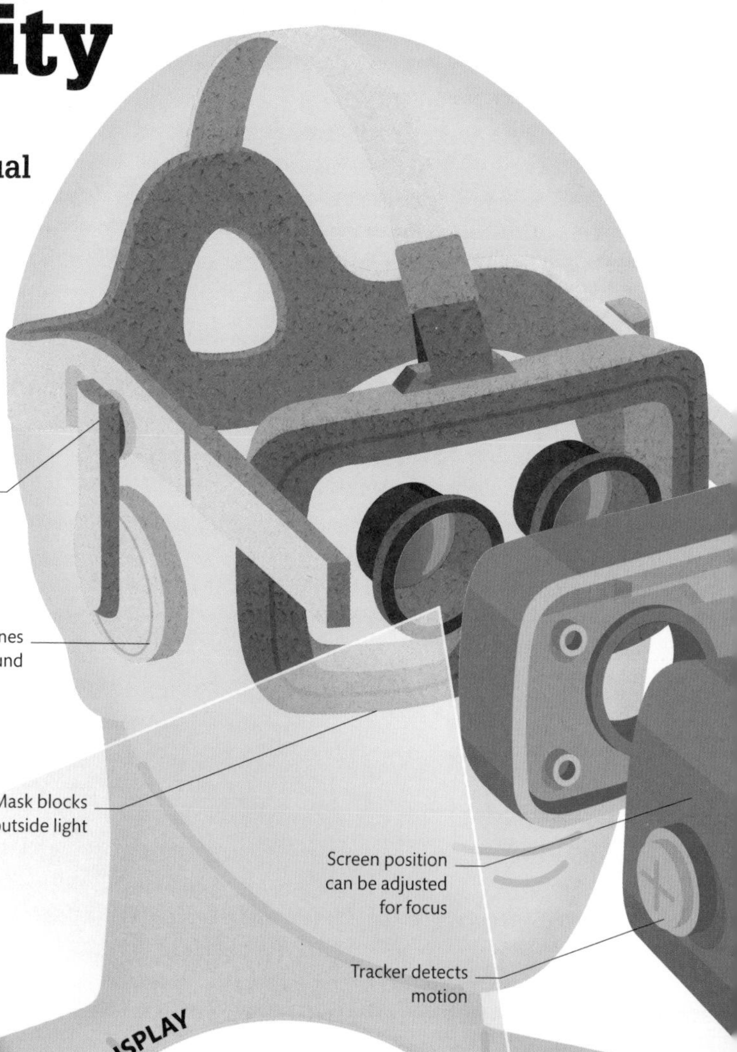
Head strap holds screen firmly in place
Headphones provide sound
Mask blocks outside light
Screen position can be adjusted for focus
Tracker detects motion

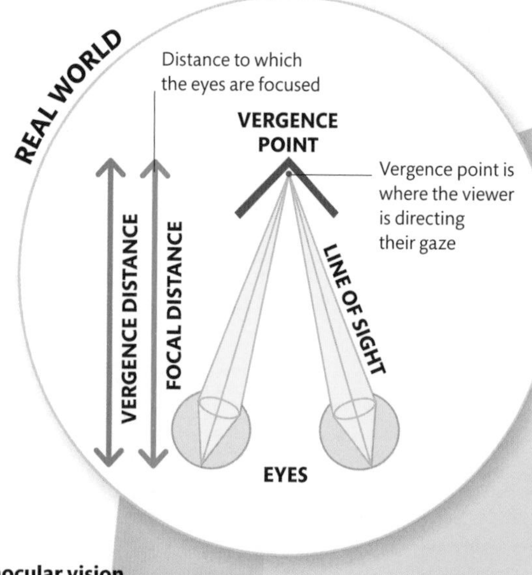
REAL WORLD
Distance to which the eyes are focused
VERGENCE POINT
Vergence point is where the viewer is directing their gaze
VERGENCE DISTANCE
FOCAL DISTANCE
LINE OF SIGHT
EYES

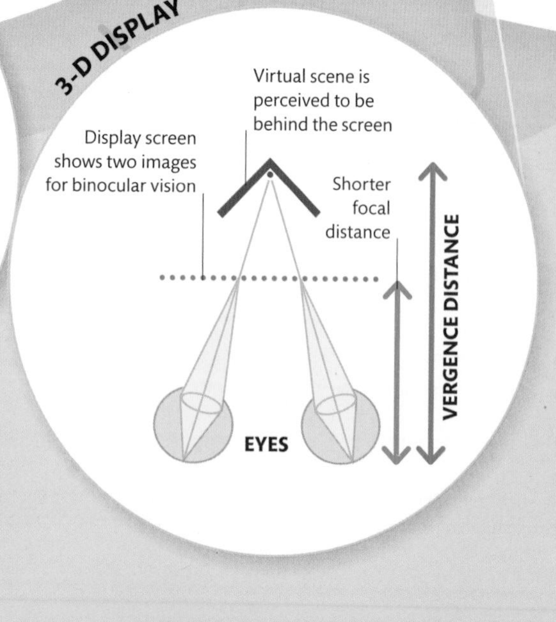
3-D DISPLAY
Virtual scene is perceived to be behind the screen
Display screen shows two images for binocular vision
Shorter focal distance
VERGENCE DISTANCE
EYES

Binocular vision
The VR screen displays two images, one for each eye. The right eye sees an image that is shifted slightly to the right compared to the left eye. This system is called stereoscopy, and it emulates true vision to create the illusion of a 3-D virtual scene.

Tracking

To make the VR experience more immersive, the headset tracks the motion of the user's head and eyes and alters the displayed scene accordingly. This means the user can look around the virtual space in a natural way. To track the motion of the user's arms and legs, a separate device can be used that bounces beams of infrared light off the body. This allows users to interact more with their virtual surroundings.

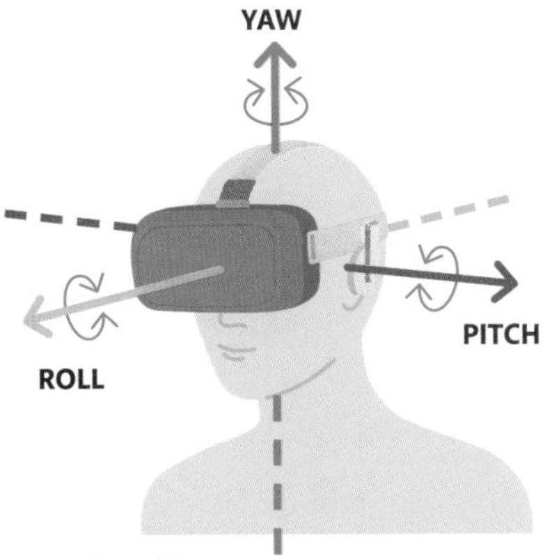
YAW
ROLL
PITCH

Head tracking
Sensors on the headset, similar to the ones in smartphones, track the motion of the user's head in three axes. This information is used for large-scale adjustments to the virtual scene.

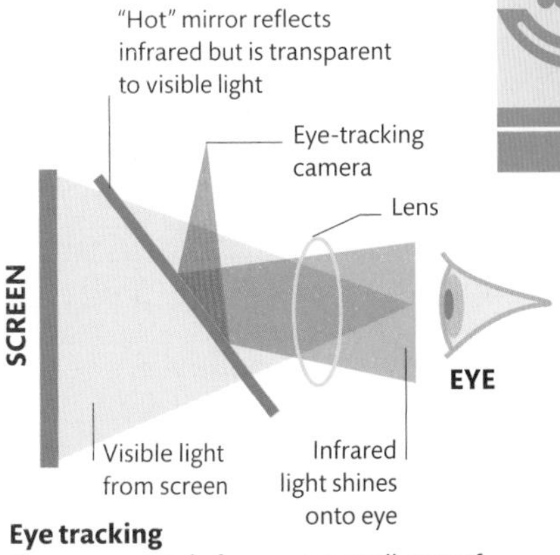
"Hot" mirror reflects infrared but is transparent to visible light
Eye-tracking camera
Lens
SCREEN
EYE
Visible light from screen
Infrared light shines onto eye

Eye tracking
Human eyes only focus on a small part of a scene, so some VR displays present the sharpest image at that point. Infrared light is shone onto the eye, and a camera analyses the reflections to track the direction of sight.

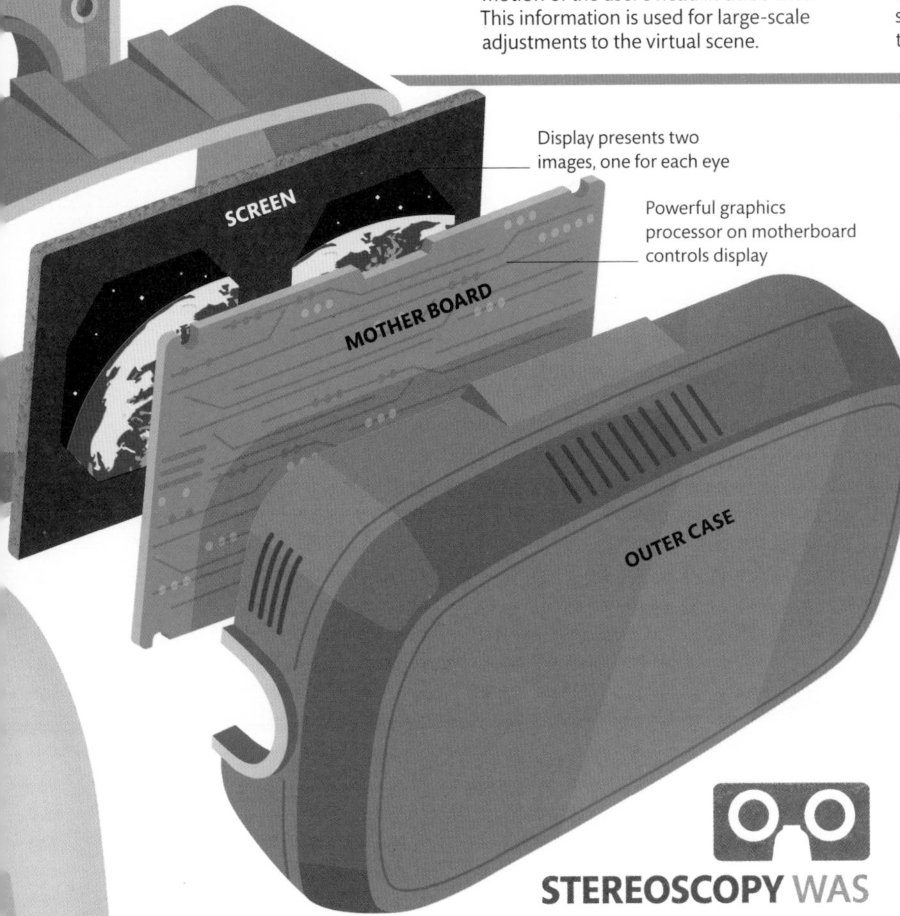
Display presents two images, one for each eye
Powerful graphics processor on motherboard controls display
SCREEN
MOTHER BOARD
OUTER CASE

Altering perceptions
VR headsets trick a user's perceptions so they experience being in a computer-rendered 3-D space. As well as using imagery and sound, "haptic" devices on gloves or other body parts allow users to feel virtual objects.

STEREOSCOPY WAS INVENTED **IN 1838–** EVEN **BEFORE PHOTOGRAPHY**

AUGMENTED REALITY

Augmented reality (AR) uses similar technology to VR, but in AR computer-generated graphics are overlaid on the real scene. AR users can view a scene through a live camera feed—on a smartphone, for example—or content can be projected onto a transparent screen such as a pair of glasses.

Nanotubes

Carbon nanotubes are cylindrical structures just a few nanometers across. At present, they are restricted to the millimeter scale, but longer nanotubes could make a material that is many times stronger than steel, while having other useful properties such as low density.

A **NANOTUBE REACHING THE MOON** WOULD ROLL INTO A BALL AS BIG AS A **POPPY SEED**

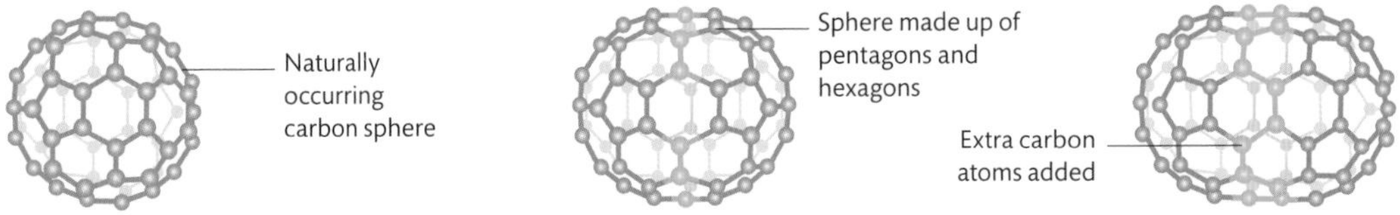

1 Growing nanotubes One way to build nanotubes is to grow them. The starting point is a naturally occurring, 60-atom carbon sphere, known as a buckyball.

2 Adding hexagons Most of the sphere is made from hexagons of carbon atoms. The length of the buckyball is increased by adding more.

3 Increasing length Successive rings of 10 carbon atoms are added to the sphere. At 1mm in length, the tube will contain over 1,000,000 atoms.

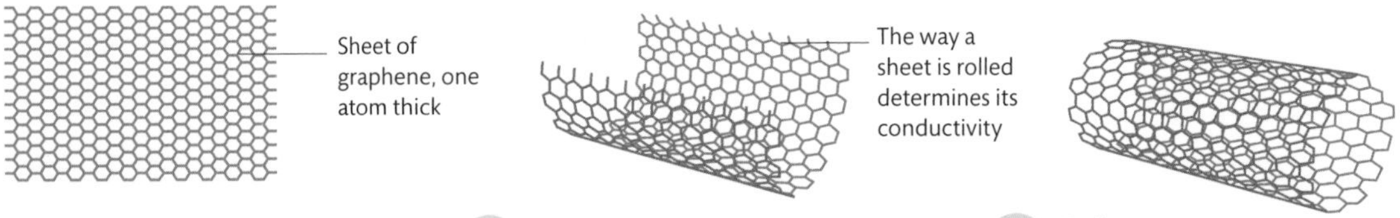

1 Rolling nanotubes Another method for making nanotubes is to roll a sheet of carbon hexagons one atom thick, called graphene.

2 Flexible and strong Graphene is very stiff in all directions, which means it can be flexed and curved into different positions—in this case, it is rolled.

3 Roll up Rolling one graphene sheet produces a single-walled nanotube. Multiwalled nanotubes are created by nesting one tube inside another.

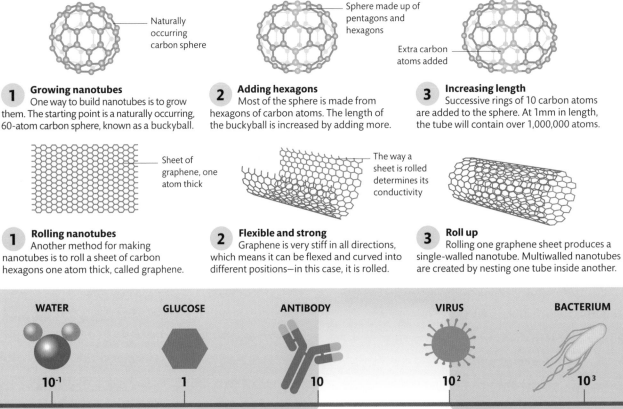

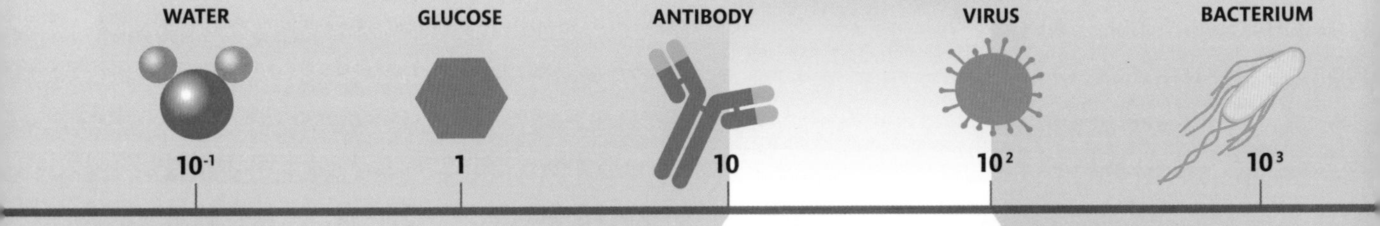

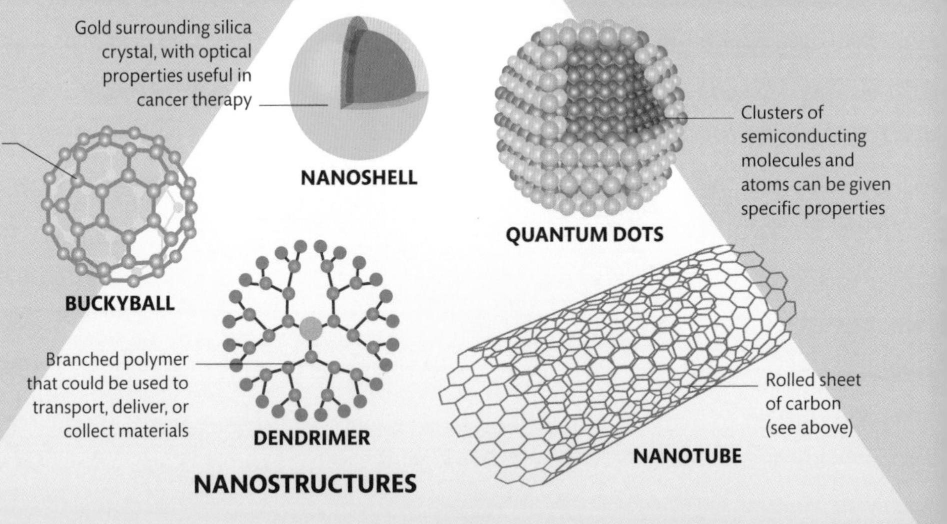

Tiny technology
For their volume, nanoparticles have a very large surface area, which means they are able to react very quickly. Nanoparticles have unique properties not shared by the same substances at other scales. There are some concerns that nanoparticles are so small they might cause damage in a person's body by entering the brain via the bloodstream.

Uses of nanotechnology

Nanotechnology is set to change the future of construction, medicine, and electronics. One theory is that tiny machines called nanobots could work in the body, delivering medicines. Another proposal is that nanoscale tools could assemble objects molecule by molecule. These technologies are decades away, but nanoscale materials are already in use. For example, scratch-resistant glass is hardened by a layer of aluminum silicate nanoparticles, which is only a few nanometers thick, and so is transparent.

Transparent sunscreen
Nanoparticles of zinc and titanium oxides are used in sunscreens. The tiny crystals scatter damaging rays away from the skin.

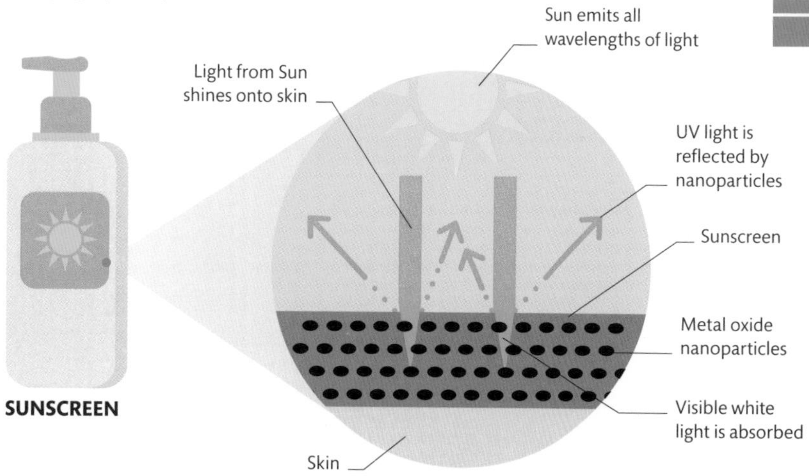

OLED TV
Organic light-emitting diode (OLED) technology makes light by electrifying a layer of molecules. OLED displays are thin and flexible.

Smaller computers
Wirelike nanotubes and quantum dots could soon be incorporated into microchips, making them smaller and more powerful.

Mega-structures
Adding nanotubes to building components makes them much stronger, which could allow for much larger structures in the future.

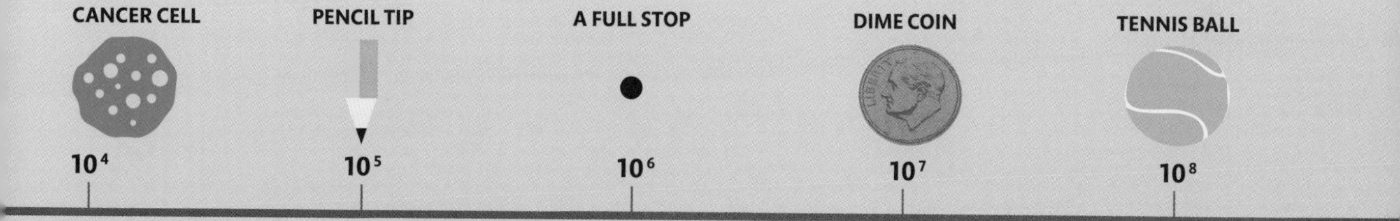

Nanotechnology

Miniaturization has long been a goal of engineering. Nanotechnology aims to build tiny machines by assembling them from individual atoms and molecules.

The nanoscale

The prefix nano means "a billionth"—there are 1 billion nanometers (nm) in 1 meter, while this full stop . is about 1 million nm across. Nanomachines, or nanobots, are theoretical machines capable of performing actions on the nanoscale, and could be between 10 and 100nm wide.

USING DNA

A useful property of DNA is that it can create copies of itself, with one DNA strand acting as a template for a new one. This property of self-replication could be manipulated to make nanoscale devices constructed of DNA, which could, in theory, change their shape and work like machines.

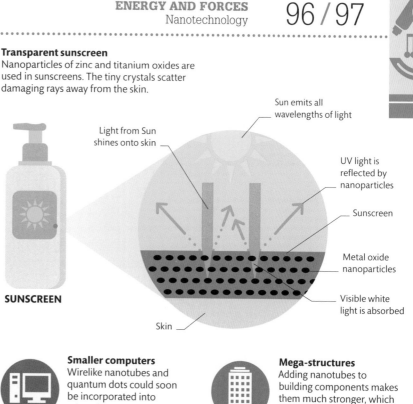

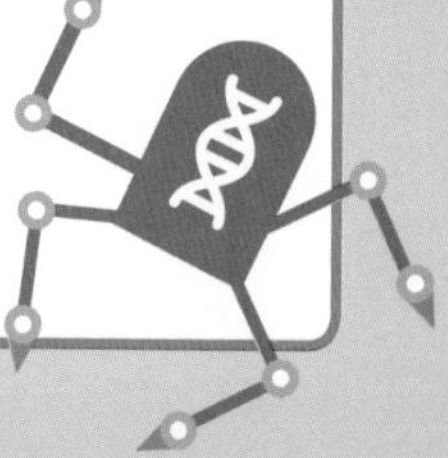

Robots and automation

A robot is a machine that is built to perform complex actions. It may be operated remotely by a human but is generally designed to work automatically.

What are robots for?

A robot's components can move independently in different directions. This allows the robot to perform certain actions required to complete a complex task that would otherwise require a human worker. Robots are mainly limited to applications where they offer obvious advantages over humans, such as working in dangerous places or performing repetitive tasks.

WILL ROBOTS REPLACE HUMANS?

Mechanical robots are designed for a small number of specific tasks, and so far we are a long way from a machine that is as versatile as a human body.

Repetitive tasks
Assembly robots are programmed to do one job over and over again. They never get tired or bored, but they are unable to modify their actions in response to an unforeseen event.

Danger zones
Robotic vehicles, such as bomb-disposal robots, are sent to places that are too dangerous for a human. They will send information back to a human operator.

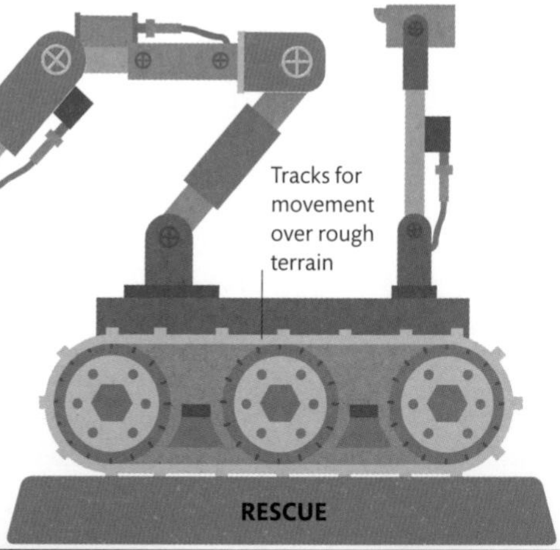

Actroids

Many engineers have attempted to build machines that emulate the human form. A recent development in this field is the Actroid, a lifelike, soft-skinned robot that recognizes and responds to speech and facial expressions. However, designers have to contend with the uncanny valley, a phenomenon where inanimate human replicas look strange, even frightening, as they get closer to resembling living humans.

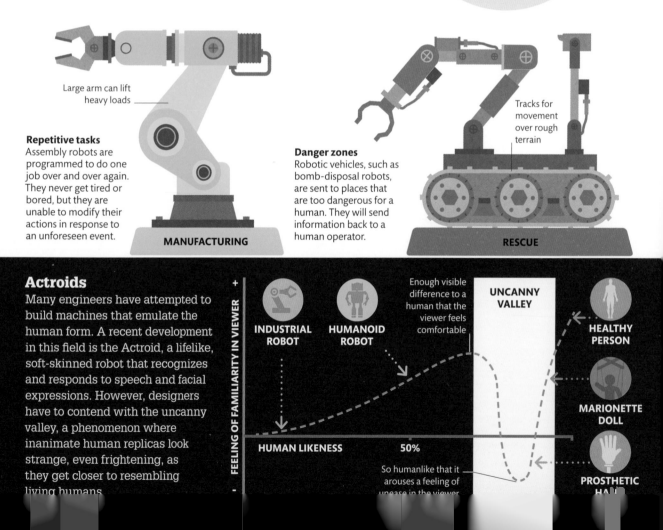

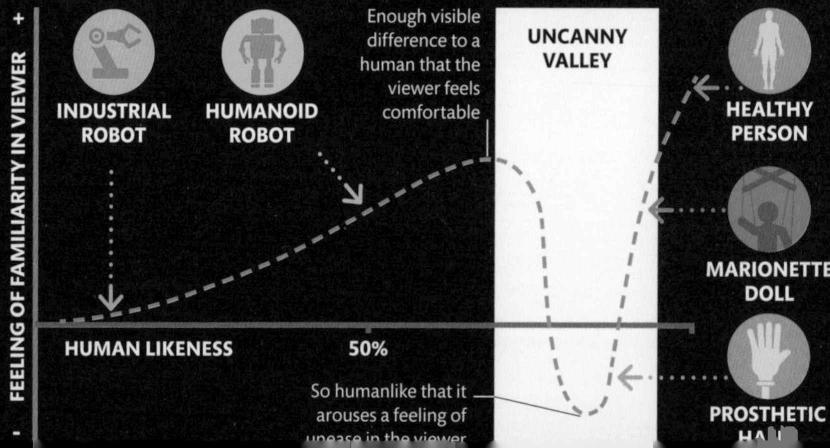

STEPPER MOTOR

Robotic joints that bend or swivel rely heavily on a kind of motor called a stepper motor. The motor uses a series of electromagnets, each of which moves an axle by a few degrees at a time. As a result, the motor can be made to make very precise turns.

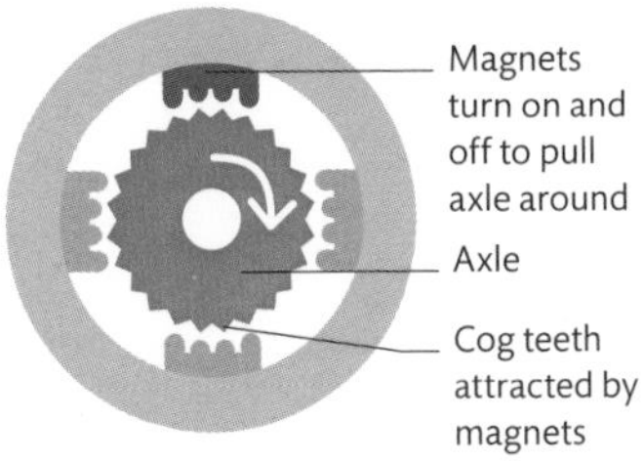

THE MARS ROVER CURIOSITY CAN VAPORIZE SAMPLES FOR ANALYSIS FROM 23FT (7M) AWAY

Other worlds
Mobile science labs, such as Mars exploration rovers, follow routes sent to them by operators but can also respond autonomously to hazards.

Precision required
Surgical robots are able to carry out very precise incisions and procedures either directed by a human doctor or following a preplanned sequence.

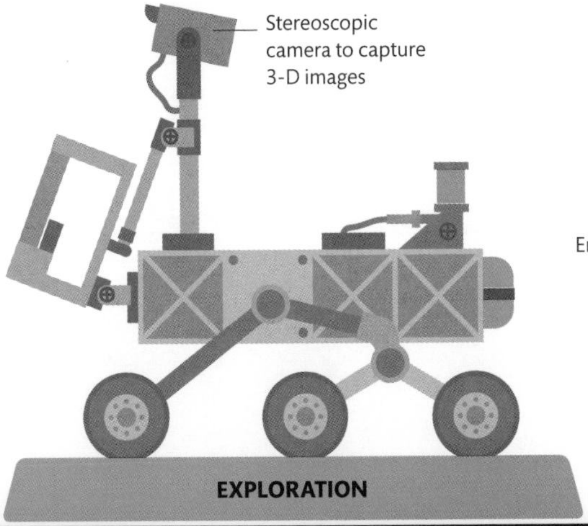

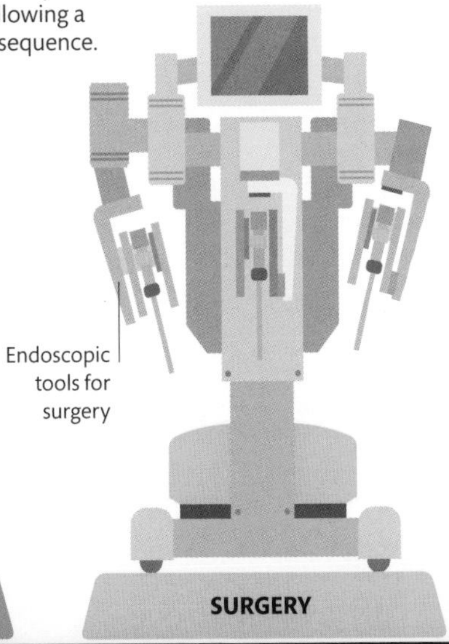

Low-status role
Robots that clean and carry may one day replace human carers, although designing a robot that can do this work is very hard.

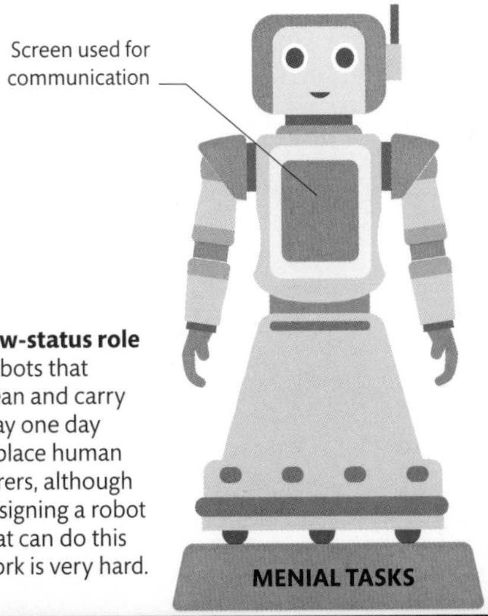

Driverless cars

Cars that automatically navigate along roads and respond to their surroundings are a type of robot. The robotic elements operate the steering and throttle, but the success of a driverless car is in its ability to interpret where it is and what is happening around it. Different detection systems are used to create a full picture of its surroundings.

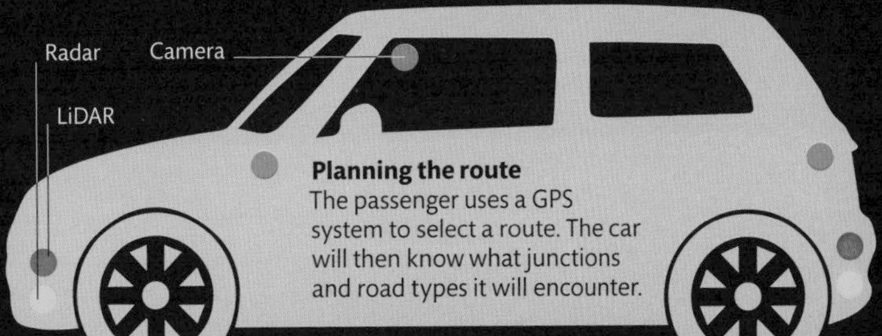

Planning the route
The passenger uses a GPS system to select a route. The car will then know what junctions and road types it will encounter.

Camera
Detects the roadway, signs, and other

Radar
Picks up the direction and speed of moving or stationary objects

LiDAR
Laser-based detector works out the size and shape of objects

Artificial intelligence

Intelligence can be thought of as the ability to make decisions about what is appropriate for the prevailing conditions. A goal of computer science is to make devices that use an artificial intelligence (AI).

Weak or strong?
Most AIs are weak; they are unable to function outside criteria set by their human creators. A strong AI is potentially more versatile—it could do almost anything a human brain can. It would be clever enough to know that it doesn't know something and then learn about it.

COULD AI TAKE OVER?

AI is unlikely to be cleverer than us anytime soon, but we will rely on AIs to make decisions for us—and not understand how they do it.

WEAK

Narrow AI
A suggestion engine, such as a social media news feed, is a narrow AI. It is able to search and select items that relate closely to what you've already viewed.

Expert
A chess computer is an expert system. It decides on its moves by consulting a database compiled by a human expert chess player.

Voice recognition
A voice-activated assistant learns to recognize spoken words and analyze the phrases to offer the best responses. However, it has no comprehension of the meaning.

General AI
IBM's Watson is a computing system capable of solving a range of problems, from playing game shows to advising doctors, all based on the same framework. It is perhaps the closest we have to a General AI.

Quantum computing
The future of AI may lie in quantum computing, where a new kind of processor will be able to handle vastly more data than even today's supercomputers.

STRONG

Types of artificial intelligence

The most common idea of AI is a nonhuman device that has an intelligence akin to our own. However, an AI is unlikely to work like that for some time (if ever). The AIs at work today focus on a narrow range of very specific tasks. Nevertheless, they are able to perform those tasks faster and more accurately than a human intelligence.

Machine learning

Allowing a computer system to learn to adjust its own behavior in response to new situations is known as machine learning. This involves an artificial neural network, inspired by the interconnected cells in animal brains, which learns by processing information and using it to make informed guesses. When it is wrong, it adjusts its guess so it can do better next time.

Trial and error
During supervised machine learning, the system is told, by its human creator, if its outputs are correct or not. The system applies and changes weights, or biases, to nodes in the network to achieve the correct output.

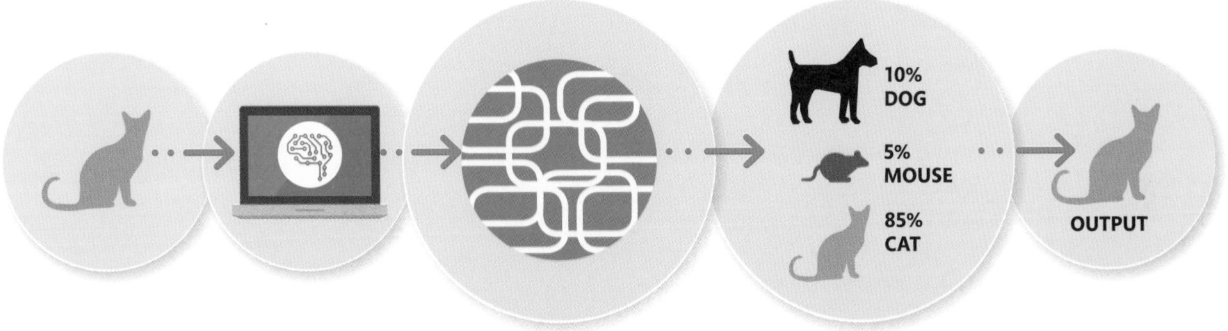

1 Input
The system enters an image into the neural network as a pattern of pixels of different color intensities.

2 Learning
The computer's aim is to recognize the patterns among the pixels it associates with different animals. At first, it will simply guess at random.

3 Analysis
The data about the pixels passes through layers within the neural network. Each of the layers learns about the pixel patterns in more and more detail.

4 Machine learning
After many learning attempts (this can be anything from hundreds to billions), the neural network gets better at recognizing the patterns of pixels that could represent a dog, a cat, or a mouse.

5 In use
After the AI system has learned its task, it can be used to analyze pictures—or do another learned task—automatically.

Turing test

One of the pioneers of computer science, Alan Turing, formulated a test for whether a computer is intelligent. A human judge holds a text conversation with a computer and a human test subject. If the judge cannot tell which is the human and which is the computer, the computer has passed the Turing test.

Blind trial
Judges are not able to see who they are speaking to. In more advanced tests, the judge will show pictures and talk to test subjects.

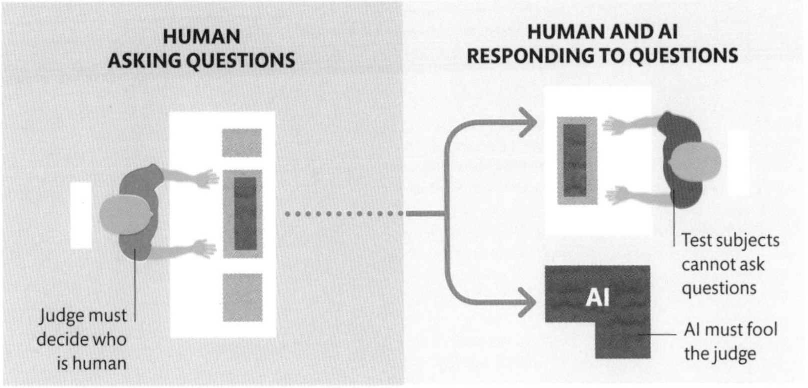

QUANTUM BITS

Classical computers use binary digits (bits), which store one piece of data at a time—a 1 or a 0. Quantum computers use quantum bits, or qubits, which have a certain chance of being 1 or 0, and so hold two bits of data at once. The power of quantum computing comes from using qubits together; a 32-qubit processor handles 4,294,967,296 bits at once.

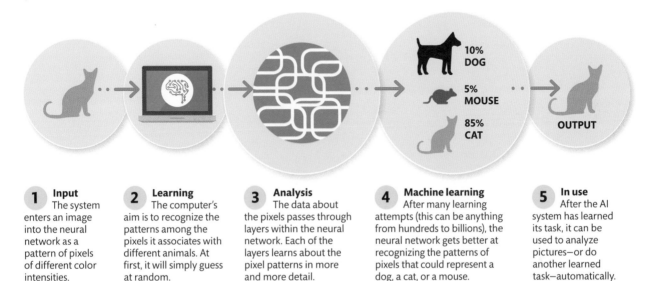

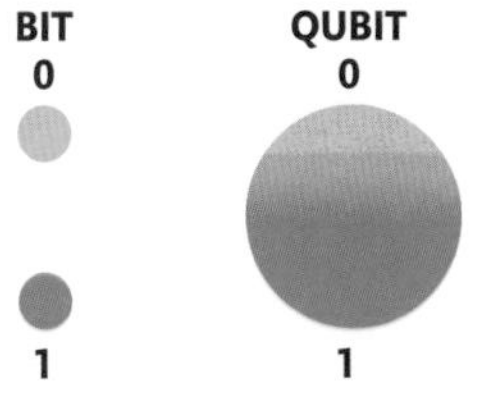

Waves

Waves are oscillations—or rhythmic fluctuations—found in nature. Light and sound are both examples of waves. Although they take different forms, there are some features and behaviors that all waves share.

WHERE DO OCEAN WAVES COME FROM?

The wind can create waves on the ocean by blowing over the surface of open water. Friction pushes the water into crests, which in turn catch more wind.

Types of waves

A wave is an example of energy moving from one place to another. Waves all exhibit the same basic behaviors due to their oscillatory motion, and that motion can arise in three forms. Sound is a longitudinal wave. Light and other kinds of radiation are transverse waves and do not require a medium to travel through. Ocean waves are an example of a complex third form known as surface waves, or seismic waves.

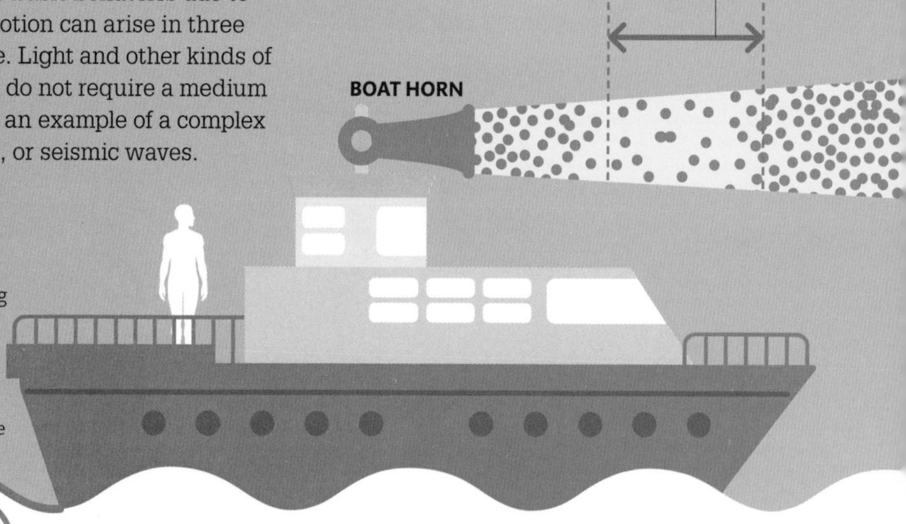

Air molecules rarified, or spread out, into low-pressure region
BOAT HORN

Surface wave
The water in a surface wave does not move forward with the wave itself. Instead, the water near the surface turns in loops, causing peaks and troughs of equal height along the level of the water during calm conditions.

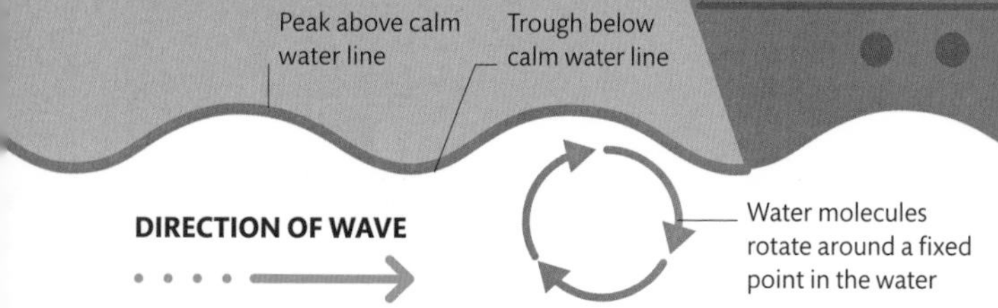

Peak above calm water line
Trough below calm water line
DIRECTION OF WAVE
Water molecules rotate around a fixed point in the water

Measuring waves

All waves, of whatever form, can be measured using the same set of dimensions. Wavelength is the distance filled by one complete oscillation of the wave. It is easiest to measure this from one wave peak to the next peak. A wave's frequency is the number of wavelengths that occur every second, and is measured in hertz (Hz). Amplitude equates to the height of the wave and indicates the power of the wave, or how much energy is being transferred over time.

Wave relationship
If the wave speed is constant, increasing the wavelength reduces the frequency and vice versa.

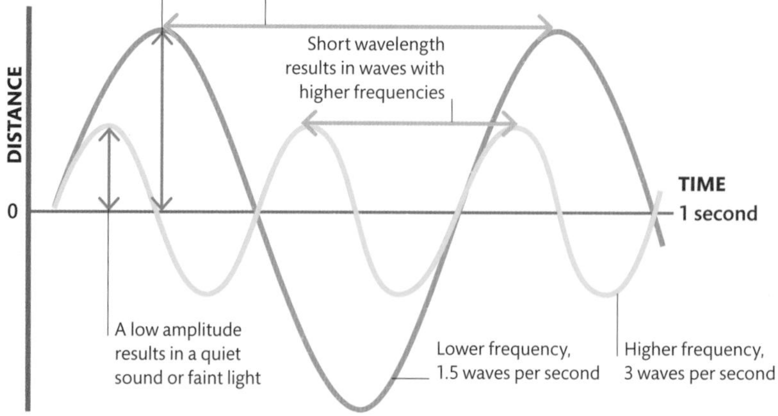

Amplitude is measured from a central line around which the wave oscillates
Longer wavelengths take longer to complete
Short wavelength results in waves with higher frequencies
DISTANCE
0
TIME
1 second
A low amplitude results in a quiet sound or faint light
Lower frequency, 1.5 waves per second
Higher frequency, 3 waves per second

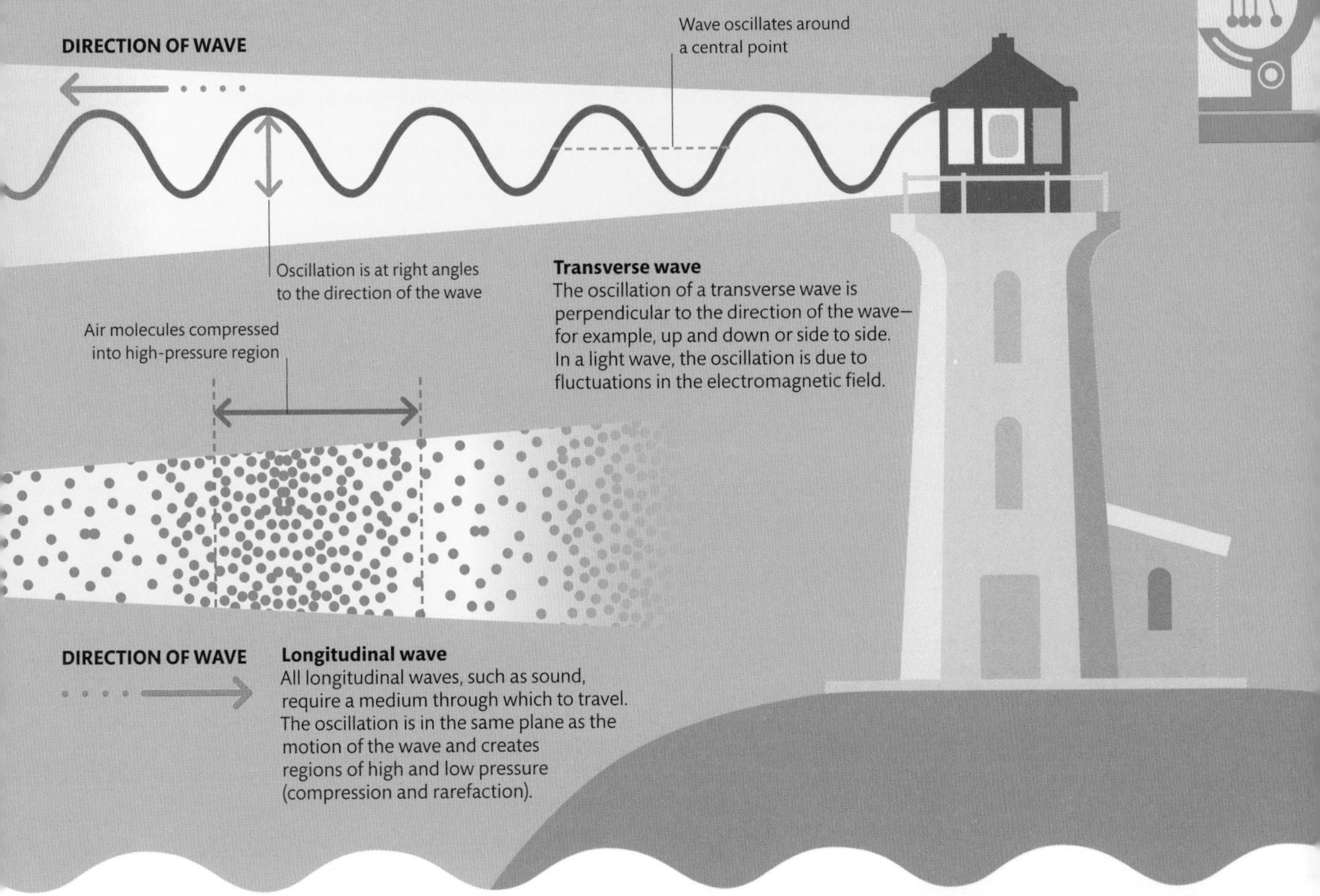
DIRECTION OF WAVE
Wave oscillates around a central point
Oscillation is at right angles to the direction of the wave
Air molecules compressed into high-pressure region
DIRECTION OF WAVE

Transverse wave
The oscillation of a transverse wave is perpendicular to the direction of the wave—for example, up and down or side to side. In a light wave, the oscillation is due to fluctuations in the electromagnetic field.

Longitudinal wave
All longitudinal waves, such as sound, require a medium through which to travel. The oscillation is in the same plane as the motion of the wave and creates regions of high and low pressure (compression and rarefaction).

Propagation of waves

Waves spread out from a source in all directions, as long as nothing is blocking the way. A wave's intensity, or the concentration of energy in it, reduces as it moves away from the source. The reduction in intensity—which makes sounds grow quieter and light appear fainter—follows what is called an inverse square law. For example, for every doubling of distance, the intensity of the wave goes down by a factor of four.

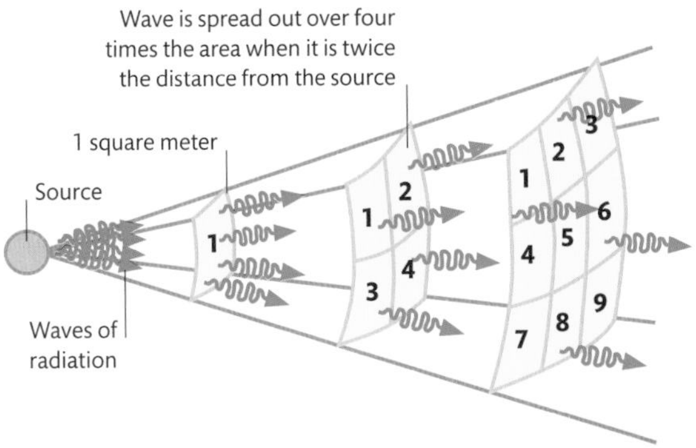
Wave is spread out over four times the area when it is twice the distance from the source
1 square meter
Source
Waves of radiation
1
1
2
3
4
1
2
3
4
5
6
7
8
9

Less effect
Reduction of wave intensity is very rapid. At triple the distance from the source, the intensity has become 9 times lower. At 100 times the distance, it is reduced by a factor of 10,000.

BREAKING WAVES

Ocean waves break when the sea becomes too shallow for the water to circulate in a loop (see p.233). As the wave enters the shallows, the rotating water surges up into a taller, elongated crest. The wave becomes top heavy and breaks.

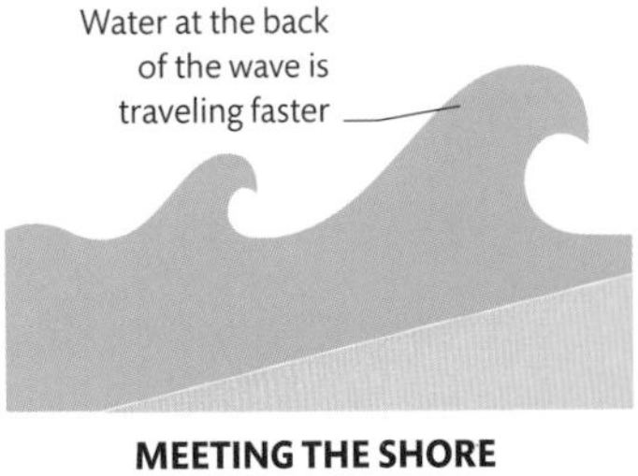
Water at the back of the wave is traveling faster

MEETING THE SHORE

From radio waves to gamma rays

Everything we see around us is a pattern of visible light that reaches our eyes in the form of waves. But these visible rays are just part of a broad spectrum of electromagnetic waves that carry energy from place to place.

ARE MICROWAVES DANGEROUS?

Strong microwaves could burn you, but weak ones are harmless. Microwave ovens are designed so the microwaves are always contained within the oven.

Electromagnetic radiation

Energy can be transferred by electromagnetic radiation. This takes the form of a wave that ripples both from side to side and up and down. The two components of the wave oscillate in phase—their peaks and troughs occur in a regular movement and are aligned with each other. The length of the wave can vary, but the wave will always travel through empty space at the speed of light.

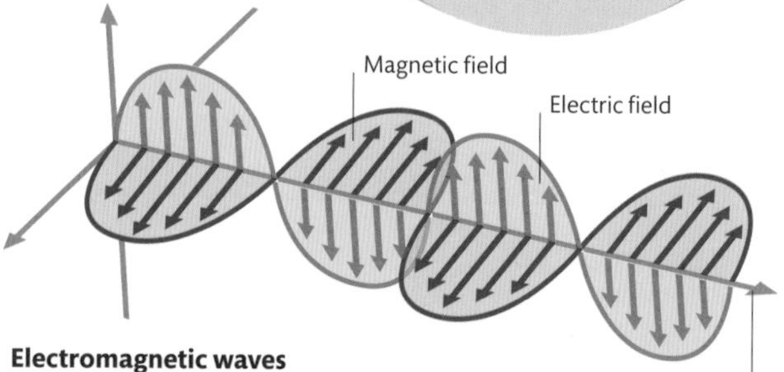

Electromagnetic waves
These are made up of two matching waves at right angles—one is an oscillating electric field, and the other is an oscillating magnetic field.

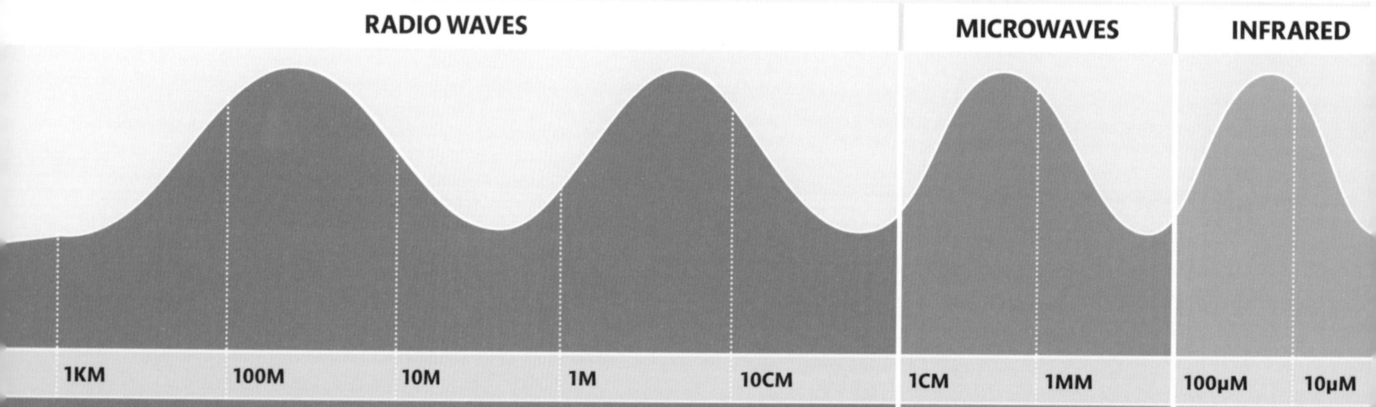

The electromagnetic spectrum

We perceive some electromagnetic waves as visible light. This consists of a spectrum of colors, each with its own wavelength ranging from red to violet. But the electromagnetic spectrum extends far beyond visible light. The longer wavelengths range from infrared rays that carry heat energy, to microwaves and radio waves. The shorter wavelengths extend from ultraviolet and X-rays to gamma rays.

Radio telescope
A dish antenna can be used to detect radio waves emitted by distant stars.

Microwave oven
Food heats up when high-energy microwaves excite the water molecules inside.

Remote control
A remote control use pulses of infrared radiation to transmit digital control codes.

Digital radio

Analog radio transmitters broadcast signals that are essentially ripples added to normal radio waves. Different radio waves can interfere with each other, and this can distort an analog broadcast. Digital radio converts sound into a digital code, so as long as the digits forming the code get through, the transmission can be converted into a clear signal.

THE SPEED OF LIGHT IN A VACUUM IS 983,571,056FT (299,792,458M) PER SECOND

High-quality sound
Sound waves are converted into a stream of numbers before transmission. A digital receiver then decodes the numbers, turning them back into a form that can drive a loudspeaker.

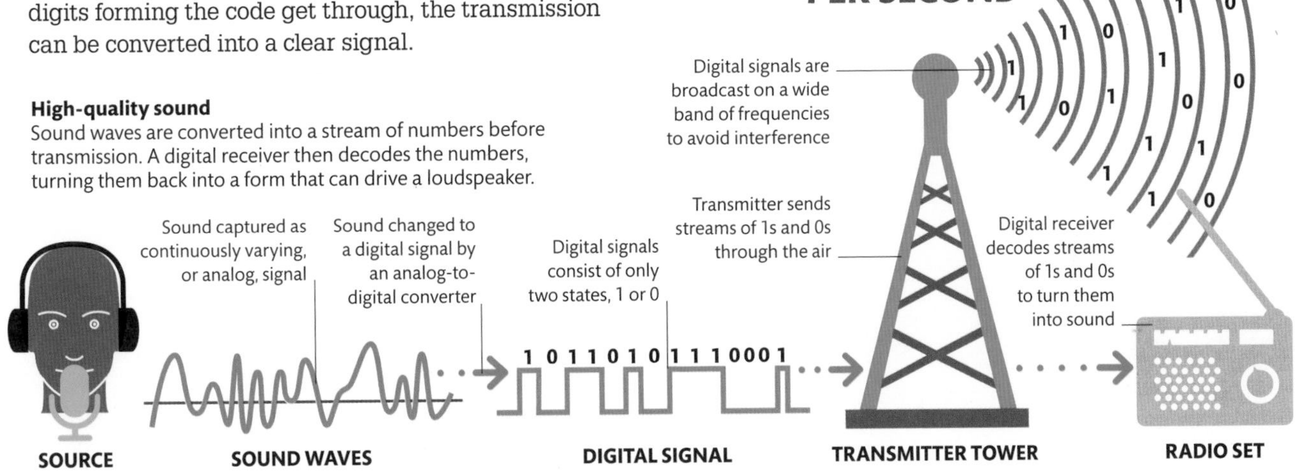
Digital signals are broadcast on a wide band of frequencies to avoid interference
Transmitter sends streams of 1s and 0s through the air
Digital receiver decodes streams of 1s and 0s to turn them into sound
Sound captured as continuously varying, or analog, signal
Sound changed to a digital signal by an analog-to-digital converter
Digital signals consist of only two states, 1 or 0
1 0 1 1 0 1 0 1 1 1 0 0 0 1
SOURCE
SOUND WAVES
DIGITAL SIGNAL
TRANSMITTER TOWER
RADIO SET

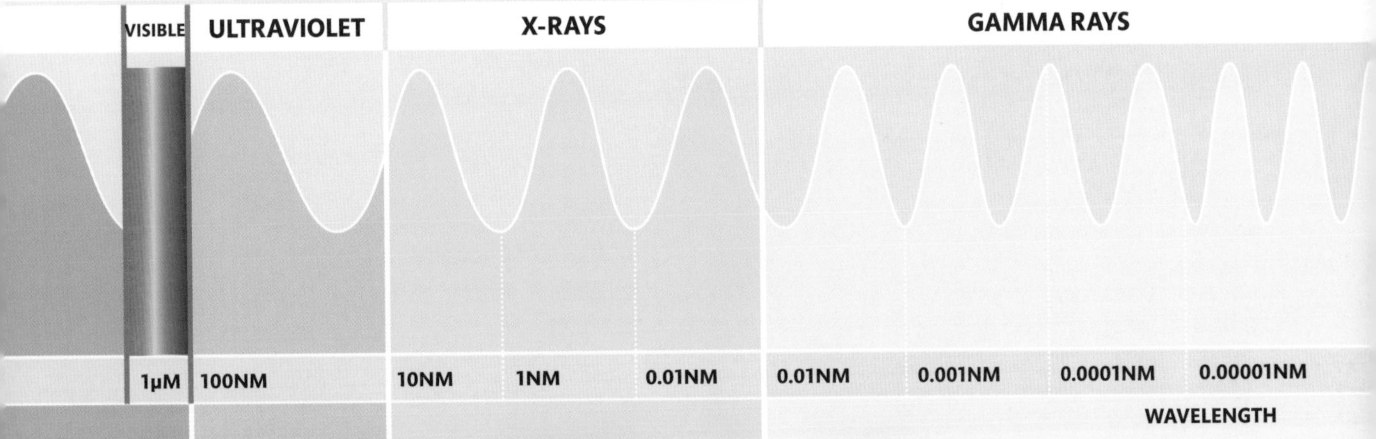
VISIBLE
ULTRAVIOLET
X-RAYS
GAMMA RAYS
1μM
100NM
10NM
1NM
0.01NM
0.01NM
0.001NM
0.0001NM
0.00001NM
WAVELENGTH

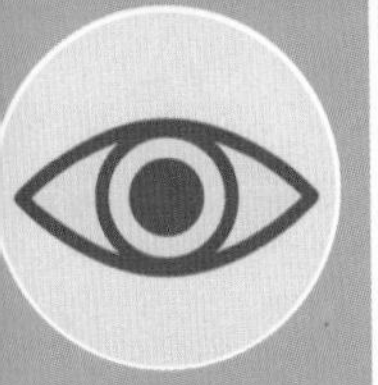

Human eye
Our eyes detect a narrow range of wavelengths as the color spectrum.

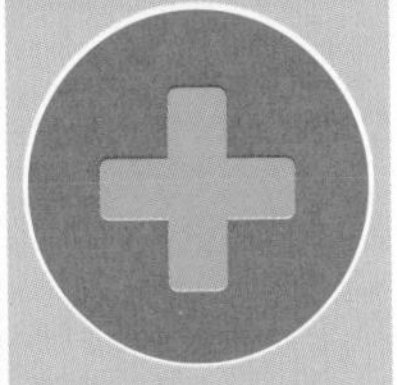

Disinfection
Certain wavelengths of UV light can be used to kill bacteria and sanitize objects.

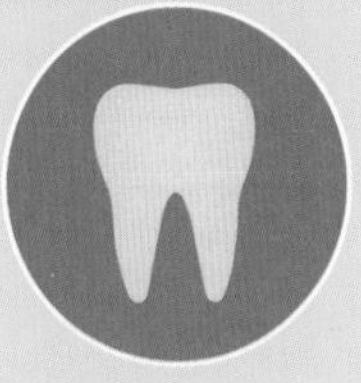

Dental X-ray
Short-wavelength X-rays pass through tissue to reveal the teeth underneath.

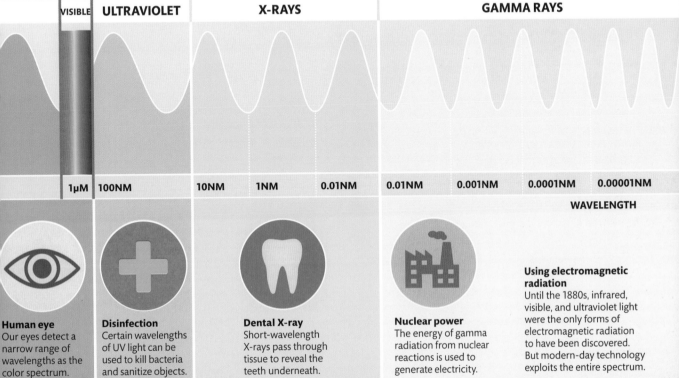

Nuclear power
The energy of gamma radiation from nuclear reactions is used to generate electricity.

Using electromagnetic radiation
Until the 1880s, infrared, visible, and ultraviolet light were the only forms of electromagnetic radiation to have been discovered. But modern-day technology exploits the entire spectrum.

Color

Color is a phenomenon generated by our eyes and vision system to enable us to see different wavelengths of light. The colors we perceive depend on the wavelengths of light that our eyes detect.

WHY CAN'T WE SEE COLORS WELL AT NIGHT?

It is too dark for the eyes' color-sensitive cone cells to work. Instead, more sensitive rod cells make images as areas of light and dark.

Visible spectrum

The eye can detect light with wavelengths ranging from about 400 to 700 nanometers. Light that contains all of these wavelengths appears as white. When the light is split into individual wavelengths, the brain assigns each one a specific color from the full color spectrum. Red light has the longest wavelengths and violet has the shortest.

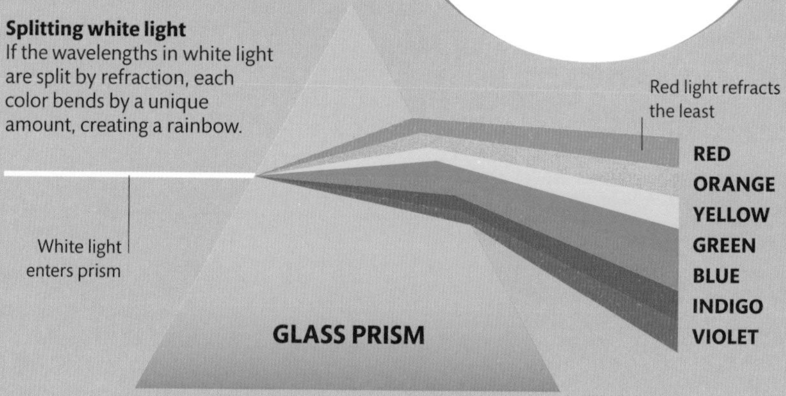

Splitting white light
If the wavelengths in white light are split by refraction, each color bends by a unique amount, creating a rainbow.

Color vision

The human eye creates images from light using three types of light-detecting cells, known as cones because of their shape. The cones in the retina contain chemical pigments that are sensitive to specific wavelengths of light. When triggered, they fire off a nerve signal. The brain receives signals for red, green, and blue light entering the eye and creates the perception of colors from them. For example, signals from both a green and a red cone create the perception of yellow. Signals from all cones makes white, while no signals from any cells make black.

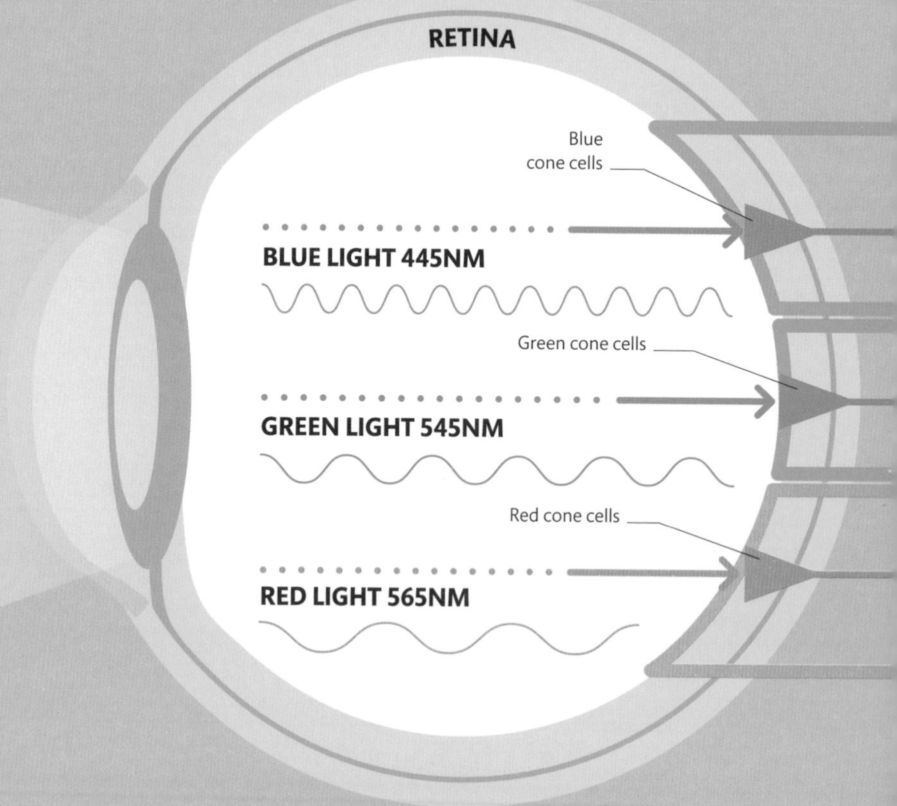

Light sensors
All parts of the retina have cone cells of all three types, although most of the cones are in the central portion directly behind the pupil. This is where the most detailed parts of images are formed.

BLUE SKY

The sky looks blue because, compared with other colors, blue light has a shorter wavelength and bounces more strongly off air molecules, scattering in all directions, before it shines into our eyes. Violet light also scatters, but there is less of it and our eyes are more sensitive to blue light.

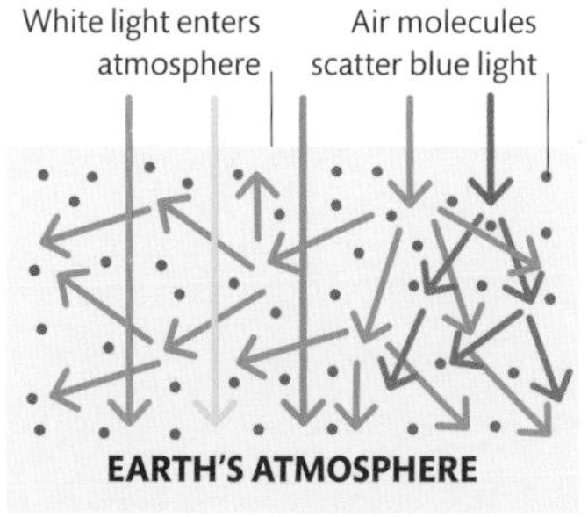
White light enters atmosphere
Air molecules scatter blue light
EARTH'S ATMOSPHERE

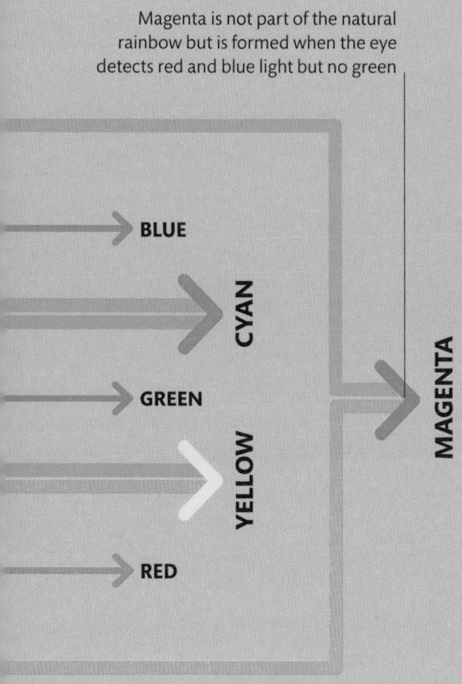
Magenta is not part of the natural rainbow but is formed when the eye detects red and blue light but no green
BLUE
CYAN
GREEN
YELLOW
RED
MAGENTA

Mixing colors

When light falls on an object, it can be either absorbed or reflected. The brain allots a particular color to an object according to the light it reflects. For example, a banana reflects yellow light and absorbs all other colors. This is called subtractive mixing and is used to manufacture colored inks and dyes. Mixing colors directly from a light source, such as in stage lighting, requires an opposite approach, called additive mixing.

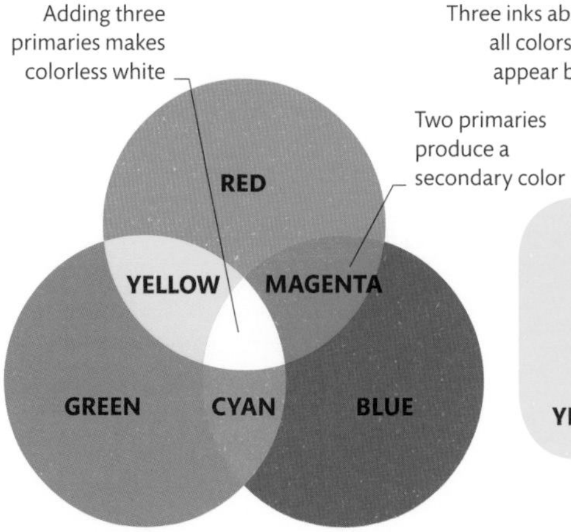
Adding three primaries makes colorless white
Two primaries produce a secondary color
RED
YELLOW
MAGENTA
GREEN
CYAN
BLUE

Additive mixing
Transmitted light is altered using the additive system. Red, green, and blue are the three primary colors. Secondary colors are made by combining two primaries. Adding all primaries makes white light.

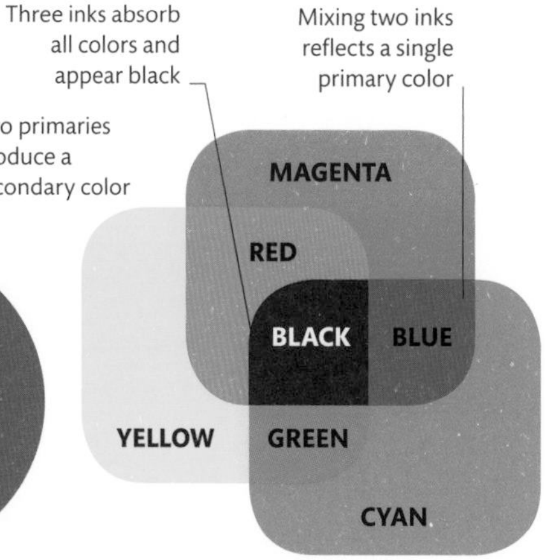
Three inks absorb all colors and appear black
Mixing two inks reflects a single primary color
MAGENTA
RED
BLACK
BLUE
YELLOW
GREEN
CYAN

Subtractive mixing
Cyan, magenta, and yellow pigments are used to build reflected color. Each absorbs one primary color and reflects two. Adding another pigment reduces the reflected light to just one primary.

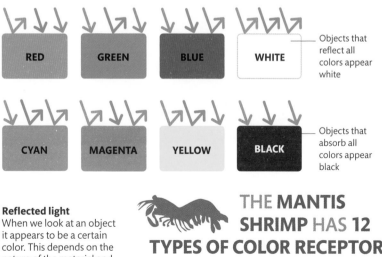

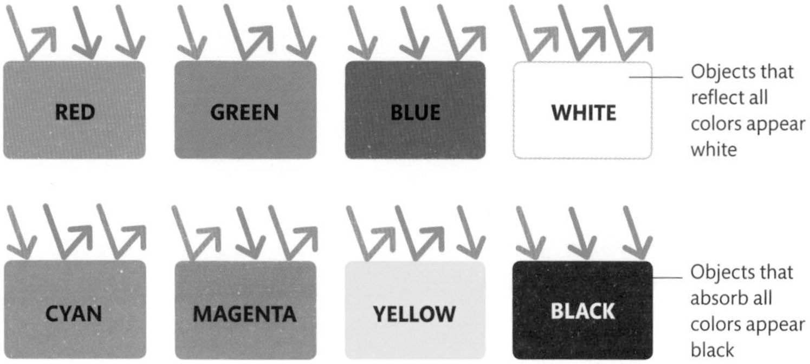
RED
GREEN
BLUE
WHITE
Objects that reflect all colors appear white
CYAN
MAGENTA
YELLOW
BLACK
Objects that absorb all colors appear black

Reflected light
When we look at an object it appears to be a certain color. This depends on the nature of the material and which wavelengths of light it either absorbs or reflects into our eyes.

THE MANTIS SHRIMP HAS 12 TYPES OF COLOR RECEPTORS AND CAN SEE UV AND NEAR-INFRARED LIGHT

Mirrors and lenses

Beams of light always travel in straight lines, but they can change direction due to phenomena such as reflection and refraction. These two processes are used to control light when using mirrors and lenses.

MIRAGES

Mirages are optical illusions that can be seen on hot days. In deserts, mirages appear to show water shimmering in the distance. The "water" is actually bright light from the sky, which is being refracted back up to the eye by a layer of hot air.

Reflecting light

The angle of a reflected beam is always the same as the angle of the incident (arriving) beam. The angles are measured from the normal, an imaginary line at right angles to the surface. Light reflecting from most objects will scatter off in all directions because the beams are hitting the rough, uneven surface at different angles. A mirror is very smooth, so the reflected beams keep their original alignments and create an image.

WHY DO DIAMONDS SPARKLE?

Cut diamonds sparkle because the angles of their surfaces ensure that any light shining into them reflects around inside and only leaves through the top.

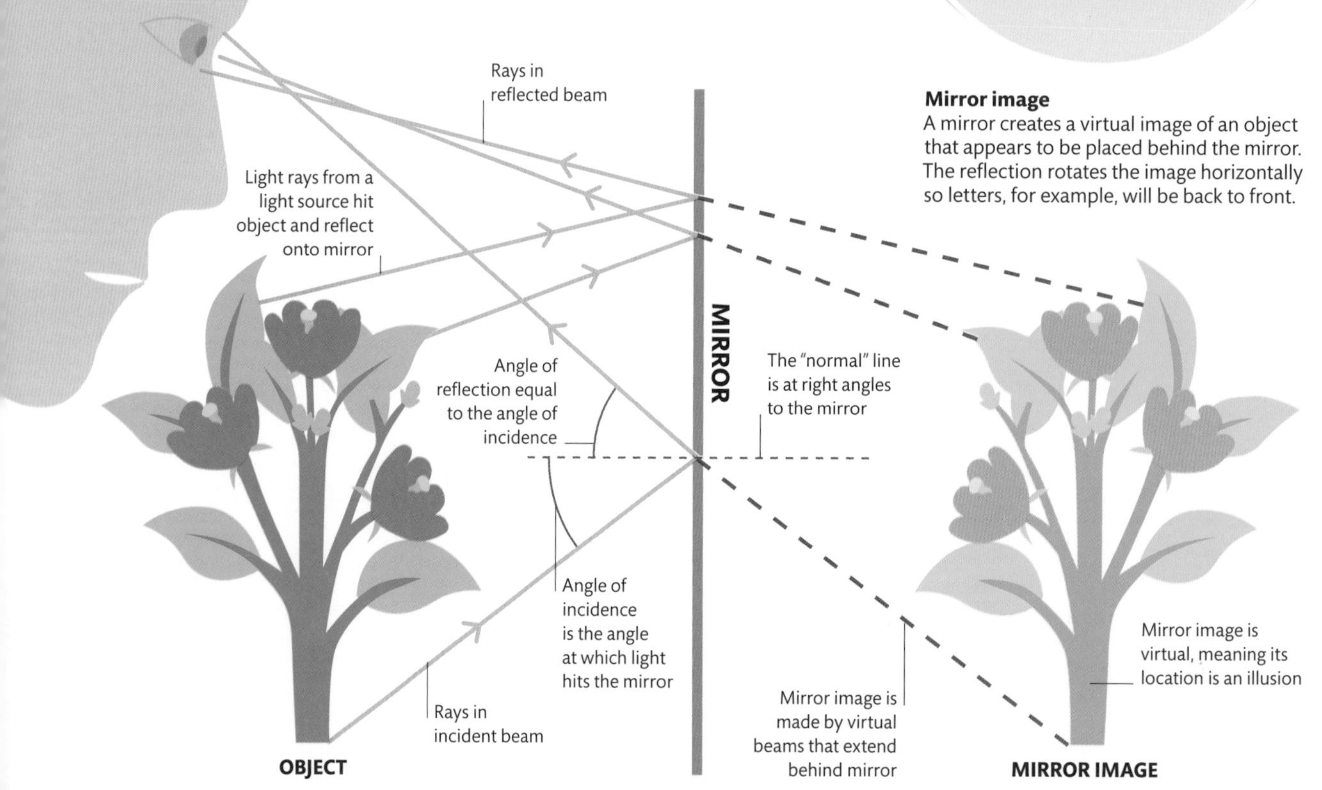

Rays in reflected beam
Light rays from a light source hit object and reflect onto mirror
MIRROR
Angle of reflection equal to the angle of incidence
The "normal" line is at right angles to the mirror
Angle of incidence is the angle at which light hits the mirror
Rays in incident beam
Mirror image is made by virtual beams that extend behind mirror
Mirror image is virtual, meaning its location is an illusion
OBJECT
MIRROR IMAGE

Mirror image
A mirror creates a virtual image of an object that appears to be placed behind the mirror. The reflection rotates the image horizontally so letters, for example, will be back to front.

Refracting light

Light waves travel at different speeds through different mediums. If light enters a new transparent medium at an angle, the change of speed also results in a small change in direction. This is known as refraction. Different parts of the light beam slow at different times, which deflects the path of the light.

RAINBOWS FORM WHEN **LIGHT** IS **REFLECTED, REFRACTED,** AND **DISPERSED** BY **RAINDROPS**

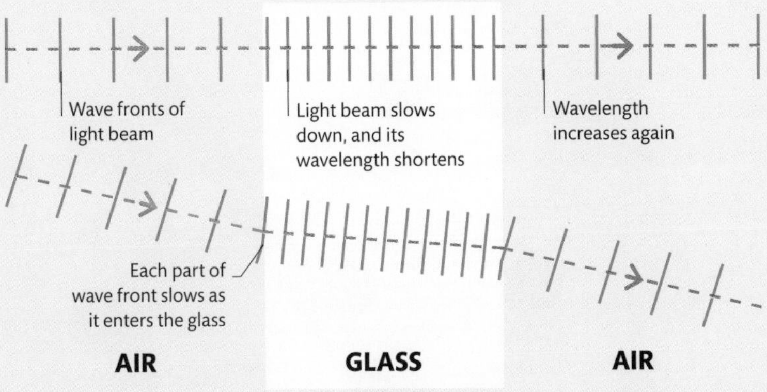
Wave fronts of light beam
Light beam slows down, and its wavelength shortens
Wavelength increases again
Each part of wave front slows as it enters the glass
AIR
GLASS
AIR

Actual and apparent

Refraction can trick the eye. The light from a fish is refracted as it moves from water to air. The brain, assuming that light travels in straight lines, sees the fish in line with the light that hits the eye.

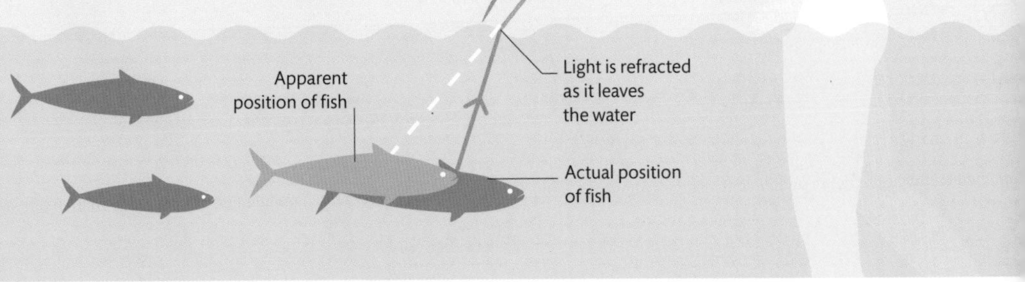
Apparent position of fish
Light is refracted as it leaves the water
Actual position of fish

Focusing light

A lens is a piece of transparent glass that uses refraction to change the direction of light. It has a curved surface, which means that light beams hit the lens at a series of different angles and, as a result, the beams are all refracted by different amounts. There are two main types of lens. A converging (convex) lens bends light inward and a diverging (concave) lens spreads light out.

Converging lens

Light beams shining through a convex lens converge at a focal point on the opposite side. The distance between the lens and the focal point is the focal length. A converging lens can be used to magnify small objects (see p.113).

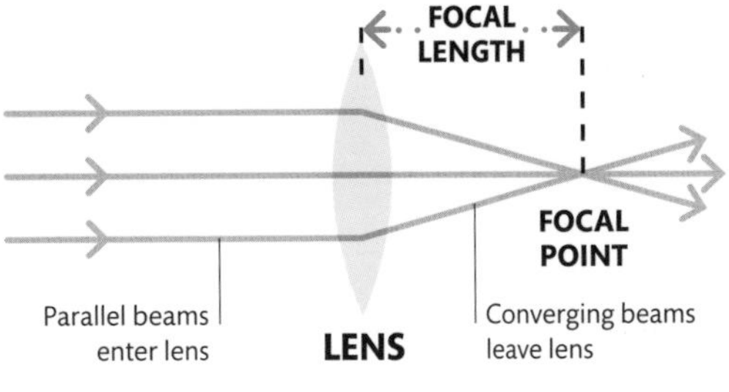
FOCAL LENGTH
FOCAL POINT
Parallel beams enter lens
LENS
Converging beams leave lens

Diverging lens

A concave lens causes light beams to spread out so that they appear to come from a focal point behind the lens. These lenses are used in glasses for short-sightedness.

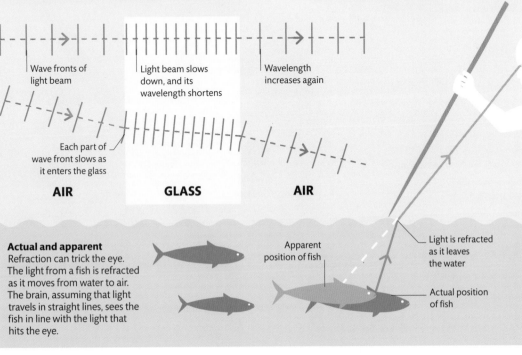

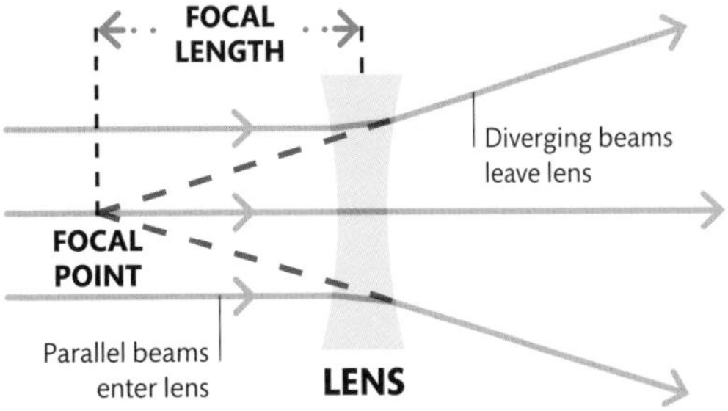
FOCAL LENGTH
Diverging beams leave lens
FOCAL POINT
Parallel beams enter lens
LENS

How lasers work

A laser is a device that produces an intense beam of light that is parallel and coherent, meaning the light waves in the beam are lined up and in step with one another. This gives the beam precision and power.

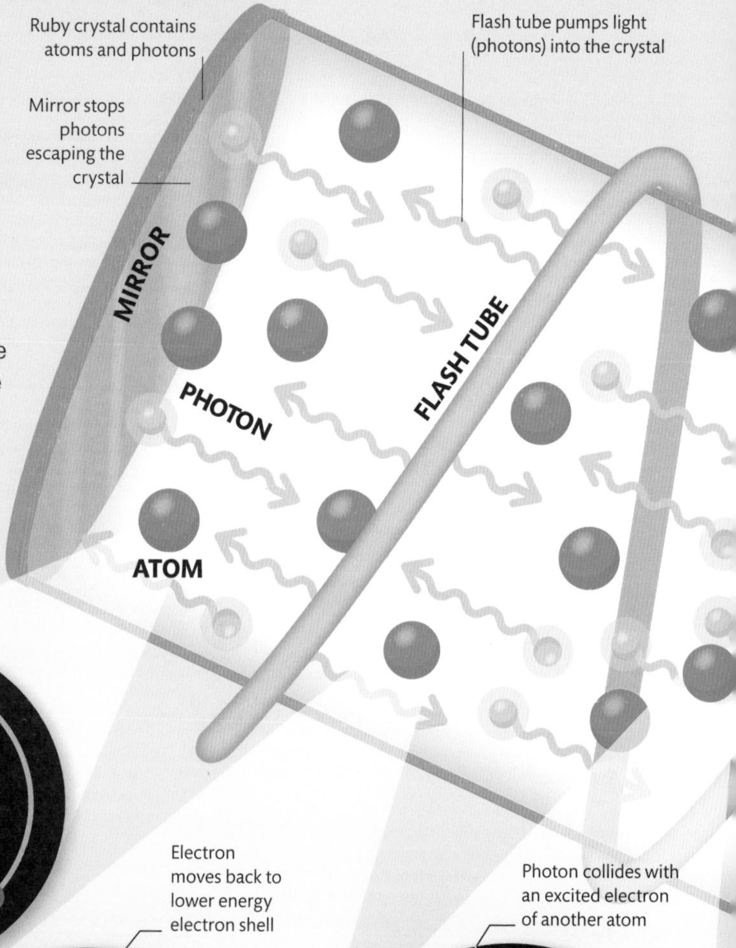

Ruby crystal contains atoms and photons
Flash tube pumps light (photons) into the crystal
Mirror stops photons escaping the crystal
MIRROR
PHOTON
FLASH TUBE
ATOM

Energizing light

In a crystal laser, light is directed into a tube made of artificial crystal, such as ruby. The atoms inside soak up the energy and re-emit the light, causing nearby atoms to give off photons of light, too, all at a very specific wavelength. The photons surge back and forth between mirrors in the tube until the light is intense enough to escape the tube as a narrow beam, which can be powerful enough to drill into diamond.

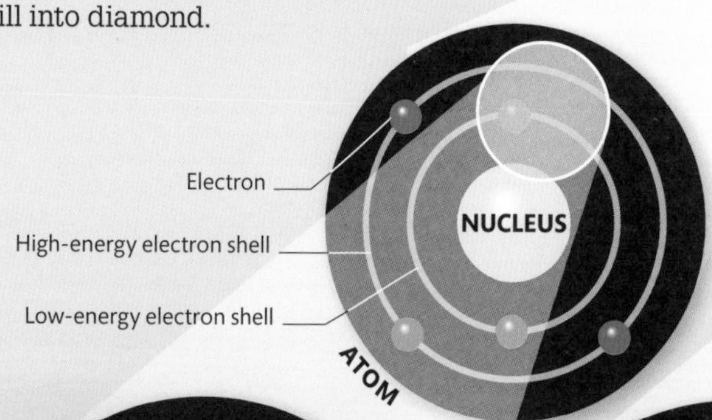

Electron
High-energy electron shell
Low-energy electron shell
NUCLEUS
ATOM

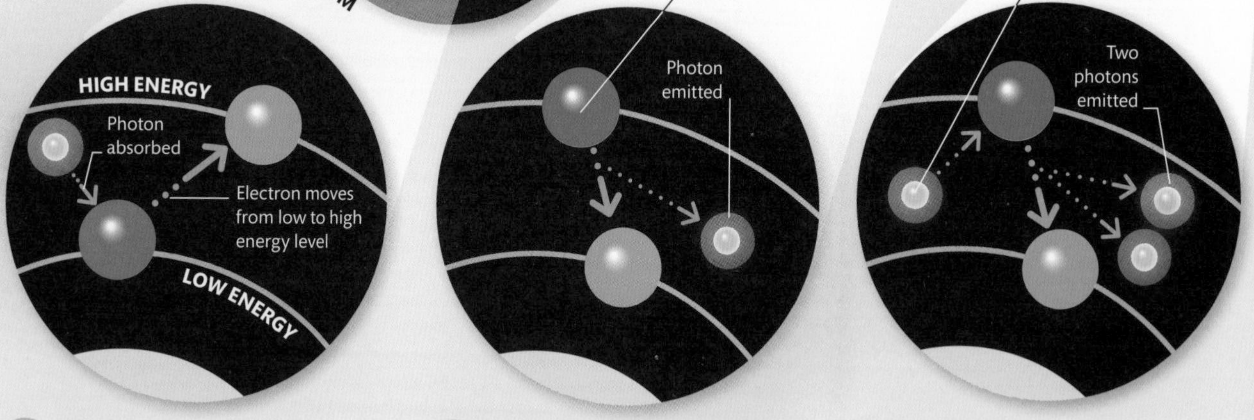

HIGH ENERGY
Photon absorbed
Electron moves from low to high energy level
LOW ENERGY
Electron moves back to lower energy electron shell
Photon emitted
Photon collides with an excited electron of another atom
Two photons emitted

1 Getting excited When an atom absorbs a photon, one of its electrons jumps from a low energy level to a higher one. An atom is unstable in its excited state.

2 Excess energy The electron stays excited for only a few milliseconds, then releases its absorbed photon. The photon released by the electron is of a particular wavelength.

3 Let it all out Already-excited electrons are hit by some photons, causing them to release two photons instead of one. This is called stimulated emission.

Using laser light

Lasers have proved themselves to be among the most versatile inventions of modern times. Today, they have a wide range of everyday and unique uses, from satellite communication to reading barcodes at the supermarket.

LASER LIGHT INTENSITY: LOW — MEDIUM — HIGH

Laser printing Lasers spread static electricity on paper to attract ink.

Burning data Data is encoded by etching patterns on optical discs.

Lighting effects Theatrical events feature controlled laser displays.

Medical Surgeons can use lasers, instead of scalpels, to cut or destroy tissue.

Material cutting A stronger laser can cut through tough materials.

Astronomy Precise lasers are used to measure distances exactly.

HOW POWERFUL CAN LASERS BE?

The world's most powerful laser can provide a beam of 2 petawatts, for a trillionth of a second—almost as much as the average power consumption of the entire world.

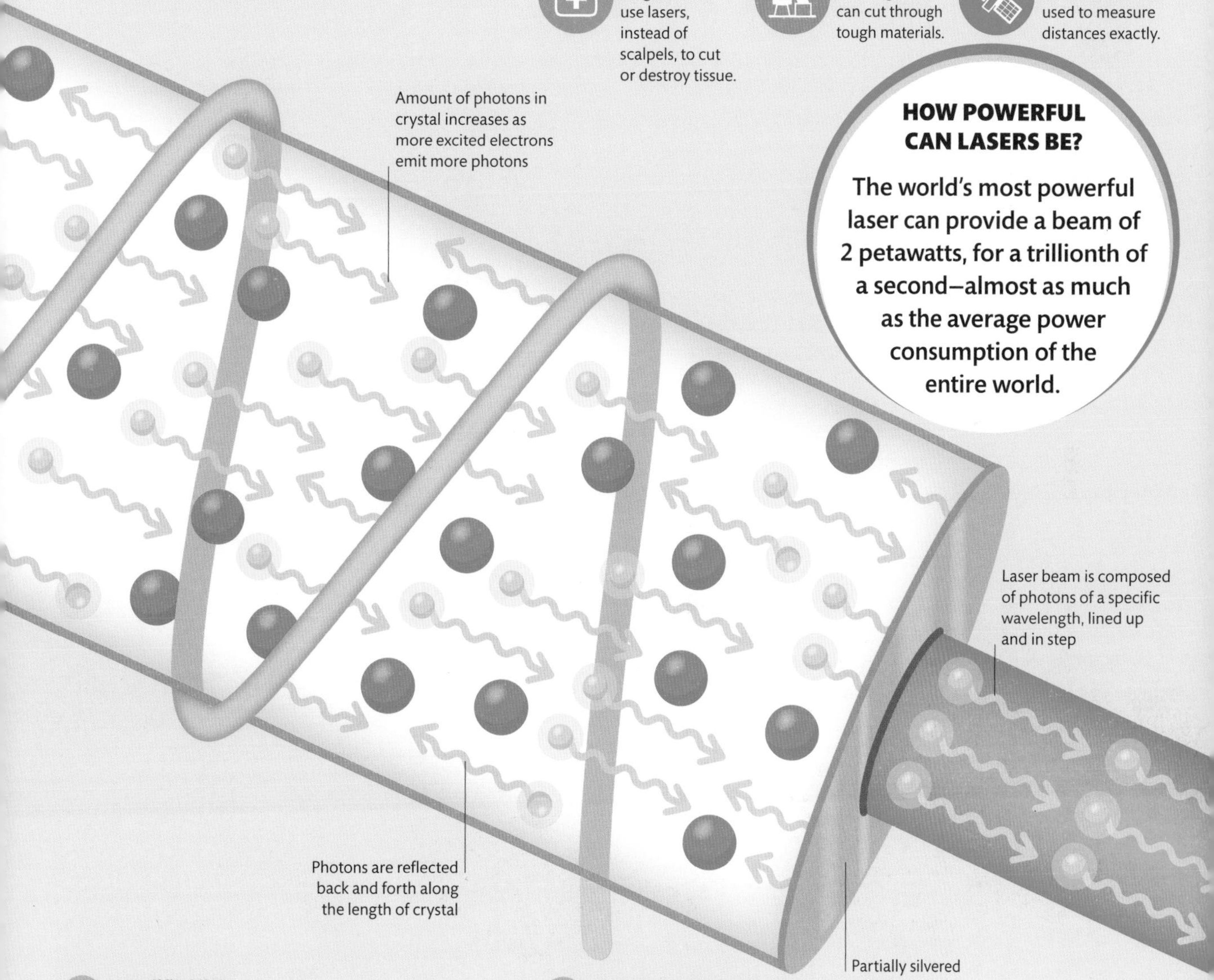
Amount of photons in crystal increases as more excited electrons emit more photons
Laser beam is composed of photons of a specific wavelength, lined up and in step
Photons are reflected back and forth along the length of crystal
Partially silvered mirror

4 Amplified light
Each time one photon stimulates the release of two photons, the light is amplified. "Laser" stands for "light amplification by stimulated emission of radiation." The light bounces up and down the tube.

5 Laser beam escapes
A partially silvered mirror lets some photons escape the crystal as a very concentrated beam of powerful, coherent light.

Using optics

Optics is the study of light. The optical behavior of light beams, such as reflection and refraction, have some powerful applications that allow us to see beyond the limits of the human eye.

Optics in action

The human eye can only see objects larger than 0.1mm across. Optical instruments can be used to view objects that are smaller than this—or to make out the detail of objects that are very far away. They do this by collecting the beams of light coming from the object. This light forms a faint image that is too small to see. The instrument collects more of the light from the object to make the image brighter, and then magnifies it with a lens.

Optical fibers

These super-fast cables send signals as coded flashes of laser light inside flexible fibers of glass. The light travels by reflecting off the inside surface of the fiber. The angle at which the laser hits the glass is crucial: too steep and it will not reflect but refract out of the fiber.

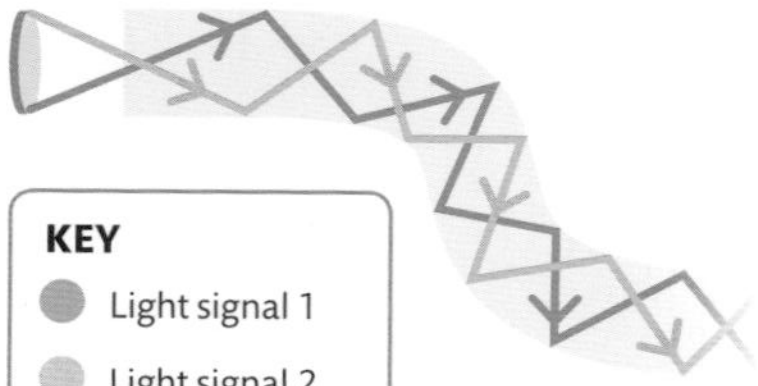

KEY
- Light signal 1
- Light signal 2

Multiplexing
One fiber can carry several signals using different-colored lasers.

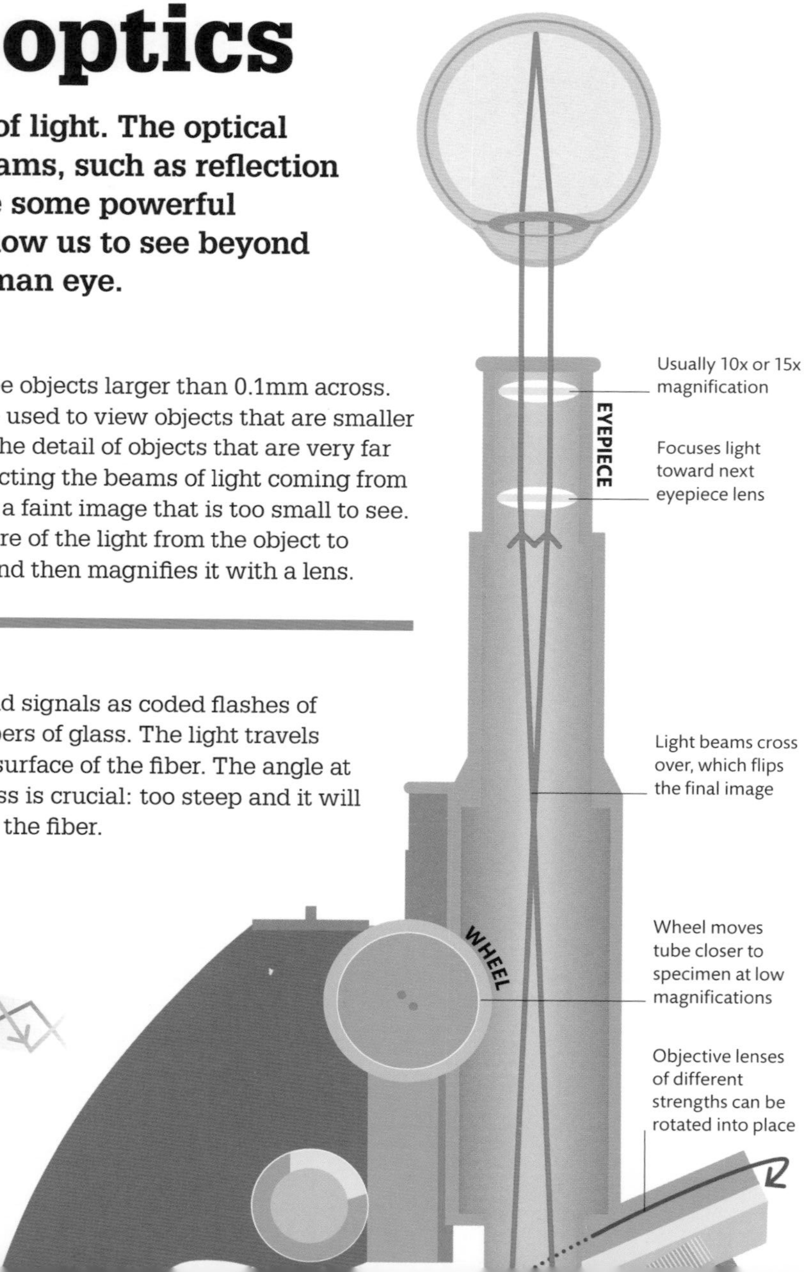

EYEPIECE
Usually 10x or 15x magnification
Focuses light toward next eyepiece lens
Light beams cross over, which flips the final image
WHEEL
Wheel moves tube closer to specimen at low magnifications
Objective lenses of different strengths can be rotated into place

INTERFERENCE

Like all kinds of waves, light waves will interfere with each other. When two light waves meet, they combine into one. If the wavelengths are in phase—they are the same length with the peaks and troughs moving in unison—they will make a more powerful wave. Waves that are exactly out of phase will cancel each other out. Interference creates patterns, such as the color swirls seen on oil.

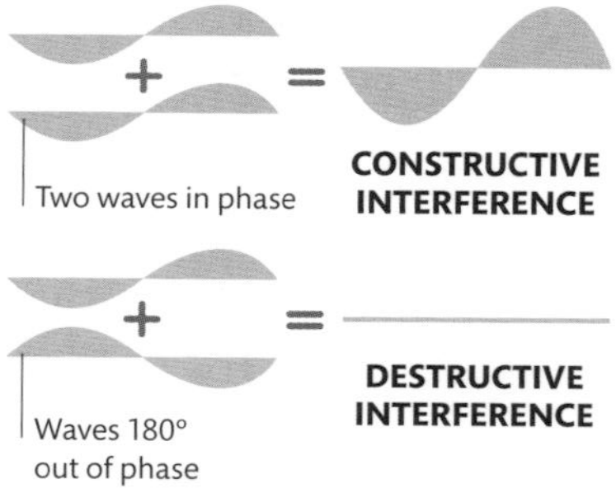

Two waves in phase
CONSTRUCTIVE INTERFERENCE
Waves 180° out of phase
DESTRUCTIVE INTERFERENCE

WILL GLASSES MAKE MY VISION WORSE?

Poor eyesight is caused by the shape of the eye and flexibility of the lens. Wearing glasses will have no effect on these—but should make you see better.

Light microscope
A microscope collects and magnifies light that has passed through a specimen. The light from the specimen enters through the chosen objective lens.

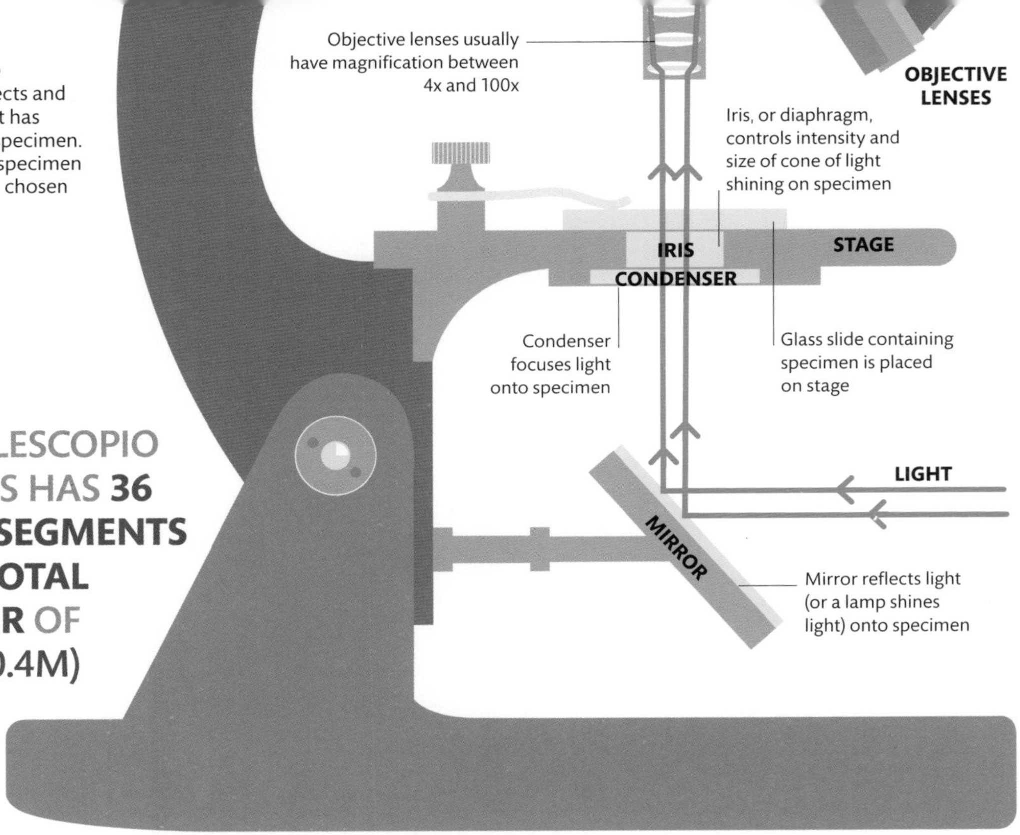
Objective lenses usually have magnification between 4x and 100x
OBJECTIVE LENSES
Iris, or diaphragm, controls intensity and size of cone of light shining on specimen
IRIS
STAGE
CONDENSER
Condenser focuses light onto specimen
Glass slide containing specimen is placed on stage
LIGHT
MIRROR
Mirror reflects light (or a lamp shines light) onto specimen

GRAN TELESCOPIO CANARIAS HAS **36 MIRROR SEGMENTS** WITH A **TOTAL DIAMETER** OF **34.1FT** (10.4M)

Telescope
Astronomical telescopes use lenses and mirrors to collect light from a distant object. Telescopes for use on Earth use lenses to flip the image the right way up.

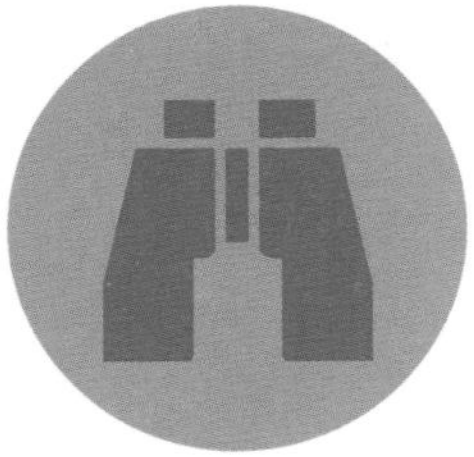

Binoculars
Light enters through wide main lenses. It is then reflected inward by mirrors and redirected through smaller magnifying lenses into each eye.

How magnification works

Most of the lenses fitted in a microscope are convex lenses (see p.109), used to form an enlarged view of a specimen. If an object is placed in between the lens and its focal point, the light beams from the object converge on the other side of the lens. Increasing the curvature of the lens will increase the focal length and, as a result, the magnification strength of the lens.

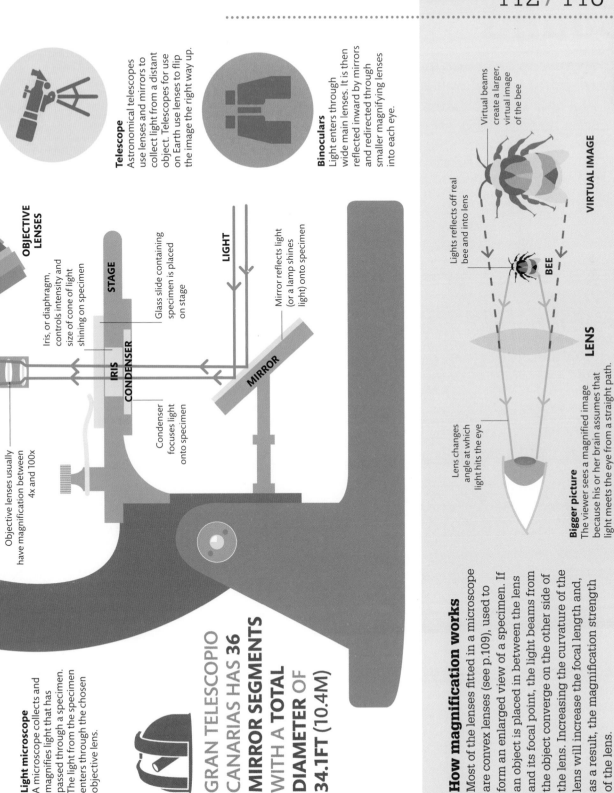

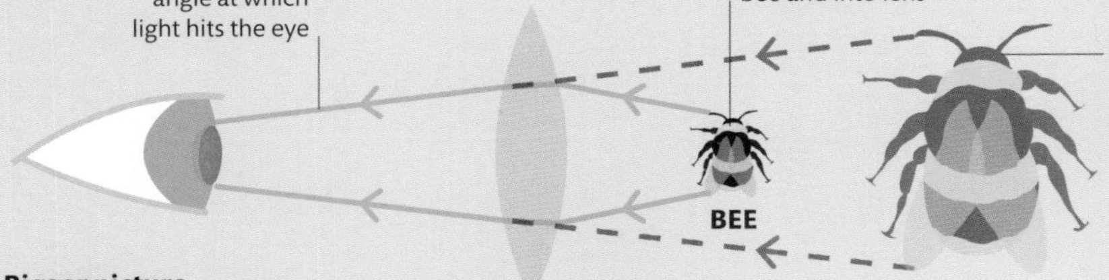
Lens changes angle at which light hits the eye
Lights reflects off real bee and into lens
Virtual beams create a larger, virtual image of the bee
BEE
LENS
VIRTUAL IMAGE

Bigger picture
The viewer sees a magnified image because his or her brain assumes that light meets the eye from a straight path.

Sound

All the sounds that reach our ears travel through a medium—such as air—in the form of waves (see pp.102–03). But sound waves are not like light or radio waves. They are ripples of compression that travel longitudinally away from the source.

Pressure waves

A sound wave is created by a push-pull mechanism, such as the cone of a loudspeaker. An electrical signal makes the cone move forward and back at high speed, and this pushes against the air and then pulls back. Each push creates a ripple of compression that moves away through the air. The farther the speaker cone moves during each cycle, the more pressure it exerts—the higher the pressure, the more compressed the air molecules are and the louder the sound.

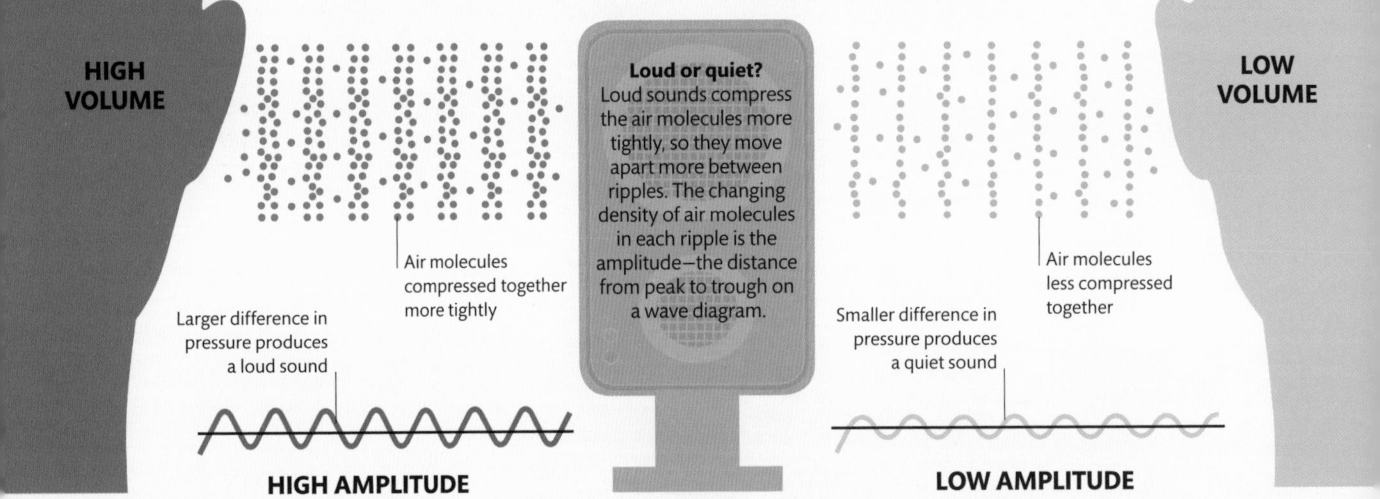

Loud or quiet?
Loud sounds compress the air molecules more tightly, so they move apart more between ripples. The changing density of air molecules in each ripple is the amplitude—the distance from peak to trough on a wave diagram.

How loud?
Sound volume is measured in decibels (dB), which increase in power exponentially. Making a sound 10dB louder actually makes it 10 times more powerful.

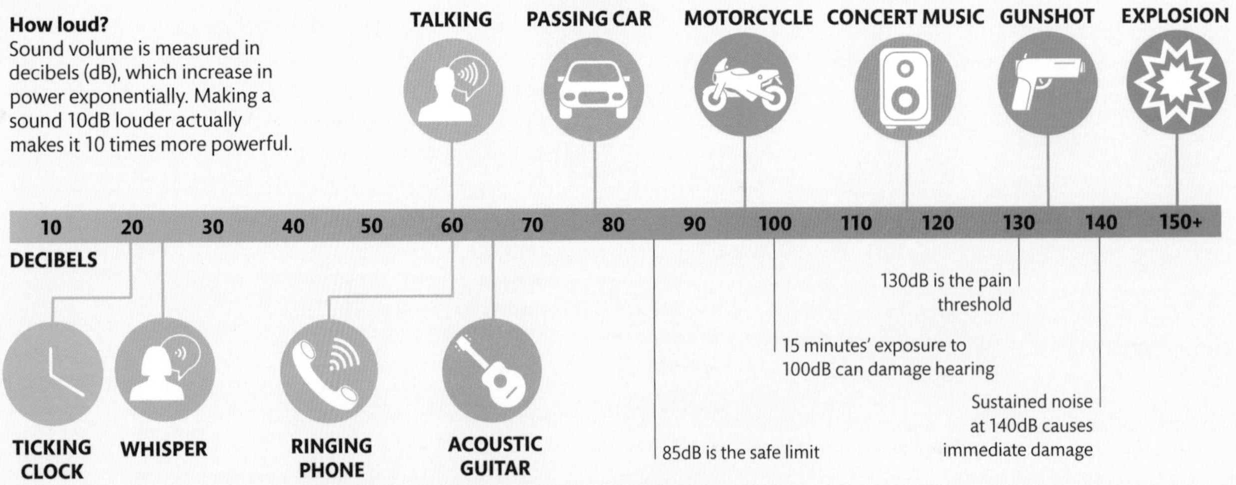

Doppler effect

Sound waves travel through air at around 770 miles (1,238km) per hour. That's fast, but even fast waves are affected by the speed of their source. If a loud vehicle is moving toward a listener, the ripples of pressure of the sound waves end up closer together, raising their frequency and pitch. As the vehicle passes, the waves stretch out, lowering their pitch.

Racing the waves
Sound waves dispersing ahead of this race car are squeezed together because its loud engine is moving a little closer to each wave before sending out the next.

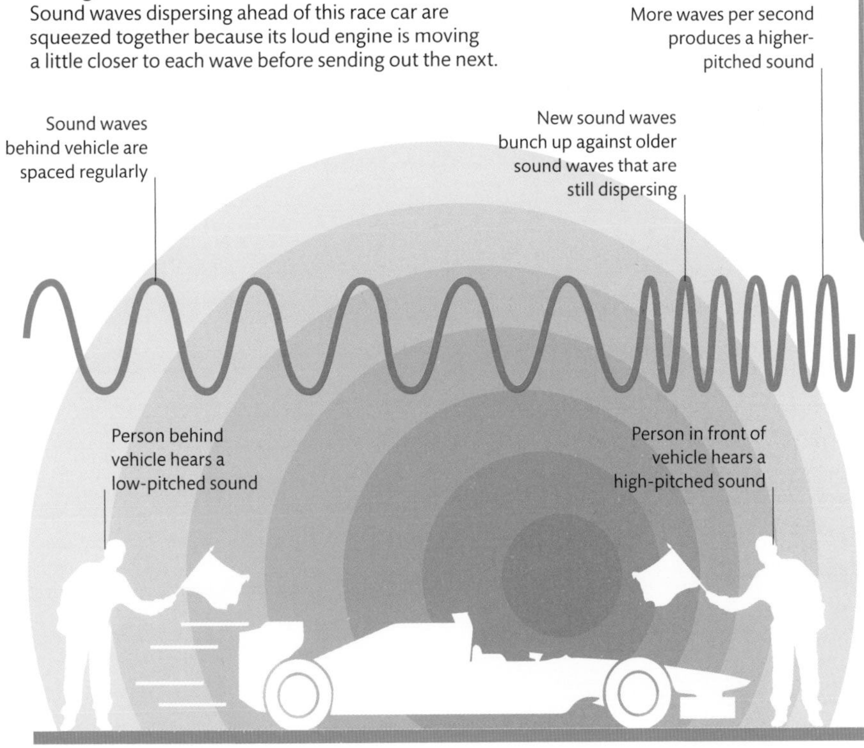
More waves per second produces a higher-pitched sound
Sound waves behind vehicle are spaced regularly
New sound waves bunch up against older sound waves that are still dispersing
Person behind vehicle hears a low-pitched sound
Person in front of vehicle hears a high-pitched sound

PITCH

A sound's pitch is related to the frequency of its wave: the higher the frequency, the higher the pitch. Measured in hertz (Hz), frequency is the number of wave peaks and troughs (or cycles) that pass a point each second.

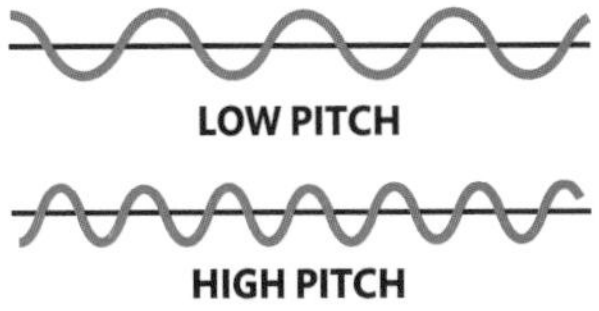
LOW PITCH
HIGH PITCH

WHY, IN SPACE, CAN NO ONE HEAR YOU SCREAM?

Sound is transmitted by pressure waves passing through a medium, such as air molecules. In the vacuum of space, there is no air.

Supersonic

Many jet aircraft travel faster than the speed of sound, passing overhead before anyone hears them coming. The sound waves are so compressed that they create a loud sonic boom.

BLUE WHALE VOICES CAN HAVE A VOLUME GREATER THAN 180dB

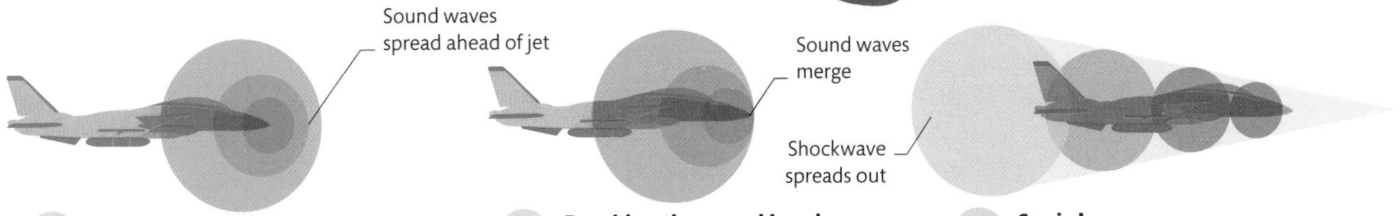
Sound waves spread ahead of jet
Sound waves merge
Shockwave spreads out

1 Speeding up
When the jet is accelerating from low speed, sound waves can disperse ahead of it, but they are squeezed more and more closely together by the Doppler effect.

2 Breaking the sound barrier
At 770 miles (1,238km) per hour, the aircraft breaks the sound barrier. It catches up with the compressed sound waves, making them merge into a single shockwave.

3 Sonic boom
The shockwave spreads out behind the aircraft like an expanding cone. Where it hits the ground, it is heard as a loud boom, which follows the flight path.

Heat

Hot objects have a lot of internal energy, which makes their atoms and molecules move. It is also known as thermal energy. An object with a high thermal energy is hot, and its heat spreads to colder places with less thermal energy.

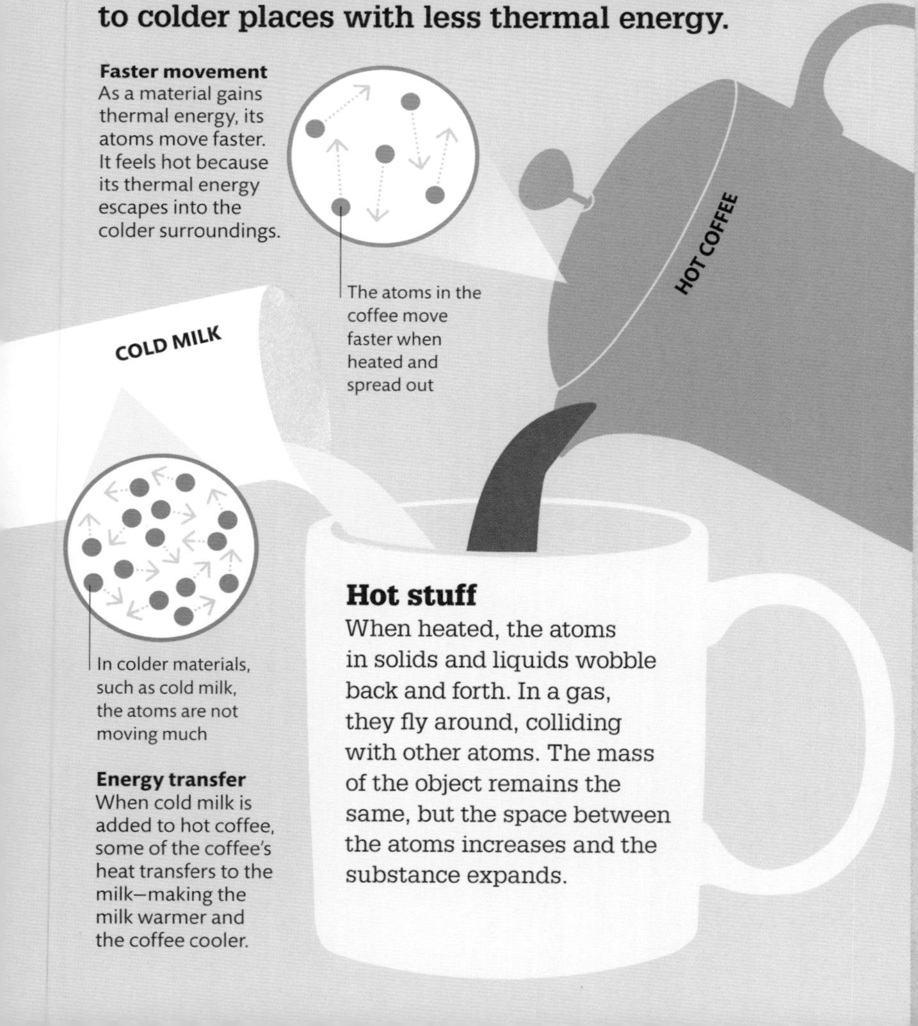

Faster movement
As a material gains thermal energy, its atoms move faster. It feels hot because its thermal energy escapes into the colder surroundings.

The atoms in the coffee move faster when heated and spread out

In colder materials, such as cold milk, the atoms are not moving much

Energy transfer
When cold milk is added to hot coffee, some of the coffee's heat transfers to the milk—making the milk warmer and the coffee cooler.

Hot stuff

When heated, the atoms in solids and liquids wobble back and forth. In a gas, they fly around, colliding with other atoms. The mass of the object remains the same, but the space between the atoms increases and the substance expands.

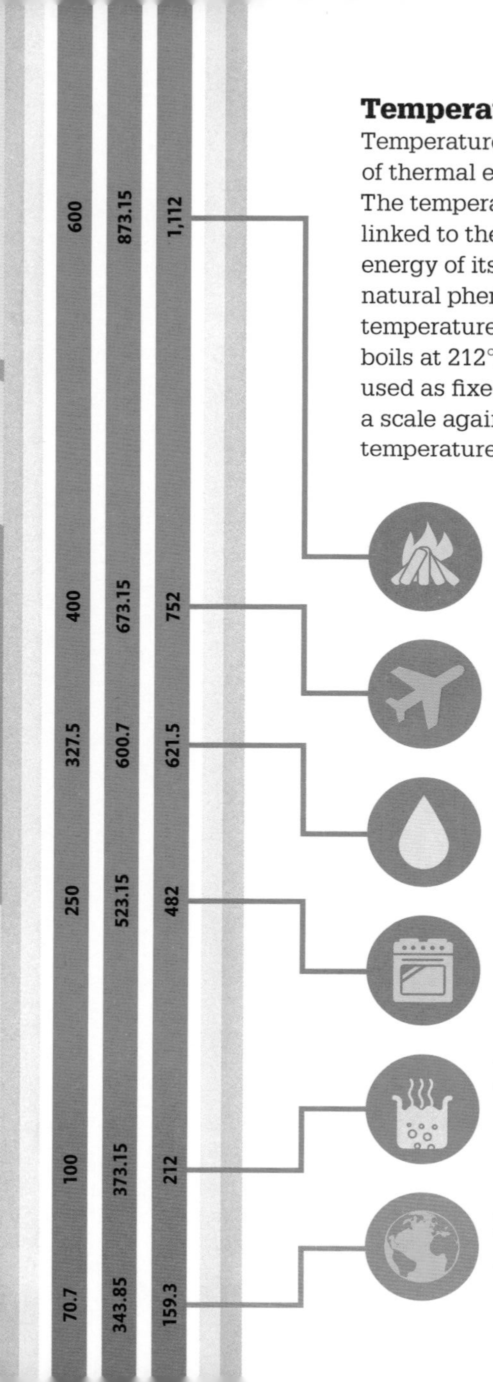

Temperature

Temperature indicates the amount of thermal energy in a substance. The temperature of a substance is linked to the average amount of energy of its particles. Certain natural phenomena occur at set temperatures—for example, water boils at 212°F (100°C). These are used as fixed points to create a scale against which other temperatures can be compared.

Wood fire
A well-ventilated wood fire is just hot enough for smelting ores into pure metals.

Airliner jet exhaust
The thrust of a jet engine comes from the rapid motion of energetic gas molecules.

Melting point of lead
Lead was the first metal to be refined due to its relatively low melting point.

Maximum home oven temperature
Prolonged cooking at this temperature gradually damages metal racks.

Water boils
The upper fixed point of the Celsius scale, chosen because it is easy to replicate.

Hottest temperature on Earth
This was recorded in 2005 during a satellite study of land surface temperatures in Iran's Lut desert.

Latent heat

As thermal energy is added to a substance, the increased movement of atoms and molecules will eventually break the bonds holding it together. The substance changes state (see pp.22–23)—it boils, for example. During such a change, heating does not make the substance get any hotter. Instead, the energy is working as hidden, or latent, heat.

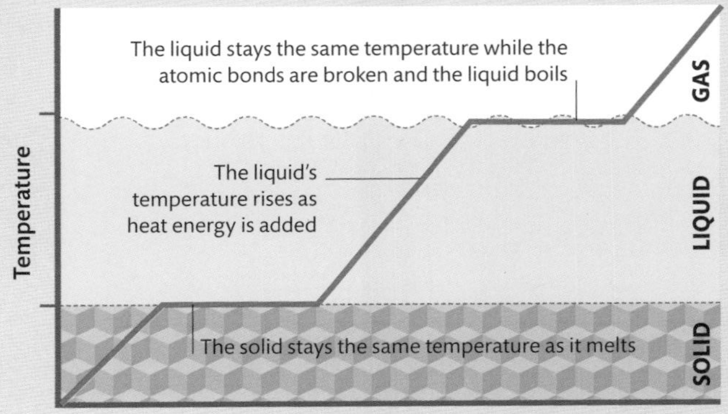

Hidden effect
Instead of increasing the movement of the atoms and molecules, the latent heat energy is used to break the bonds between them. So the temperature will briefly stay constant during a change of state even though thermal energy is being added. Once the bonds are broken, the temperature rises again.

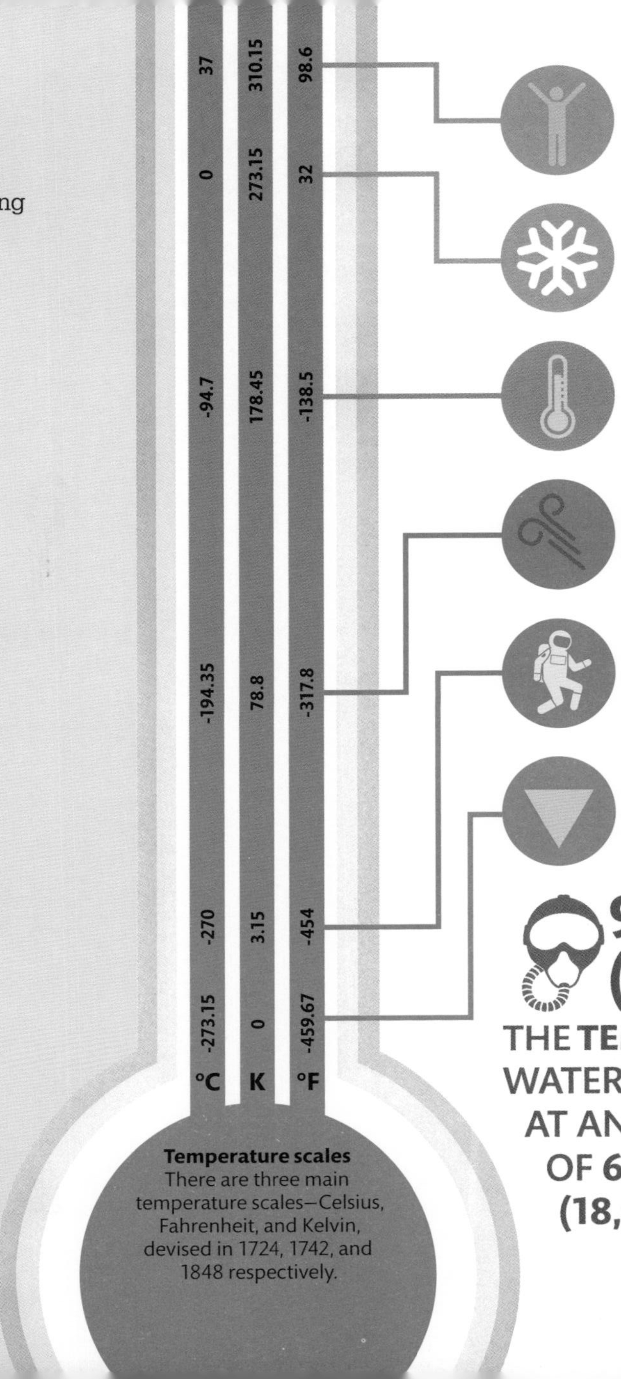

Temperature scales
There are three main temperature scales—Celsius, Fahrenheit, and Kelvin, devised in 1724, 1742, and 1848 respectively.

Normal body temperature
This was originally chosen as the upper fixed point for the Fahrenheit scale.

Water freezes
The zero point of the Celsius scale, creating 100 degrees between the freezing and boiling points of water.

Coldest temperature on Earth
This was measured in eastern Antarctica in 2010.

Air liquefies
Most of the gases in air become liquids at this temperature.

Outer space
The coldest temperature in interstellar space.

Absolute zero
The lowest theoretical temperature, though it is impossible for an object to become that cold.

99°F (37°C)
THE **TEMPERATURE** WATER BOILS AT, AT AN ALTITUDE OF **60,000FT (18,000M)**

ENERGY VS. TEMPERATURE

A sparkler burns at about 1,800°F (1,000°C). However, the hot sparks that the sparkler emits will not burn a person's skin, whereas the burning sparkler itself will. While the tiny spark has a high temperature, its low mass means that the total energy it contains is very small, so it is mostly harmless.

Sparks are burning grains of iron, magnesium, aluminum, and other metals

Heat transfer

Heat can be transferred from one object to another by three processes: conduction, convection, and radiation. The way it moves depends on the object's atomic structure.

WHICH MATERIAL CONDUCTS HEAT BEST?

Diamond is regarded as the best heat conductor–it is more than twice as effective as copper, and over four times better than aluminum.

Convection

Heat moves through fluids—liquids and gases—by convection. This process works on the principle that hot fluids rise and cold fluids sink. Heat makes the atoms and molecules in a fluid spread out, so its volume increases and its density decreases. This allows the hot fluid to float upward while the colder one sinks, and creates a convection current that moves heat energy with it.

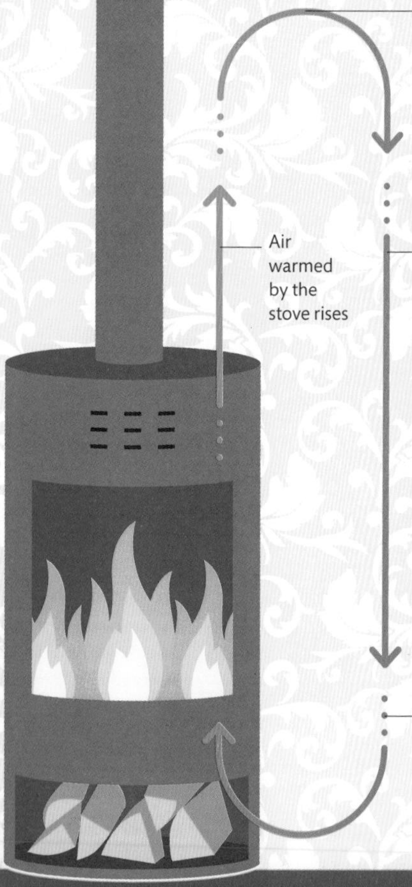
Warm air spreads through room, giving out its heat to the surroundings
Air warmed by the stove rises
Cooler air sinks to make way for rising warm air
Sinking air is drawn closer to the stove, where it is warmed and then rises

Heating space
Space heaters, such as a wood stove, use convection to spread heat through a room. The radiators of a central heating system do the same thing.

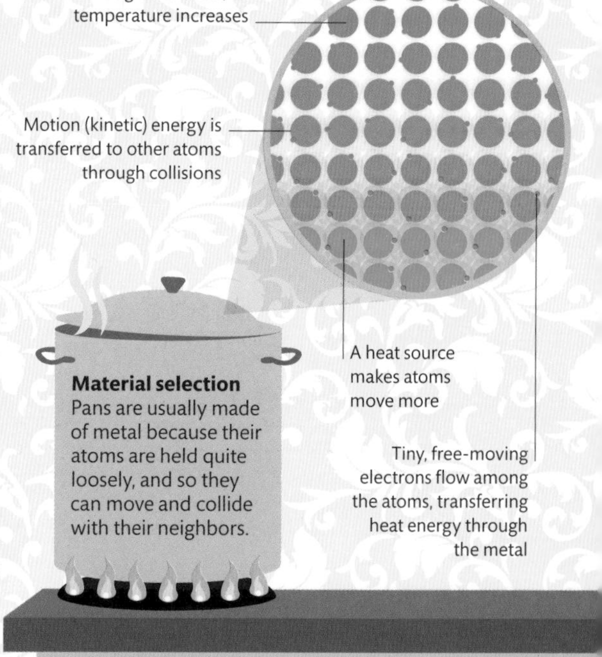
As motion energy spreads through the metal, its temperature increases
Motion (kinetic) energy is transferred to other atoms through collisions
A heat source makes atoms move more
Tiny, free-moving electrons flow among the atoms, transferring heat energy through the metal

Material selection
Pans are usually made of metal because their atoms are held quite loosely, and so they can move and collide with their neighbors.

Conduction

Solids transmit heat by conduction. The atoms in a hot part of the solid vibrate a lot, and they regularly collide with neighboring atoms. These collisions transfer motion energy to nearby atoms, making them warmer. This process continues until the heat has moved all the way through the material.

INFRARED RADIATION TRAVELS AT THE SPEED OF LIGHT AND CAN MOVE THROUGH THE VACUUM OF SPACE

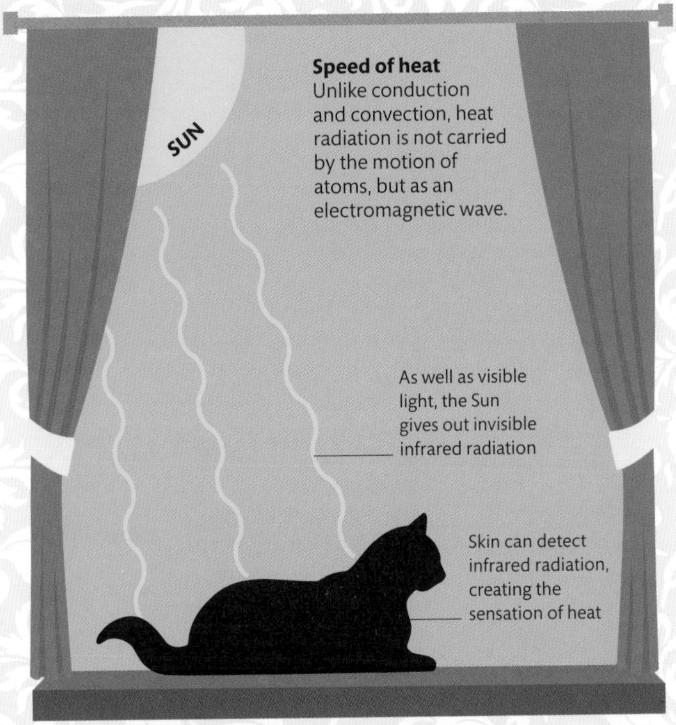
SUN
Speed of heat
Unlike conduction and convection, heat radiation is not carried by the motion of atoms, but as an electromagnetic wave.
As well as visible light, the Sun gives out invisible infrared radiation
Skin can detect infrared radiation, creating the sensation of heat

INSULATION

Thermal insulators work by preventing heat transfer. A gas like air is a poor conductor, so some insulators are filled with pockets of air. Clothes keep us warm by trapping air close to our bodies. Body heat cannot conduct through the air, so it stays inside. Double-glazed windows comprise two panes of glass separated by a cavity filled under vacuum conditions with an inert gas or dehydrated air. These windows are even better insulators because they block radiation and convection.

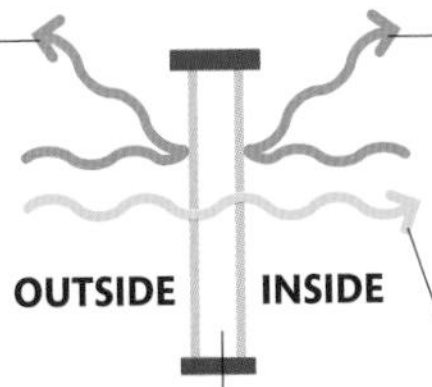
Infrared radiation from outside is reflected away
Infrared radiation from inside is reflected back
OUTSIDE
INSIDE
Visible light shines through the double-glazed window and into the house
Cavity filled with inert gas or dehydrated air makes it impossible for convection currents to form and transfer heat through the window

Radiation

The third means of heat transfer is radiation. Heat energy is carried by an invisible form of radiation called infrared—so-called because its frequency is below that of visible red light (and above that of radio waves). Infrared is given out by all hot objects, but perhaps most notably the Sun. An object with a large surface area in relation to its volume radiates heat—and cools down—much more quickly than an object with a relatively small surface area.

THERMAL EQUILIBRIUM

When two objects are in physical contact with each other, heat moves from the hot object to the cold one, and never the other way around. The heat will keep being transferred until both objects have the same temperature. This state is called thermal equilibrium, where no further heat transfer takes place.

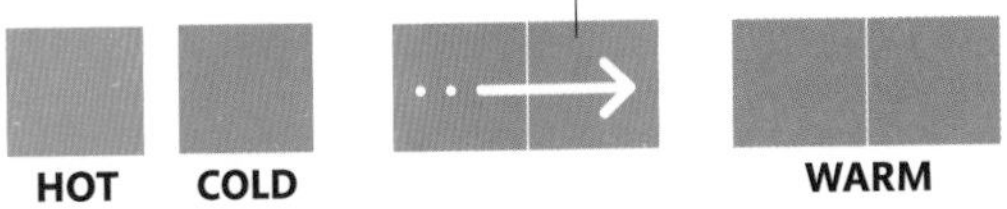
Heat energy spreads out until it is evenly distributed
HOT
COLD
WARM

Forces

Motion is created by a force acting on a mass. Forces affect objects differently, depending on the object's mass. Force is measured in newtons (N). One newton of force (1N) will accelerate an object with a mass of 1kg to 1 meter per second in 1 second.

WHY DO SOME OBJECTS BOUNCE WHILE OTHERS SMASH?

Flexible objects deform when they strike a surface, but brittle objects barely change shape when a force is applied and are likely to smash.

Transfer of energy

When two objects collide, their atoms come close together. The negatively charged electrons around atoms repel each other, so the objects do not merge into one, but are forced apart. That force transfers energy from one object to the other, but the total amount of energy stays the same. By moving energy between objects, the force creates a change in the status quo, such as altering motion or changing the shape of the objects involved.

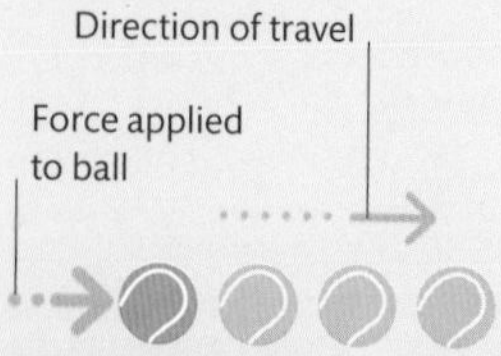
Direction of travel
Force applied to ball

Speeding up
A force acting on a tennis ball makes it accelerate so it begins to move and increase its speed.

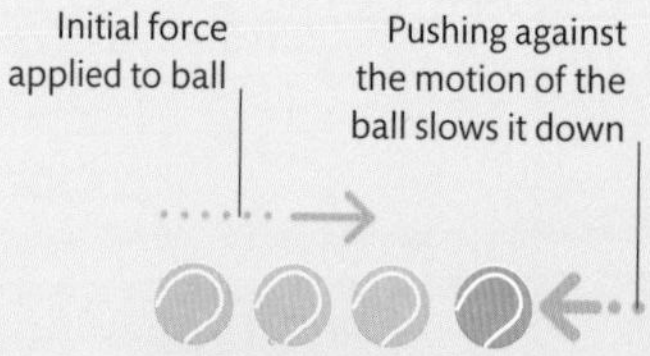
Initial force applied to ball
Pushing against the motion of the ball slows it down

Slowing down
A force pushing against the motion of the ball makes it slow down.

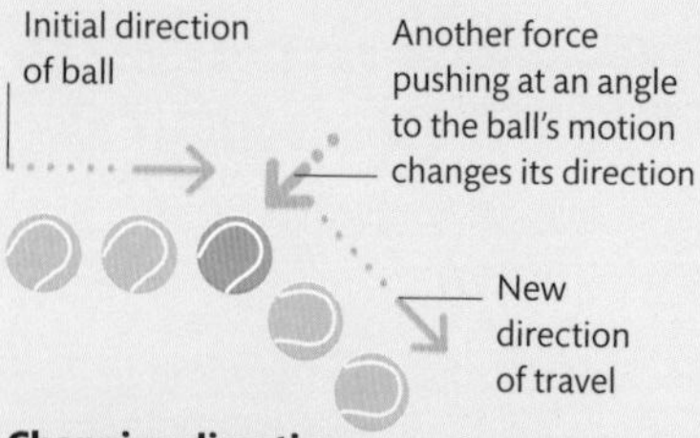
Initial direction of ball
Another force pushing at an angle to the ball's motion changes its direction
New direction of travel

Changing direction
Another force acting at a different angle to the original force is required to make the tennis ball change its direction.

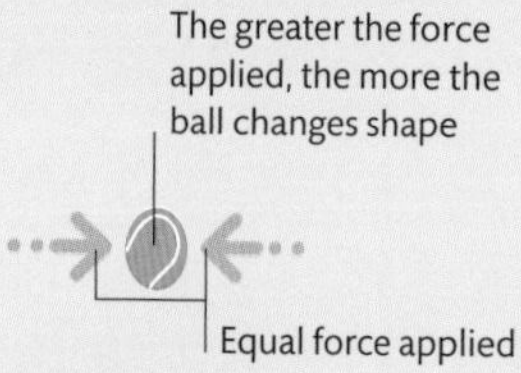
The greater the force applied, the more the ball changes shape
Equal force applied

Changing shape
The ball changes shape because it is compressed by two opposing equal forces.

THE FASTEST TENNIS SERVE IN HISTORY IS 163.7 MILES (263.4KM) PER HOUR

Projectile motion
A tennis ball, or any other projectile, follows a curved path because of the combination of forces acting on the ball. The ball's kinetic energy is transformed into gravitational potential energy (energy stored as a result of its vertical position) and then back to kinetic energy as it falls.

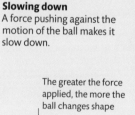

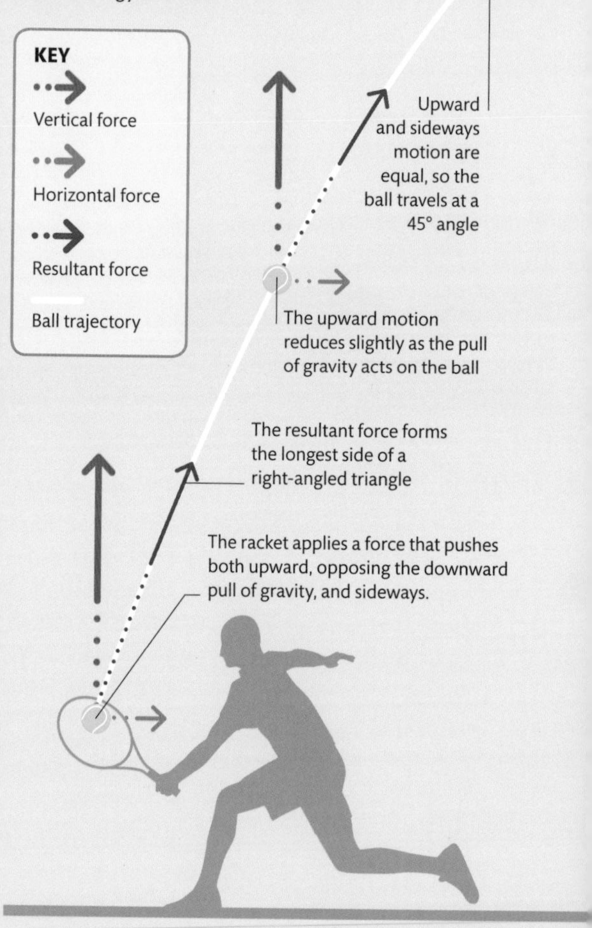
KEY
Vertical force
Horizontal force
Resultant force
Ball trajectory
Upward and sideways motion are equal, so the ball travels at a 45° angle
The upward motion reduces slightly as the pull of gravity acts on the ball
The resultant force forms the longest side of a right-angled triangle
The racket applies a force that pushes both upward, opposing the downward pull of gravity, and sideways.

Inertia

Inertia is a property of matter that resists a change in its state of motion, whether it is at rest or traveling at a constant speed. An external force is needed to overcome inertia. A larger mass has more inertia than a smaller one, so the larger mass requires a larger force to alter its state of motion.

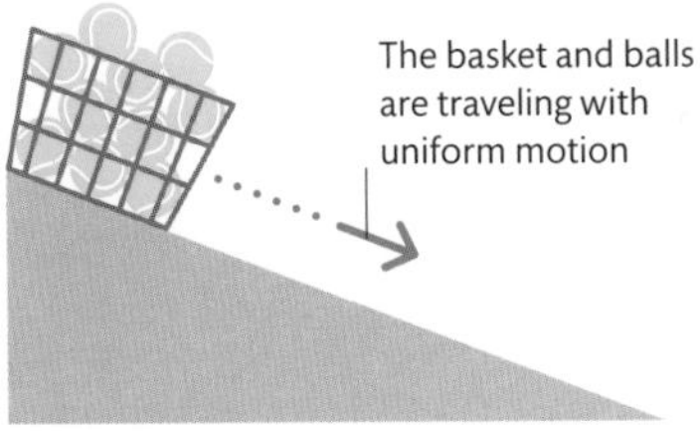
The basket and balls are traveling with uniform motion

Equal motion
The basket and balls are moving at the same speed in the same direction. Only a force can change their motion.

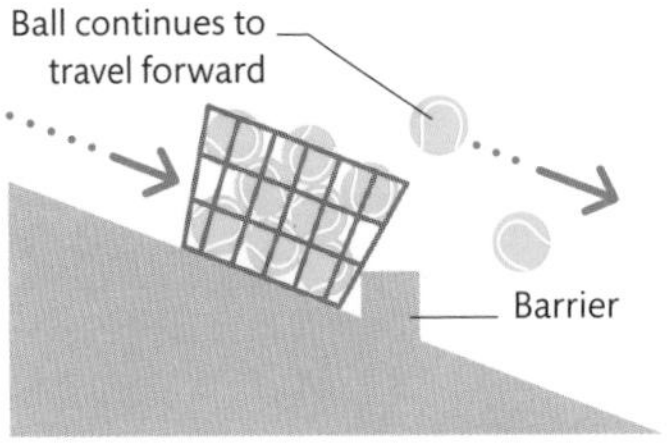
Ball continues to travel forward
Barrier

Inertial shift
A force (a barrier) stops the basket from moving, but this force hardly affects the balls, so their inertia keeps them moving.

Resultant forces

There is nearly always more than one force acting on an object, pushing it in different directions by different amounts. These individual forces combine into a single resultant force. A resultant force is calculated using the Pythagorean theorem, with two forces represented by the short sides of a right-angled triangle, and the resultant force having a size and direction equivalent to the longest side, or hypotenuse.

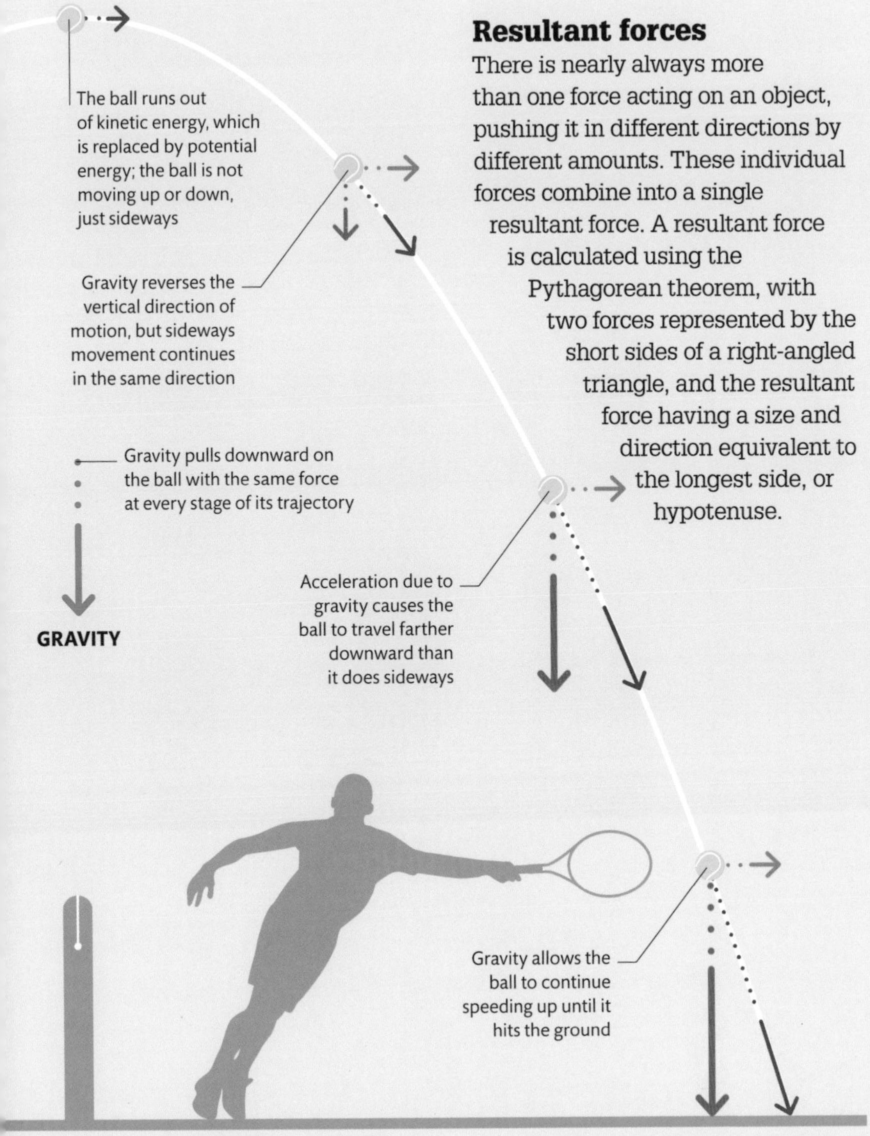
The ball runs out of kinetic energy, which is replaced by potential energy; the ball is not moving up or down, just sideways
Gravity reverses the vertical direction of motion, but sideways movement continues in the same direction
Gravity pulls downward on the ball with the same force at every stage of its trajectory
GRAVITY
Acceleration due to gravity causes the ball to travel farther downward than it does sideways
Gravity allows the ball to continue speeding up until it hits the ground

HOW AIRBAGS WORK

The inertia of passengers is one of the main dangers during a car crash because their bodies keep moving when the car suddenly stops. Airbags harness inertia to sense a crash and inflate, slowing the passengers at a safer rate.

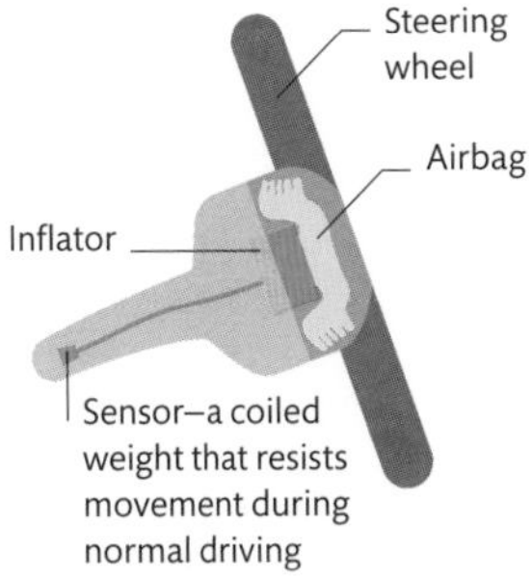
Steering wheel
Airbag
Inflator
Sensor—a coiled weight that resists movement during normal driving

AIRBAG BEFORE IMPACT

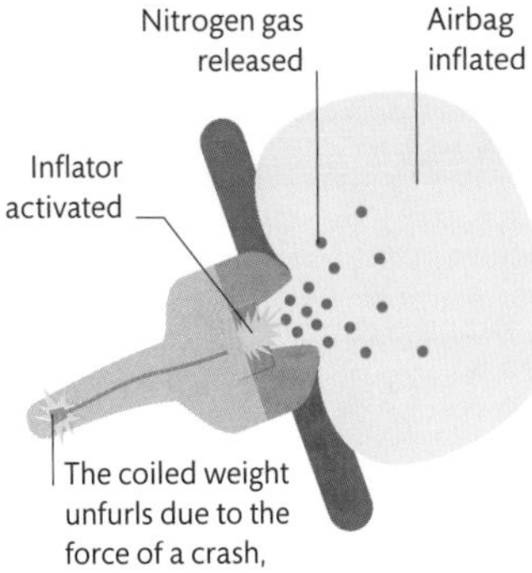
Nitrogen gas released
Airbag inflated
Inflator activated
The coiled weight unfurls due to the force of a crash, triggering the inflator

AIRBAG AFTER IMPACT

Velocity and acceleration

Velocity is the speed at which an object is traveling in a particular direction. A change in an object's velocity requires a force to be applied, and the rate of change in velocity is measured as an acceleration.

Velocity

Speed is a measure of distance over time—for example, how far a car travels in 1 hour. Velocity is a measure of speed but also includes the direction of motion. Cars traveling in opposite directions may move at the same speed but have different velocities. Every moving object has a relative velocity compared to other moving objects that is different from their actual speed.

Zero difference
These two cars have the same velocity in terms of their speed and direction. Therefore, their relative velocity is zero, and they will remain at a fixed distance from each other.

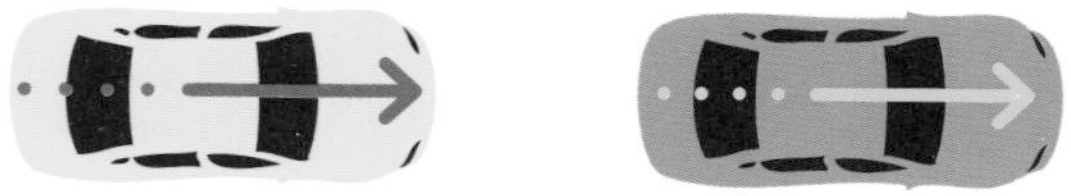

CAR TRAVELING AT 19MPH (30KPH)

CAR TRAVELING AT 19MPH (30KPH)

Catching up
The yellow car is moving 19mph (30kph) faster than the green car, so it can be said to have a relative velocity of 19mph (30kph) with respect to the green car.

CAR TRAVELING AT 37MPH (60KPH)

CAR TRAVELING AT 19MPH (30KPH)

Head on
The cars are traveling at the same speed but in opposite directions. Their relative velocities with respect to one another are both 37mph (60kph).

CAR TRAVELING AT 19MPH (30KPH)

CAR TRAVELING AT 19MPH (30KPH)

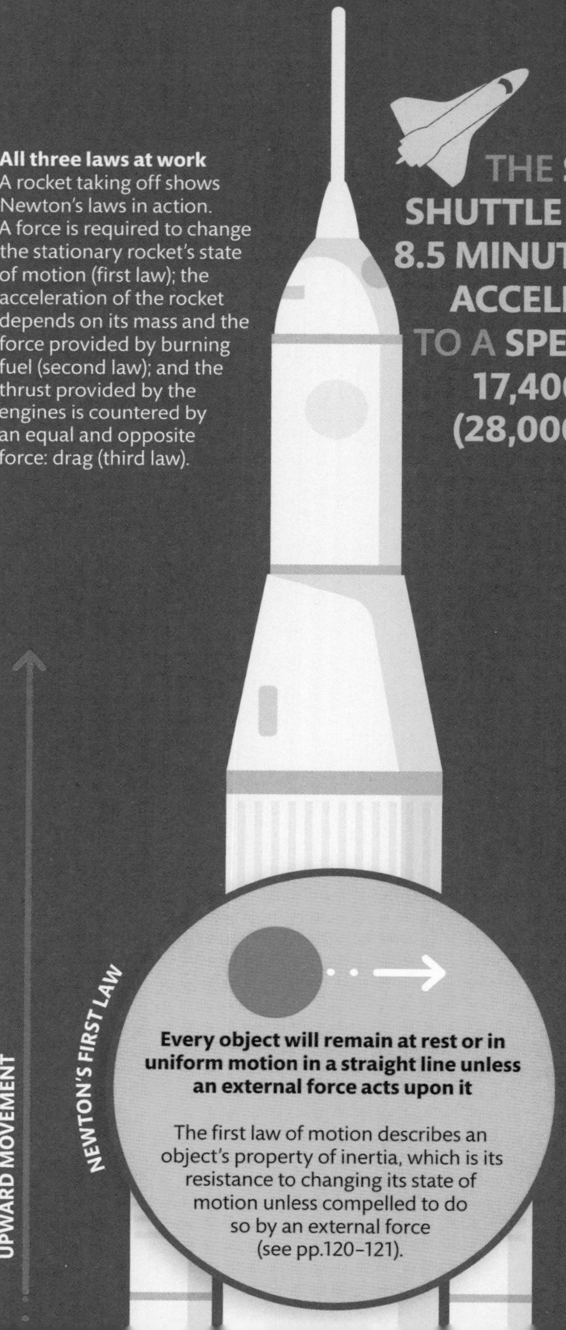
THE SPACE SHUTTLE TOOK 8.5 MINUTES TO ACCELERATE TO A SPEED OF 17,400MPH (28,000KPH)
UPWARD MOVEMENT
NEWTON'S FIRST LAW
Every object will remain at rest or in uniform motion in a straight line unless an external force acts upon it
The first law of motion describes an object's property of inertia, which is its resistance to changing its state of motion unless compelled to do so by an external force (see pp.120–121).

All three laws at work
A rocket taking off shows Newton's laws in action. A force is required to change the stationary rocket's state of motion (first law); the acceleration of the rocket depends on its mass and the force provided by burning fuel (second law); and the thrust provided by the engines is countered by an equal and opposite force: drag (third law).

Acceleration

Acceleration is a change in velocity and is measured in meters per second per second (m/s^2). Slowing down is also an acceleration, but one where velocity is decreasing. Acceleration is calculated by subtracting the initial velocity from the final velocity, and dividing this figure by the time elapsed.

Acceleration
If this car doubled its speed in 1 minute, its acceleration can be calculated by finding the velocity change (6m/s) and dividing by the time elapsed in seconds (60). This works out as 0.1m/s per second, or $0.1m/s^2$.

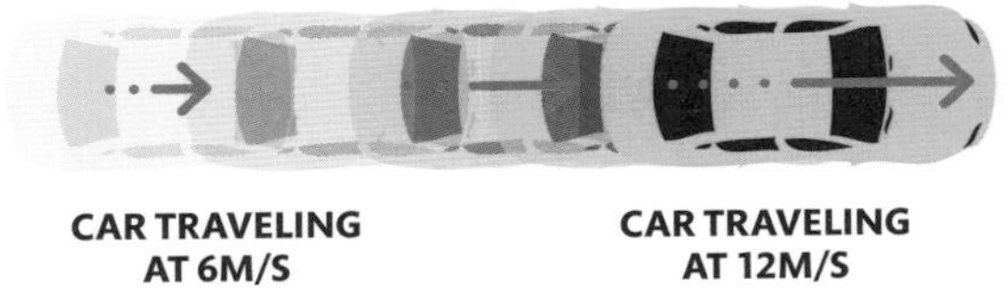

Changing direction
A change in direction, like a turn, is a change in velocity. Because a force is needed to do that, the turn is an acceleration even though the speed does not change.

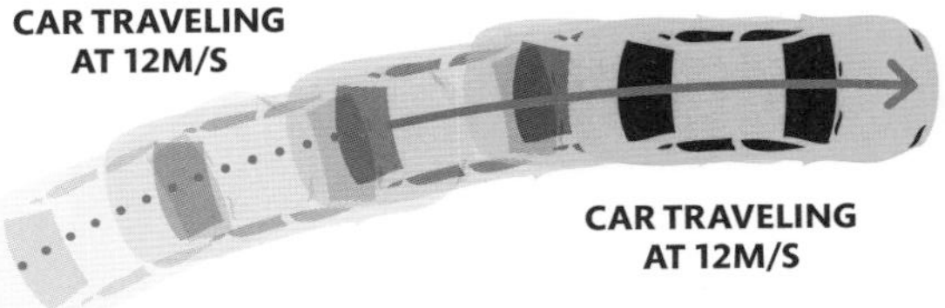

Deceleration
If this car halved its speed in 1 minute, its acceleration would be $-0.1m/s^2$. It is a negative value because the final velocity (6m/s) is lower than the initial velocity (12m/s).

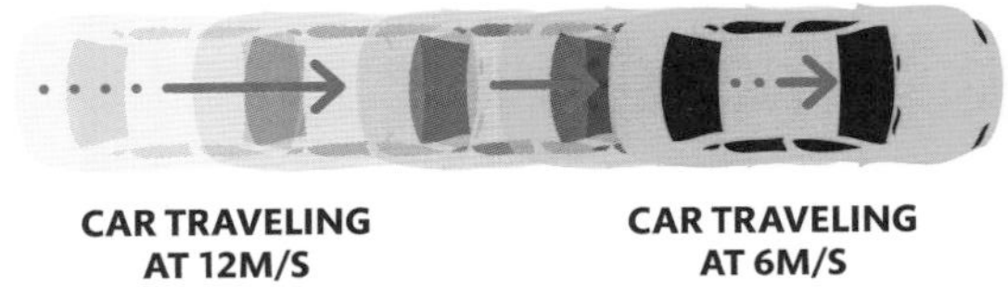

SLIPSTREAMS

As an object moves through air, it pushes the air out of the way. The air pushes back, creating drag. The effect of drag can be reduced by moving into the slipstream, the area with reduced drag, behind a moving object. This enables a trailing car to travel at the same velocity but use less fuel.

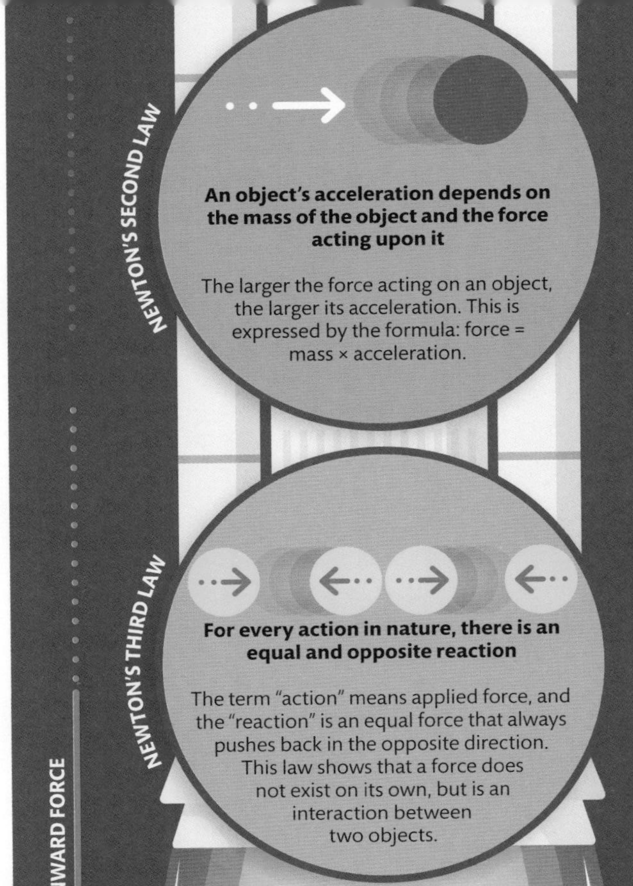

An object's acceleration depends on the mass of the object and the force acting upon it

The larger the force acting on an object, the larger its acceleration. This is expressed by the formula: force = mass × acceleration.

For every action in nature, there is an equal and opposite reaction

The term "action" means applied force, and the "reaction" is an equal force that always pushes back in the opposite direction. This law shows that a force does not exist on its own, but is an interaction between two objects.

Laws of motion

All motion is governed by three laws that show the relationship between an object's mass, the forces acting on it, and the accelerations that result. The laws of motion were published by Isaac Newton in 1687. While they are accurate enough for most applications, Albert Einstein famously theorized in 1905 that the laws fall down when objects approach the speed of light (see pp.140–141).

Machines

Simple machines are devices that turn one type of force into another. There are six simple machines, some of which do not seem like machines at all.

THE AERIAL SCREW WAS THE NAME **LEONARDO DA VINCI** GAVE TO HIS EARLY **HELICOPTER DESIGN**

Six simple machines

Like most mechanical devices, a bicycle is a combination of simple machines. Some, such as the chain mechanism and brake levers, have a clear mechanical function. Others are less obvious because they are used for adjustment, repair, or even to make cycling uphill possible. Altogether, riding and maintaining a bicycle makes use of all six simple machines: the lever, pulley, wheel and axle, screw, wedge, and inclined plane.

SCREW

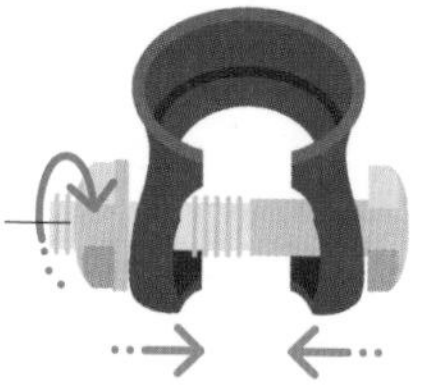
Nut tightened onto screw thread

Tightening the screw that secures the saddle transforms a lot of rotation into a small amount of very powerful compression. It is essentially a long, spiral wedge.

WEDGE

Forcing a tool under a tire to remove it from a wheel uses the wedge principle. A pushing force is turned into a stronger separating force that acts over a shorter distance.

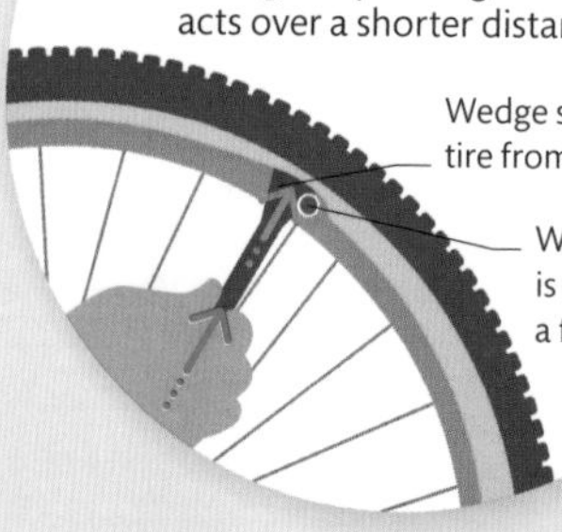
Wedge separates tire from rim
Wheel rim is used as a fulcrum

PULLEY

Smaller wheel turns faster

The bicycle chain is basically a pulley system—one wheel drives another by pulling a type of cable. The relative sizes of the wheels determine their relative speed and power.

WHEEL AND AXLE

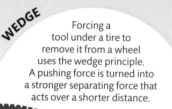

Rim moves faster
Axle bearing moves slowly

A wheel turns on a fixed axle, overcoming friction (see pp.126–27) by acting as a lever. It converts a lot of rim movement into a small but powerful turning movement at the axle.

Mechanical advantage

All machines apply the principle of mechanical advantage, which is a measure of force amplification. This means that they allow you to convert a large movement into a smaller movement with greater power, like levering the lid off a can of paint. But it can also work in reverse, such as when an angler applies strength to the butt of a fishing rod to swing the tip through a wide arc. More movement gives less power, and vice versa.

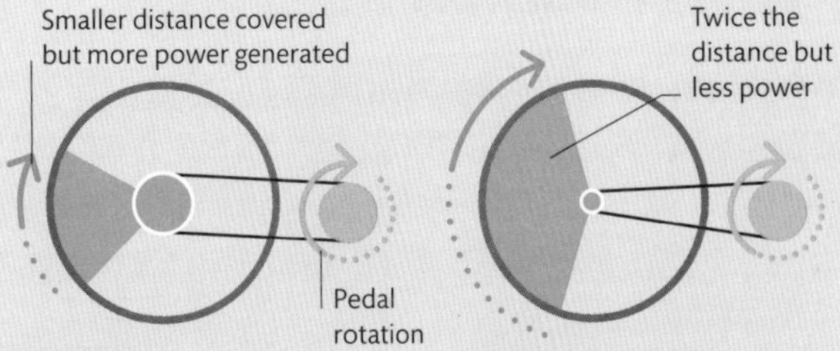
Smaller distance covered but more power generated
Pedal rotation
Twice the distance but less power

Low gear
On a bicycle, a lower gear converts pedal rotation into more power for climbing hills, at the cost of speed.

High gear
Changing into a higher gear when reaching the top of a hill increases the speed.

KEY ··→ Effort force (input force) ··→ Load (output force) ● Fulcrum

LEVER

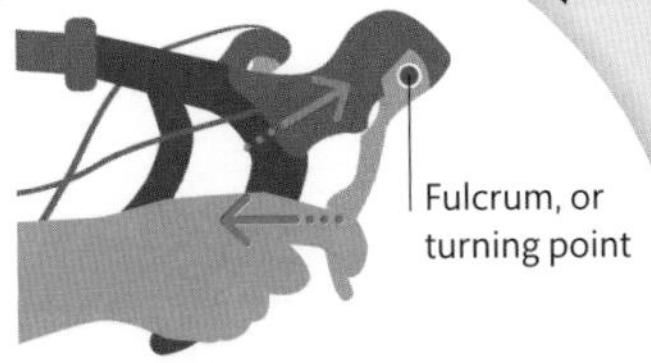
Fulcrum, or turning point

The brakes are operated by a lever pivoting on a turning point. The lever magnifies a small force into a larger one because the former acts over a greater distance. Pulling the lever tightens a cable, forcing calipers to clamp down on the wheel rim.

Lever classes

There are three classes of lever, depending on where the load and effort are located relative to the fulcrum. They can be chosen to increase either power or movement, in different directions.

First-class
The load and the effort are situated on opposite sides of the fulcrum. Examples include a pair of scissors and pliers.

Second-class
The load is located between the effort and the fulcrum. An example of a second-class lever is a pair of nutcrackers.

Third-class
The effort is applied between the load and the fulcrum, such as a pair of tongs or tweezers.

INCLINED PLANE

Shorter distance is harder work

Cycling straight up a wall would not be possible. An inclined plane or ramp solves the problem at the cost of increasing the distance the cyclist has to pedal.

Gear ratio

Power in the form of turning force, or torque, is often transmitted through the interlocking "teeth" of gears. If the larger driver gear has three times as many teeth as the smaller gear, it will make the small one rotate three times as fast. Several gears together are often called gear trains.

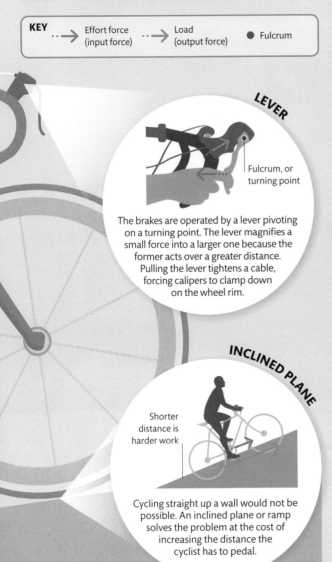

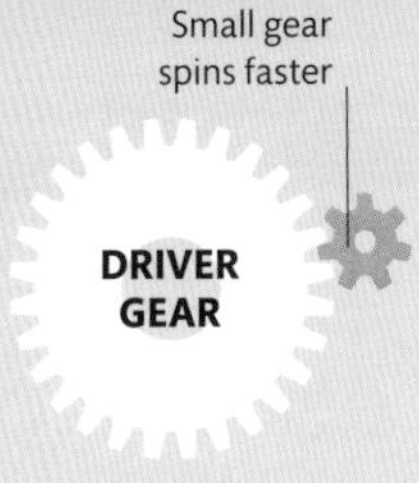
Small gear spins faster
DRIVER GEAR

Gear ratio
A large gear driving a small one raises speed. The reverse gives more power.

Friction

Friction is a resistive force that occurs when two objects or substances rub together, because friction pushes against the direction of motion. When an object pushes through a liquid or gas, it causes a form of friction known as drag.

Opposing forces

Friction is generated when the surfaces of two materials meet. At the microscopic level, surfaces are never smooth, and the small indentations snag each other as the surfaces move in opposite directions. Each snag applies a tiny force, but together they add up to a resistance force that slows or stops motion. When two surfaces move together, the friction between them converts kinetic energy into thermal energy, or heat.

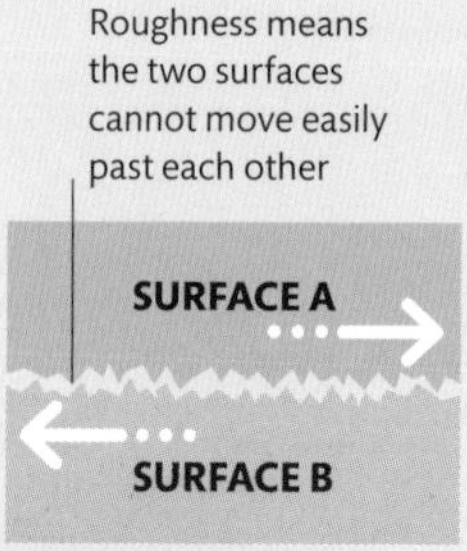

Rubbing together
Friction is related to the roughness of the surfaces. The close contact between the surfaces is caused by the weight of one object pushing down on the other.

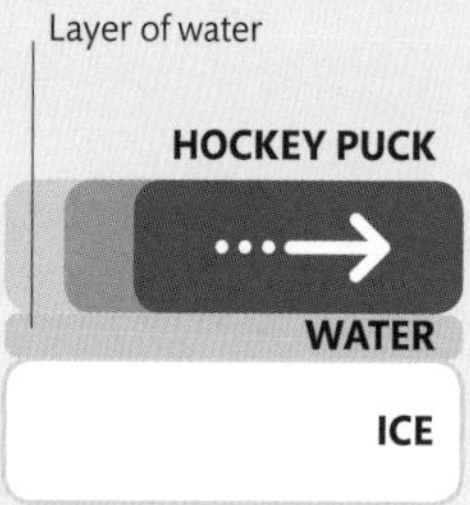

Smoothly sliding
Ice is slippery because a fine layer of liquid water separates it from other surfaces, so they make very little contact. Therefore, the friction force is small.

LUBRICATION

The friction between moving parts in a machine causes damage as the components rub together, wearing each other away. To reduce this effect, mechanics often cover the components with oil-based lubricants. This provides a slippery barrier between surfaces and is sticky enough to cover the parts for a long time.

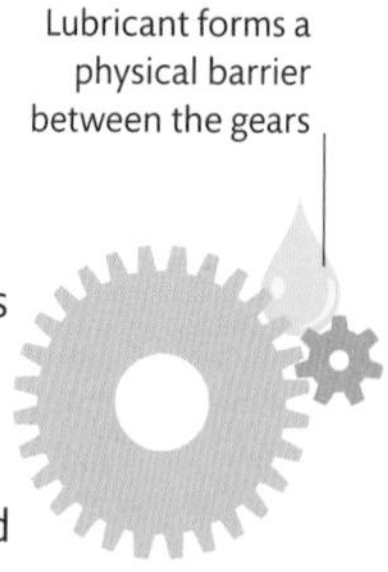

MAGLEV TRAINS CUT OUT **FRICTION** BETWEEN THE **TRAIN** AND **TRACK** BY ALLOWING THE TRAINS TO **LEVITATE**

Gripping the road
The surface of a tire is covered in indentations. This "grip" makes the tire rougher so it connects to more parts of the rough road surface. Grooves in the tire's surface force water away from the car. Adhesion and deformation help the tires grip the road, but too much stress deforms the rubber beyond elastic recovery and the surface tears.

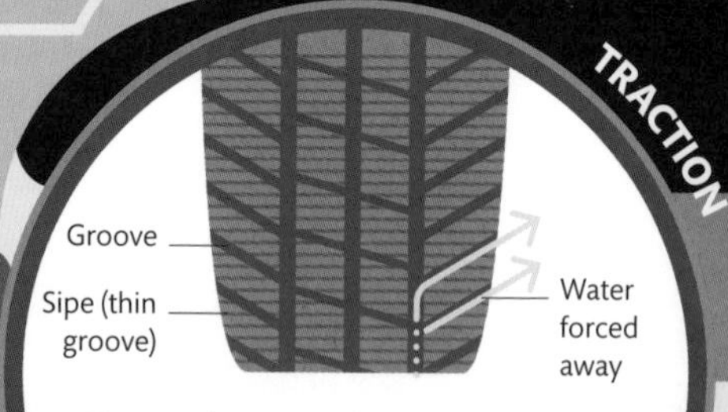

The tread pattern of a tire is designed to maximize traction in certain conditions, such as rain or snow. Regular road tires funnel rainwater away so it does not reduce contact between the tire and the road and create traction problems.

Grip and traction

The tires of a car are designed to grip the road, creating a large amount of friction with the road's surface. This friction gives the wheels traction so they can push against the road as they turn, propelling the car forward. Without sufficient grip, the wheels spin and skid.

Increased contact
A heavy load pushes the tire into the ground more, increasing the area of contact and thus increasing the frictional force.

SMALL VERTICAL LOAD

Less contact with the road surface

LARGE VERTICAL LOAD

More contact with the surface

USING FRICTION TO LIGHT FIRES

Some of the most common ways to make fire use friction, such as scraping flint on a hard surface to create a spark. A bow drill involves moving a bow rapidly left and right, which forces a pointed hardwood spindle into a groove filled with sawdust in a piece of fireboard. The heat from the friction sets the sawdust on fire.

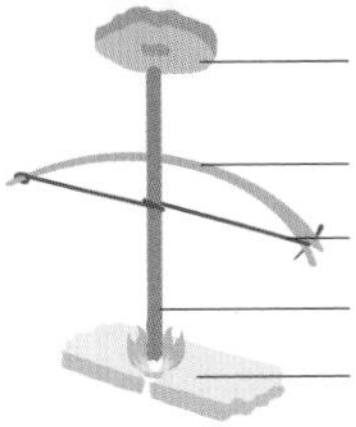

Reducing drag

Drag is the friction caused by objects moving through fluids, such as water, and air. Airplane wings and ship hulls are designed to reduce drag. Some hulls, such as those of trimarans and hydrofoils, limit the area of contact with the water. Plane wingtips control turbulent air flow to reduce drag.

Outrigger provides stability

Limited water contact
A trimaran has three small hulls, which have a relatively small total surface area to reduce drag.

Hydrofoil lifts hull out of water

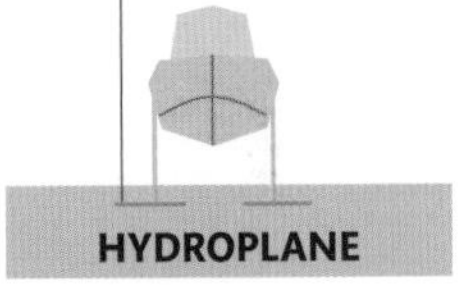

Lifting surfaces
A hydroplane uses winglike hydrofoils to lift the vessel above the water, reducing drag hugely.

Wingtip vortices
Wingtips create vortices in flight, which reduces fuel efficiency. Adding a winglet reduces the size of the tip of the wing and thus reduces drag.

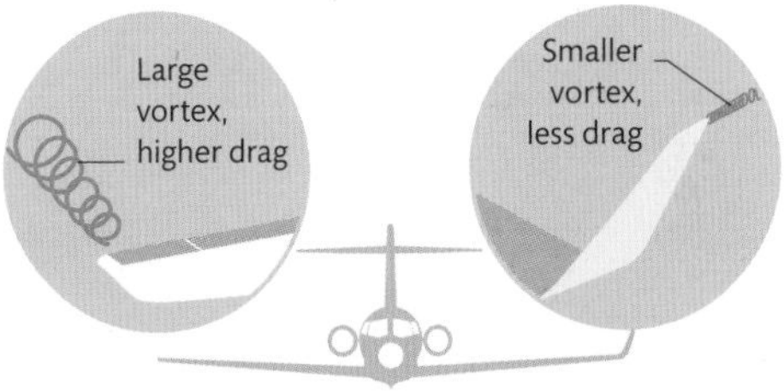

REGULAR WINGTIP **BLENDED WINGLET**

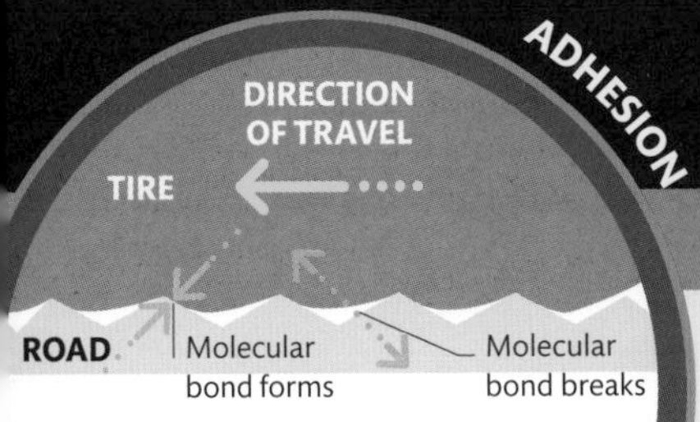

The surface of rubber contains molecules with spare chemical bonds. When the rubber contacts the road, it forms weak bonds with it, making the materials stick together briefly before separating as the bonds break.

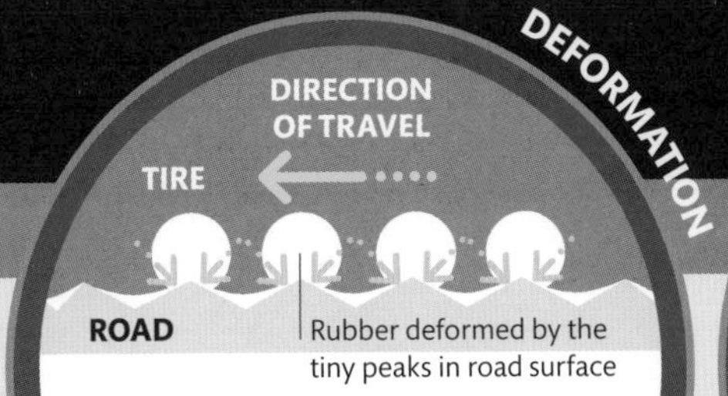

A tire's rubber is flexible, although its shape is stiffened by the high-pressure air inside. The tire deforms under the weight of the car around the lumps in the road. This focuses the car's weight onto these lumps, increasing grip.

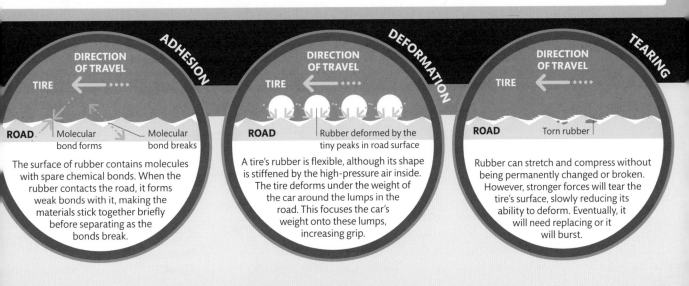

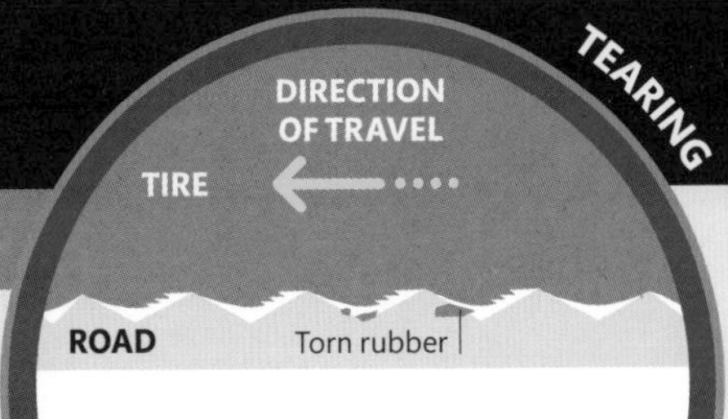

Rubber can stretch and compress without being permanently changed or broken. However, stronger forces will tear the tire's surface, slowly reducing its ability to deform. Eventually, it will need replacing or it will burst.

Springs and pendulums

A spring is an elastic object that returns to its original position when compressed or stretched. It is subject to a force called a restoring force, a key aspect of simple harmonic motion, in which a mass moves around a central point, or oscillates. This is a feature it shares with the way a pendulum moves.

WHERE ARE SPRINGS FOUND IN THE HOME?

We use springs in hundreds of everyday items without realizing it, from mattresses and watches to light switches, toasters, vacuum cleaners, and door hinges.

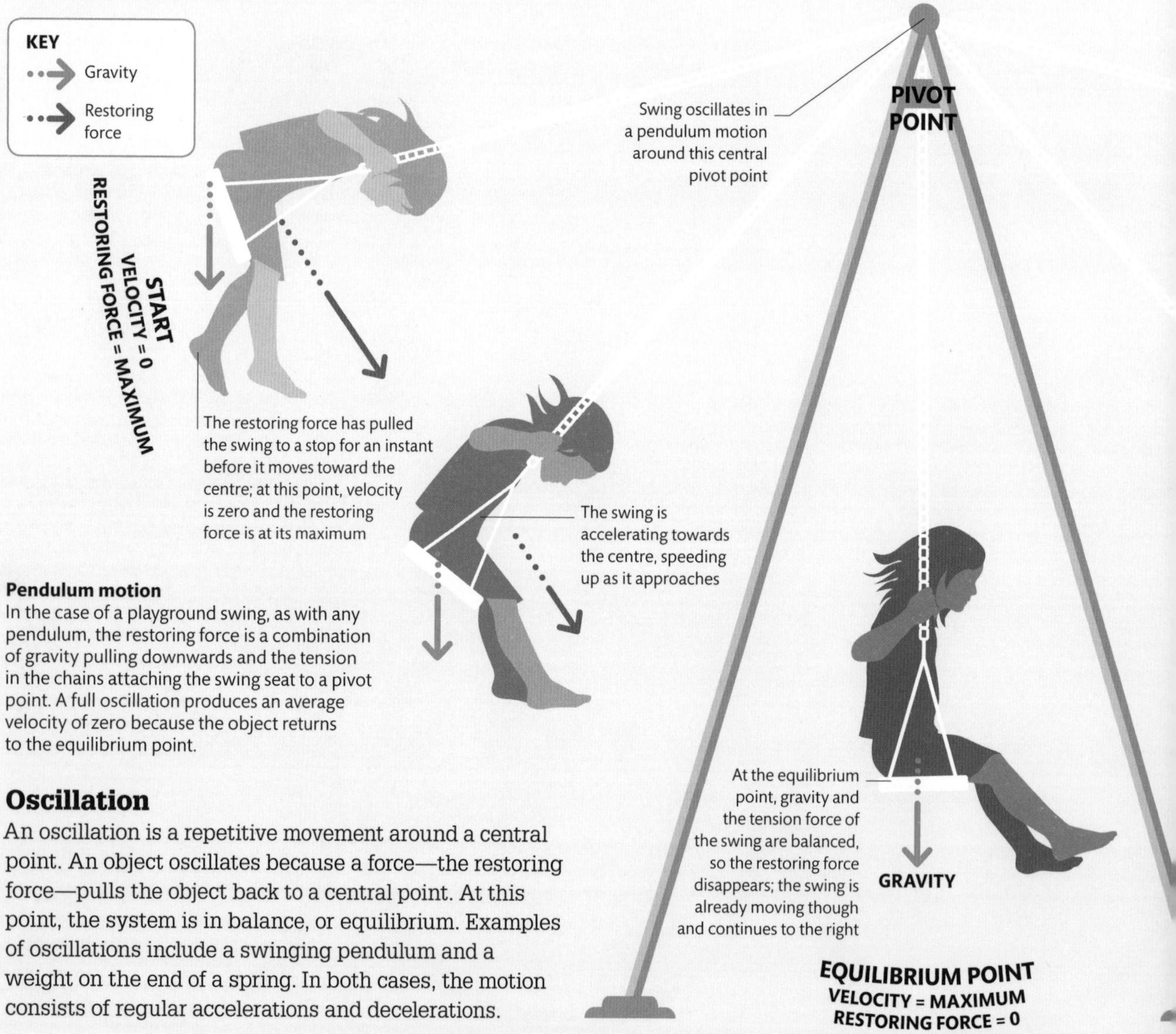
KEY
Gravity
Restoring force
START
VELOCITY = 0
RESTORING FORCE = MAXIMUM
Swing oscillates in a pendulum motion around this central pivot point
PIVOT POINT
The restoring force has pulled the swing to a stop for an instant before it moves toward the centre; at this point, velocity is zero and the restoring force is at its maximum
The swing is accelerating towards the centre, speeding up as it approaches
At the equilibrium point, gravity and the tension force of the swing are balanced, so the restoring force disappears; the swing is already moving though and continues to the right
GRAVITY
EQUILIBRIUM POINT
VELOCITY = MAXIMUM
RESTORING FORCE = 0

Pendulum motion
In the case of a playground swing, as with any pendulum, the restoring force is a combination of gravity pulling downwards and the tension in the chains attaching the swing seat to a pivot point. A full oscillation produces an average velocity of zero because the object returns to the equilibrium point.

Oscillation

An oscillation is a repetitive movement around a central point. An object oscillates because a force—the restoring force—pulls the object back to a central point. At this point, the system is in balance, or equilibrium. Examples of oscillations include a swinging pendulum and a weight on the end of a spring. In both cases, the motion consists of regular accelerations and decelerations.

Elastic forces

A spring is a particularly elastic object, meaning it is able to change shape temporarily before bouncing back. When a mass pulls on it, the spring extends. The extension creates a restoring force in the spring that pulls it back to its original shape. When the restoring force equals the force deforming the spring, the extension will stop.

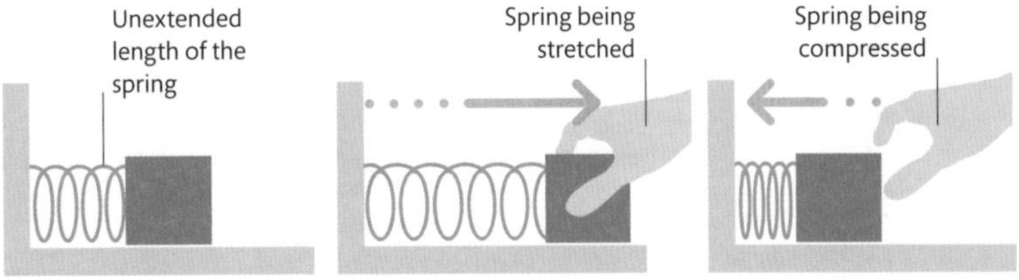
Unextended length of the spring
Spring being stretched
Spring being compressed

Resting state
The mass on the end of the spring exerts no force on the spring. This is called the equilibrium point.

Stretching force
Moving the mass creates a restoring force in the spring, which pulls it back to the equilibrium point.

Compressing force
Pushing the spring and letting it go makes it overshoot the equilibrium point, but the restoring force pulls it back.

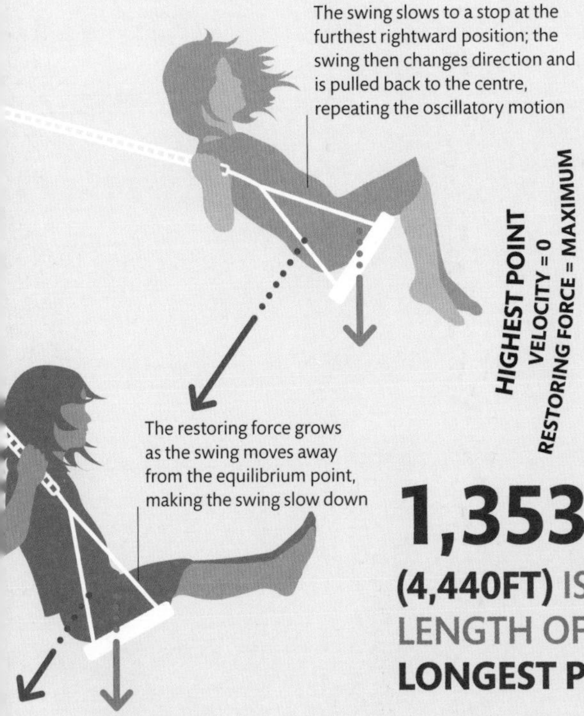
The swing slows to a stop at the furthest rightward position; the swing then changes direction and is pulled back to the centre, repeating the oscillatory motion
HIGHEST POINT
VELOCITY = 0
RESTORING FORCE = MAXIMUM
The restoring force grows as the swing moves away from the equilibrium point, making the swing slow down

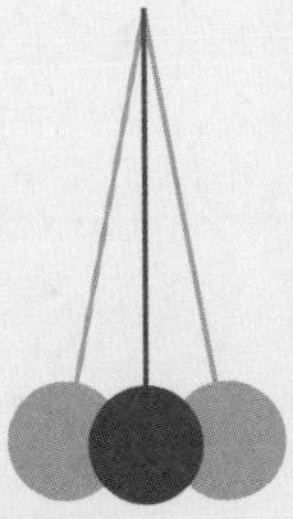

1,353M
(4,440FT) IS THE LENGTH OF THE **WORLD'S LONGEST PENDULUM**

Young's modulus

Engineers need to know how stiff a substance is, so they can learn how to construct with it. The elasticity of a substance is measured as its Young's modulus, which shows how much force is needed to deform it. It is measured in pascals, the unit of pressure. A high Young's modulus shows a material is stiff and barely changes shape when stretched. A low value shows a substance can undergo large elastic deformations.

Substance	Young's modulus (pascals)
Rubber	0.01–0.1
Wood	11
High-strength concrete	30
Aluminum	69
Gold	78
Glass	80
Tooth enamel	83
Copper	117
Stainless steel	215.3
Diamond	1050–1210

Deformation

Some forces can change the shape of a material. At first, a stretching force causes elastic deformation. When the force is removed, a restoring force pulls it back to its original shape. If the stretching force increases, the material will reach its elastic limit, so any change of shape will be permanent.

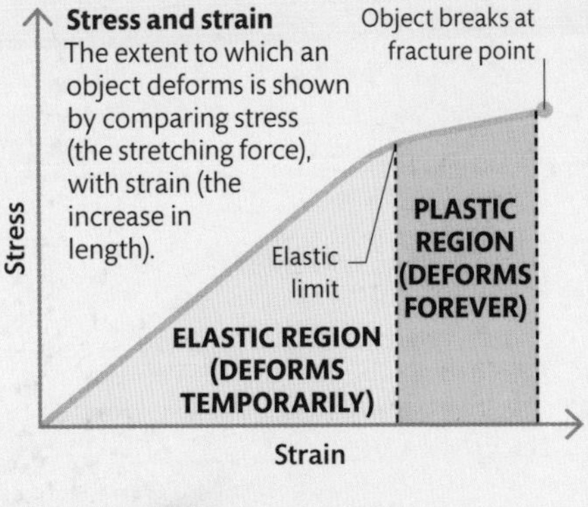
Object breaks at fracture point
Stress
Elastic limit
PLASTIC REGION (DEFORMS FOREVER)
ELASTIC REGION (DEFORMS TEMPORARILY)
Strain

Stress and strain
The extent to which an object deforms is shown by comparing stress (the stretching force), with strain (the increase in length).

Pressure

Pressure is the force applied against a surface divided by the area of the surface. It can be applied to or by any medium, including water and air.

Pressure in gases

When a force is applied, a gas is compressed into a smaller volume. The molecules become more tightly packed together, until they stop behaving as gas molecules and change into a liquid. This is why a pressurized gas cylinder clearly contains liquid. Relaxing the pressure by opening the valve allows the liquid to revert back to a gas.

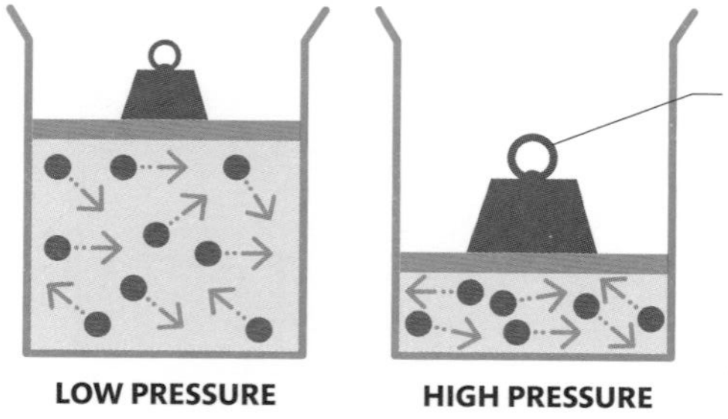

Higher density
Compressing a gas, such as air, decreases its volume while its mass remains the same, which increases the gas's density.

HOW A PRESSURE COOKER WORKS

At atmospheric pressure, water boils at 212°F (100°C). The resulting steam usually escapes, but a pressure cooker keeps it in, raising the pressure. This increases the water's boiling point and temperature, cooking food faster.

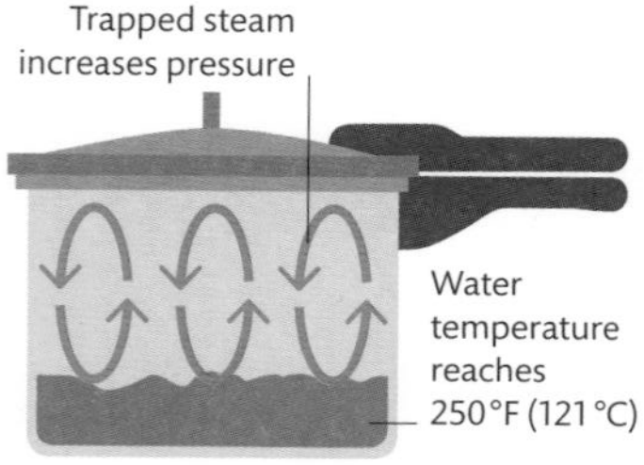

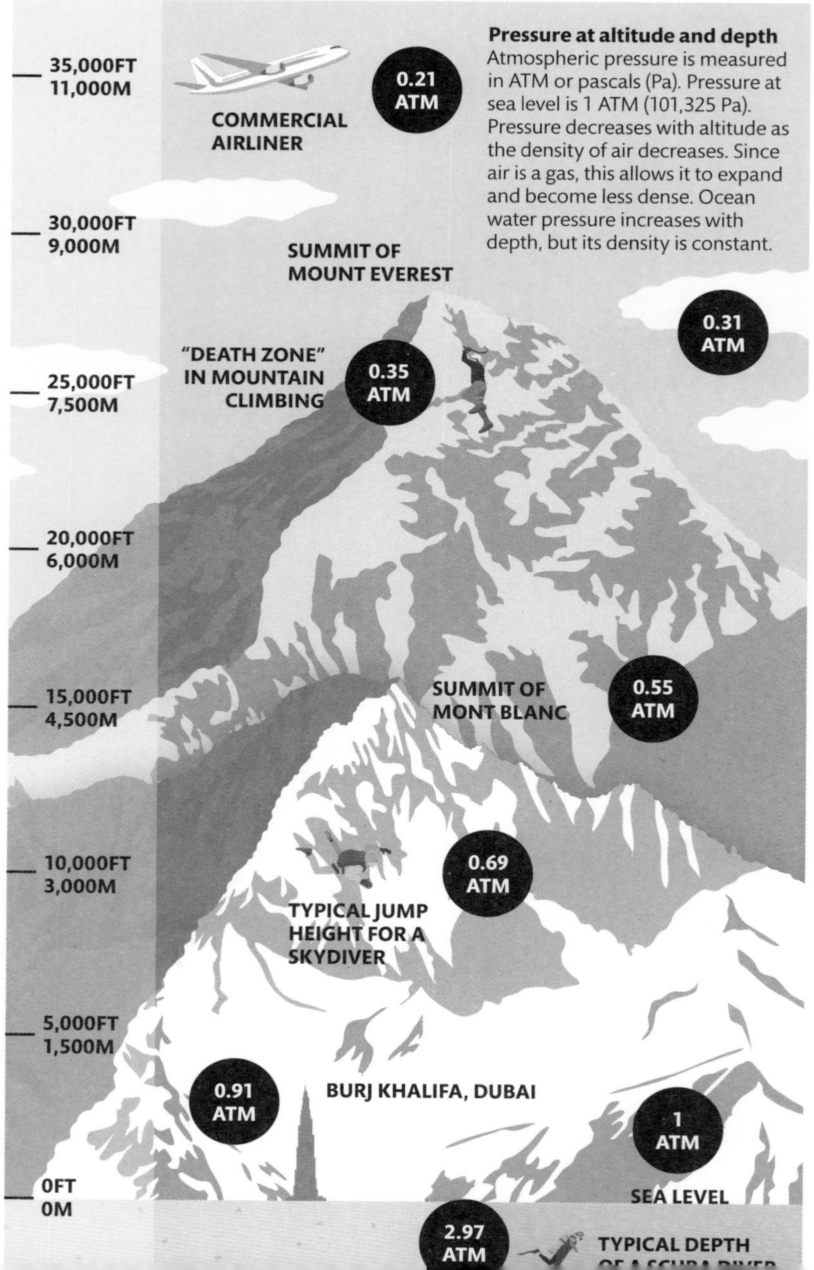

Pressure at altitude and depth
Atmospheric pressure is measured in ATM or pascals (Pa). Pressure at sea level is 1 ATM (101,325 Pa). Pressure decreases with altitude as the density of air decreases. Since air is a gas, this allows it to expand and become less dense. Ocean water pressure increases with depth, but its density is constant.

Pressure in liquids

Unlike gases, liquids are very hard to squeeze into a smaller volume by pressure. Any pressure applied to a liquid is transferred through it. For example, if a liquid is inside a pipe, a pressure applied at one end will be passed all the way to the far end. Pressure increases with depth because of the weight of water above it, which is why dams have to be thicker at their base. Pressure is also affected by density. The more dense the liquid, the higher the pressure it exerts.

The leaking bucket
The pressure increase at depth is shown by the rate at which water leaks from three identically sized holes in a bucket.

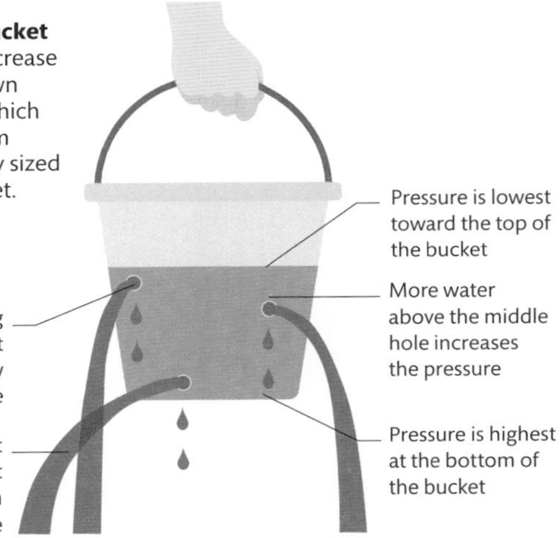
Pressure is lowest toward the top of the bucket
More water above the middle hole increases the pressure
Pressure is highest at the bottom of the bucket
Water leaking from the highest hole is under low pressure
Water spurts at the highest pressure from the lowest hole

HYDRAULICS

The near-incompressibility of liquids enables them to transfer pressure through pipe networks to operate machines. By using a pipe to link a pumping cylinder to a lifting mechanism with twice the area, the force exerted will double, although the pressure stays the same.

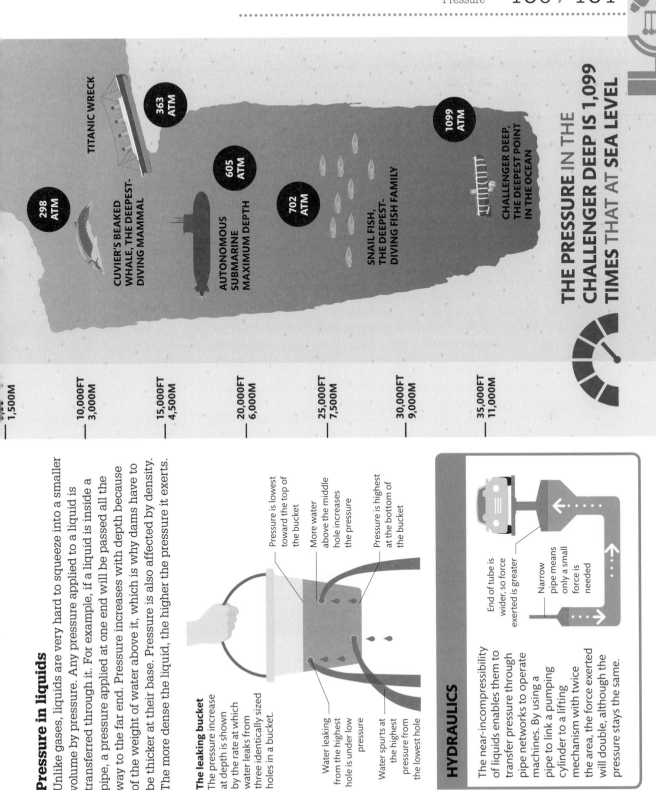

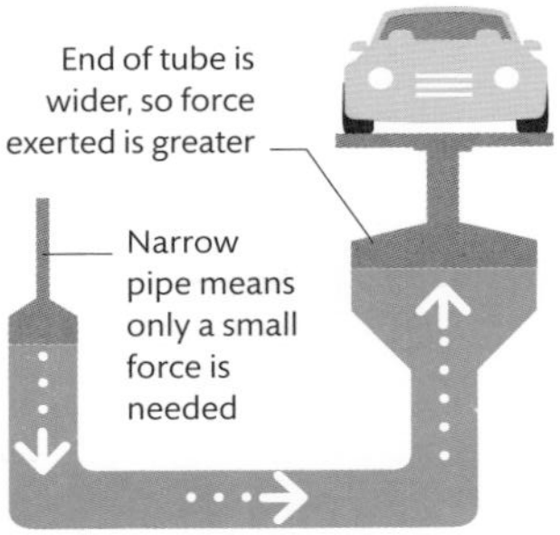
End of tube is wider, so force exerted is greater
Narrow pipe means only a small force is needed

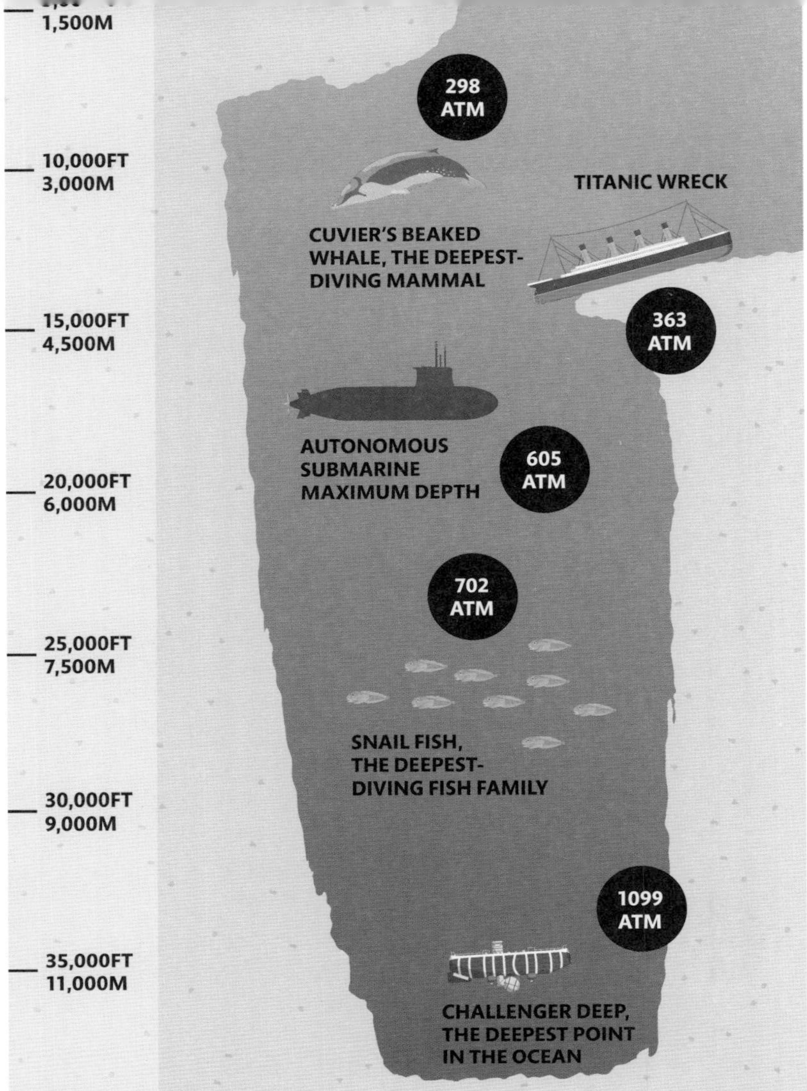
1,500M
10,000FT
3,000M
15,000FT
4,500M
20,000FT
6,000M
25,000FT
7,500M
30,000FT
9,000M
35,000FT
11,000M
298 ATM
CUVIER'S BEAKED WHALE, THE DEEPEST-DIVING MAMMAL
TITANIC WRECK
363 ATM
AUTONOMOUS SUBMARINE MAXIMUM DEPTH
605 ATM
702 ATM
SNAIL FISH, THE DEEPEST-DIVING FISH FAMILY
1099 ATM
CHALLENGER DEEP, THE DEEPEST POINT IN THE OCEAN

THE PRESSURE IN THE **CHALLENGER DEEP IS 1,099 TIMES** THAT AT **SEA LEVEL**

Flight

The technology of flight makes use of two very different principles. Balloons and airships rely on the fact that hot air and gases, such as hydrogen and helium, float upward. All other aircraft depend on the generation of lift using wings and rotors.

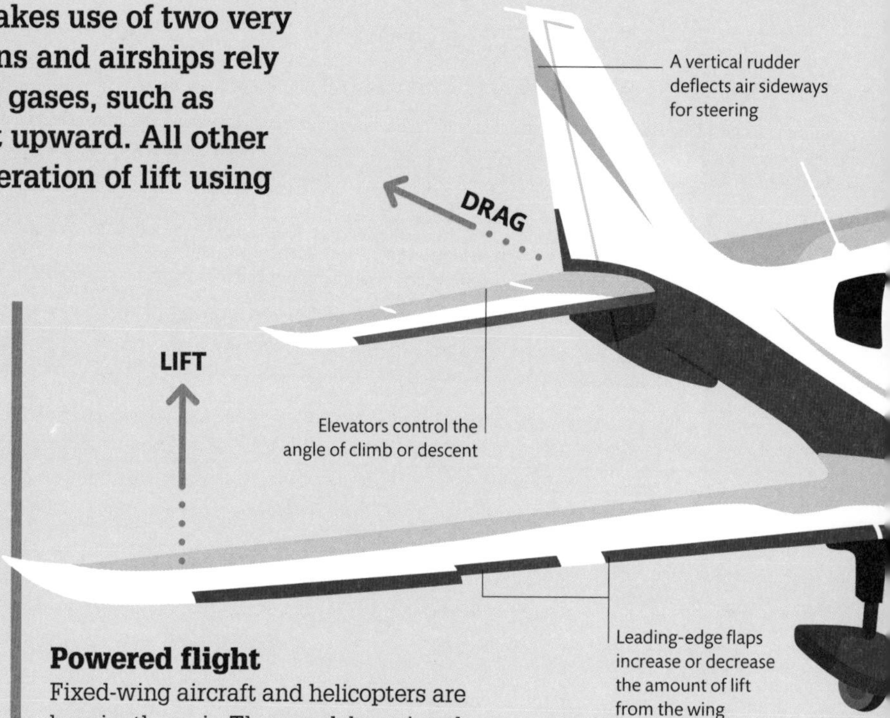

A vertical rudder deflects air sideways for steering
DRAG
LIFT
Elevators control the angle of climb or descent
Leading-edge flaps increase or decrease the amount of lift from the wing

Lighter than air

A regular balloon rises into the sky because it is filled with a gas that is lighter than the air outside it. Most manned balloons achieve this by heating air to make it expand; this makes the air less dense and therefore lighter than cool air. Airships usually contain hydrogen or helium. Helium is also used to inflate party balloons. Hydrogen is twice as light as helium but is dangerously flammable, whereas helium is nonflammable.

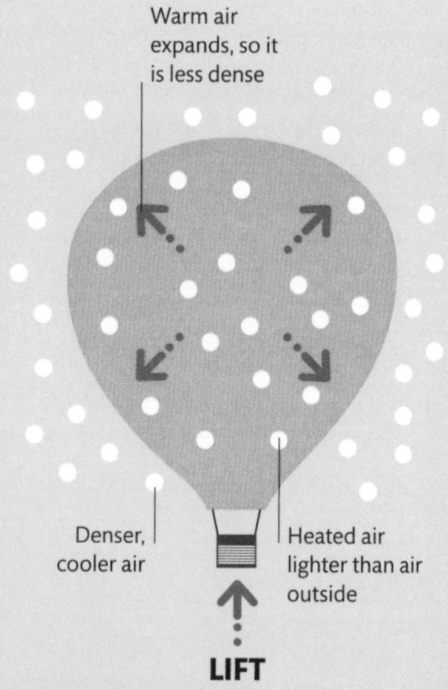

Warm air expands, so it is less dense
Denser, cooler air
Heated air lighter than air outside
LIFT

Lift in a hot air balloon
When air is heated, its molecules move apart, so it expands. Since there are now fewer gas molecules occupying the same volume, the air inside the balloon is less dense.

Powered flight

Fixed-wing aircraft and helicopters are heavier than air. They work by using the motion of their specially profiled wings or rotors to deflect the air, reducing the pressure above them. The angle between the wings and the oncoming air—the angle of attack—is critical. For takeoff, flaps on the plane's wings are extended to increase the angle of attack and curvature of the wing, providing the maximum lift possible.

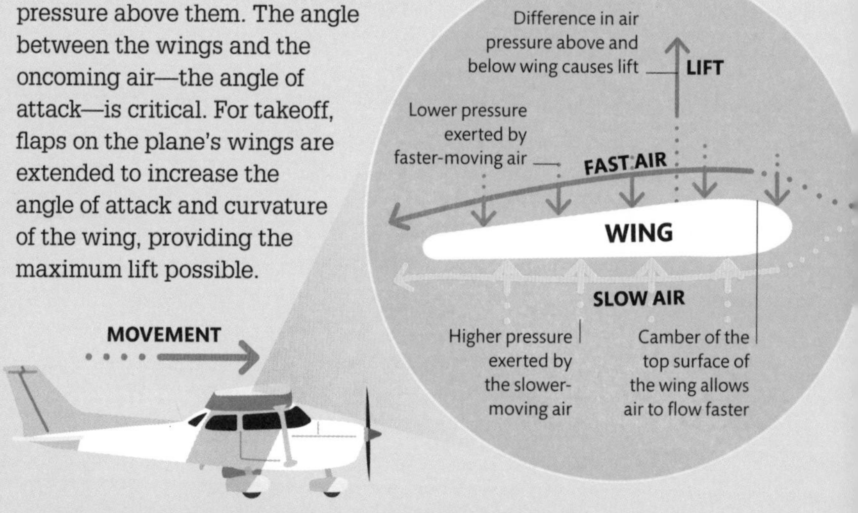

MOVEMENT
Difference in air pressure above and below wing causes lift
LIFT
Lower pressure exerted by faster-moving air
FAST AIR
WING
SLOW AIR
Higher pressure exerted by the slower-moving air
Camber of the top surface of the wing allows air to flow faster

1 Preparing for takeoff
An aircraft relies on forward motion to force air over its wings and create lift for takeoff. It uses a powerful engine to accelerate, while adjustable flaps increase lift at low speed.

2 The Bernoulli effect
Pressure varies depending on the motion of a medium—this is called the Bernoulli effect. The top surface of the wing has a longer curve than the bottom, so the air flows faster over it. This reduces pressure above the wing, creating lift.

APPROXIMATELY **9,250 PASSENGER PLANES** ARE IN THE AIR AT **ANY GIVEN MOMENT**

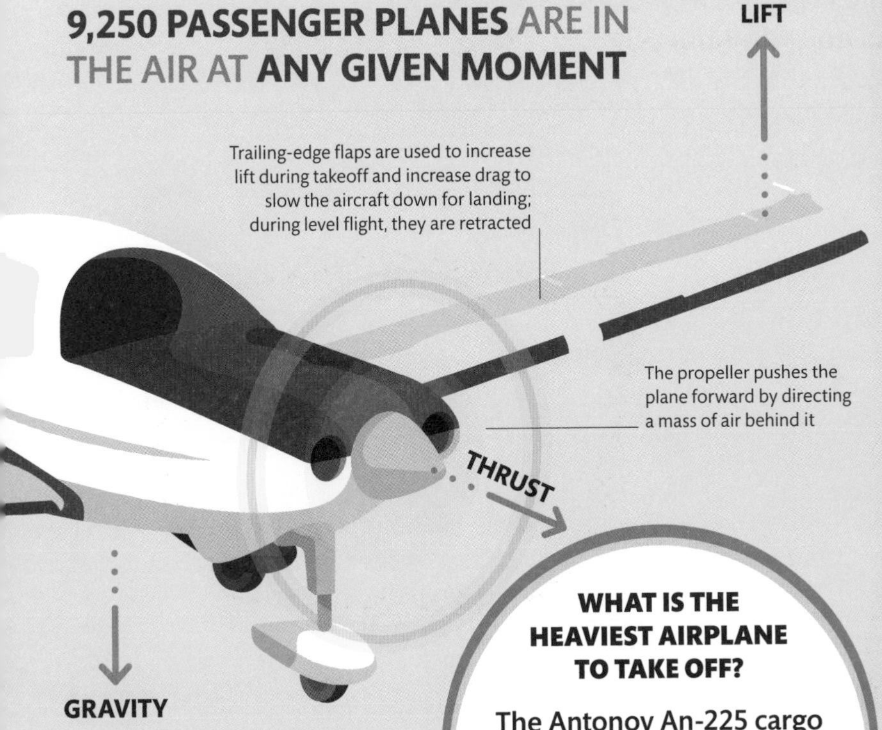
LIFT
Trailing-edge flaps are used to increase lift during takeoff and increase drag to slow the aircraft down for landing; during level flight, they are retracted
The propeller pushes the plane forward by directing a mass of air behind it
THRUST
GRAVITY

WHAT IS THE HEAVIEST AIRPLANE TO TAKE OFF?

The Antonov An-225 cargo aircraft was produced in 1985. It has a maximum weight of 705 tons (640 tonnes) and is powered by six turbofan engines.

3 Level flight
Lift provided by the wings offsets the force of gravity, but only if the thrust from the engine keeps the aircraft moving forward fast enough. This thrust must also overcome drag created by the lift forces.

HOW A HELICOPTER GENERATES LIFT

A helicopter's fast-moving rotor blades generate the lift that keeps it in the air. Moving the cyclic control forward alters the angle of the rotors, propelling the helicopter through the air.

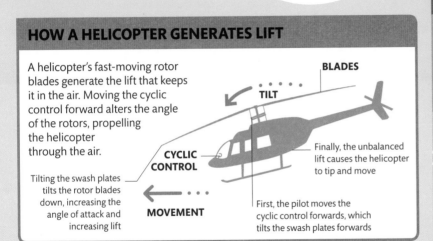

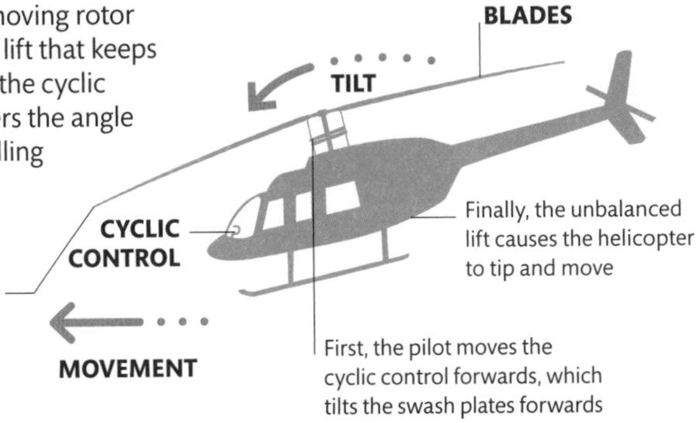
BLADES
TILT
CYCLIC CONTROL
Finally, the unbalanced lift causes the helicopter to tip and move
Tilting the swash plates tilts the rotor blades down, increasing the angle of attack and increasing lift
MOVEMENT
First, the pilot moves the cyclic control forwards, which tilts the swash plates forwards

Kármán line

Air density decreases with altitude. This reduces drag, allowing aircraft to fly faster, but it also forces them to fly faster to generate lift. Above an altitude of 62 miles (100km), called the Kármán line, air-supported flight is not possible. The Kármán line is regarded as the boundary between Earth's atmosphere and space.

THERMOSPHERE
50–370 MILES (80–600KM)

Into orbit
To keep flying above the Kármán line, an object must move at orbital velocity—the speed at which centrifugal force offsets gravity (see pp.214–15).

18,000mph (29,000kph)

KÁRMÁN LINE 62 miles (100km)

Speed at which an aircraft would need to fly to stay aloft

MESOSPHERE
30–50 miles (50–80km)

STRATOSPHERE
10–30 miles (16–50km)

Speed at which a commercial airliner needs to fly to stay aloft at an altitude of 7.5 miles (12km)

TROPOSPHERE
0–10 miles (0–16km)

560mph (900kph)

How buoyancy works

Buoyancy is an upward force exerted by liquids and gases on solids. However, buoyancy works in balance with density. If an object is too dense, buoyancy is not enough to stop that sinking feeling.

What is buoyancy?

When placed in a fluid—either a liquid or gas—an object will push aside, or displace, a volume of fluid equal to its own volume. If the object is denser than the fluid, the volume displaced will weigh less than the object itself, so the object will sink. But an object that's less dense than the fluid will float because the buoyancy force will balance its weight.

WHY DID ARCHIMEDES SHOUT "EUREKA"?

He discovered that objects heavier than water displace their volume of water—useful for measuring the volume of objects of any shape.

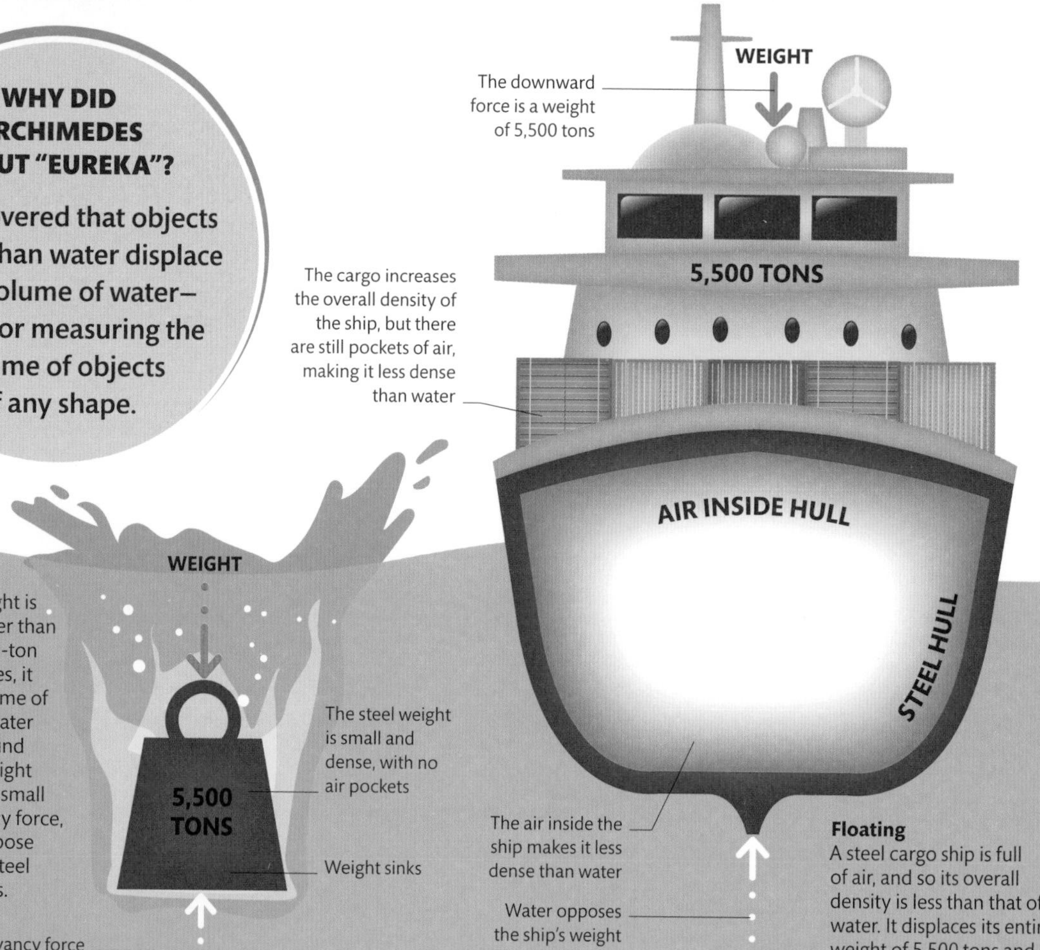

WEIGHT
The downward force is a weight of 5,500 tons
5,500 TONS
The cargo increases the overall density of the ship, but there are still pockets of air, making it less dense than water
AIR INSIDE HULL
STEEL HULL
The air inside the ship makes it less dense than water
Water opposes the ship's weight with an equal buoyancy force of 5,500 tons
BUOYANCY
WEIGHT
5,500 TONS
The steel weight is small and dense, with no air pockets
Weight sinks
The buoyancy force is 690 tons—not enough to stop the weight sinking
BUOYANCY

Sinking
A solid steel weight is eight times denser than water. As a 5,500-ton weight submerges, it displaces its volume of water, but that water weighs only around 690 tons. The weight of water exerts a small upward buoyancy force, but it cannot oppose the force of the steel weight, so it sinks.

Floating
A steel cargo ship is full of air, and so its overall density is less than that of water. It displaces its entire weight of 5,500 tons and remains afloat, buoyed up by the upward force of 5,500 tons of ocean.

SWIM BLADDER

Like submarines, some fish rise in water by releasing gas dissolved in their blood, via gas glands, into a swim bladder. This increases the volume of the bladder, making the fish less dense and causing it to rise. To sink, the gas is dissolved in the blood again, making the bladder shrink.

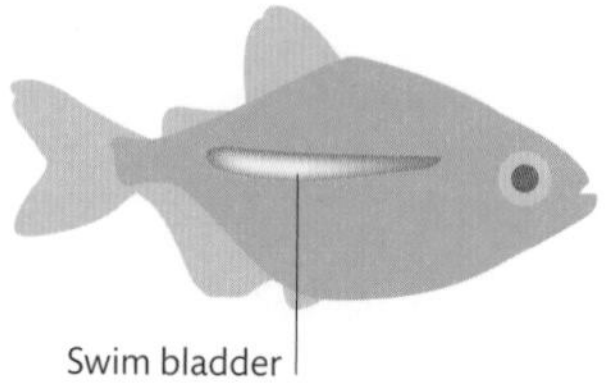

Swim bladder

Weight and density

As a ship is loaded, its air spaces are filled with heavier-than-air cargo and so its overall density increases. Each time a container is loaded, the ship sits lower in the water, because its greater weight pushes more water aside until it finds a new equilibrium between weight and buoyancy. The waterline for the highest safe load—a Plimsoll line—is painted on the ship's hull.

ALL FLOATING OBJECTS DISPLACE AN EQUAL VOLUME OF WATER

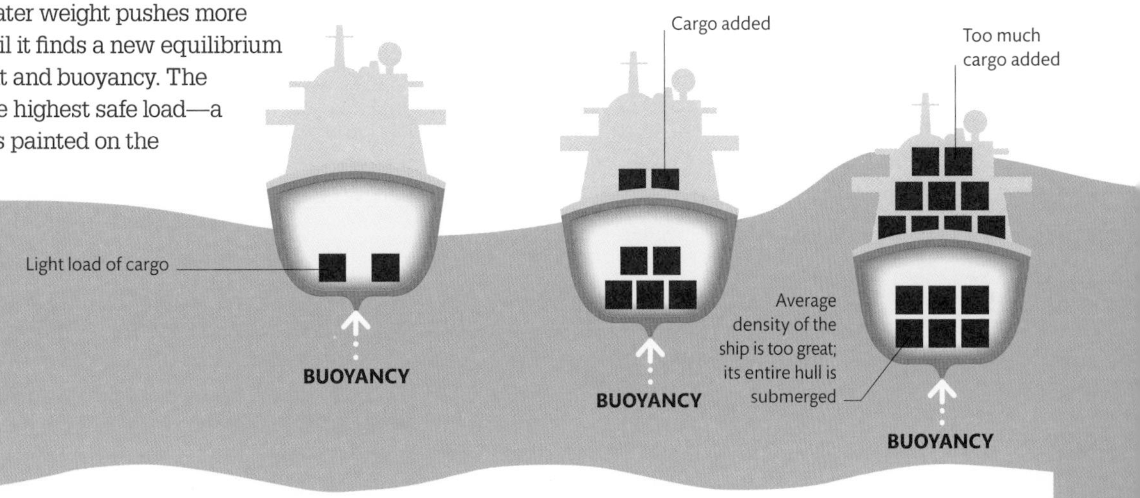

Submarines

To dive and surface at will, submarines manipulate their average density using tanks of compressed air. As long as they have a power source, they can do this indefinitely because, while at the surface, they can pump in fresh air from the atmosphere and compress it into their tanks ready for their next surfacing.

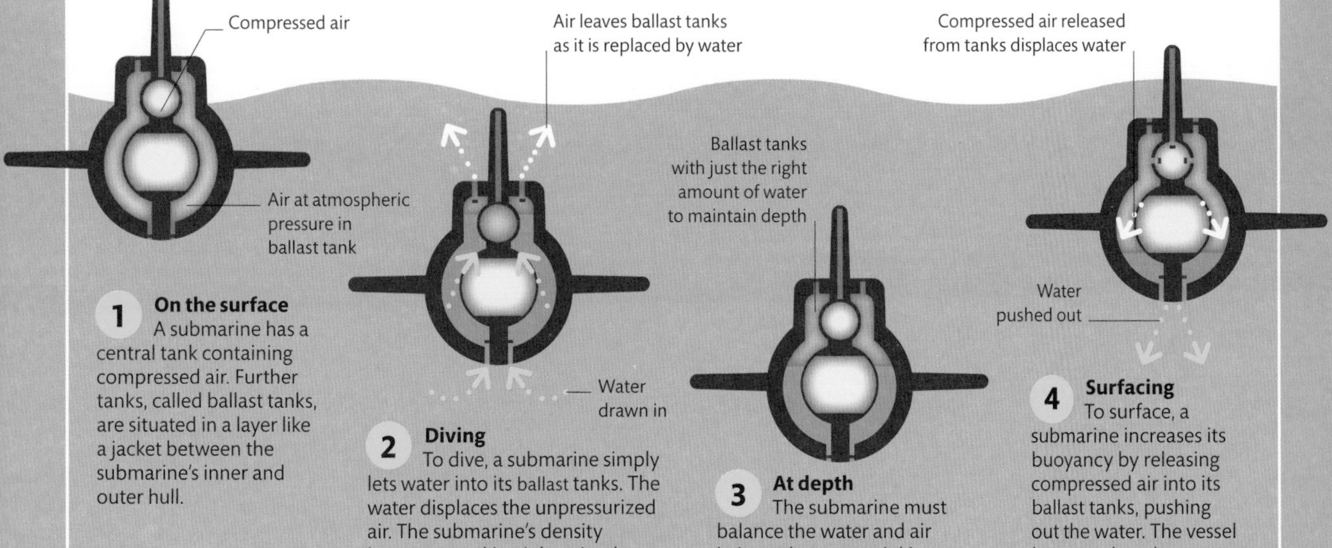

1 On the surface A submarine has a central tank containing compressed air. Further tanks, called ballast tanks, are situated in a layer like a jacket between the submarine's inner and outer hull.

2 Diving To dive, a submarine simply lets water into its ballast tanks. The water displaces the unpressurized air. The submarine's density increases, making it heavier than water, so it sinks.

3 At depth The submarine must balance the water and air in its tanks to stop sinking farther than it wants to.

4 Surfacing To surface, a submarine increases its buoyancy by releasing compressed air into its ballast tanks, pushing out the water. The vessel becomes less dense than water, and rises.

Vacuums

A perfect vacuum is a region of empty space that contains no material of any kind. This has never been observed in practice—even outer space contains some matter, which exerts a measurable pressure—so vacuums in the real world are called partial vacuums.

HOW A VACUUM FLASK WORKS

A vacuum flask uses a vacuum to stop warm liquids cooling and prevent cold ones warming up. The liquid is in a chamber surrounded by a vacuum, which blocks convection currents that might transfer heat to the outside. The vessel is silvered to both reflect heat back inside and reflect it away from outside.

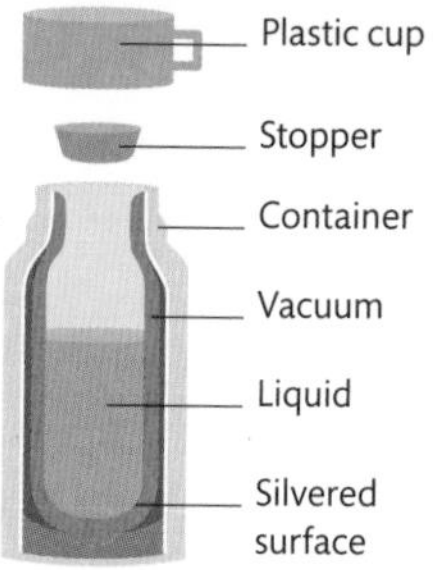

What is a vacuum?

In the 17th century, it became possible to create a vacuum using a pump to suck the air out of a vessel. Experiments showed that a flame went out and that sound could not pass through it, because sound needs a medium, such as air, to travel through. Light does not need a medium and can pass through a vacuum.

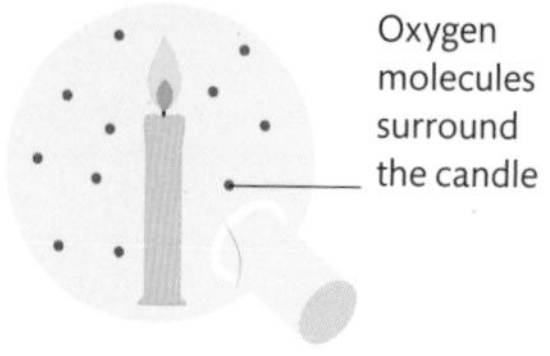

Flame in air
A candle burns inside a vessel that is filled with air. The oxygen in the air reacts with the wax to release heat and light.

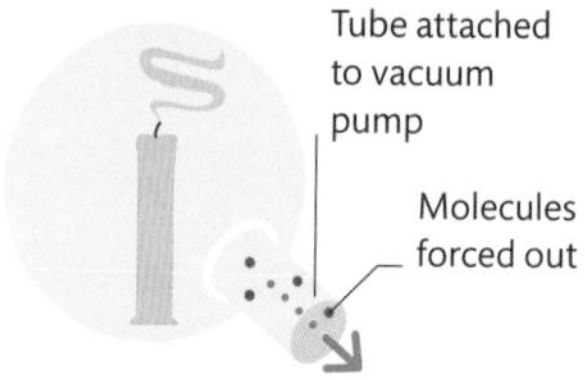

Flame extinguished
Sucking out the air to make a vacuum results in the flame going out. This is because combustion requires oxygen.

Environment	Pressure (pascals)	Molecules per cubic centimeter
Standard atmosphere	101,325	2.5×10^{19}
Vacuum cleaner	approx. 80,000	1×10^{19}
Earth's thermosphere	1–0.0000007	10^{7}–10^{14}
The Moon's surface	1–0.000000009	400,000
Interplanetary space		11
Intergalactic space		0.000006

Inside a vacuum

Material will always spread out to fill any empty spaces. This process is what creates the "suck" in a vacuum cleaner, because air from outside rushes into the vacuum created inside. The molecules in materials placed in a vacuum, especially those of liquids, will also break free of their bonds to form a gas that fills the emptiness.

No resistance
Objects falling in a vacuum experience no air resistance, which would otherwise slow their descent. A hammer and a feather fall at different rates through air, but they plummet side by side in a vacuum.

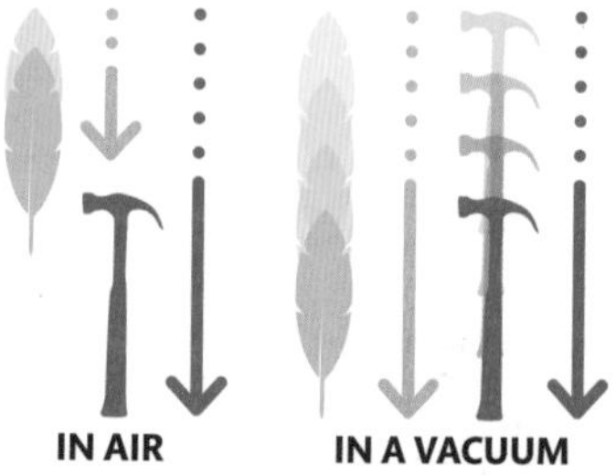

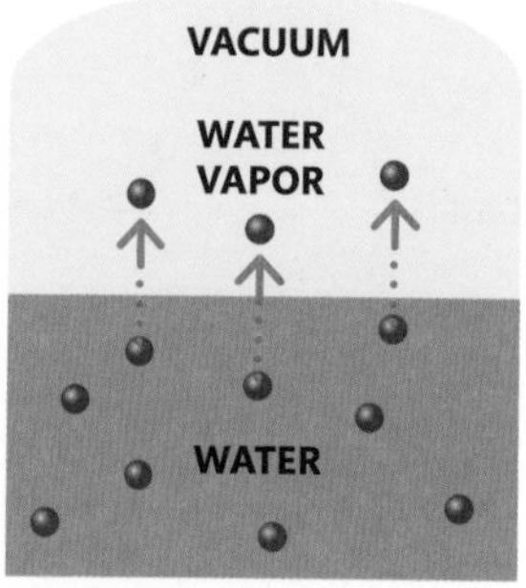

Perfect vacuum
When exposed to a vacuum, water molecules become vapor, filling the space. Very few of them rejoin the liquid.

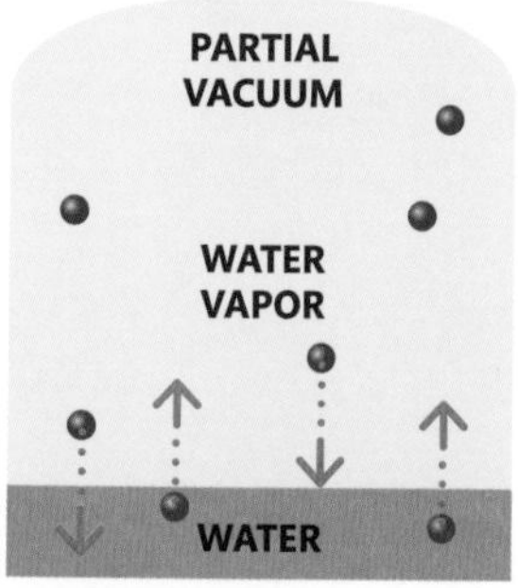

Partial vacuum
The water evaporates, increasing the pressure. The system reaches equilibrium as water molecules move equally in both directions.

Exposure to a vacuum

Outer space is a near-perfect vacuum. Space walkers must wear space suits to protect them from radiation, sunlight, and the cold of empty space, but also to create a pressurized atmosphere around the body. If the suit or visor fails, a quick death is almost certain, but it would not be as dramatic as is often depicted in science fiction.

TARDIGRADES ARE MICROSCOPIC ANIMALS THAT CAN **SURVIVE** IN THE **VACUUM OF SPACE**

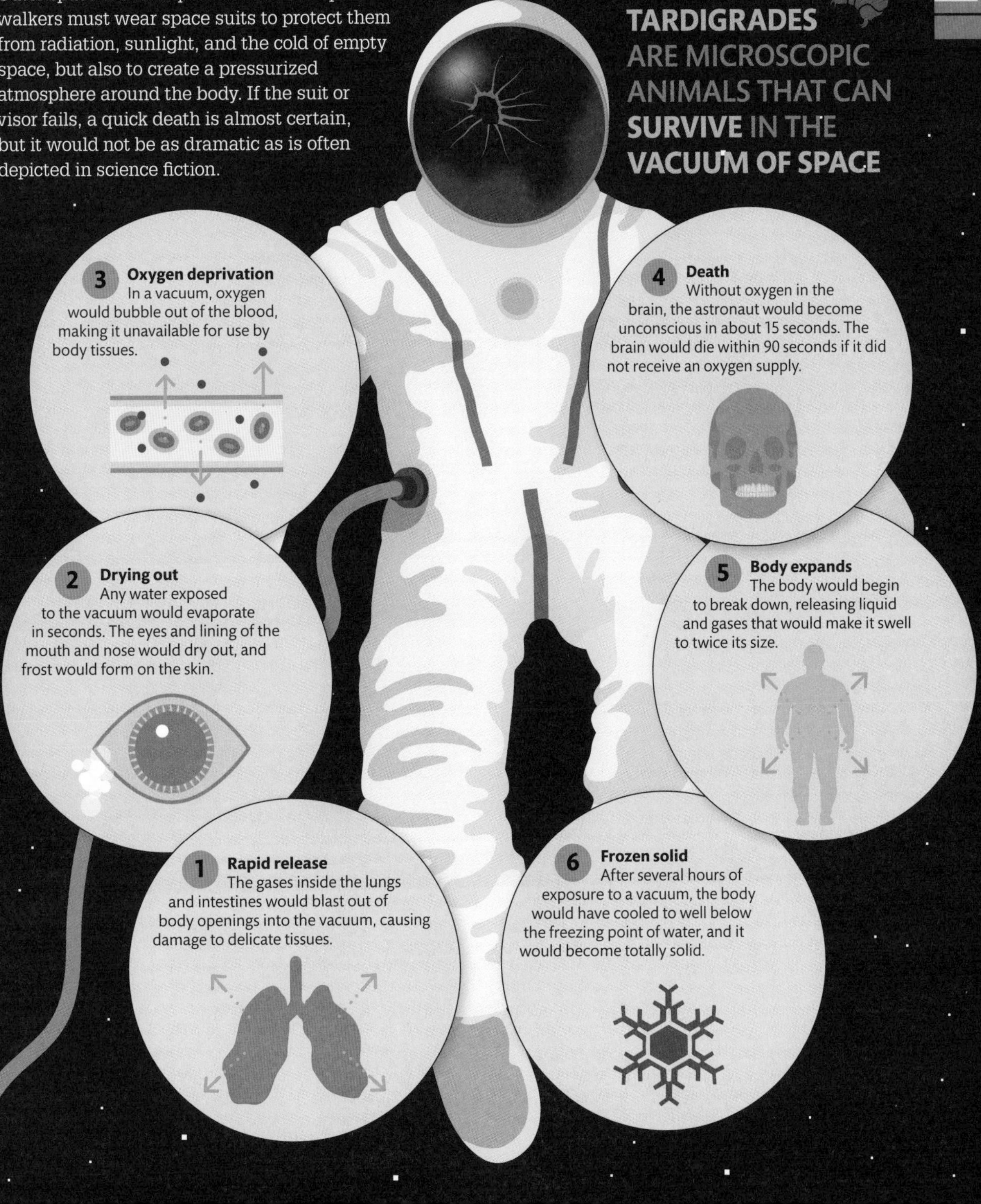

1 Rapid release
The gases inside the lungs and intestines would blast out of body openings into the vacuum, causing damage to delicate tissues.

2 Drying out
Any water exposed to the vacuum would evaporate in seconds. The eyes and lining of the mouth and nose would dry out, and frost would form on the skin.

3 Oxygen deprivation
In a vacuum, oxygen would bubble out of the blood, making it unavailable for use by body tissues.

4 Death
Without oxygen in the brain, the astronaut would become unconscious in about 15 seconds. The brain would die within 90 seconds if it did not receive an oxygen supply.

5 Body expands
The body would begin to break down, releasing liquid and gases that would make it swell to twice its size.

6 Frozen solid
After several hours of exposure to a vacuum, the body would have cooled to well below the freezing point of water, and it would become totally solid.

Gravity

We can think of the gravitational force as a force of attraction. It pulls falling objects to the ground and holds Earth in orbit around the Sun. Isaac Newton first described the gravitational force with mathematics in the 1600s.

Features of gravity

The gravitational force is an attractive force that pulls matter together. As set out in Newton's universal law of gravitation, the size of the attraction depends on two factors: the size of the masses involved and the distance between them. Gravity is the weakest of the four fundamental forces of nature (see p.27). However, huge masses such as stars and galaxies still produce large gravitational forces that act over long distances.

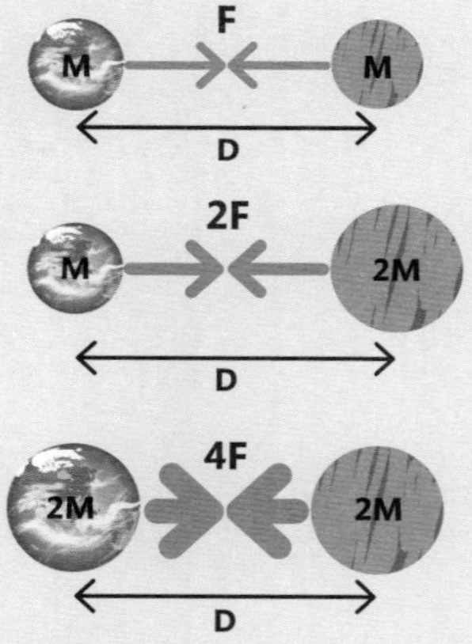

Gravity and mass
Assuming the distance (D) between two objects stays the same, the pull of gravity (F) is directly proportional to their masses (M). Doubling one object's mass (2M) doubles the attraction (2F) between the objects. Doubling both masses increases the pull by a factor of four (4F).

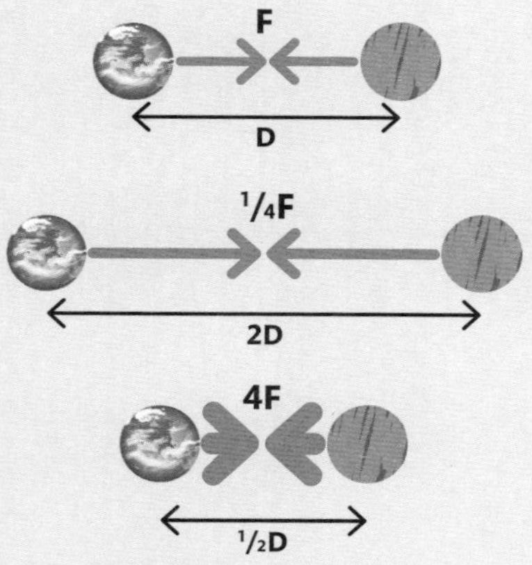

Gravity and distance
The force of gravity is inversely proportional to the square of the distance (D) between two masses, assuming the mass of the objects remains the same. Doubling the distance (2D) decreases the attraction by a factor of four (¼F). Halving the distance (½D) increases the gravity fourfold (4F).

Terminal velocity

Gravity makes falling objects accelerate, speeding up to greater velocities as they approach the ground. However, objects that fall for a long time will reach a maximum, or terminal, velocity. This happens when the downward pull of gravity is matched by the upward push of air resistance.

WHAT IS G-FORCE?

G-force is a change in an object's motion that can make a person feel heavier when they accelerate. The force of gravity when a person stands on the ground is 1g.

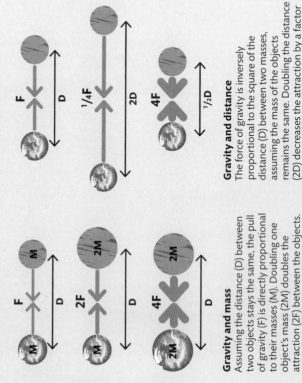

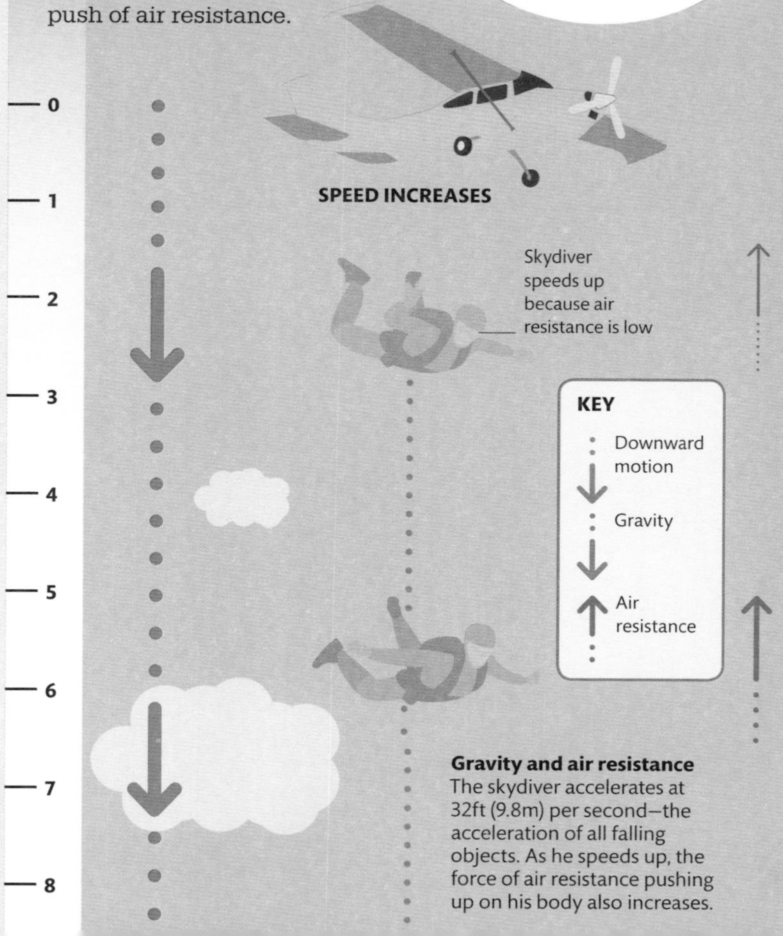

Gravity and air resistance
The skydiver accelerates at 32ft (9.8m) per second—the acceleration of all falling objects. As he speeds up, the force of air resistance pushing up on his body also increases.

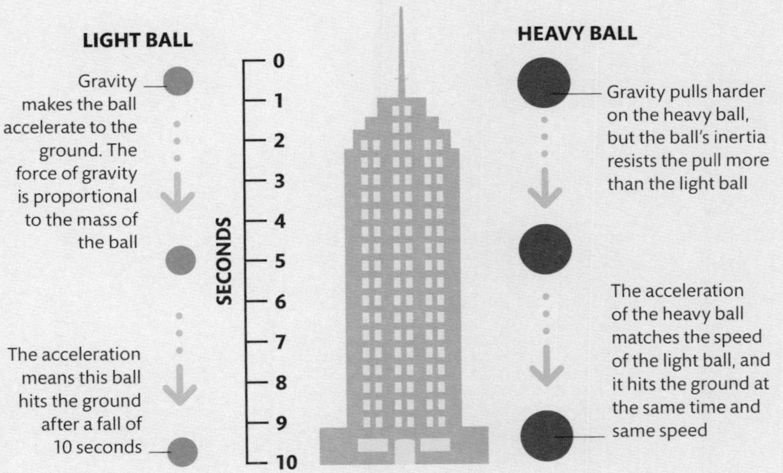
LIGHT BALL
Gravity makes the ball accelerate to the ground. The force of gravity is proportional to the mass of the ball
The acceleration means this ball hits the ground after a fall of 10 seconds
SECONDS
0
1
2
3
4
5
6
7
8
9
10
HEAVY BALL
Gravity pulls harder on the heavy ball, but the ball's inertia resists the pull more than the light ball
The acceleration of the heavy ball matches the speed of the light ball, and it hits the ground at the same time and same speed

Mass and weight

When describing objects on Earth, mass and weight are interchangeable terms. However, mass is an independent measure of the material and energy in an object, whereas weight is the force applied to a mass by Earth's gravitational field.

Landing simultaneously
A light and a heavy ball fall (accelerate) at the same rate. The gravitational force on the heavy ball is greater, because this force is proportional to mass. This larger force causes the heavy ball to accelerate as quickly as the light ball.

WHAT WE WEIGH ON OTHER PLANETS

While an object's mass stays the same, its weight depends on the pull of gravity, which varies from planet to planet. On smaller planets, a person would weigh less than they do on Earth, but they would weigh far more on larger planets, such as Jupiter.

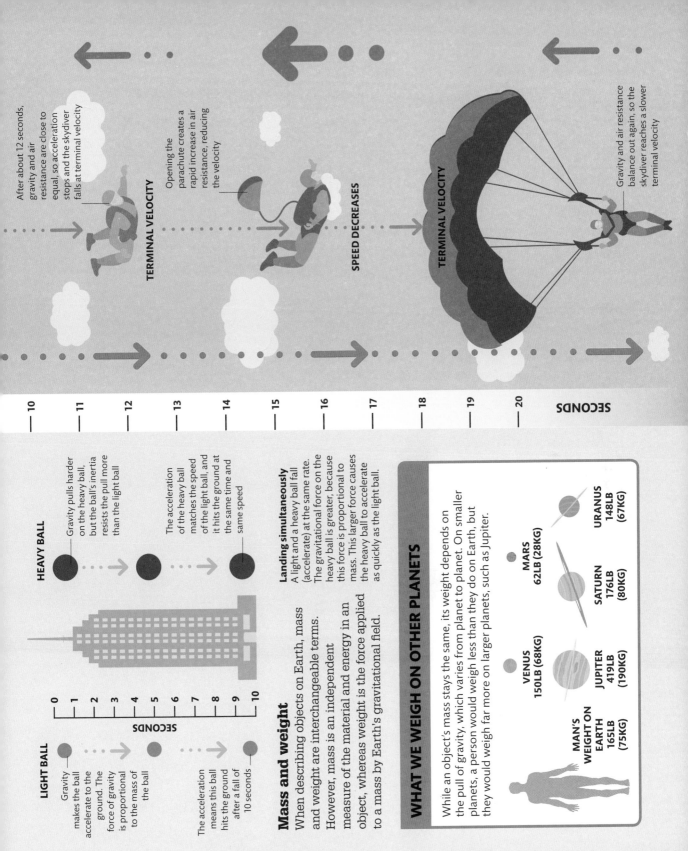

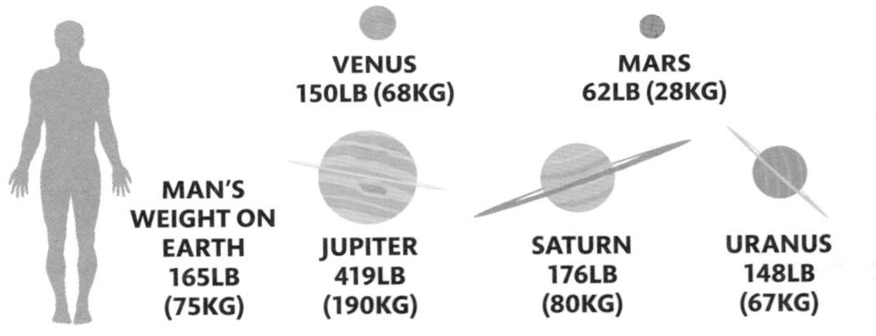
MAN'S WEIGHT ON EARTH 165LB (75KG)
VENUS 150LB (68KG)
JUPITER 419LB (190KG)
MARS 62LB (28KG)
SATURN 176LB (80KG)
URANUS 148LB (67KG)

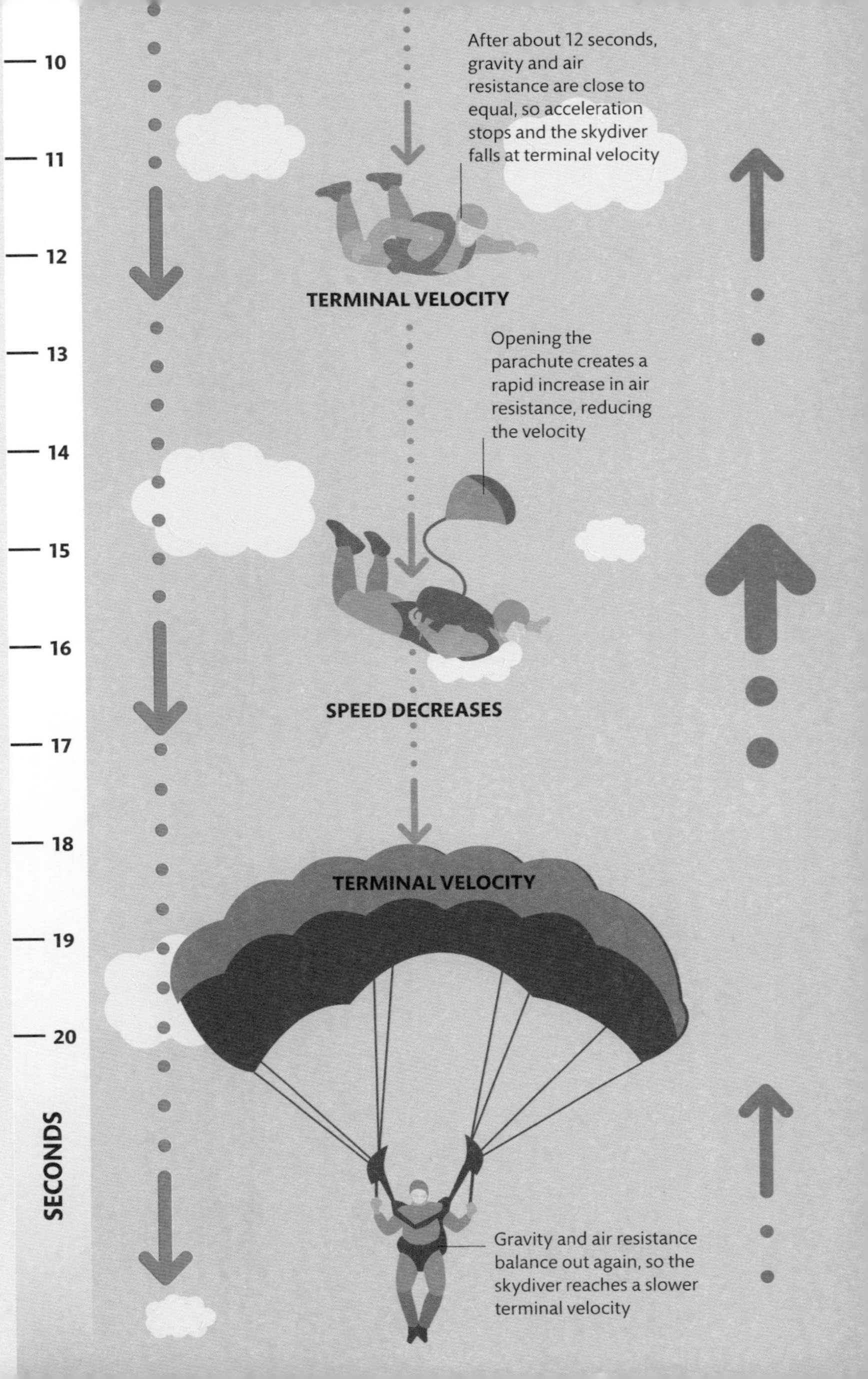
10
11
12
13
14
15
16
17
18
19
20
SECONDS
After about 12 seconds, gravity and air resistance are close to equal, so acceleration stops and the skydiver falls at terminal velocity
TERMINAL VELOCITY
Opening the parachute creates a rapid increase in air resistance, reducing the velocity
SPEED DECREASES
TERMINAL VELOCITY
Gravity and air resistance balance out again, so the skydiver reaches a slower terminal velocity

Special relativity

In 1905, Albert Einstein proposed a revolutionary way of understanding how motion, space, and time work together. He called it the special theory of relativity, and its purpose was to solve the biggest problem in physics at the time—the contradiction between the different ways that light and objects move through space.

Contradictory laws

The laws of motion say that every object's velocity is relative to the motion of other objects. However, according to the rules of electromagnetism, light travels at a fixed speed. Light always reaches an observer at that speed, whether the light's source is stationary or moving toward or away from them.

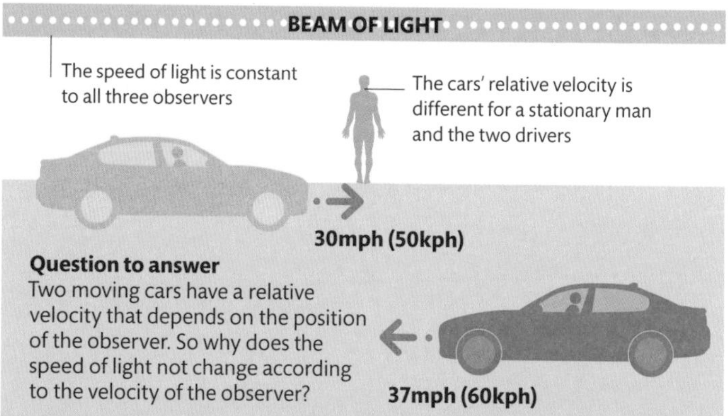
BEAM OF LIGHT
The speed of light is constant to all three observers
The cars' relative velocity is different for a stationary man and the two drivers
30mph (50kph)
37mph (60kph)

Question to answer
Two moving cars have a relative velocity that depends on the position of the observer. So why does the speed of light not change according to the velocity of the observer?

LENGTH CONTRACTION

As well as slowing in time, the space around a moving object contracts. It is not possible to measure the contraction because measuring devices also contract by the same amount. As the object approaches the speed of light, space is so contracted from an observer's point of view, and time so dilated, that the object appears to stop moving altogether.

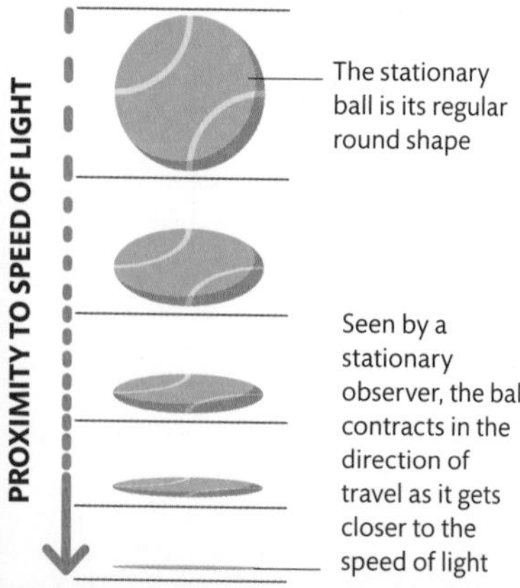
PROXIMITY TO SPEED OF LIGHT
The stationary ball is its regular round shape
Seen by a stationary observer, the ball contracts in the direction of travel as it gets closer to the speed of light

Time dilation

Einstein explained the contradiction between the velocity of light and that of other objects by theorizing that as an object moves faster through space, it also moves more slowly through time. This means that time passes at different speeds for observers moving at different velocities. For a stationary observer, time moves faster than it does for an observer moving at close to the speed of light.

Explaining light's constant speed
Inside a spaceship moving at close to light speed, an astronaut using a clock to measure the speed of light finds that light travels a relatively short distance in a short time. For a stationary observer, light moves a longer distance in a longer time. But both observers measure light moving at the same speed.

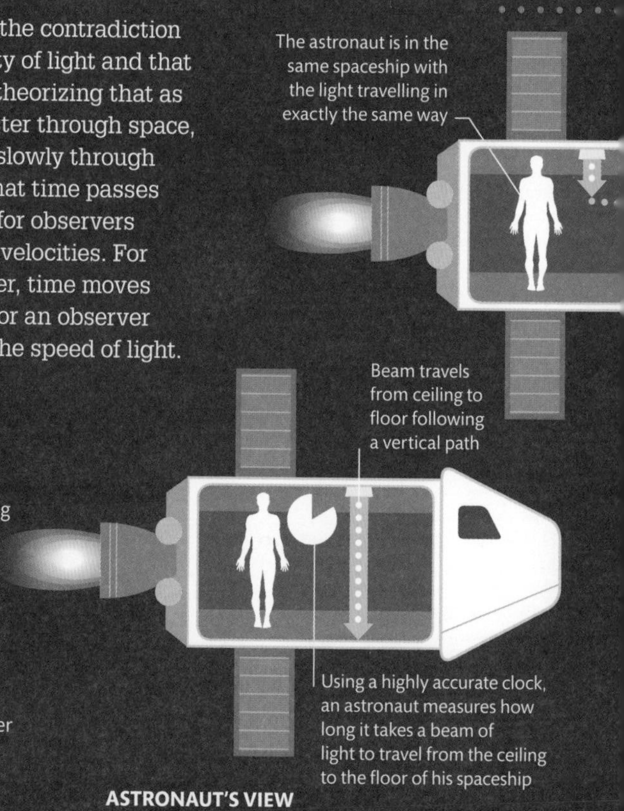
The astronaut is in the same spaceship with the light travelling in exactly the same way
Beam travels from ceiling to floor following a vertical path
Using a highly accurate clock, an astronaut measures how long it takes a beam of light to travel from the ceiling to the floor of his spaceship
ASTRONAUT'S VIEW

Mass and energy

As Einstein pondered how light always traveled at a fixed speed, he also examined the nature of mass and energy. He realized that mass and energy are equivalent and related the two properties using the famous equation $E=mc^2$ where E stands for energy, m for mass, and c the speed of light. Adding energy to a stationary object can make it move. Since energy and mass are equivalent, the motion makes the object act as if it is heavier than when it is stationary. At low speeds, this effect is negligible, but close to the speed of light, an object's mass approaches infinity.

$$E = mc^2$$

The amount of energy locked into matter in the form of mass is enormous—during nuclear explosions, small amounts of mass are converted into huge amounts of both heat and light

Mass is a property of matter that describes its resistance to changes in motion; the greater the mass, the more energy that can potentially be released

Light is carried by massless particles, and so it travels at the fastest speed possible—the speed of light

WHEN WAS THE EXPRESSION "SPECIAL THEORY OF RELATIVITY" FIRST USED?

Einstein only described it this way 10 years after it was published, to distinguish it from his general theory. The paper was originally called *On the Electrodynamics of Moving Bodies.*

TRAVELING AT CLOSE TO THE SPEED OF LIGHT

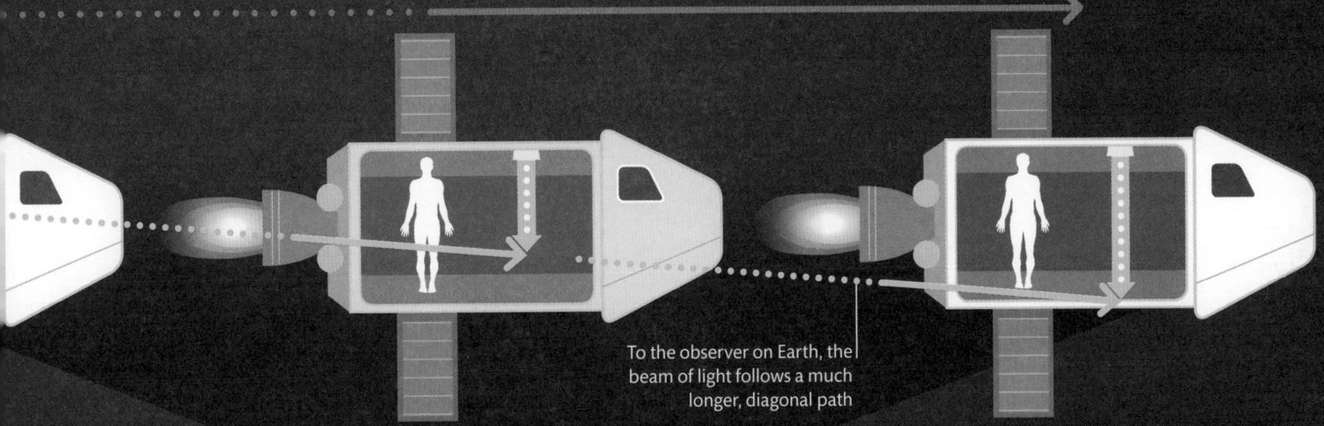

To the observer on Earth, the beam of light follows a much longer, diagonal path

Perception of motion from outside
The activity inside the speeding spaceship appears different to an observer watching from a different reference frame, such as Earth's surface. To the observer on Earth, the beam of light traces a diagonal instead of vertical line.

OBSERVER'S VIEW FROM EARTH

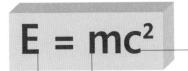

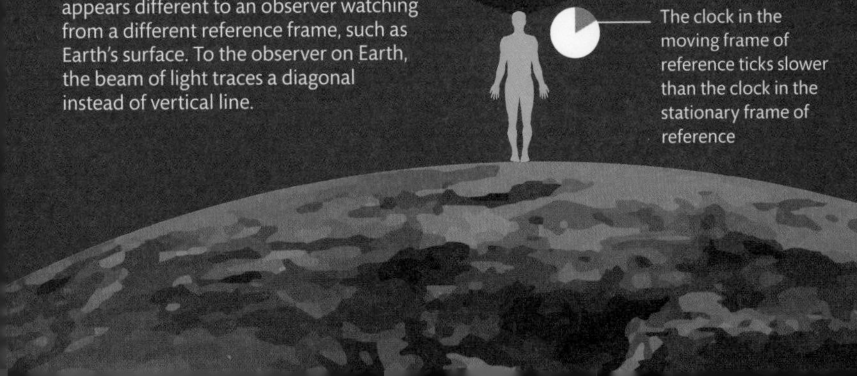

The clock in the moving frame of reference ticks slower than the clock in the stationary frame of reference

EARLY MEASUREMENTS OF THE **SPEED OF LIGHT,** MADE IN **THE 1600S,** WERE ABOUT **26 PER CENT** TOO LOW

General relativity

Gravity, as described by Isaac Newton in 1687, appeared to be incompatible with Albert Einstein's special theory of relativity. So, in 1916, Einstein unified gravity and his relativistic ideas of space and time in his general theory of relativity.

HOW WAS THE THEORY PROVED?

In 1919, astronomer Arthur Eddington observed deflected starlight during a total solar eclipse. This demonstrated the effects of warped space-time and made Einstein world-famous.

Space-time

Special relativity describes how objects experience space and time differently, depending on their motion. An important implication of special relativity is that space and time are always linked. General relativity describes them in a four-dimensional continuum called space-time, which is warped by massive objects. Mass and energy are equivalent to each other, and the warping they cause in space-time creates the effects of gravity, such as the Moon orbiting Earth.

GENERAL RELATIVITY EXPLAINS THE MOTION OF THE PLANETS AROUND THE SUN

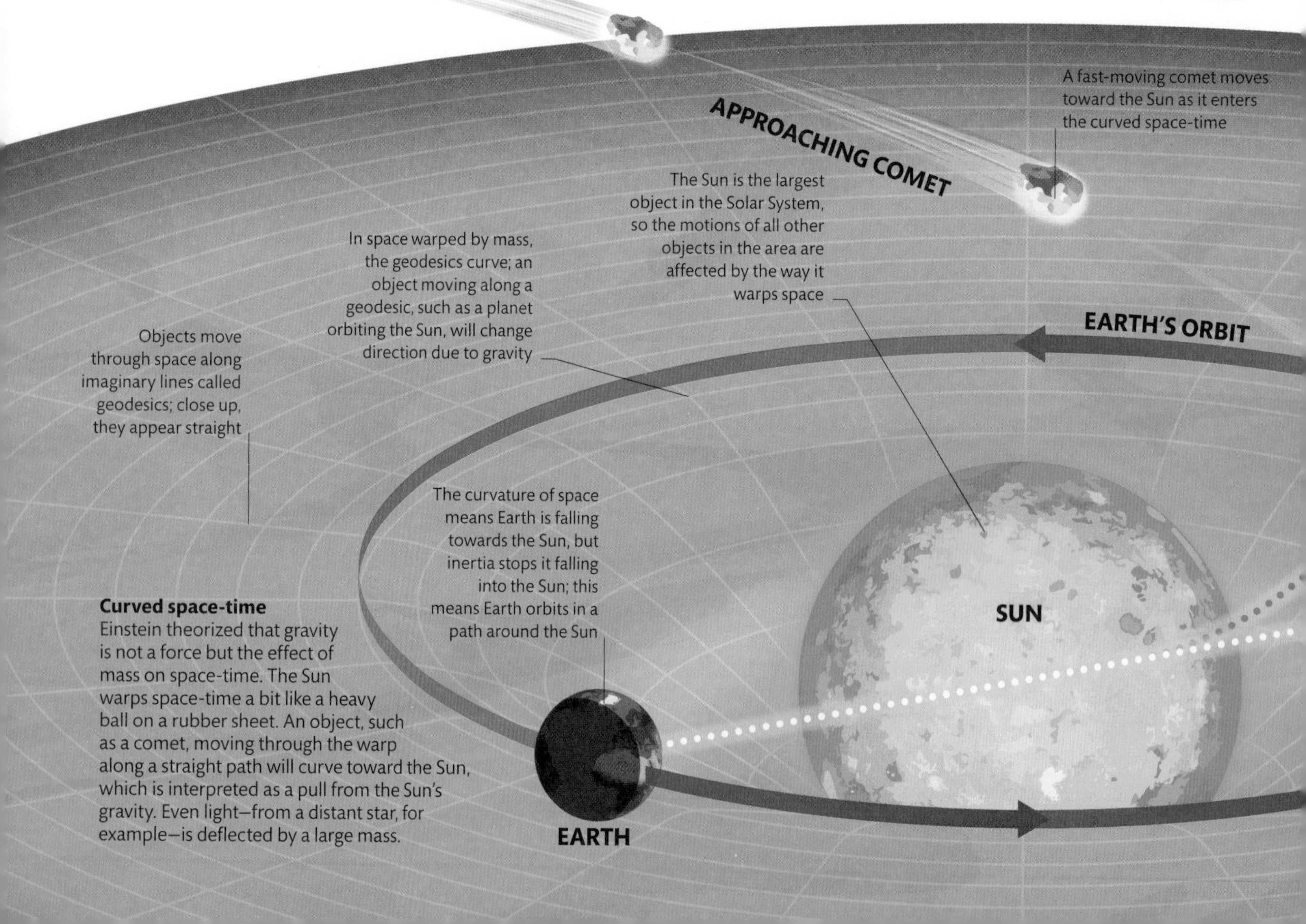

A fast-moving comet moves toward the Sun as it enters the curved space-time
APPROACHING COMET
The Sun is the largest object in the Solar System, so the motions of all other objects in the area are affected by the way it warps space
In space warped by mass, the geodesics curve; an object moving along a geodesic, such as a planet orbiting the Sun, will change direction due to gravity
EARTH'S ORBIT
Objects move through space along imaginary lines called geodesics; close up, they appear straight
The curvature of space means Earth is falling towards the Sun, but inertia stops it falling into the Sun; this means Earth orbits in a path around the Sun
SUN
EARTH

Curved space-time
Einstein theorized that gravity is not a force but the effect of mass on space-time. The Sun warps space-time a bit like a heavy ball on a rubber sheet. An object, such as a comet, moving through the warp along a straight path will curve toward the Sun, which is interpreted as a pull from the Sun's gravity. Even light—from a distant star, for example—is deflected by a large mass.

The equivalence principle

To understand gravity, Einstein imagined himself in an elevator and asked himself whether the force keeping him on the floor was the pull of gravity or the effect of inertia as the elevator moved upward. From inside, there is no way of telling. This is called the equivalence principle. From this idea, he began to think of himself as an observer in a still frame of reference watching the Universe move around him.

Einstein's elevator experiment
Einstein extended his elevator thought experiment by imagining what a beam of light would look like for a person inside the elevator in three different scenarios. The person inside is not able to describe the motion of the elevator fully but can see the behavior of the light beam. The experiment reveals that when traveling very fast or being pulled by powerful gravity, space—and the beam of light—becomes curved.

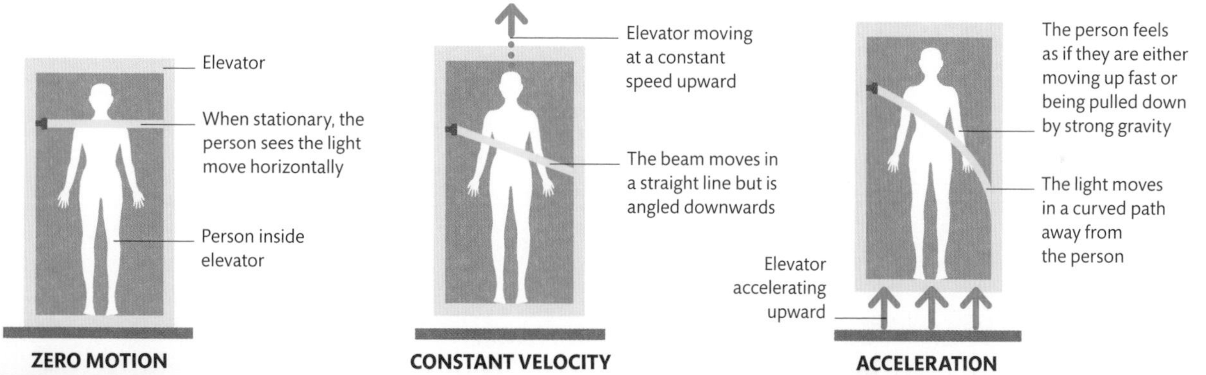
Elevator
When stationary, the person sees the light move horizontally
Person inside elevator
ZERO MOTION
Elevator moving at a constant speed upward
The beam moves in a straight line but is angled downwards
CONSTANT VELOCITY
The person feels as if they are either moving up fast or being pulled down by strong gravity
The light moves in a curved path away from the person
Elevator accelerating upward
ACCELERATION

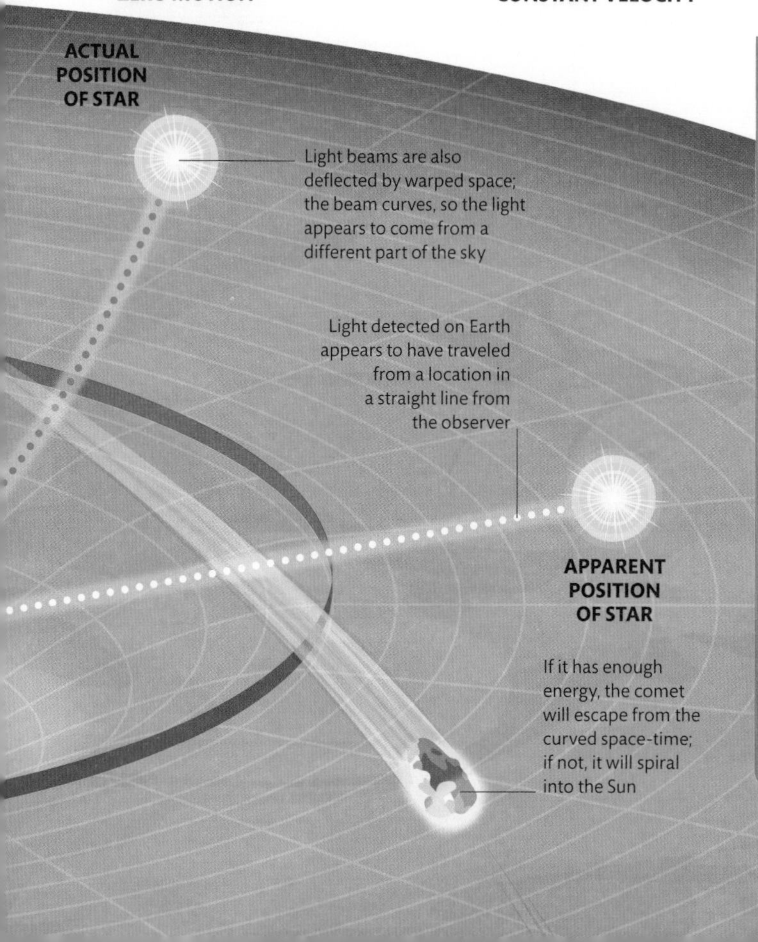
ACTUAL POSITION OF STAR
Light beams are also deflected by warped space; the beam curves, so the light appears to come from a different part of the sky
Light detected on Earth appears to have traveled from a location in a straight line from the observer
APPARENT POSITION OF STAR
If it has enough energy, the comet will escape from the curved space-time; if not, it will spiral into the Sun

GPS NAVIGATION

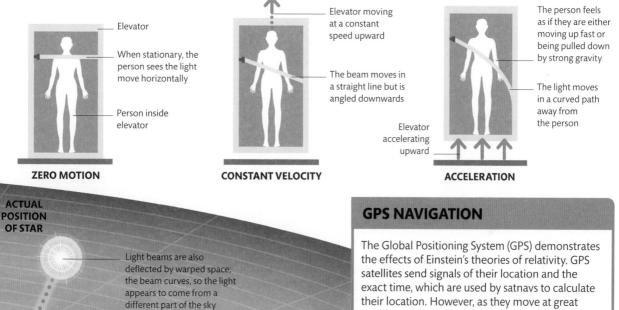

The Global Positioning System (GPS) demonstrates the effects of Einstein's theories of relativity. GPS satellites send signals of their location and the exact time, which are used by satnavs to calculate their location. However, as they move at great speed, the satellites' onboard clocks run more slowly than on Earth, so this relativistic effect has to be accounted for by the satnavs.

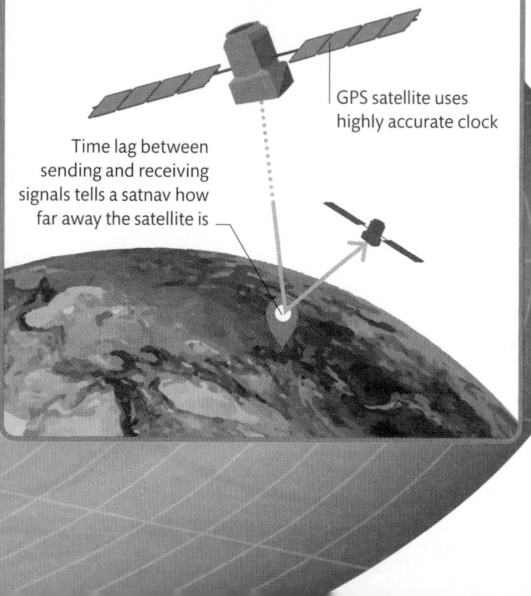
GPS satellite uses highly accurate clock
Time lag between sending and receiving signals tells a satnav how far away the satellite is

Gravitational waves

The theory of general relativity predicted that objects moving through space create ripples in space-time called gravitational waves. In 2015, these waves were detected for the first time.

WORMHOLES

In the 1930s, Albert Einstein and Nathan Rosen described how space-time could be warped so that two distant locations could be linked by a shortcut. Such a bridge, or wormhole, would create shortcuts for long journeys, but no evidence for their existence has been found.

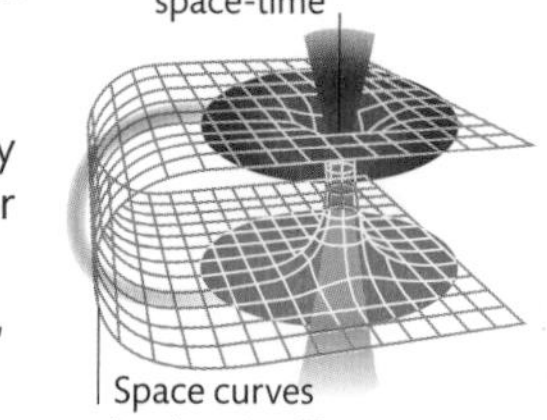

What are they?

Matter accelerating through space in certain ways creates gravitational waves. The largest gravitational events produce low-frequency waves with huge wave periods. For example, waves from the Big Bang are thought to be many millions of light-years long. Gravitational waves offer a way of imaging the Universe that does not rely on light. They may reveal things that are currently invisible to us, like exactly what happens inside a black hole.

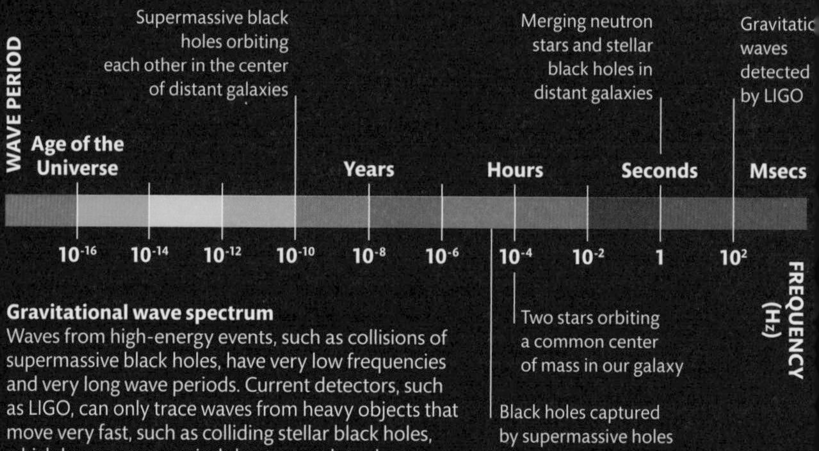

Gravitational wave spectrum
Waves from high-energy events, such as collisions of supermassive black holes, have very low frequencies and very long wave periods. Current detectors, such as LIGO, can only trace waves from heavy objects that move very fast, such as colliding stellar black holes, which have a wave period short enough to detect.

How gravitational waves form

The first gravitational waves detected by LIGO were generated by waves traveling toward Earth from two colliding black holes nearly 1.3 billion light-years away. The black holes were pulled together by their mutual gravitational attraction.

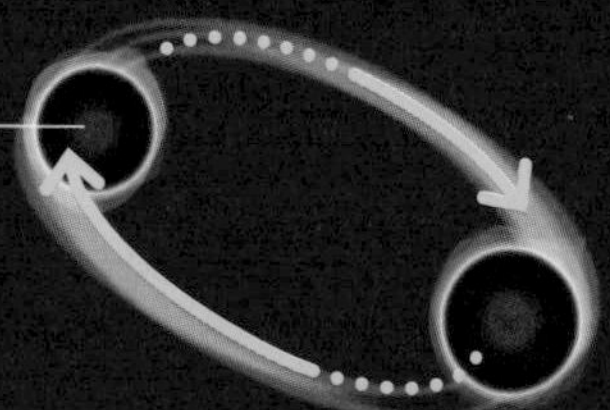

1 Colliding black holes
The two black holes were pulled together by the force of gravity. Regular oscillations detected by LIGO indicated that the orbits of the black holes were near-perfect circles and that they orbited each other at a rate of over 15 times per second.

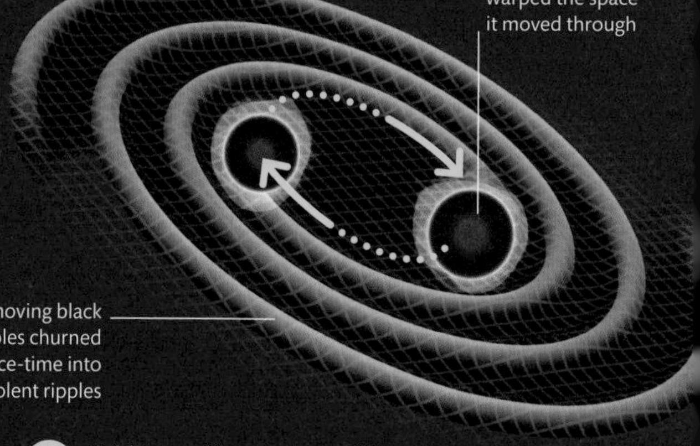

2 Orbit speed increases rapidly
As the black holes got nearer, their spiral orbits got smaller and smaller, and they both sped up to close to the speed of light. All this mass moving at such great speed produced powerful gravitational waves spreading in all directions.

How LIGO detects waves

LIGO, or the Laser Interferometer Gravitational-Wave Observatory, detects gravitational waves by their effects on laser beams fired down two 2.5-mile- (4-km-) long tubes. One beam travels half a wavelength farther than the other. This means that when the beams meet, they cancel each other out, making the light vanish. A gravitational wave alters the distance covered by the lasers, so when the beams meet, they create a signal of flickering light.

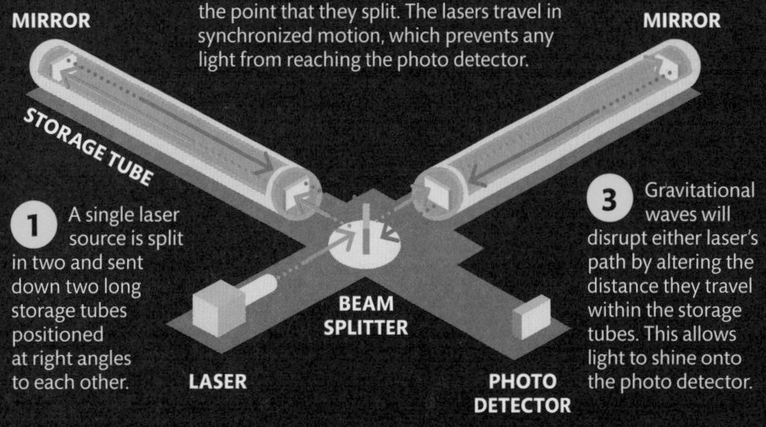
2 The lasers bounce back and forth off the mirrors, and then recombine at the point that they split. The lasers travel in synchronized motion, which prevents any light from reaching the photo detector.
MIRROR
MIRROR
STORAGE TUBE
1 A single laser source is split in two and sent down two long storage tubes positioned at right angles to each other.
BEAM SPLITTER
LASER
PHOTO DETECTOR
3 Gravitational waves will disrupt either laser's path by altering the distance they travel within the storage tubes. This allows light to shine onto the photo detector.

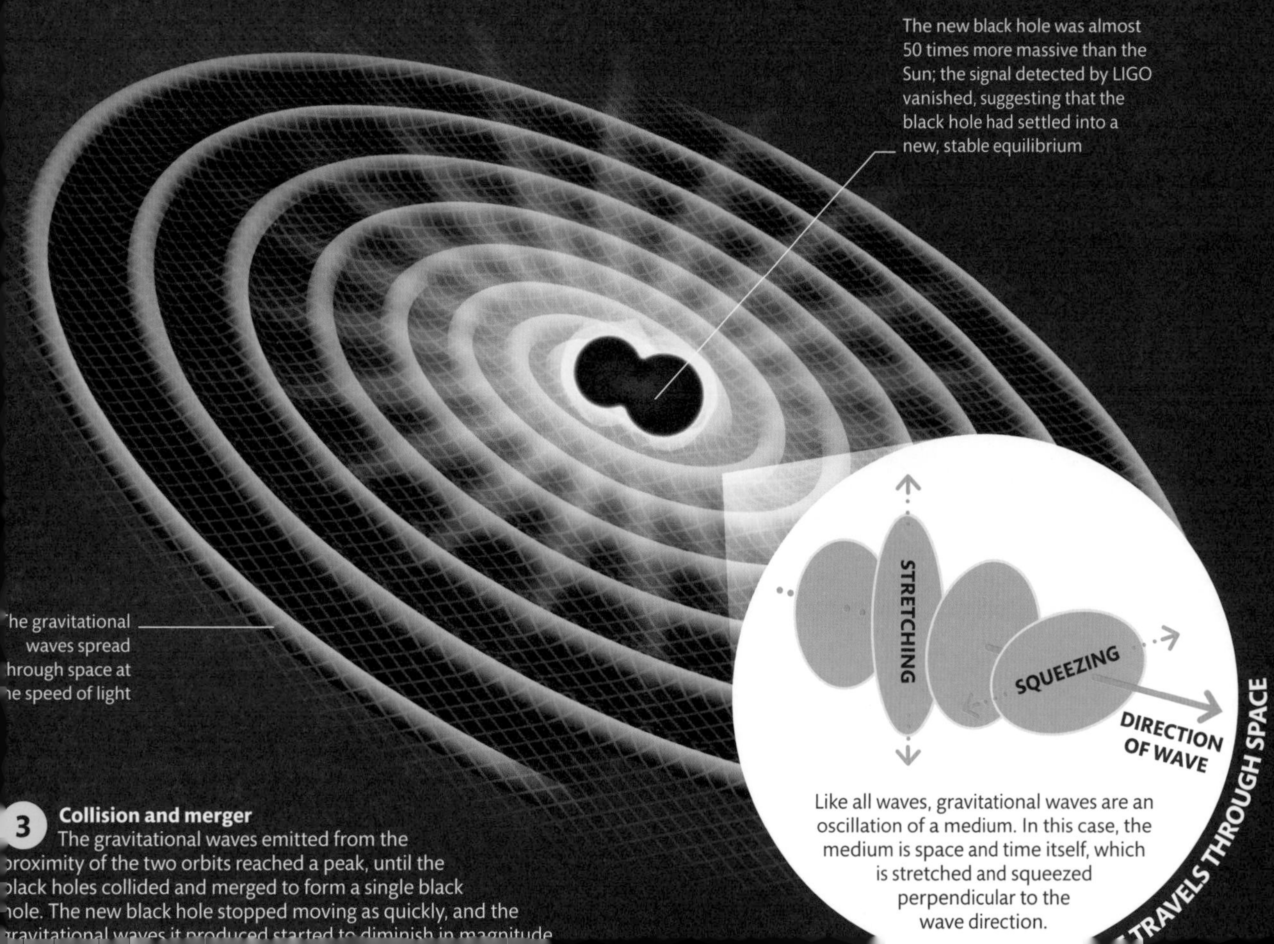
The new black hole was almost 50 times more massive than the Sun; the signal detected by LIGO vanished, suggesting that the black hole had settled into a new, stable equilibrium
The gravitational waves spread through space at the speed of light
STRETCHING
SQUEEZING
DIRECTION OF WAVE
Like all waves, gravitational waves are an oscillation of a medium. In this case, the medium is space and time itself, which is stretched and squeezed perpendicular to the wave direction.
TRAVELS THROUGH SPACE

3 Collision and merger
The gravitational waves emitted from the proximity of the two orbits reached a peak, until the black holes collided and merged to form a single black hole. The new black hole stopped moving as quickly, and the gravitational waves it produced started to diminish in magnitude

String theory

String theory is an attempt to solve the biggest problems in physics, such as how gravity works at an incredibly small scale. It proposes that all particles are one-dimensional "strings" and are part of a universal framework.

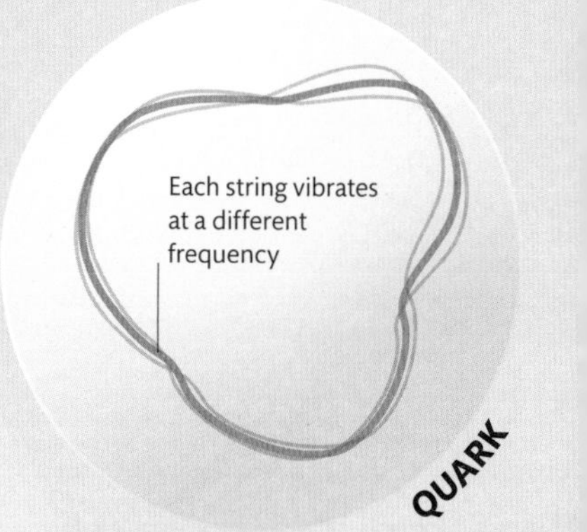

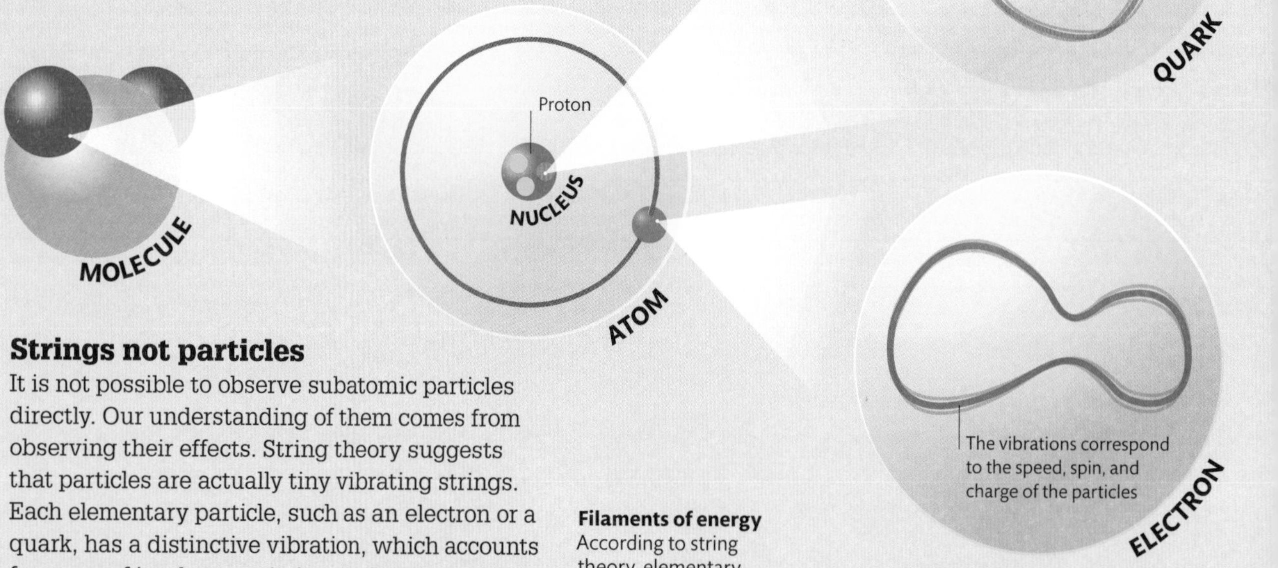

Strings not particles

It is not possible to observe subatomic particles directly. Our understanding of them comes from observing their effects. String theory suggests that particles are actually tiny vibrating strings. Each elementary particle, such as an electron or a quark, has a distinctive vibration, which accounts for many of its characteristics, such as mass, charge, and momentum. No one has figured out how to test string theory; for the time being, it is a mathematical system that seems to fit with the way quantum particles behave.

Filaments of energy
According to string theory, elementary particles, such as electrons or quarks, which make up protons, are strings or filaments of energy with their own distinctive vibration.

WHY DOES THERE HAVE TO BE A THEORY OF EVERYTHING?

The Universe follows a series of rules that work at the smallest and largest scales. They must be linked, and a theory of everything seeks to explain how.

Quantum gravity

Quantum gravity theory is intended to link general relativity, which describes the gravitation of huge structures like planets, with quantum mechanics, which shows how the other three fundamental forces act on an atomic scale. Quantum gravity's effects may operate at a scale measured in a unit of distance called the Planck length.

The Planck length
It is not possible to determine the location of two objects less than a Planck length apart, making it the smallest unit with any physical meaning.

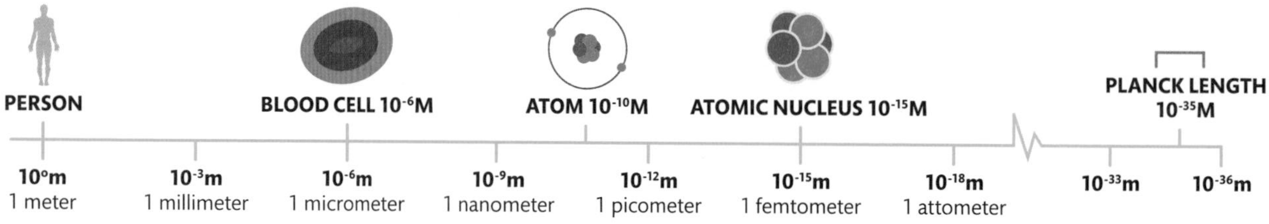

Many dimensions

String theorists suggest that strings do not just vibrate in the three visible dimensions (length, width, and depth) but in at least seven other dimensions that are hidden from us. The dimensions are described as "compact," which means that they only appear at the smallest subatomic scales. These other spatial dimensions may be all around us and might be a way of explaining mysterious phenomena such as dark matter and dark energy (see pp.206–07).

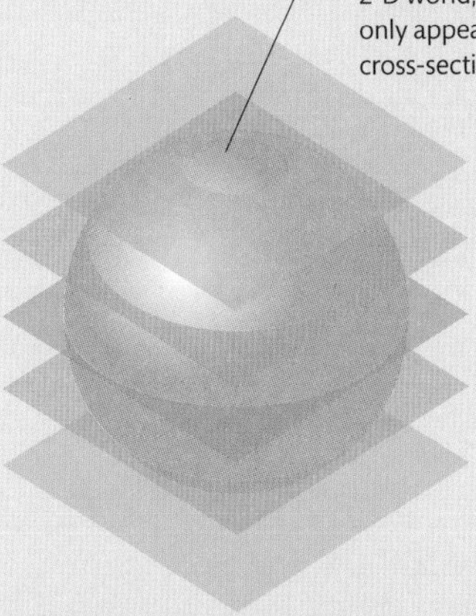
If a 3-D sphere were to pass through a 2-D world, it would only appear as 2-D cross-sections

3-D shapes in a 2-D world
Imagining a 3-D shape viewed in two dimensions helps us to understand higher spatial dimensions. Only circular slices of a 3-D sphere can be seen when viewed in 2-D.

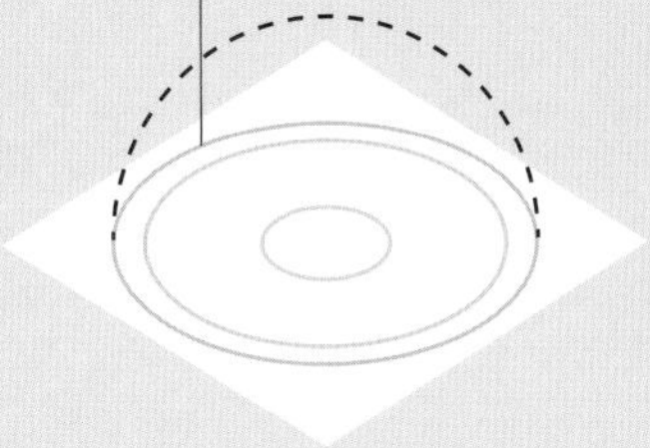
To a 2-D observer, the cross-sections, or slices of the sphere, would appear as concentric rings as each segment passes through the 2-D surface

View of a 2-D observer
A 2-D being, who cannot see up or down, sees a sphere growing and shrinking as it moves up and down. This strange behavior is due to the invisible spatial dimension.

Calabi-Yau manifold
According to some string theorists, the extra dimensions that are invisible to us could be curled up in geometric structures called Calabi-Yau manifolds. This shows a 2-D cross-section of a six-dimensional manifold called a Calabi-Yau quintic.

The manifold is split into 25 regions or "patches," with each represented by a different color

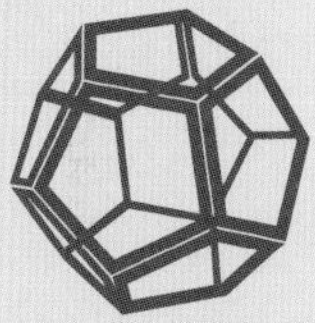

IN ONE VERSION OF STRING THEORY, THERE ARE 10 DIMENSIONS IN SPACE

Sparticles

Some forms of string theory suggest that matter is just the lowest vibration of energy, and that there are other strings vibrating in higher octaves, like the harmonies of music. The higher vibrations represent super particles, or sparticles, each of which theoretically partners a regular elementary particle. Some string theorists predict that sparticles may have masses up to 1,000 times greater than that of their corresponding particles.

MATTER PARTICLES AND PROPOSED SPARTICLES		FORCE PARTICLES AND PROPOSED SPARTICLES	
Particle	**Sparticle**	**Particle**	**Sparticle**
Quark	Squark	Graviton	Gravitino
Neutrino	Sneutrino	W bosons	Winos
Electron	Selectron	Z^{0}	Zino
Muon	Smuon	Photon	Photino
Tau	Stau	Gluon	Gluino
		Higgs boson	Higgsino

LIFE

What is "alive"?

Life is the most complex thing in the known universe. Its molecular building blocks and the collaboration of its working parts are more intricate than those of any computer. We need to strip down the biology of an organism to its basic functions to appreciate what makes something alive.

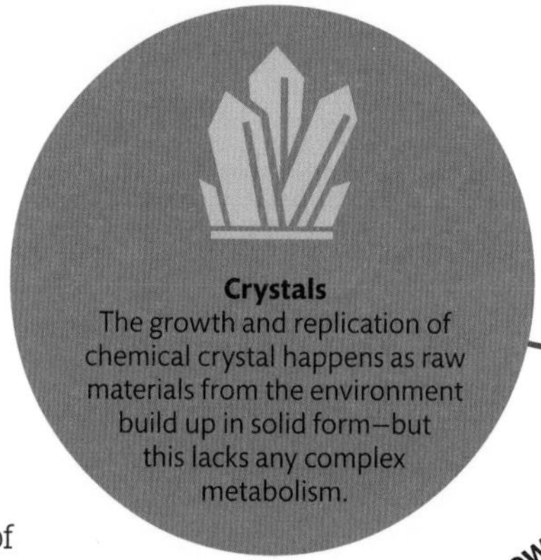

Crystals
The growth and replication of chemical crystal happens as raw materials from the environment build up in solid form—but this lacks any complex metabolism.

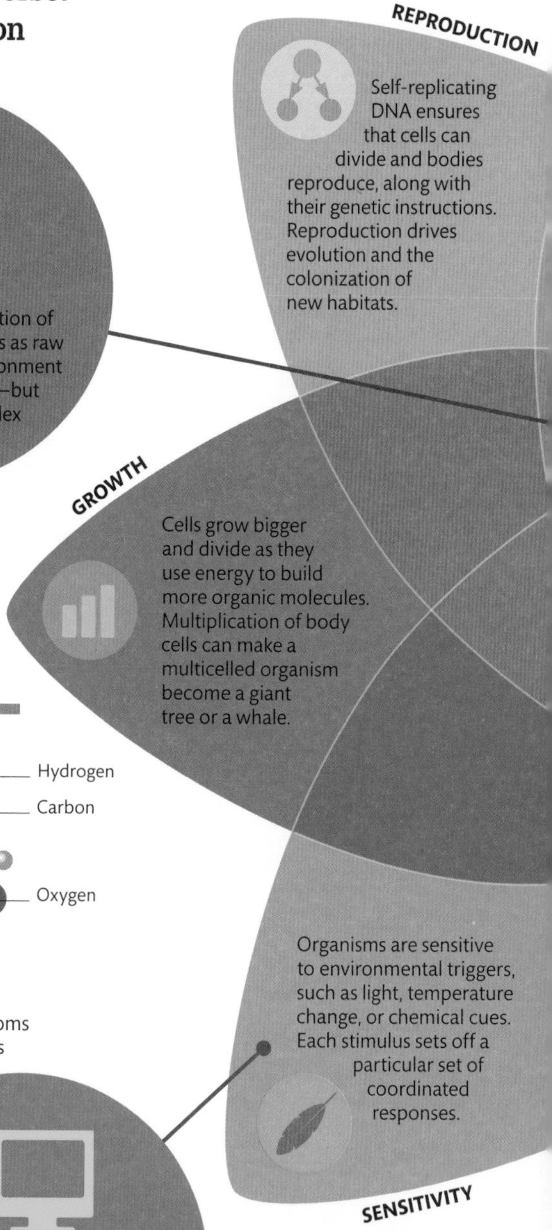
REPRODUCTION
Self-replicating DNA ensures that cells can divide and bodies reproduce, along with their genetic instructions. Reproduction drives evolution and the colonization of new habitats.
GROWTH
Cells grow bigger and divide as they use energy to build more organic molecules. Multiplication of body cells can make a multicelled organism become a giant tree or a whale.
Organisms are sensitive to environmental triggers, such as light, temperature change, or chemical cues. Each stimulus sets off a particular set of coordinated responses.
SENSITIVITY

Venn diagram of life
Even though there is astonishing variation across the range of organisms, the same seven basic functions are shared by living things as different as bacteria, plants, and animals.

Signs of life

Millions of species of organism share a combination of traits: the characteristics of life. Only when all these characteristics come together can something be called alive. A living thing uses food, respires to release energy, and excretes waste. It moves, senses its surroundings, grows, and reproduces. Nonliving things may have one or two of these functions but never the complete set.

Building complexity

The complex chemicals that make up life are built around a framework of carbon atoms and are among the biggest molecules known. Chains of DNA or cellulose can be many centimeters long. Plants make these organic molecules from simple ingredients, such as carbon dioxide and water. Animals acquire them by eating food—either other organisms or their waste. These food molecules act as both fuel and building materials.

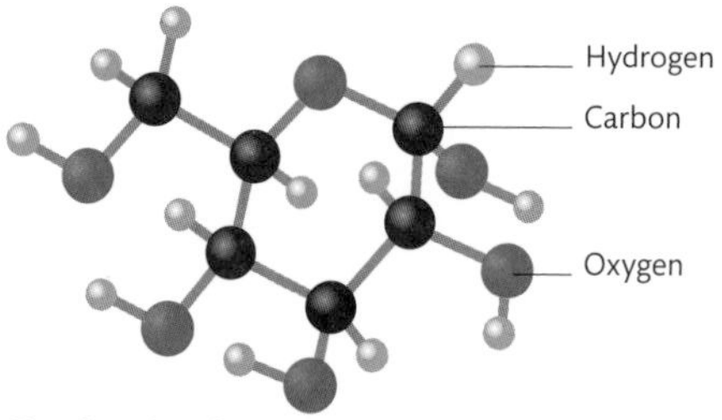
Hydrogen
Carbon
Oxygen

Food molecule
A glucose molecule is made of 24 atoms and is among the simplest molecules used as food. As in other biomolecules, carbon atoms form the framework.

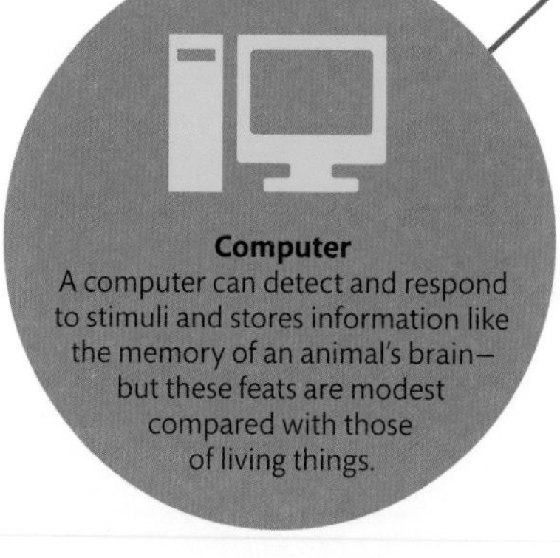

Computer
A computer can detect and respond to stimuli and stores information like the memory of an animal's brain—but these feats are modest compared with those of living things.

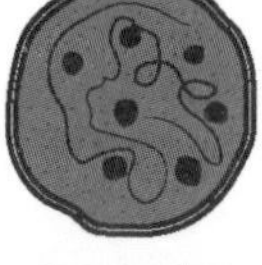

AMONG THE **SIMPLEST** LIVING **ORGANISMS** ARE PNEUMONIA-CAUSING **BACTERIA** THAT HAVE ONLY **687 GENES**

NUTRITION

All life needs an ongoing supply of energy and raw materials. Many organisms need to acquire these in the form of organic food molecules, such as proteins and carbohydrates.

MOVEMENT

From the steady streaming of fluid and constituents in their microscopic cells to the powerful contraction of animal muscles, all organisms can move to a greater or lesser degree.

DOES LIFE HAVE TO BE CARBON-BASED?

Science fiction writers have speculated that silicon could form an alternative biology. However, carbon is the only element that combines with so many other kinds of atoms to form such complex molecules—and, therefore, life.

LIVING ORGANISM

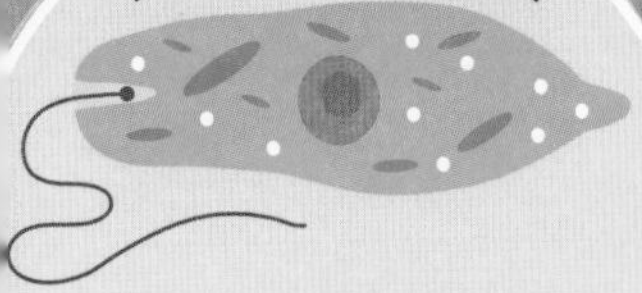

Euglena—a microscopic single-celled inhabitant of ponds—can photosynthesize as plants do or consume food the way an animal does.

EXCRETION

The continuous chemical reactions in the cells of an organism produce waste products, such as carbon dioxide. Excretion is how this metabolic waste is removed from the body.

RESPIRATION

Much of an organism's organic food is broken down in chemical reactions equivalent to the burning of fuel in an engine. This respiration releases energy that is used by the body.

Internal combustion engine
By taking in and burning fuel to cause movement and "excreting" waste fumes, an engine has four of the signs of life. But it lacks sensitivity, growth, and reproduction.

WHAT IS METABOLISM?

Countless chemical reactions—called metabolism—underpin life. Molecules are changed in sequences of reactions, each step driven by a specific protein catalyst called an enzyme. The unique metabolism of each organism depends on a set of enzymes determined by DNA's genetic instructions.

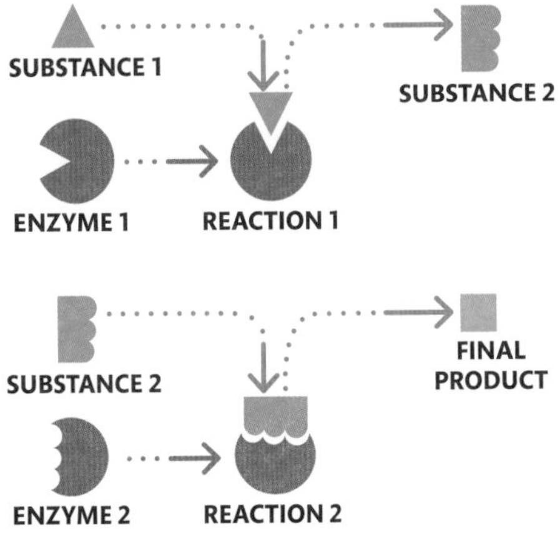

Types of living thing

We classify things to make sense of the world. And when it comes to organizing the variety of life, modern scientific classification has an extra goal—to chart the physical and genetic similarities between species in ways that reflect their evolutionary relationships.

The tree of life

Similarities among organisms as different as bacteria and animals, especially in their cells and genes, are strong evidence that all life stems from a single ancestor. Over billions of years, living things evolved into a vast family tree. Scientists classify them in a series of ever-smaller groups, which mirrors the way branches have split from larger limbs during evolution. The tree's oldest branches mark the foundations of life's kingdoms; the outermost twigs are the millions of species that have ever lived.

A SINGLE TEASPOON OF SOIL COULD CONTAIN MORE THAN 100,000 SPECIES OF MICROBES

SCIENTIFIC NAMES

Each species is given a unique scientific name. This makes each name unambiguous—something that common names such as tree heath or giant heather (both of which refer to the same species, *Erica arborea*) rarely achieve. Scientific names are often descriptive (*arborea* meaning "treelike") and always have two parts. The first, such as *Erica*, defines a group of related species called the genus. With the second part included (*Erica cinerea* or *E. arborea*), the name defines the species.

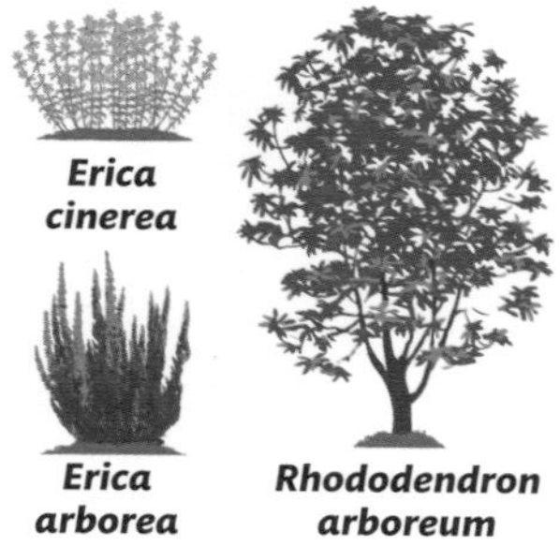
Erica cinerea
Erica arborea
Rhododendron arboreum

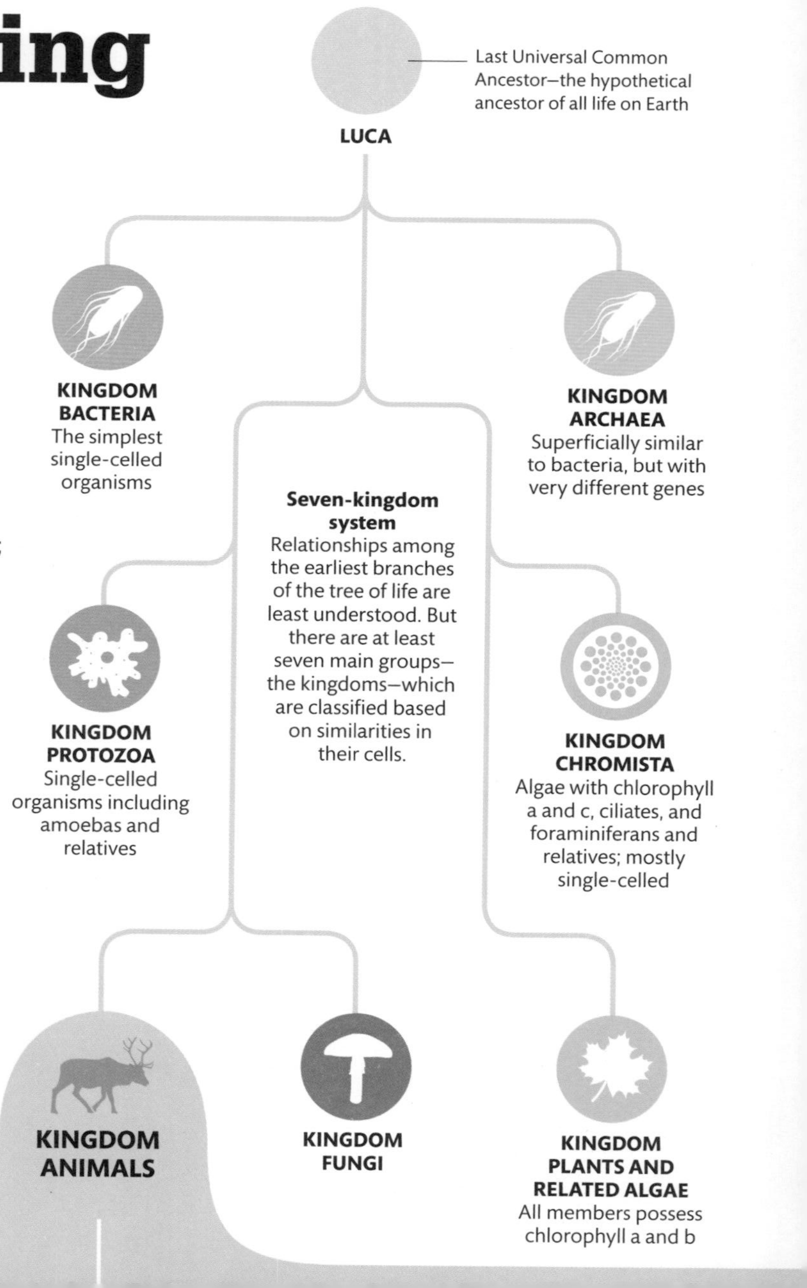
LUCA
Last Universal Common Ancestor—the hypothetical ancestor of all life on Earth
KINGDOM BACTERIA
The simplest single-celled organisms
KINGDOM ARCHAEA
Superficially similar to bacteria, but with very different genes
Seven-kingdom system
Relationships among the earliest branches of the tree of life are least understood. But there are at least seven main groups—the kingdoms—which are classified based on similarities in their cells.
KINGDOM PROTOZOA
Single-celled organisms including amoebas and relatives
KINGDOM CHROMISTA
Algae with chlorophyll a and c, ciliates, and foraminiferans and relatives; mostly single-celled
KINGDOM ANIMALS
KINGDOM FUNGI
KINGDOM PLANTS AND RELATED ALGAE
All members possess chlorophyll a and b

Natural and unnatural groups

Many organisms share features by a coincidence of evolution. Birds and insects evolved wings separately, so it would not be natural to group them as "flying animals." Natural groups, or clades, include all descendants of a common ancestor—a forking point in life's tree. Mammals and birds are both clades. The animal groups we call fish and invertebrates are not, because they do not include all descendants. "Fish," for example, excludes their offshoot, land vertebrates.

Groups within groups
If we classify strictly according to relatedness, our system must reflect that birds descended from a group of theropods—upright dinosaurs that included *Tyrannosaurus*. This means they are classified as a subgroup of dinosaurs, nested within the reptiles.

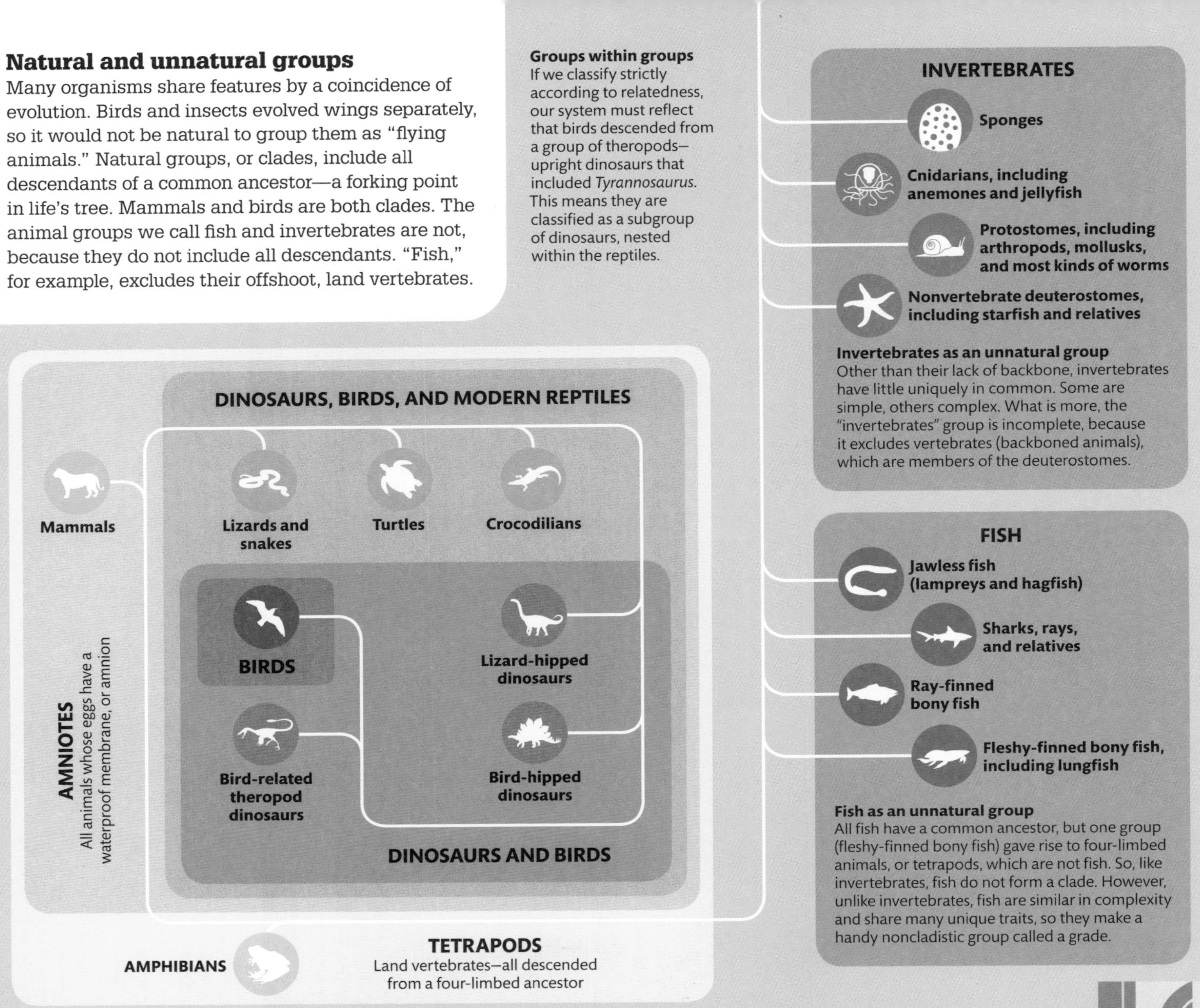

Invertebrates as an unnatural group
Other than their lack of backbone, invertebrates have little uniquely in common. Some are simple, others complex. What is more, the "invertebrates" group is incomplete, because it excludes vertebrates (backboned animals), which are members of the deuterostomes.

Fish as an unnatural group
All fish have a common ancestor, but one group (fleshy-finned bony fish) gave rise to four-limbed animals, or tetrapods, which are not fish. So, like invertebrates, fish do not form a clade. However, unlike invertebrates, fish are similar in complexity and share many unique traits, so they make a handy noncladistic group called a grade.

Viruses

Viruses provide a vivid demonstration of the drive to replicate. Not truly alive themselves, these infectious particles—little more than tiny packages of genes—sabotage living cells to scatter their replicas through their host. Some cause little harm, but others cause Earth's most feared diseases.

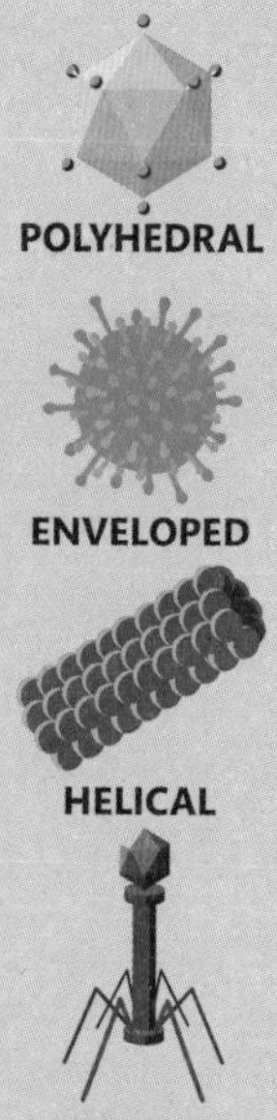
POLYHEDRAL
ENVELOPED
HELICAL
COMPLEX

Types of virus

Viruses come in different shapes but share the same essential parts: a cluster of genes wrapped up in a protein coat. Some are based on DNA. Others instead have RNA—the substance otherwise used as a go-between for making protein in true cells (see pp.158–59). Most remarkably of all, many virus genes are more related to those of their hosts than they are to other virus genes—evidence that viruses may actually be parcels of renegade genes that have escaped from host chromosomes.

A virus cycle

All viruses are parasites, transmitted through contact, across air, or in infected food. They are not true living organisms (see pp.150–51), partly because they rely on the inner workings of a cell to replicate. Like the living organisms they use, their behavior is encoded in their genes—helping them infect their host's body in ways that maximize multiplication. Each kind of virus has its own effects: from the mild common cold of a rhinovirus to the complete system shutdown of Ebola.

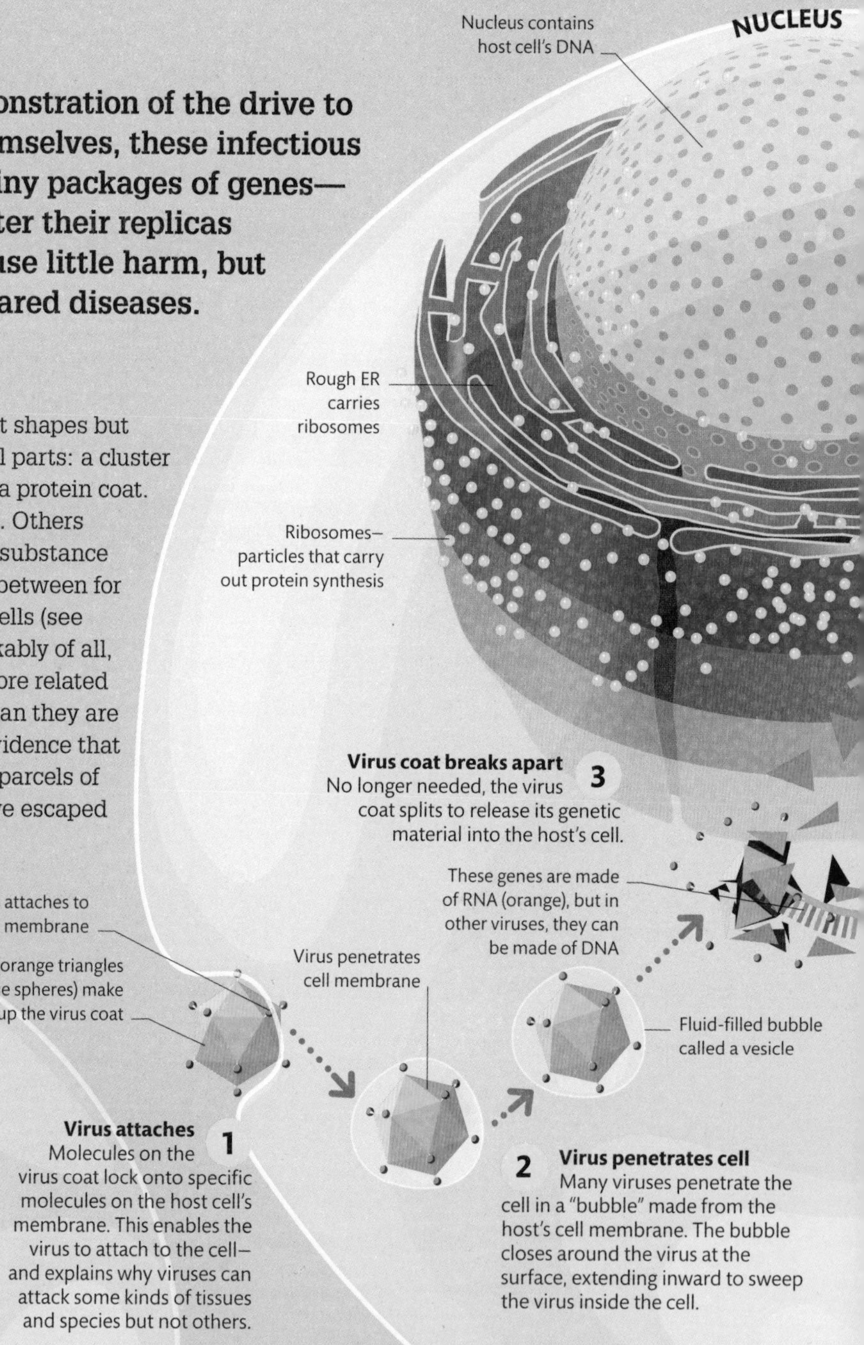
NUCLEUS
Nucleus contains host cell's DNA
Rough ER carries ribosomes
Ribosomes—particles that carry out protein synthesis
Virus coat breaks apart 3
No longer needed, the virus coat splits to release its genetic material into the host's cell.
These genes are made of RNA (orange), but in other viruses, they can be made of DNA
Virus attaches to cell membrane
Proteins (orange triangles and blue spheres) make up the virus coat
Virus penetrates cell membrane
Fluid-filled bubble called a vesicle
Virus attaches 1
Molecules on the virus coat lock onto specific molecules on the host cell's membrane. This enables the virus to attach to the cell—and explains why viruses can attack some kinds of tissues and species but not others.
2 **Virus penetrates cell**
Many viruses penetrate the cell in a "bubble" made from the host's cell membrane. The bubble closes around the virus at the surface, extending inward to sweep the virus inside the cell.

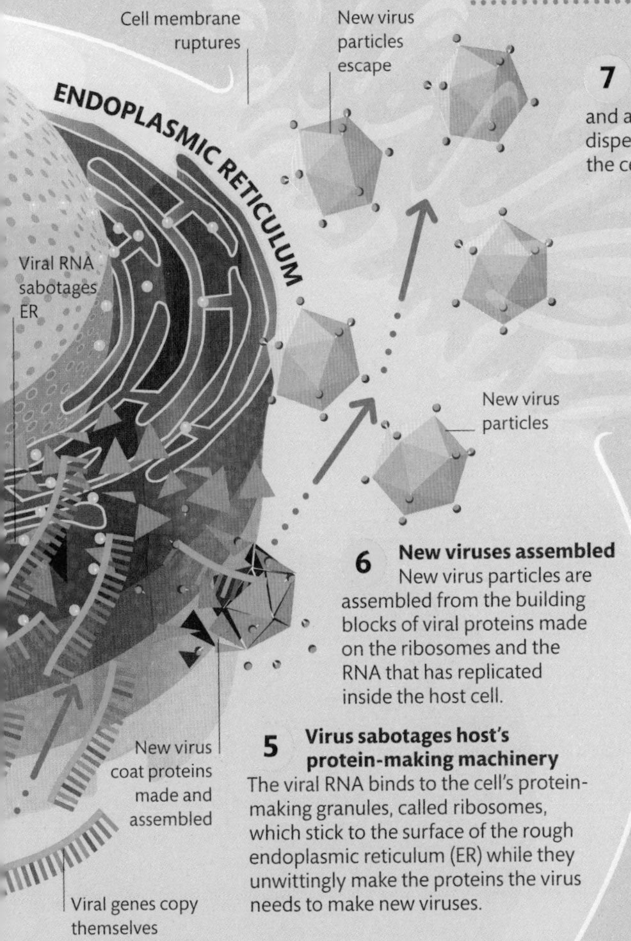

7 New virus particles released
Virus particles escape from the cell and are ready to infect other cells or disperse to a new host. This may rupture the cell membrane, killing the host cell.

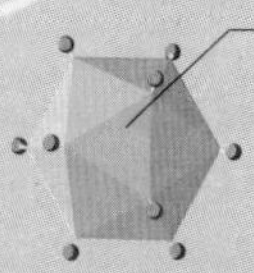

6 New viruses assembled
New virus particles are assembled from the building blocks of viral proteins made on the ribosomes and the RNA that has replicated inside the host cell.

5 Virus sabotages host's protein-making machinery
The viral RNA binds to the cell's protein-making granules, called ribosomes, which stick to the surface of the rough endoplasmic reticulum (ER) while they unwittingly make the proteins the virus needs to make new viruses.

4 Viral genes replicate
The virus's genetic material produces many identical copies. Viruses with RNA carry their own enzymes to make DNA first, or simply to replicate directly. Although not shown here, DNA-based viruses go straight into the host's nucleus, where they insert themselves into the host's DNA.

CELL MEMBRANE

Fighting viruses

Faced with a virus attack, the body mobilizes the white blood cells of its immune system. Some release proteins called antibodies that bind to viruses, disabling them. Others, named "killer cells," sacrifice cells that have already been infected. Viruses cannot be treated with antibiotics, which work only on microbes, such as bacteria. At the front line of virus control, however, are vaccines, which prime the immune system with a "fake" infection.

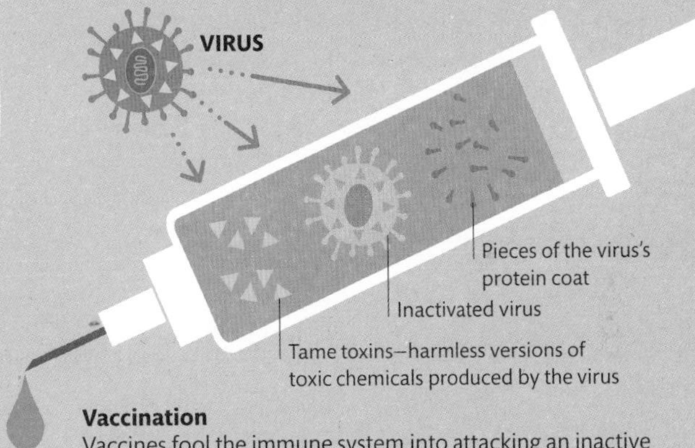

Vaccination
Vaccines fool the immune system into attacking an inactive version of the infection—enough to trigger the immune response but not enough to cause disease. Once primed in this way, the immune system recognizes the true virus, if encountered, and can mount a fast and powerful response.

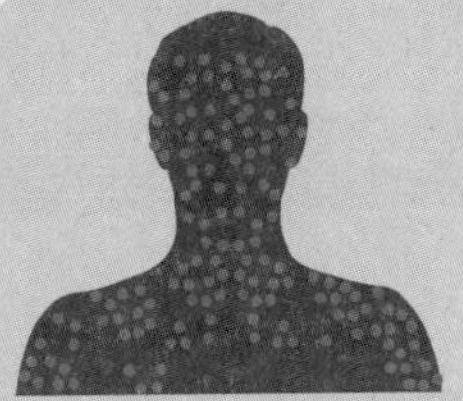

SMALLPOX IS THE ONLY INFECTIOUS DISEASE THAT HAS BEEN ELIMINATED BY VACCINATION

VIRUSES FOR GOOD

Viruses can be genetically modified to carry drugs to specific cells, helping to target cancers. DNA viruses can also deliver "healthy" genes into cells for gene therapy (as shown here). Other viruses even have the potential for fighting disease-causing bacteria, offering an alternative to antibiotics for treating infections.

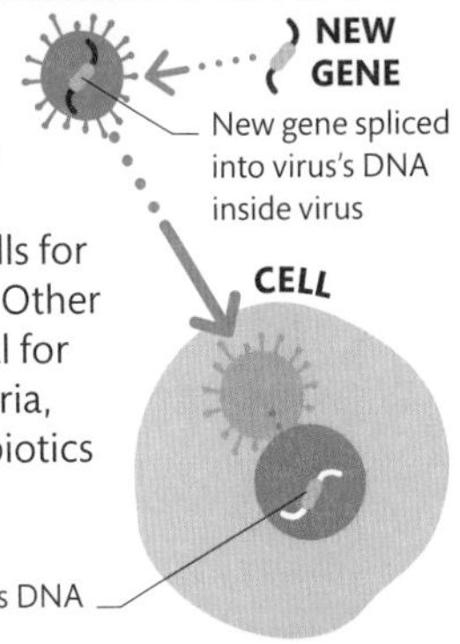

Cells

Nearly every part of any organism's body consists of living units called cells. Cells process food and energy, sense their surroundings, and grow and repair themselves—all within a space five times smaller than a full stop.

How cells work

A cell is full of tiny structures called organelles. Like organs in a body, each carries out one or more specialty tasks vital to the working of the cell. All cells gather materials from their surroundings to make a wealth of complex substances.

1 **Protein manufacture**
Most substances the cell needs are specific proteins, which are made according to genetic instructions (see pp.158–59) at sites called ribosomes, which stud the complex surface of an organelle called the rough endoplasmic reticulum.

2 **Packaging**
The proteins travel in vesicles—little cellular bubbles that drift to the Golgi body. This organelle acts as the cell's mail room, packaging and labeling the proteins, which determines where they are sent next.

3 **Shipping**
The Golgi body places the proteins into different vesicles depending on their label. The vesicles bud off, and those destined for outside of the cell fuse with the cell membrane and release the proteins outside.

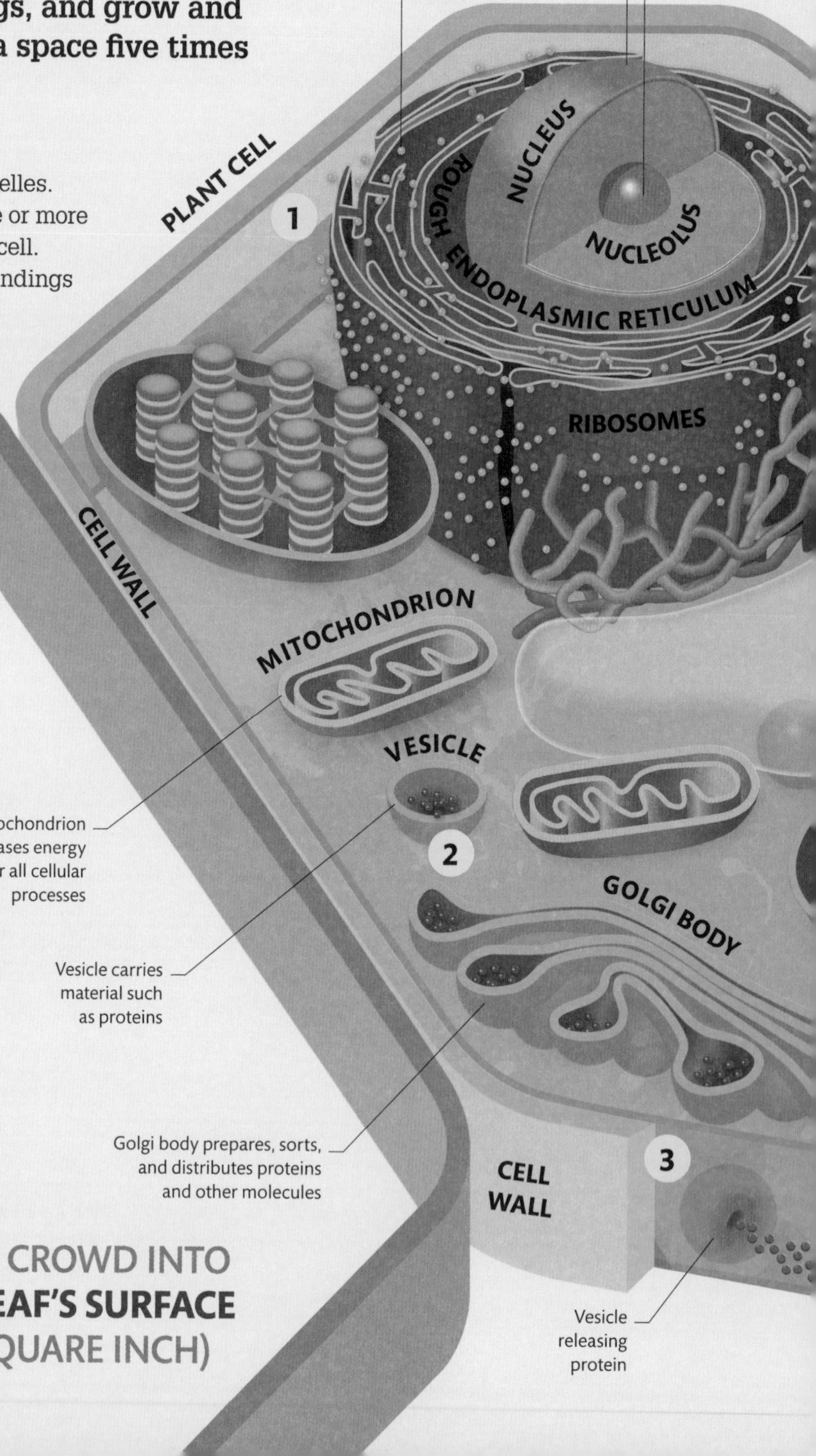

Ribosomes stud the rough endoplasmic reticulum, making it look rough
Nucleus stores DNA, which acts as a library of instructions for making proteins
Nucleolus helps to make ribosomes
PLANT CELL
1
NUCLEUS
NUCLEOLUS
ROUGH ENDOPLASMIC RETICULUM
RIBOSOMES
CELL WALL
MITOCHONDRION
VESICLE
Mitochondrion releases energy for all cellular processes
2
GOLGI BODY
Vesicle carries material such as proteins
Golgi body prepares, sorts, and distributes proteins and other molecules
CELL WALL
3
Vesicle releasing protein

800,000

CHLOROPLASTS MAY CROWD INTO EVERY **SQUARE MM OF A LEAF'S SURFACE** (THAT'S 500 MILLION PER SQUARE INCH)

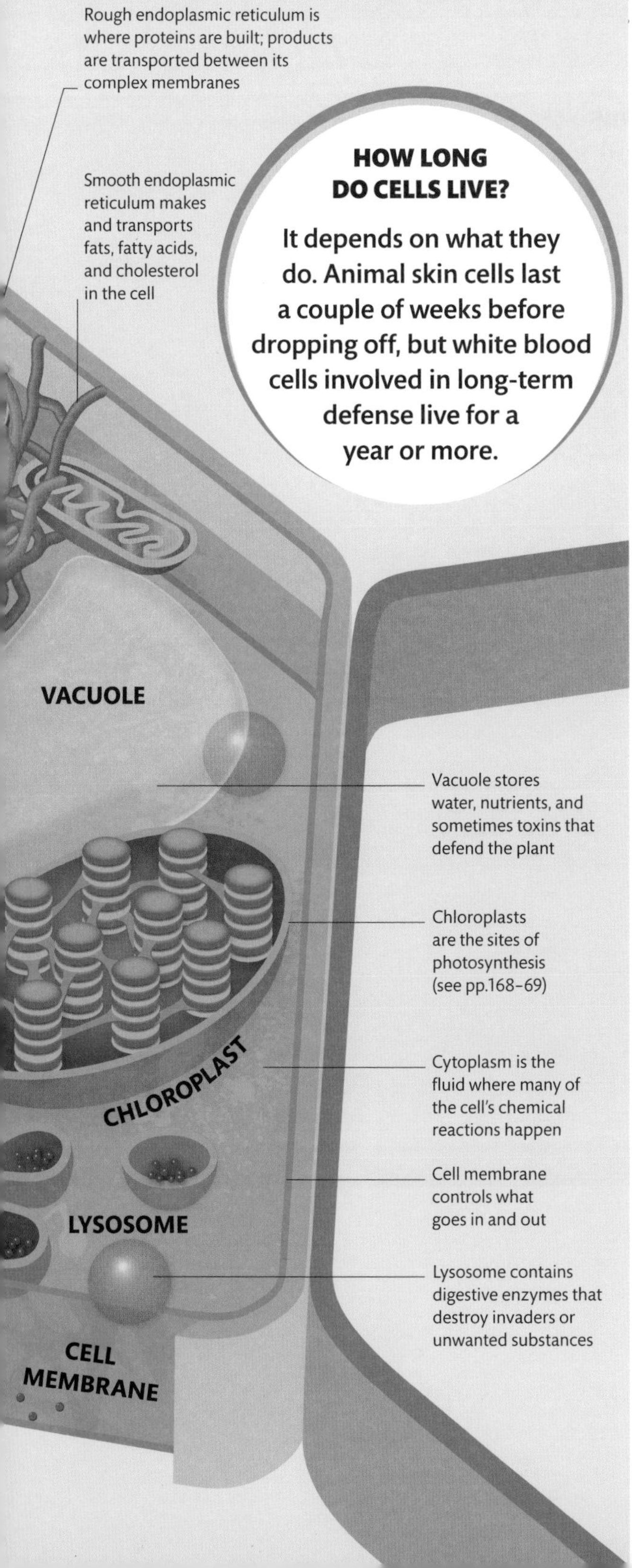
Rough endoplasmic reticulum is where proteins are built; products are transported between its complex membranes
Smooth endoplasmic reticulum makes and transports fats, fatty acids, and cholesterol in the cell
VACUOLE
Vacuole stores water, nutrients, and sometimes toxins that defend the plant
Chloroplasts are the sites of photosynthesis (see pp.168–69)
CHLOROPLAST
Cytoplasm is the fluid where many of the cell's chemical reactions happen
Cell membrane controls what goes in and out
LYSOSOME
Lysosome contains digestive enzymes that destroy invaders or unwanted substances
CELL MEMBRANE

HOW LONG DO CELLS LIVE?

It depends on what they do. Animal skin cells last a couple of weeks before dropping off, but white blood cells involved in long-term defense live for a year or more.

The diversity of cells

Animal cells are unlike plant cells. Unbound and unsupported by a cell wall, animal cells cannot grow as big. But, like those of plants, they vary in shape depending on what they do. Because animals are more energetic than plants, many of their cells have more mitochondria. But they always lack photosynthesizing chloroplasts, since animals consume, rather than make, their food.

Different animal cells
Flat skin cells can form a sheet, but because they are not busy making proteins, they have few mitochondria. The many mitochondria in a white blood cell, in contrast, help it leap into rapid-response activity in its defense of the body.

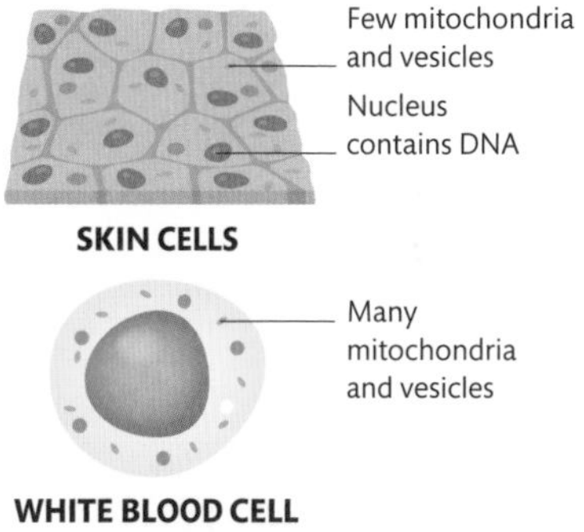
Few mitochondria and vesicles
Nucleus contains DNA
SKIN CELLS
Many mitochondria and vesicles
WHITE BLOOD CELL

Bacterial cells
Some cells are utterly unlike animal and plant cells—those of bacteria. Bacteria evolved long before animals, plants, or even single-celled algae. They have a cell wall but no nucleus to contain their DNA.

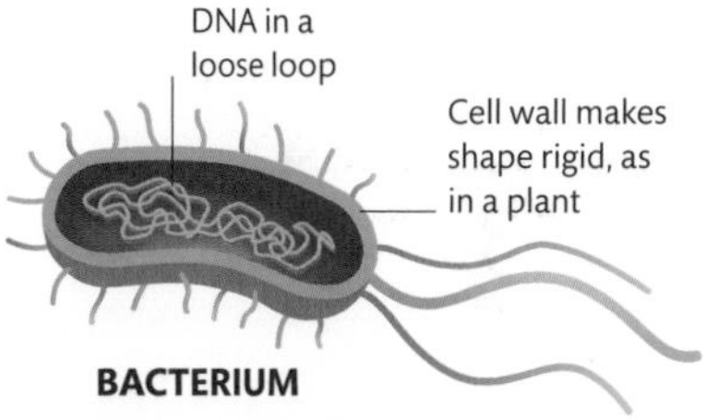
DNA in a loose loop
Cell wall makes shape rigid, as in a plant
BACTERIUM

MAKING MORE CELLS

Cells that are part of a multicellular body must duplicate themselves many times so that the body grows or renews itself. The copying process, known as mitosis, is not easy, since each cell must have its own copy of the genome—the full set of DNA instructions for the body. First, the DNA is copied in full before the cell splits into "daughter" cells.

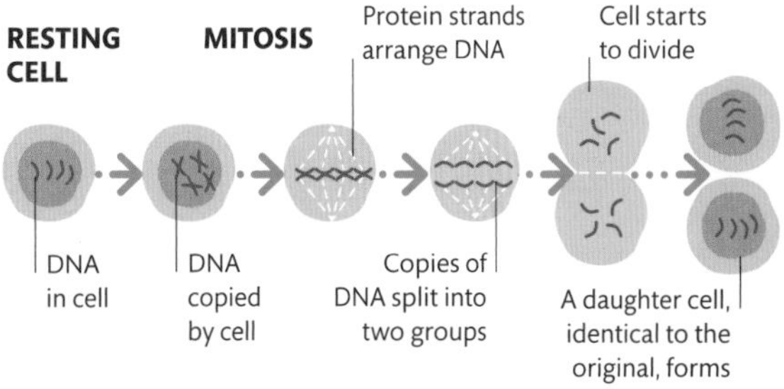
RESTING CELL
MITOSIS
Protein strands arrange DNA
Cell starts to divide
DNA in cell
DNA copied by cell
Copies of DNA split into two groups
A daughter cell, identical to the original, forms

How genes work

DNA contains coded information that controls the growth and maintenance of living things. Its instructions translate into specific proteins that an organism needs. A length of DNA containing the code for a protein is called a gene.

Building proteins

Hundreds of kinds of protein carry out the cellular processes of life. Many of them are enzymes that speed up, or catalyze, chemical reactions; others move materials across cell membranes or perform other vital tasks. All are made according to the instructions of the DNA's genes. Each gene must be copied onto a molecule called RNA that carries its instructions from the nucleus to the protein-making machinery in the cell.

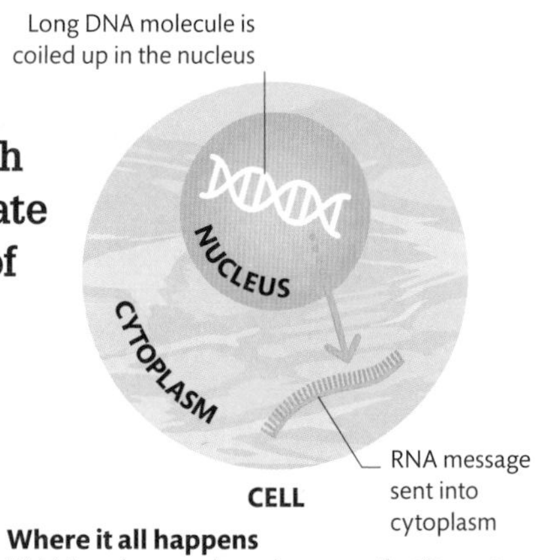

Where it all happens
DNA is so long and cumbersome that it must stay inside the nucleus. But proteins are made in the cell's cytoplasm, so copies of genes must be sent out as messenger RNA.

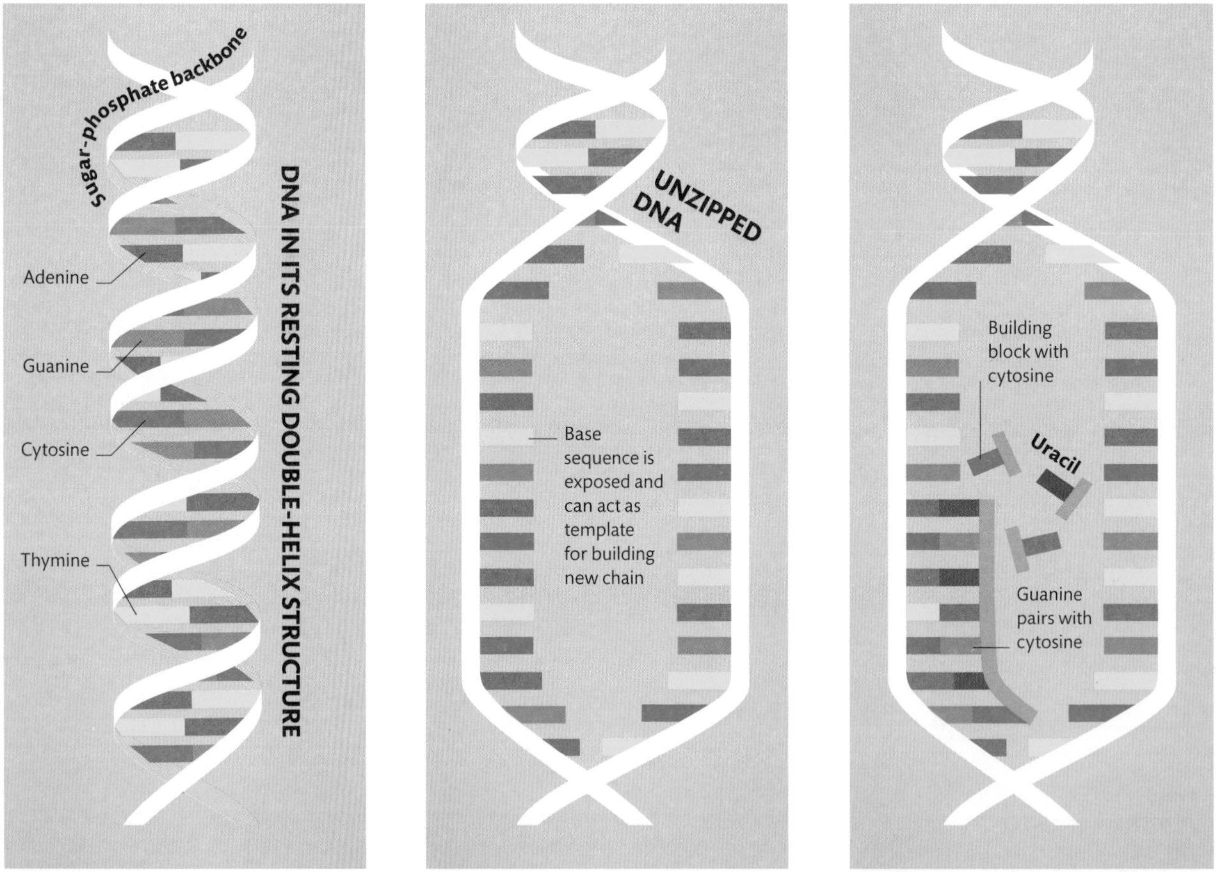

1 DNA structure
The DNA molecule is a double helix comprising two twisted strands. Four chemical units called bases pair up between the strands in a complementary way: adenine with thymine and guanine with cytosine.

2 DNA unzips
Genetic instructions are encoded in the sequence of bases along one strand. Sections called genes that contain code for specific proteins are exposed when the double helix unzips in appropriate places.

3 RNA forms on DNA template
A strand of RNA is built along the exposed gene, its sequence of bases being complementary to that of the gene's base sequence. Uracil in RNA is used to pair with adenine, instead of thymine.

THE GENETIC CODE—A UNIVERSAL LANGUAGE

Each kind of organism has its own set of genes, but the way the base sequence translates into different amino acids is the same throughout all organisms, from bacteria to plants and animals: a base triplet always translates into the same amino acid. For instance, AAA codes for the amino acid lysine, AAC is the code for asparagine, and so on.

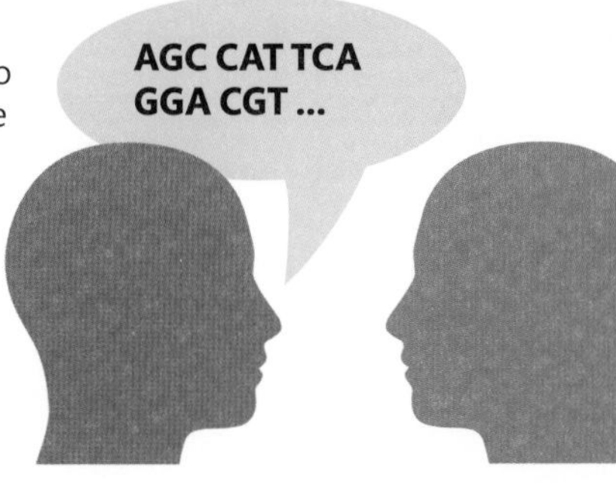
AGC CAT TCA
GGA CGT ...

50 BASES ARE ADDED EVERY SECOND WHEN DNA IS COPIED IN A **HUMAN CELL**

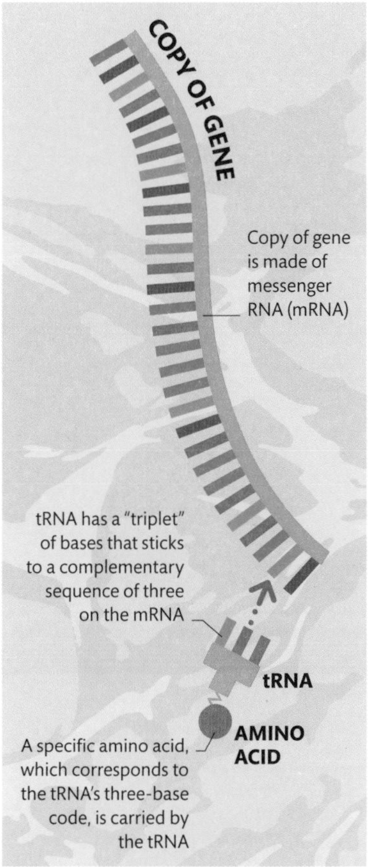
COPY OF GENE
Copy of gene is made of messenger RNA (mRNA)
tRNA has a "triplet" of bases that sticks to a complementary sequence of three on the mRNA
tRNA
AMINO ACID
A specific amino acid, which corresponds to the tRNA's three-base code, is carried by the tRNA

4 Gene leaves nucleus
The finished messenger RNA (mRNA) strand—effectively a mirror of the gene—detaches and moves into the cytoplasm of the cell. There, it attracts specific transfer RNA (tRNA) molecules.

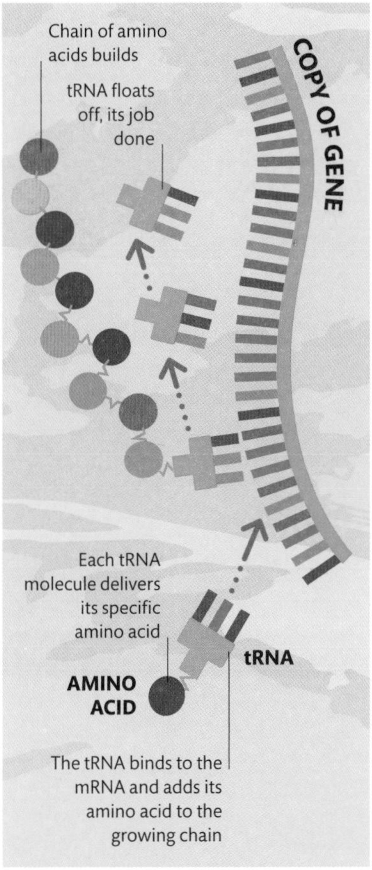
Chain of amino acids builds
tRNA floats off, its job done
COPY OF GENE
Each tRNA molecule delivers its specific amino acid
tRNA
AMINO ACID
The tRNA binds to the mRNA and adds its amino acid to the growing chain

5 Translation into amino acids
tRNA molecules recognize and stick to specific sequences on the mRNA. Each tRNA brings with it a specific amino acid, which joins a growing chain. In this way, the base sequence is translated into amino acids.

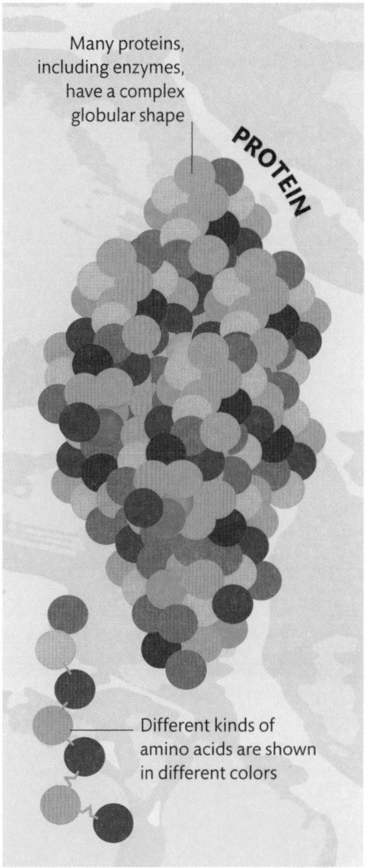
Many proteins, including enzymes, have a complex globular shape
PROTEIN
Different kinds of amino acids are shown in different colors

6 Amino acids wrap up into protein
The specific sequence of amino acids, determined by the order of bases in the gene, controls how the chain folds up into a complex protein molecule. The way the protein folds decides its shape and function.

Reproduction

As life produces more life, organisms find different ways of passing on their genes, in the greatest possible quantity, from one generation to the next. Some living things simply fragment, but for the clear majority that are sexual beings, reproduction breeds genetic variety.

Asexual reproduction

All organisms copy their DNA when cells divide. In some, replication of the entire organism's body is also a simple matter of copying (see pp.186–87). Asexual reproduction—without fertilization—results in identical offspring that are equally susceptible to disease or any ecological crisis. But its simplicity makes it ideal for proliferating quickly.

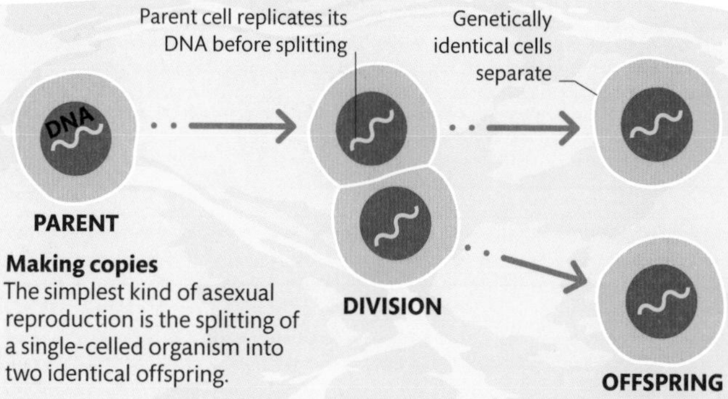

Making copies
The simplest kind of asexual reproduction is the splitting of a single-celled organism into two identical offspring.

Budding
Simple animals, such as sea anemones, can sprout new offspring from the parent body.

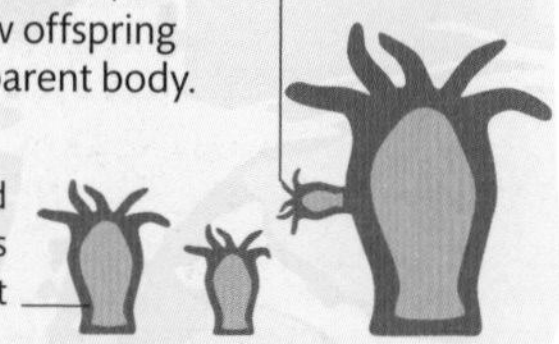

Parthenogenesis
A few animals engage in virgin birth. Aphid eggs can grow into babies inside their mother without fertilization.

Vegetative reproduction
The branching growth of many plants makes them suited to reproducing asexually from runners or side shoots.

Reproductive strategies

There are contrasting ways of investing in the next generation. Some organisms produce countless offspring to offset the fact that each one has a very slim chance of surviving. Others are much less prolific, but they are such devoted parents that each infant thrives on the care it receives.

Many offspring
Frogs can produce hundreds of eggs during a single spawning—and continue to do so year after year. But most offspring will succumb to predators.

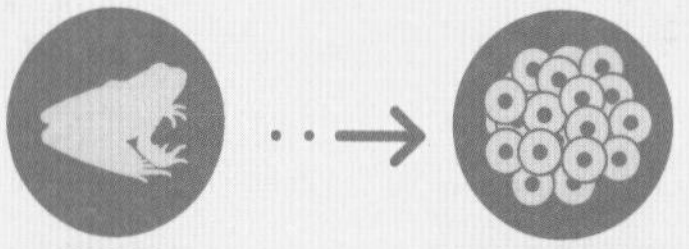

Few offspring
The California Condor—a bird of prey—only starts breeding when 8 years old and produces a maximum of a single egg every other year.

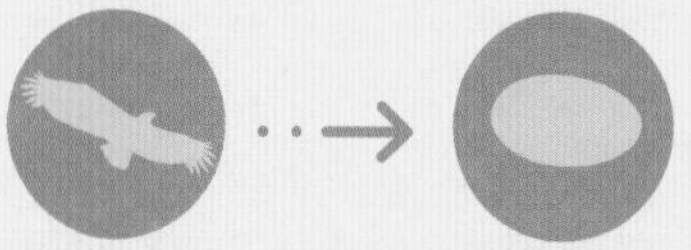

BARRIERS TO BREEDING

Different species rarely interbreed because reproductive barriers prevent it. Birds respond only to courtship songs of their own kind. Tigers and lions are separated by geography and habitat. Natural hybrids occasionally arise but fail to persist because their fertility is usually poor. However, in captivity natural barriers break down, and hybrids such as ligers are more likely.

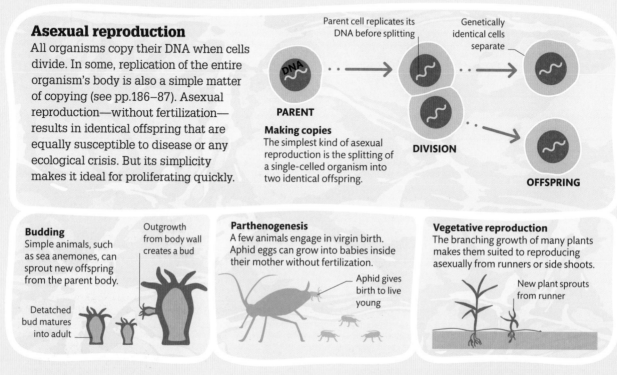

LIGER—A CROSS BETWEEN A MALE LION AND A FEMALE TIGER

Sexual reproduction

Reproduction involving sex produces offspring that are genetically different from each other and from their parents. This happens because cell division in sex organs generates sperm or eggs with unique genetic combinations. The act of fertilization mingles these combinations. This means that each new generation, as it is exposed to the vagaries of a changeable environment, is more likely to contain a winning combination.

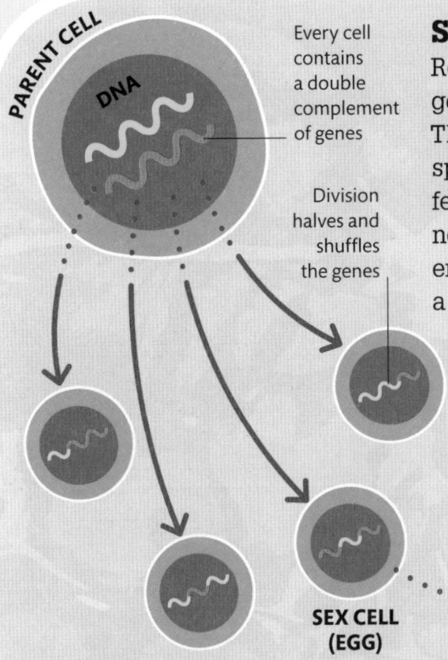
PARENT CELL
DNA
Every cell contains a double complement of genes
Division halves and shuffles the genes
SEX CELL (EGG)

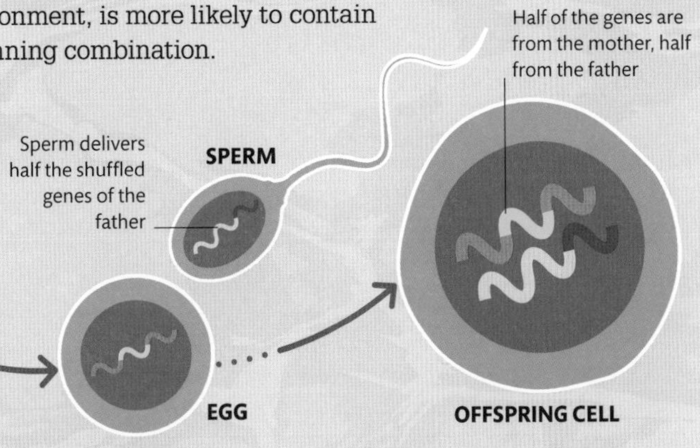
Half of the genes are from the mother, half from the father
Sperm delivers half the shuffled genes of the father
SPERM
EGG
OFFSPRING CELL

1 Reduction division
Sex cells (eggs and sperm) are produced from a type of cell division called meiosis. This halves the chromosome number but also shuffles genes around.

2 Fusion (fertilization)
Organisms typically produce many small, mobile male sex cells and fewer, bigger eggs. Once these sex cells fuse, the resulting cell is a genetic mixture of each parent.

3 New combination
Fertilization restores the double complement of genes but produces genetically unique young. The new genetic combination is replicated in every body cell in the offspring.

Sex in plants

Seed plants transfer male sex cells, in pollen grains, to female organs. Each grain sprouts a microscopic tube to deliver the male cell to an egg deep inside the flower.

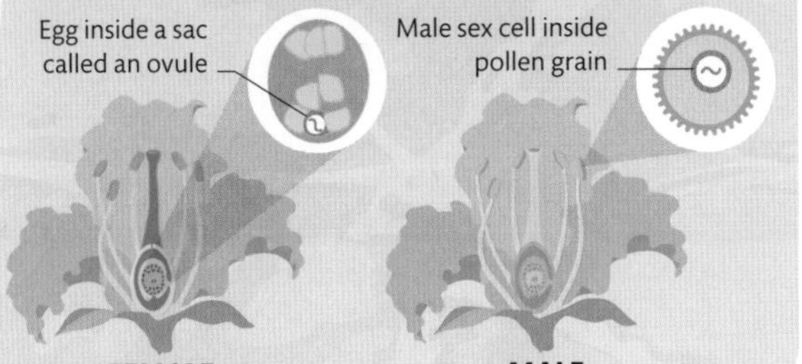
Egg inside a sac called an ovule
Male sex cell inside pollen grain
FEMALE
MALE

Sex in animals

Sperm use a whiplike tail to swim to the egg. In many aquatic animals, fertilization happens in the surrounding water. On land, sperm must enter the female's body, so fertilization is internal.

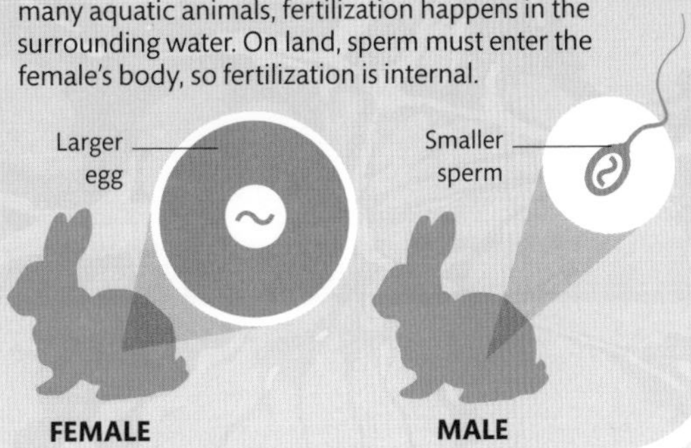
Larger egg
Smaller sperm
FEMALE
MALE

OCEAN SUNFISH LAY **300 MILLION EGGS** AT A TIME—MORE THAN ANY OTHER **VERTEBRATE**

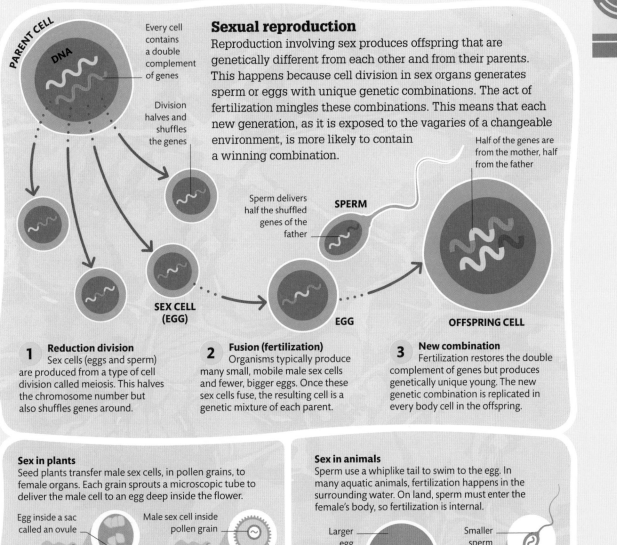

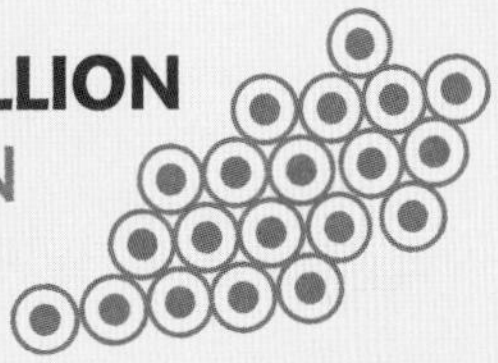

Passing on genes

Offspring inherit characteristics from their parents because those traits are influenced by genes in cells (see pp.158–59). Genes are copied whenever cells divide, and those carried in eggs and sperm are transmitted from one generation to the next. At fertilization, genes from different parents meet. The resulting combination of gene variants is the basis of inheritance.

Basic inheritance

The simplest patterns of inheritance involve a straightforward relationship between one gene and one trait. For example, tiger coat color is controlled by a single gene. The normal variant of this gene gives an orange coat; a rarer mutated version gives white. Each body cell has at least two copies of every kind of gene. But because the orange version is always read in preference, if present, two copies of the white mutation must come together to have any effect. Only then is the white version read, giving a white-coated cub.

WHITE TIGERS ARE NOT A SPECIES— NEARLY ALL ARE BENGAL TIGERS AND ALL CAN BREED WITH ORANGE MATES

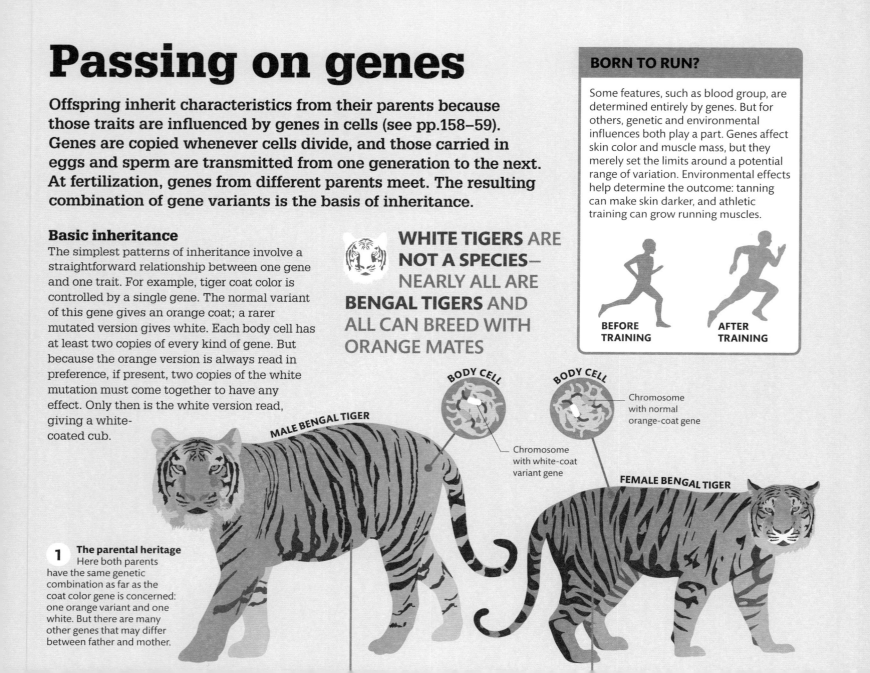

BORN TO RUN?

Some features, such as blood group, are determined entirely by genes. But for others, genetic and environmental influences both play a part. Genes affect skin color and muscle mass, but they merely set the limits around a potential range of variation. Environmental effects help determine the outcome: tanning can make skin darker, and athletic training can grow running muscles.

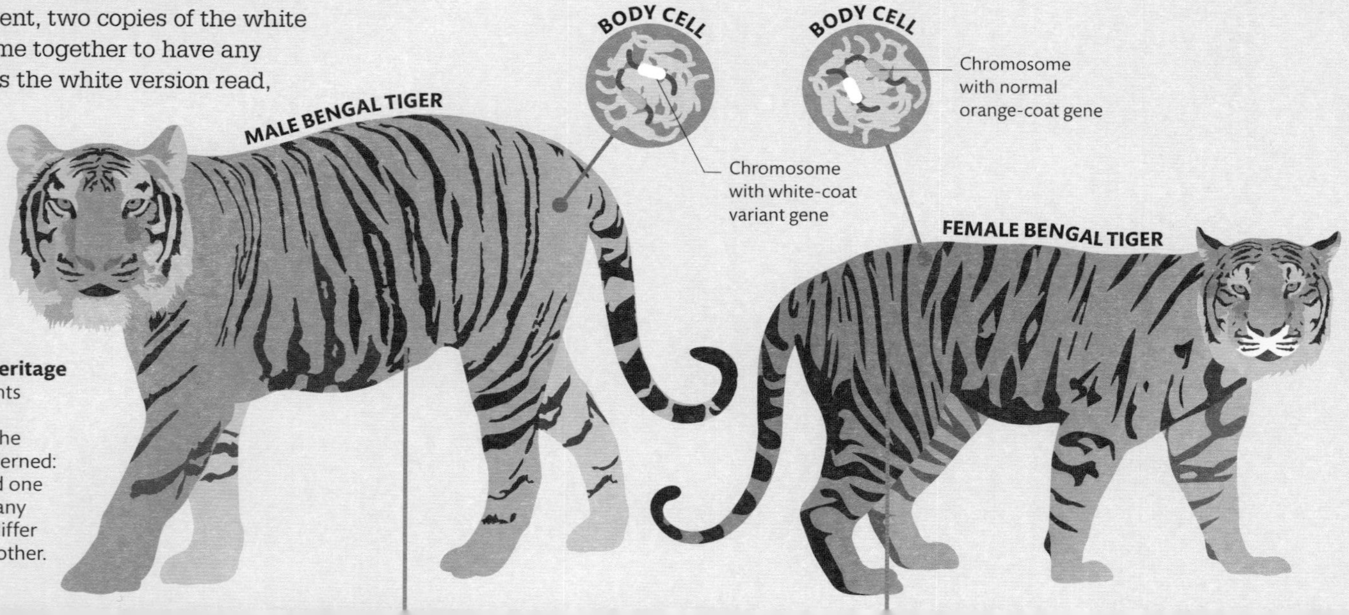

1 The parental heritage Here both parents have the same genetic combination as far as the coat color gene is concerned: one orange variant and one white. But there are many other genes that may differ between father and mother.

2 Sex cells
Formation of sex cells—sperm or eggs—involves a kind of cell division that splits up each pair of chromosomes, so separating gene pairs. It means that half of all sperm or eggs end up with the orange gene variant, and half the white version.

3 The inherited outcome
We cannot predict which sperm will fuse with which egg. But the fixed proportions of orange and white genes among sex cells means that here there is just a one-in-four chance of ending up with two white varieties coming together to form a white cub.

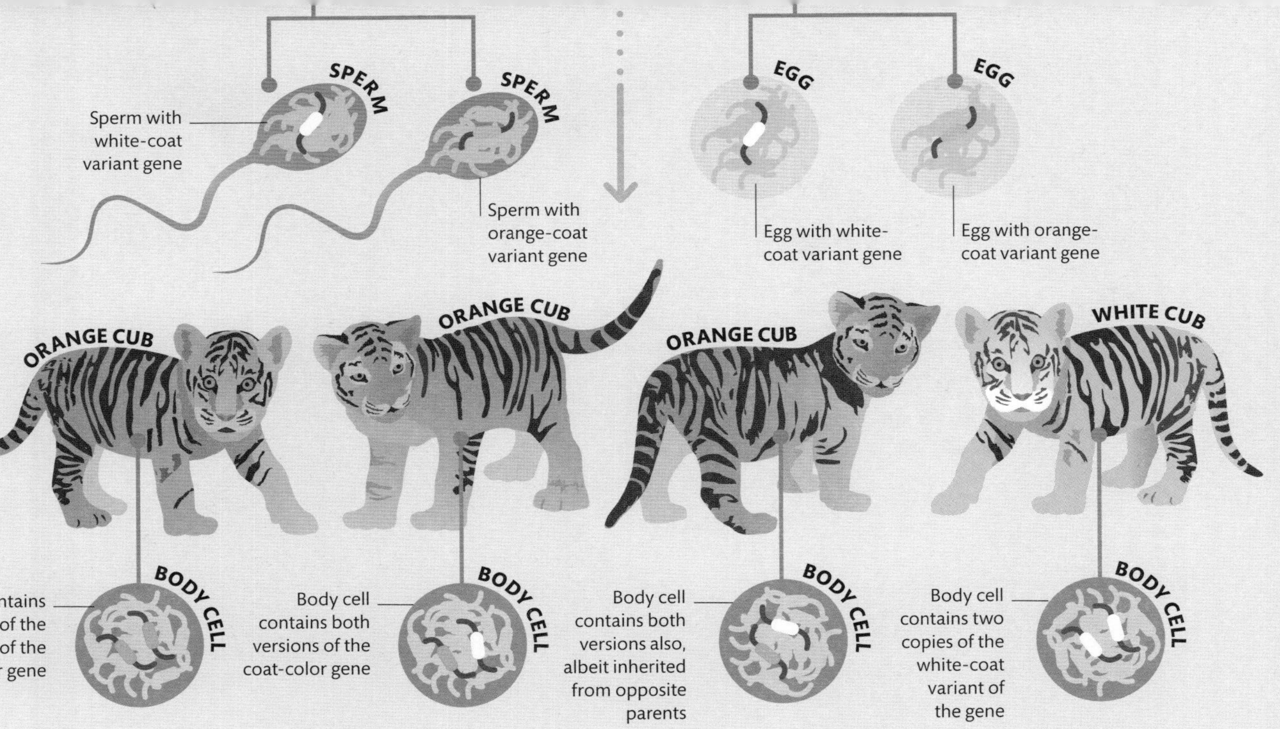

Smooth variation

Not all features are inherited to give the simple fixed proportions that we see with tiger coat colors. In fact, most traits result from multiple interacting genes. For example, human height is influenced by many genes affecting the growth of bones and muscles, making intermediate offspring with a smoother pattern of variation.

How tall will your children grow?
Not only is human height influenced by many interacting genes, but other factors, such as diet, have an influence. Overall, taller parents may have taller offspring, but we cannot predict their actual heights.

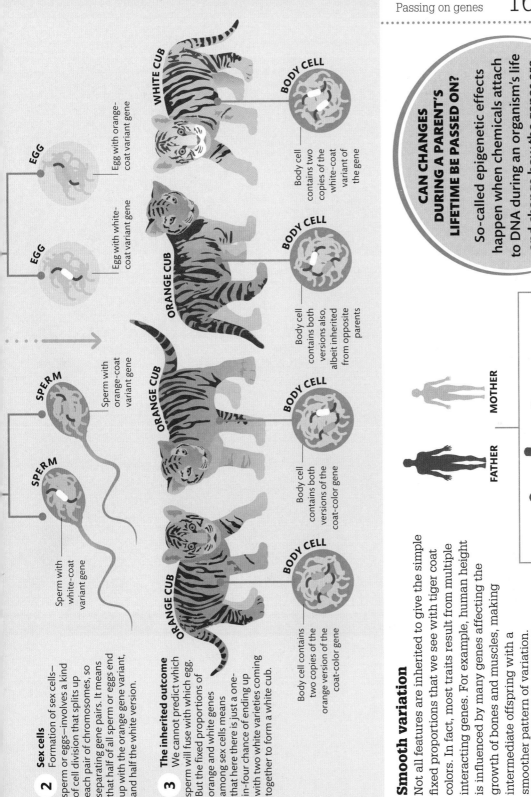

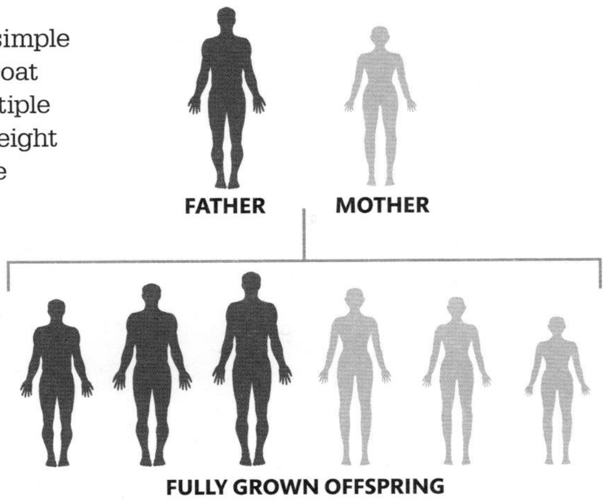

CAN CHANGES DURING A PARENT'S LIFETIME BE PASSED ON?

So-called epigenetic effects happen when chemicals attach to DNA during an organism's life and change how the genes are read. Occasionally, these changes can be passed on to its offspring.

How did life begin?

We will probably never be certain how living things emerged from nonliving matter. But clues about this momentous event are in the rocks around us and in the raw materials in organisms alive today. They suggest that conditions billions of years ago could have fostered an assembly line of increasingly complex molecules leading to the first cells.

Ingredients of life

When life emerged on Earth, the world was a violent place, very different from today. A volcanic landscape lay beneath an atmosphere of poisonous gases that could not screen out the Sun's burning rays. Experiments show that under these high-energy conditions, simple chemicals such as carbon dioxide, methane, water, and ammonia could combine to form the first organic molecules. As these building blocks of life condensed in the early oceans, it is even likely that the emergence of living things was not just fortuitous but inevitable.

VITAL SPARK

In 1952, Stanley Miller and Harold Urey at the University of Chicago tested the idea that complex organic molecules can form from simple, inorganic materials. By energizing their inorganic mixture with a spark to simulate lightning, they recreated early Earth conditions and formed simple amino acids—the building blocks of biological proteins.

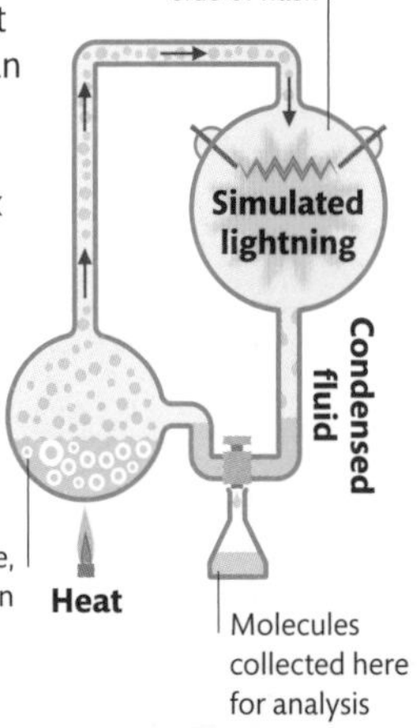

MILLER-UREY EXPERIMENT

Primordial soup
More than 4 billion years ago, Earth's crust was hot and unstable, bombarded by asteroids and in constant volcanic turmoil. But liquid water persisted in places, forming the first oceans and seas—home for the first life.

EARLY EARTH

INORGANIC INGREDIENTS

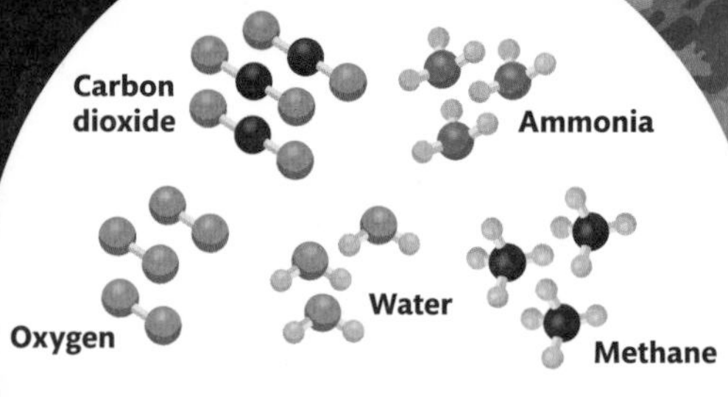

1 The early atmosphere lacked gaseous oxygen but contained a complex mixture of other gases. Carbon dioxide, ammonia, and others were the sources of life's main elements: carbon, hydrogen, oxygen, and nitrogen.

ENERGY INPUT (GEOTHERMAL HEAT AND LIGHTNING)

SIMPLEST ORGANIC MOLECULES

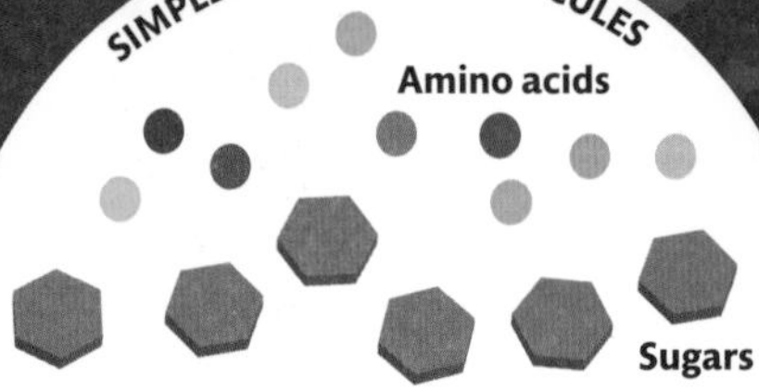

2 Charged with sufficient energy, inorganic substances reacted together to form some of the building blocks of life, such as amino acids and simple sugars. These slightly more complex molecules are called "organic" (see pp.50–51), meaning they contain carbon and have biological potential.

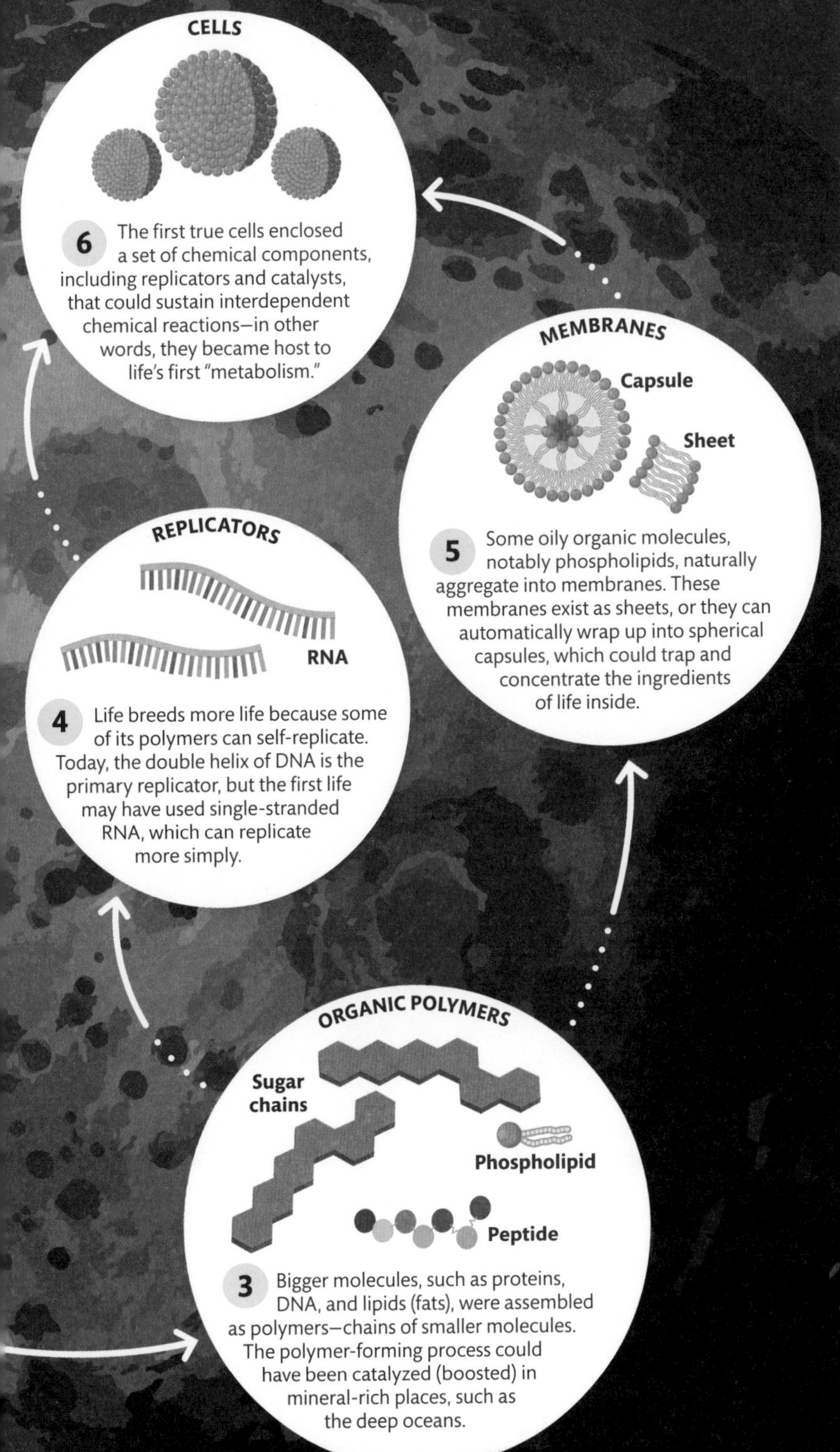
ORGANIC POLYMERS
Sugar chains
Phospholipid
Peptide
3 Bigger molecules, such as proteins, DNA, and lipids (fats), were assembled as polymers—chains of smaller molecules. The polymer-forming process could have been catalyzed (boosted) in mineral-rich places, such as the deep oceans.
REPLICATORS
RNA
4 Life breeds more life because some of its polymers can self-replicate. Today, the double helix of DNA is the primary replicator, but the first life may have used single-stranded RNA, which can replicate more simply.
MEMBRANES
Capsule
Sheet
5 Some oily organic molecules, notably phospholipids, naturally aggregate into membranes. These membranes exist as sheets, or they can automatically wrap up into spherical capsules, which could trap and concentrate the ingredients of life inside.
CELLS
6 The first true cells enclosed a set of chemical components, including replicators and catalysts, that could sustain interdependent chemical reactions—in other words, they became host to life's first "metabolism."

EARTH IS 4.54 BILLION YEARS OLD, AND ITS RECORD OF LIFE MAY STRETCH BACK AS FAR AS 4.28 BILLION YEARS

Life from nonlife

The simplest organic molecules, on their own, are not sufficient to make cells. Small organic molecules must link together into bigger ones, such as proteins and DNA. In the absence of any pre-existing hungry organisms, large molecules would have persisted long enough to get encapsulated, by chance, within oily membranes. It is thought that deep-sea volcanic vents—which, to this day, are rich in minerals capable of catalyzing chemical reactions—could have worked like "hatcheries" to form the first proto-cells in this way.

WHY IS THERE NO LIFE ELSEWHERE IN OUR SOLAR SYSTEM?

Conditions on Earth alone (including a solid surface with oceans of liquid water) are "just right" for life—something sometimes called the Goldilocks effect.

How do things evolve?

Organisms as varied as oak trees, people, and periwinkles are so similar in their genes that a far-reaching scientific conclusion is unavoidable—all life springs, like a giant family tree, from a single common ancestor. Evolution over countless generations is the process that makes this tree branch to give life its diversity.

The case of the giant Galápagos tortoises

Life evolves in particularly distinctive ways when isolated on remote islands. DNA analysis shows that the giant tortoises of the Galápagos are closely related to tortoises on the mainland—and that within a few million years, the diverse island forms emerged from a single colonization.

CAN WE SEE EVOLUTION HAPPENING?

Evolution is slow, but lab populations of fast-breeding organisms, such as fruit flies, have produced strains that cannot interbreed. These can be regarded as new species.

1 Variation
Any natural population varies due to random mutations—errors of DNA copying. Each gene mutates rarely, but mutations are inevitable and accumulate over long periods. They lead to variations in size, shape, and color in a tortoise population. This variation provides the raw material for evolution.

2 Dispersal
The largest South American tortoises—now extinct—were probably ancestors of today's Galápagos giants, drifting up the west coast of South America on the Pacific Ocean's Humboldt Current to reach the Galápagos Islands.

SOUTH AMERICA
GALÁPAGOS ISLANDS
Varying colors show a naturally varied population of tortoises
2
1
Color variation is down to natural mutation
Tortoises float on ocean currents to the Galápagos
Larger tortoises adapted to dry grassland

3 Isolation
Making landfall, the tortoises were now isolated and began evolving separately from those on the mainland. Those adapted to arid habitats survived the dry Galápagos and spread through the archipelago. In the driest places, those with a "saddleback" shell could browse on higher vegetation, so—over time—these predominated.

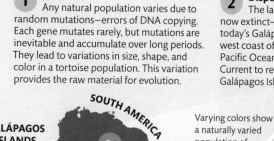

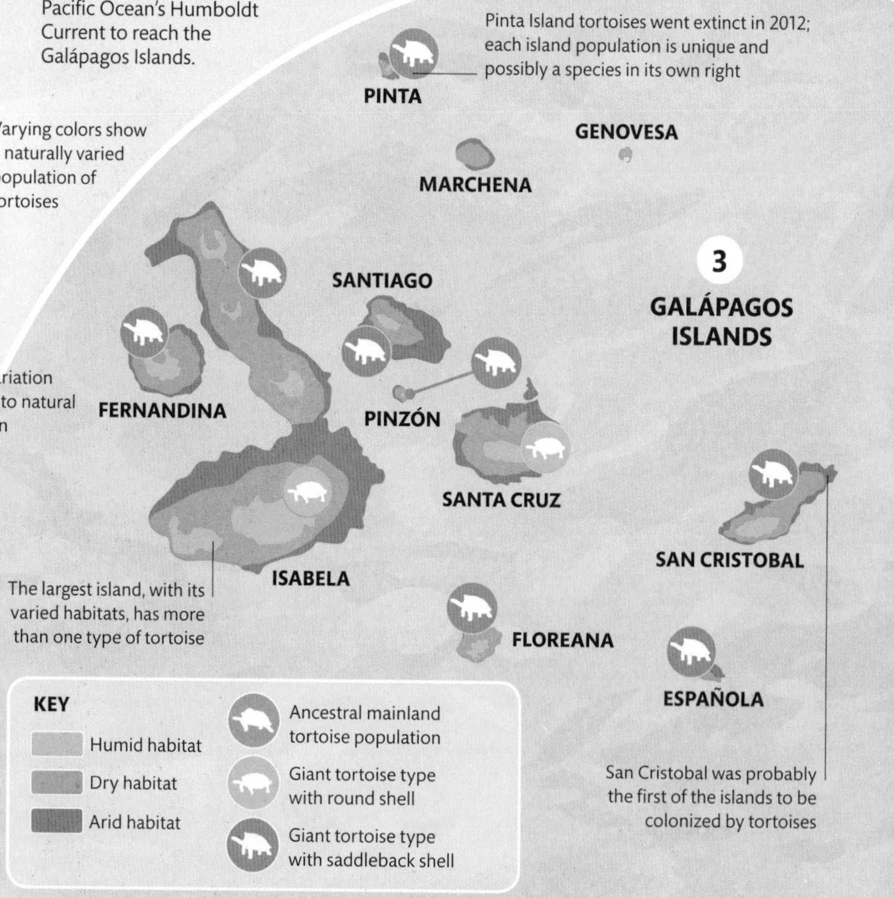

Pinta Island tortoises went extinct in 2012; each island population is unique and possibly a species in its own right
PINTA
GENOVESA
MARCHENA
3
GALÁPAGOS ISLANDS
SANTIAGO
FERNANDINA
PINZÓN
SANTA CRUZ
SAN CRISTOBAL
ISABELA
The largest island, with its varied habitats, has more than one type of tortoise
FLOREANA
ESPAÑOLA
San Cristobal was probably the first of the islands to be colonized by tortoises
KEY
Humid habitat
Dry habitat
Arid habitat
Ancestral mainland tortoise population
Giant tortoise type with round shell
Giant tortoise type with saddleback shell

Survival of the fittest

Genetic variation can make the difference between life and death. A leaf-eating insect that is leaf-green can stay hidden from insect-eaters, but color mutations spoil its camouflage. Green leaf-eaters are more likely to survive and breed, whereas other colors will not. This "natural selection" is the essence of Darwin's famous theory, which explains that species evolve because some variants are better adapted, and they survive to leave more offspring. It inspired the Victorian thinker Herbert Spencer to coin the phrase "survival of the fittest."

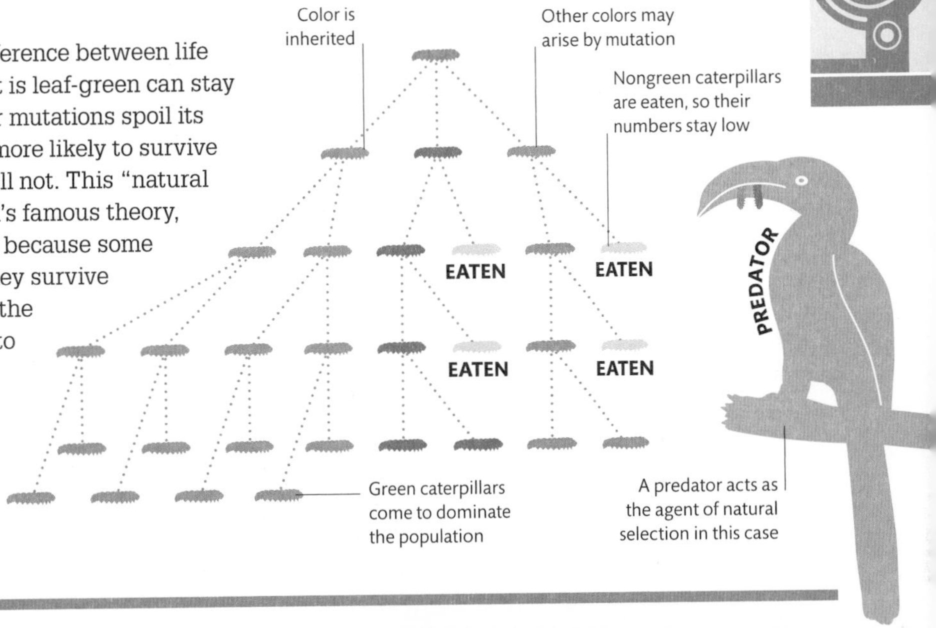
Color is inherited
Other colors may arise by mutation
Nongreen caterpillars are eaten, so their numbers stay low
PREDATOR
EATEN
EATEN
EATEN
EATEN
Green caterpillars come to dominate the population
A predator acts as the agent of natural selection in this case

Selection by predators
Green caterpillars are camouflaged from predators. Mutated gray and brown ones match their background less well, and are "selected out" of the population.

New species from old

Natural selection does not, on its own, make populations diverge into new species. For new species to form, populations have to be prevented from interbreeding, either by geographical isolation, like the Galápagos tortoises, or by behavioral or biological barriers, which often develop while populations are separated. Anything that splits populations and gives evolution enough time to generate reproductive isolation could make new species diverge.

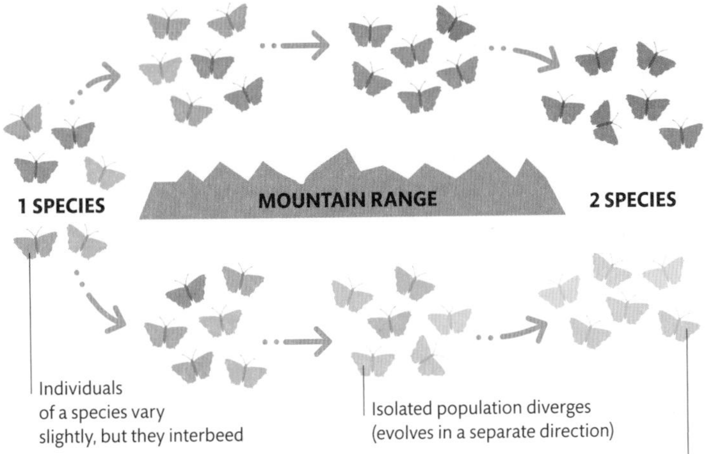
1 SPECIES
MOUNTAIN RANGE
2 SPECIES
Individuals of a species vary slightly, but they interbeed
Isolated population diverges (evolves in a separate direction)
Populations have become new species—they do not interbreed, even when reunited

How new species may form
Natural selection leads to butterflies evolving differently on either side of a new mountain range. After sufficient time, their differences may be so great that they can no longer interbreed.

MACROEVOLUTION

Tiny changes over a few generations add up to bigger ones over millions of years, so diverging species can give rise to entirely new groups of organisms. This is evolution on a grand scale, known as macroevolution, and the fossil record of extinct forms helps to show how it can produce living things as different as giant redwoods and sunflowers from the same ancestor.

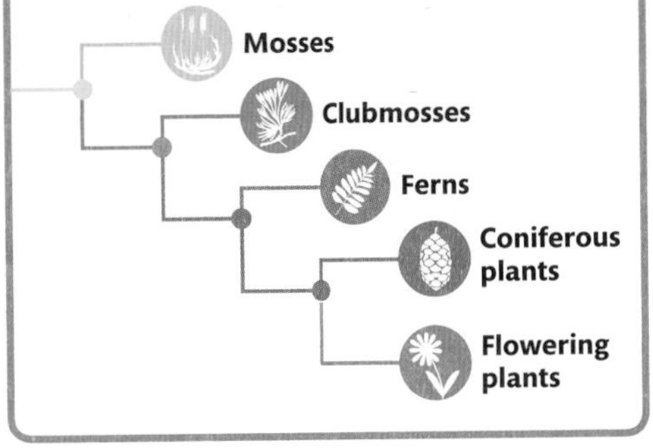
Mosses
Clubmosses
Ferns
Coniferous plants
Flowering plants

A GENE MAY MUTATE AS RARELY AS IN 1 IN A MILLION SPERM OR EGGS

How plants fuel the world

The photosynthesis that generates life-giving sugars in the green parts of plants sustains practically all food chains on the planet. Billions of microscopic solar panels in plant cells harness sunlight to make food from the simplest ingredients: water and carbon dioxide.

SUN

Light energy from sunshine is converted by photosynthesis into chemical energy in sugar

WHY IS CHLOROPHYLL GREEN?

Chlorophyll absorbs red and blue wavelengths of light, using the energy in this light for photosynthesis. The energy of green light is not used and is reflected back—and into our eyes.

Stoma is one of many pores in the leaf that admit carbon dioxide

OXYGEN RELEASED

CARBON DIOXIDE TAKEN IN

The chemical process

More than 90 percent of organic food molecules comprise the elements carbon, hydrogen, and oxygen. When plants make food, carbon dioxide absorbed from the air provides the carbon and oxygen; water absorbed from the soil provides the hydrogen. First, light energy absorbed by the green pigment chlorophyll strips high-energy hydrogen from the water. Second, this hydrogen combines with carbon dioxide to form sugars. The entire process happens in granules called chloroplasts.

Stem contains microscopic vessels that transport sugar

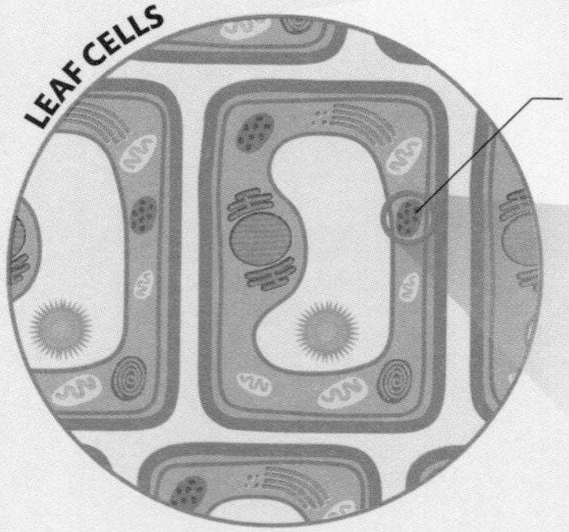

Factories of photosynthesis
Chloroplasts are concentrated in cells of the upper leaf layers, where they are angled to intercept as much light as possible. There are dozens of chloroplasts per cell and billions in a leaf.

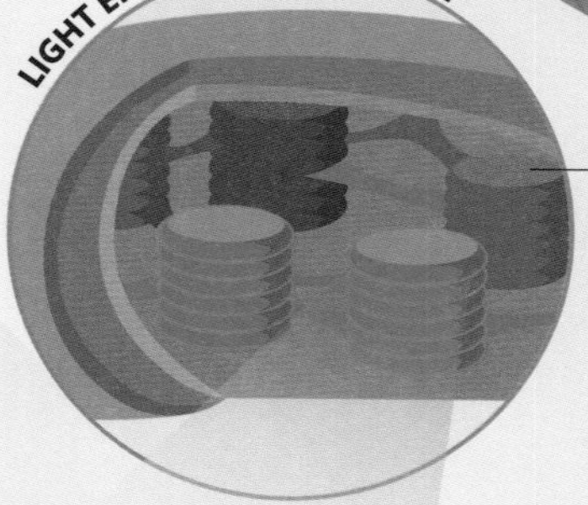

Food-making machinery
The working parts of a chloroplast consist of stacks of membranes, called thylakoids, suspended in a fluid called the stroma. Chlorophyll is bound to the thylakoids, while both membranes and fluid are rich in reaction-driving enzymes.

1 Sunlight splits water
Each disclike arrangement of thylakoid is coated with a cluster of chlorophyll molecules and enzymes needed to extract hydrogen from water. It means that energy in sunlight is efficiently transferred to the hydrogen.

2 Sugar manufacture
The energized hydrogen passes to the stroma. Here, enzymes add the hydrogen to carbon dioxide to make the sugar, glucose.

3 Building biomass
Some of the glucose is "burned" to release energy (see pp.172–73), and some is used in metabolic processes that create substances such as oils, proteins, and woody lignin. The rest is built into long chains of sugars, such as starch—the plant's energy store—and fibrous cellulose, a building material.

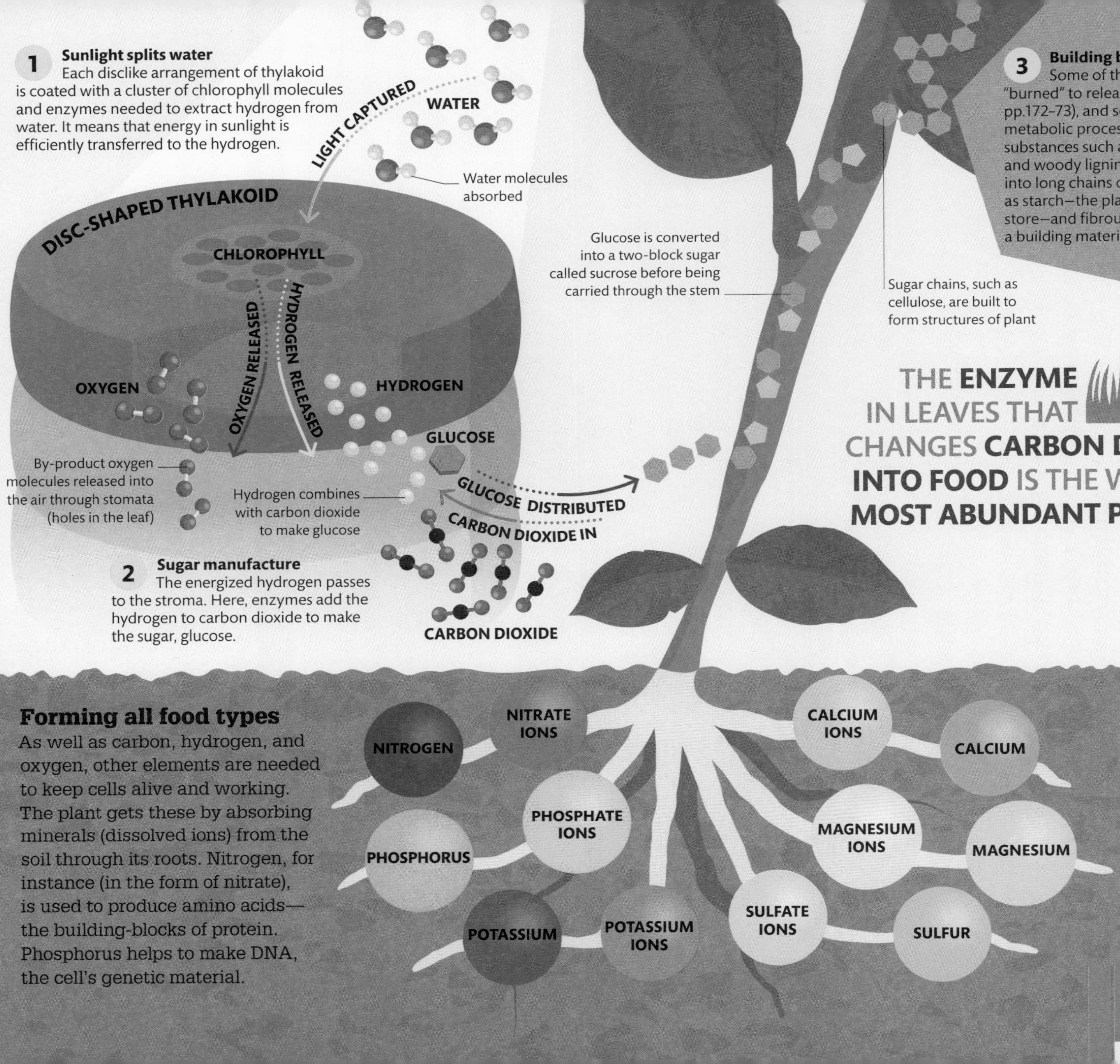

THE ENZYME IN LEAVES THAT CHANGES CARBON DIOXIDE INTO FOOD IS THE WORLD'S MOST ABUNDANT PROTEIN

Forming all food types

As well as carbon, hydrogen, and oxygen, other elements are needed to keep cells alive and working. The plant gets these by absorbing minerals (dissolved ions) from the soil through its roots. Nitrogen, for instance (in the form of nitrate), is used to produce amino acids—the building-blocks of protein. Phosphorus helps to make DNA, the cell's genetic material.

How plants grow

The lives of plants are finely regulated by substances that control every aspect of their growth, from the germination of seeds to the blooming of flowers. These growth regulators are produced in tiny amounts, but they profoundly influence the final form of a mature plant.

TREE RINGS

A plant's growth rate can vary according to temperature and rainfall. Growth is faster in summer but may almost stop in winter. These spurts give the familiar rings that run through a tree's trunk. Even in the tropics, with little or no winter to slow growth, trees often grow faster in the wet season, if there is one, so they produce rings just like those in temperate parts. If tropical trees grow steadily and continuously, no rings appear.

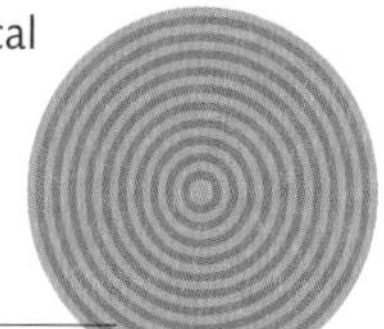
Each pale ring marks a summer's fast growth, the oldest ring at the center

CROSS-SECTION OF A TREE TRUNK

Stimulating growth

At each stage of a plant's life, different growth regulators ensure that development is coordinated. They are produced by cells in shoots, roots, or leaves and seep through tissues from their point of origin, then they are carried to other parts of the plant in the flowing sap. The outcome can depend on the balance between two or more regulators. Some counteract one another, while others are reinforcing. The same regulator may even have opposite effects in different parts of a plant.

KEY

- Water
- Gibberellin
- Auxin
- Cytokinin
- Florigen

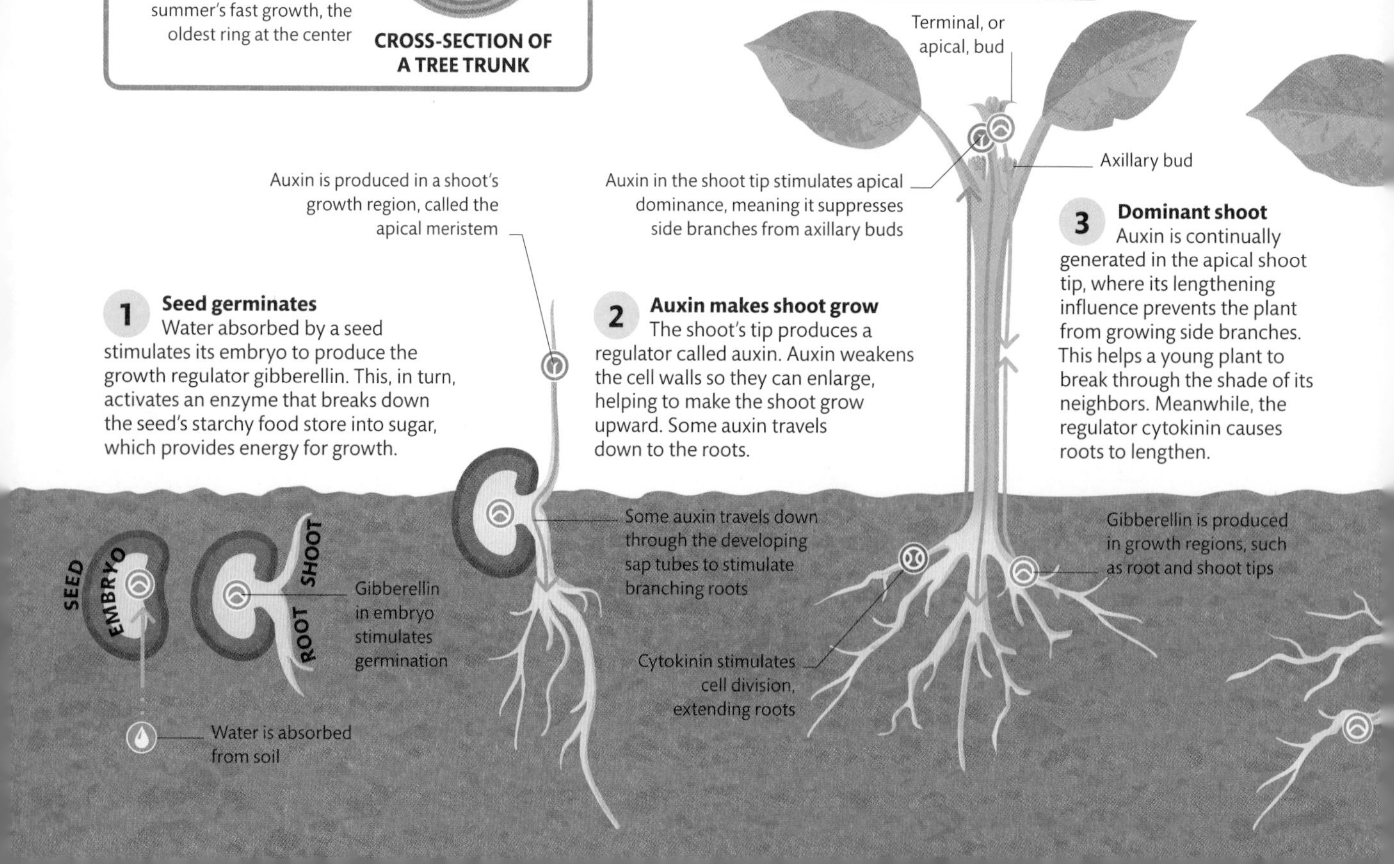
Terminal, or apical, bud
Axillary bud
Auxin in the shoot tip stimulates apical dominance, meaning it suppresses side branches from axillary buds
Auxin is produced in a shoot's growth region, called the apical meristem
SEED
EMBRYO
SHOOT
ROOT
Gibberellin in embryo stimulates germination
Water is absorbed from soil
Some auxin travels down through the developing sap tubes to stimulate branching roots
Cytokinin stimulates cell division, extending roots
Gibberellin is produced in growth regions, such as root and shoot tips

1 Seed germinates
Water absorbed by a seed stimulates its embryo to produce the growth regulator gibberellin. This, in turn, activates an enzyme that breaks down the seed's starchy food store into sugar, which provides energy for growth.

2 Auxin makes shoot grow
The shoot's tip produces a regulator called auxin. Auxin weakens the cell walls so they can enlarge, helping to make the shoot grow upward. Some auxin travels down to the roots.

3 Dominant shoot
Auxin is continually generated in the apical shoot tip, where its lengthening influence prevents the plant from growing side branches. This helps a young plant to break through the shade of its neighbors. Meanwhile, the regulator cytokinin causes roots to lengthen.

Quick response

Auxin is responsible for making plant shoots bend toward the Sun. When light shines from one direction, the auxin moves to the shadier side, making cells grow bigger there. This results in the shoot curving away from the shade toward the light, bringing leaves to face the Sun. The action can be quick enough to track the Sun across the sky.

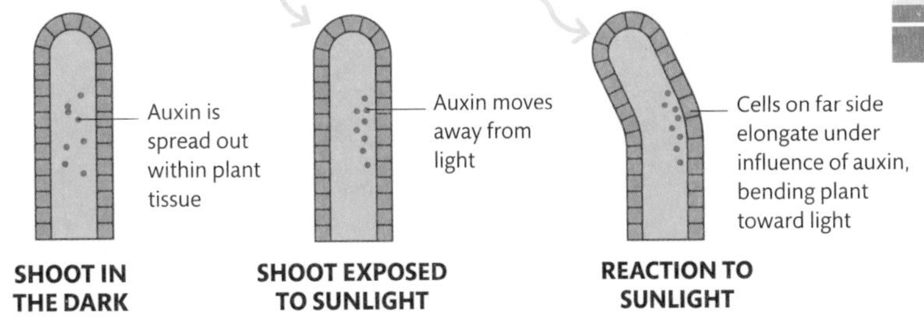

Auxin is spread out within plant tissue
SHOOT IN THE DARK
Auxin moves away from light
SHOOT EXPOSED TO SUNLIGHT
Cells on far side elongate under influence of auxin, bending plant toward light
REACTION TO SUNLIGHT

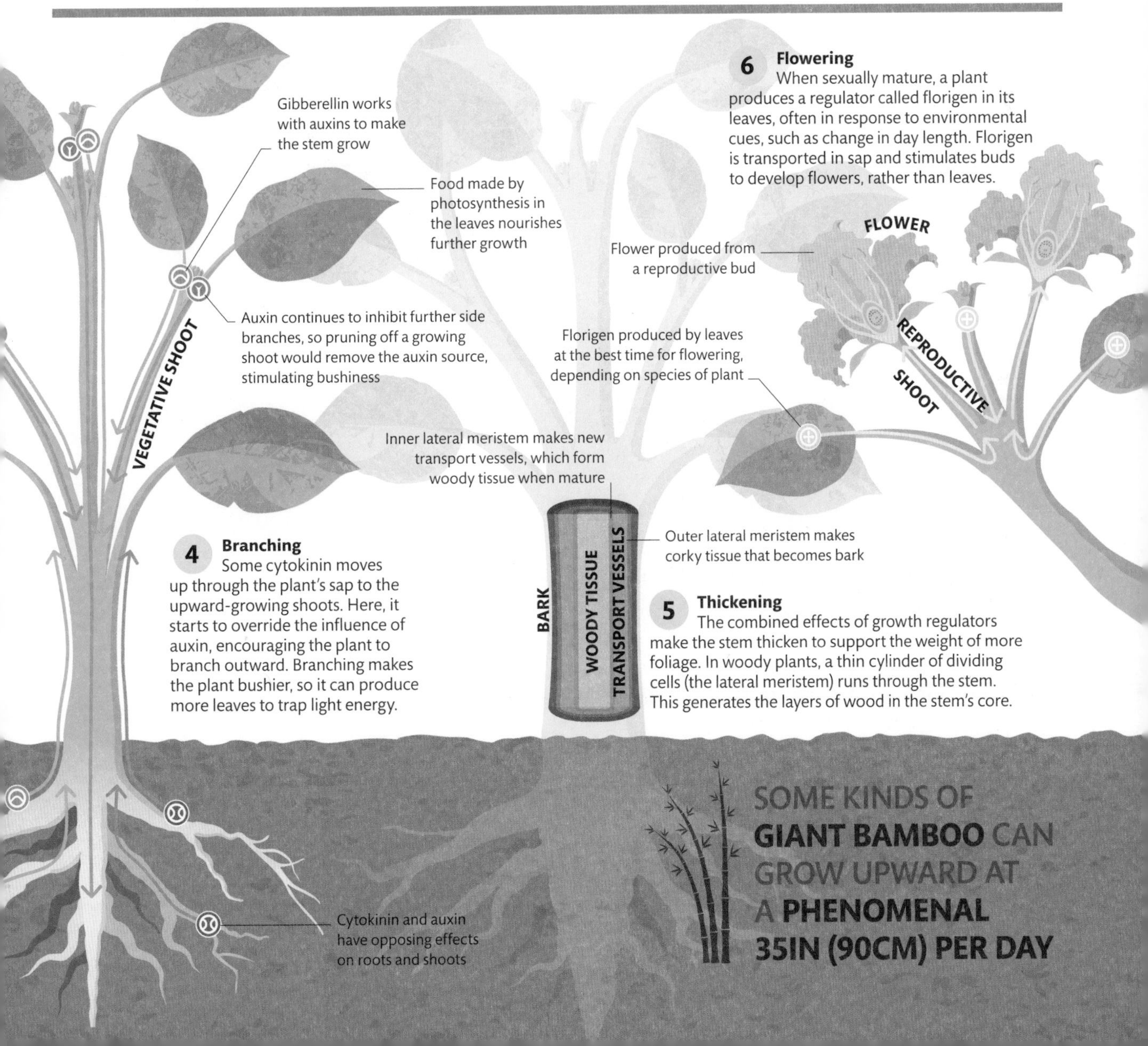

Gibberellin works with auxins to make the stem grow
Food made by photosynthesis in the leaves nourishes further growth
Auxin continues to inhibit further side branches, so pruning off a growing shoot would remove the auxin source, stimulating bushiness
VEGETATIVE SHOOT
Inner lateral meristem makes new transport vessels, which form woody tissue when mature
BARK
WOODY TISSUE
TRANSPORT VESSELS
Outer lateral meristem makes corky tissue that becomes bark
FLOWER
Flower produced from a reproductive bud
Florigen produced by leaves at the best time for flowering, depending on species of plant
REPRODUCTIVE SHOOT
Cytokinin and auxin have opposing effects on roots and shoots

6 Flowering
When sexually mature, a plant produces a regulator called florigen in its leaves, often in response to environmental cues, such as change in day length. Florigen is transported in sap and stimulates buds to develop flowers, rather than leaves.

4 Branching
Some cytokinin moves up through the plant's sap to the upward-growing shoots. Here, it starts to override the influence of auxin, encouraging the plant to branch outward. Branching makes the plant bushier, so it can produce more leaves to trap light energy.

5 Thickening
The combined effects of growth regulators make the stem thicken to support the weight of more foliage. In woody plants, a thin cylinder of dividing cells (the lateral meristem) runs through the stem. This generates the layers of wood in the stem's core.

SOME KINDS OF **GIANT BAMBOO** CAN GROW UPWARD AT A **PHENOMENAL 35IN (90CM) PER DAY**

Respiration

Life needs energy just to keep going. It is used deep inside cells, where life's microscopic machinery is hard at work processing food, growing new material, and responding to change. Chemical reactions—collectively called respiration—generate this energy in a series of steps involving the breakdown of food.

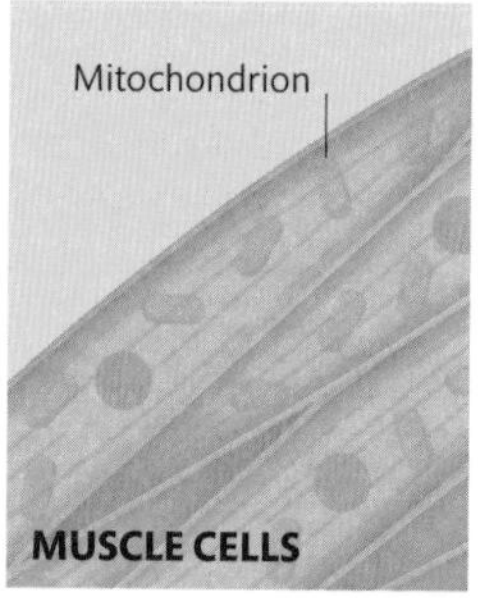
Mitochondrion
MUSCLE CELLS

Fueling the cells

Practically all life forms—from microbes to oak trees—get their energy by breaking down glucose. The most efficient way to do this is to split it completely, so each of glucose's six carbon atoms are separated into six carbon dioxide molecules. But this needs oxygen—much like the combustion of any fuel. Animals deliver glucose and oxygen to cells in their blood circulatory system. Once inside cells, a chain of reactions begins in the cell fluid and ends in mitochondria—the cell's power houses. The entire process unlocks the maximum possible amount of energy.

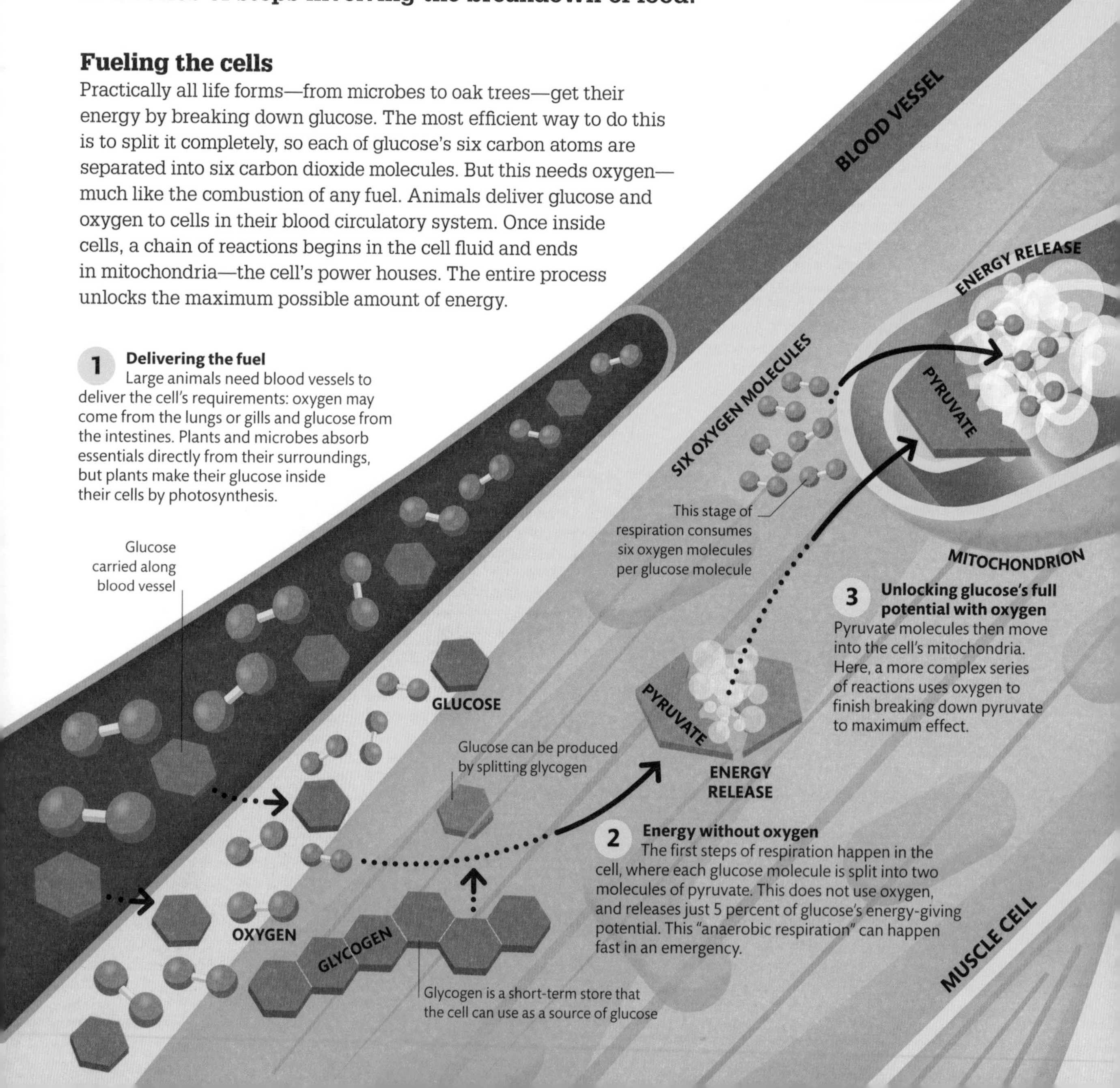
BLOOD VESSEL
ENERGY RELEASE
SIX OXYGEN MOLECULES
PYRUVATE
This stage of respiration consumes six oxygen molecules per glucose molecule
MITOCHONDRION
Glucose carried along blood vessel
GLUCOSE
Glucose can be produced by splitting glycogen
PYRUVATE
ENERGY RELEASE
OXYGEN
GLYCOGEN
Glycogen is a short-term store that the cell can use as a source of glucose
MUSCLE CELL

1 Delivering the fuel
Large animals need blood vessels to deliver the cell's requirements: oxygen may come from the lungs or gills and glucose from the intestines. Plants and microbes absorb essentials directly from their surroundings, but plants make their glucose inside their cells by photosynthesis.

2 Energy without oxygen
The first steps of respiration happen in the cell, where each glucose molecule is split into two molecules of pyruvate. This does not use oxygen, and releases just 5 percent of glucose's energy-giving potential. This "anaerobic respiration" can happen fast in an emergency.

3 Unlocking glucose's full potential with oxygen
Pyruvate molecules then move into the cell's mitochondria. Here, a more complex series of reactions uses oxygen to finish breaking down pyruvate to maximum effect.

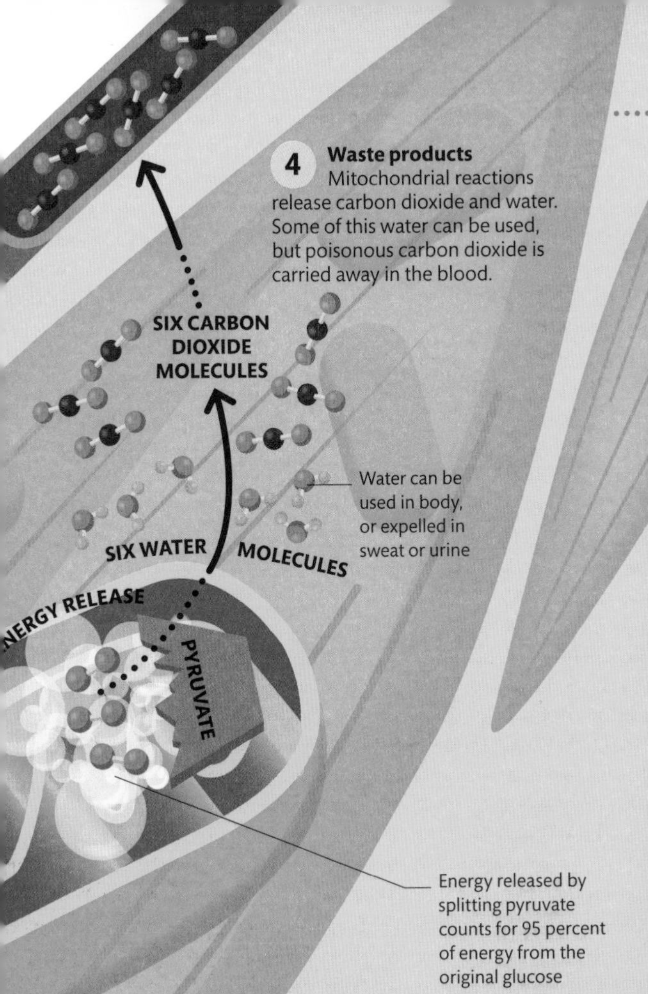

4 Waste products
Mitochondrial reactions release carbon dioxide and water. Some of this water can be used, but poisonous carbon dioxide is carried away in the blood.

Water can be used in body, or expelled in sweat or urine

Energy released by splitting pyruvate counts for 95 percent of energy from the original glucose

Where does energy go?

All organisms use energy to maintain the cell's functions: their basal metabolism. But extra work is needed to move, grow, and reproduce. Animals use proportionately more energy than plants on movement, because muscle contraction requires energy. Warm-blooded animals have the highest energy demands. Maintaining a high body temperature accounts for a large proportion of this high energy expenditure.

KEY
- Metabolism
- Reproduction
- Generating body heat
- Growth
- Movement

Plant
Although plants use light energy to make food in photosynthesis, they must still respire to release energy to power life's vital processes.

Cold-blooded snake
As in other animals, much of a snake's energy powers movement. However, respiratory energy is not used to heat their body—they rely on the Sun to do this.

Warm-blooded fully grown mouse
Small warm-blooded animals lose more heat, proportionately, so use the greatest part of their energy budget in their body's central heating.

DO PLANTS BREATHE IN CO_2?

No – in sunlight, plants absorb carbon dioxide to make sugar, but this isn't breathing. Plants also respire just like animals, taking in oxygen and releasing CO_2. This process is akin to breathing.

GAS EXCHANGE

Contrary to popular understanding, respiration does not mean the same as breathing. Energy-releasing respiration happens in all cells of organisms, but breathing is the movement of lungs in animals that have them. Technically called ventilation, breathing helps bring fresh supplies of oxygen into the blood and flush away carbon dioxide.

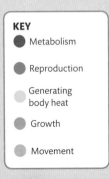

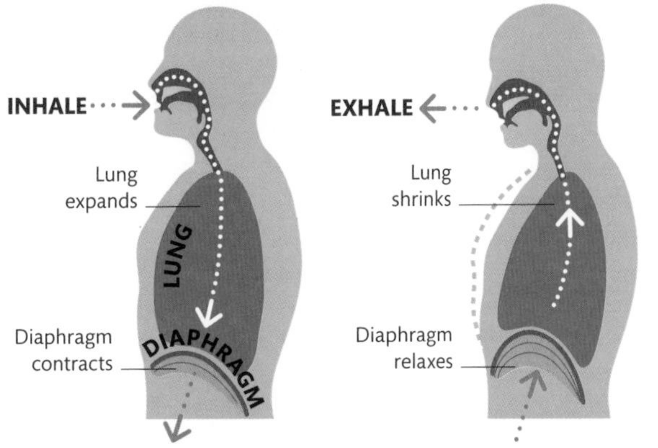

MANGROVE TREES GROW IN AIRLESS MUD SO THEIR ROOTS GROW UP TO GET THEIR OXYGEN

The carbon cycle

Carbon atoms are moved by biological and physical processes through the air, oceans, land, and the bodies of living things. Stores of carbon are called "carbon sinks"—and carbon moves between them at various rates.

Natural balance

Each year, photosynthesis concentrates carbon into plants and algae by changing carbon dioxide (CO_2) from the air into food. Respiration and natural combustion push carbon back into the air in roughly equal amounts. Much slower transitions, over millions of years, move carbon through the rocks. But when humans burn fossil fuels, CO_2 release from the ground is rapidly accelerated, leading to an extra 9 billion tons of carbon being released every year.

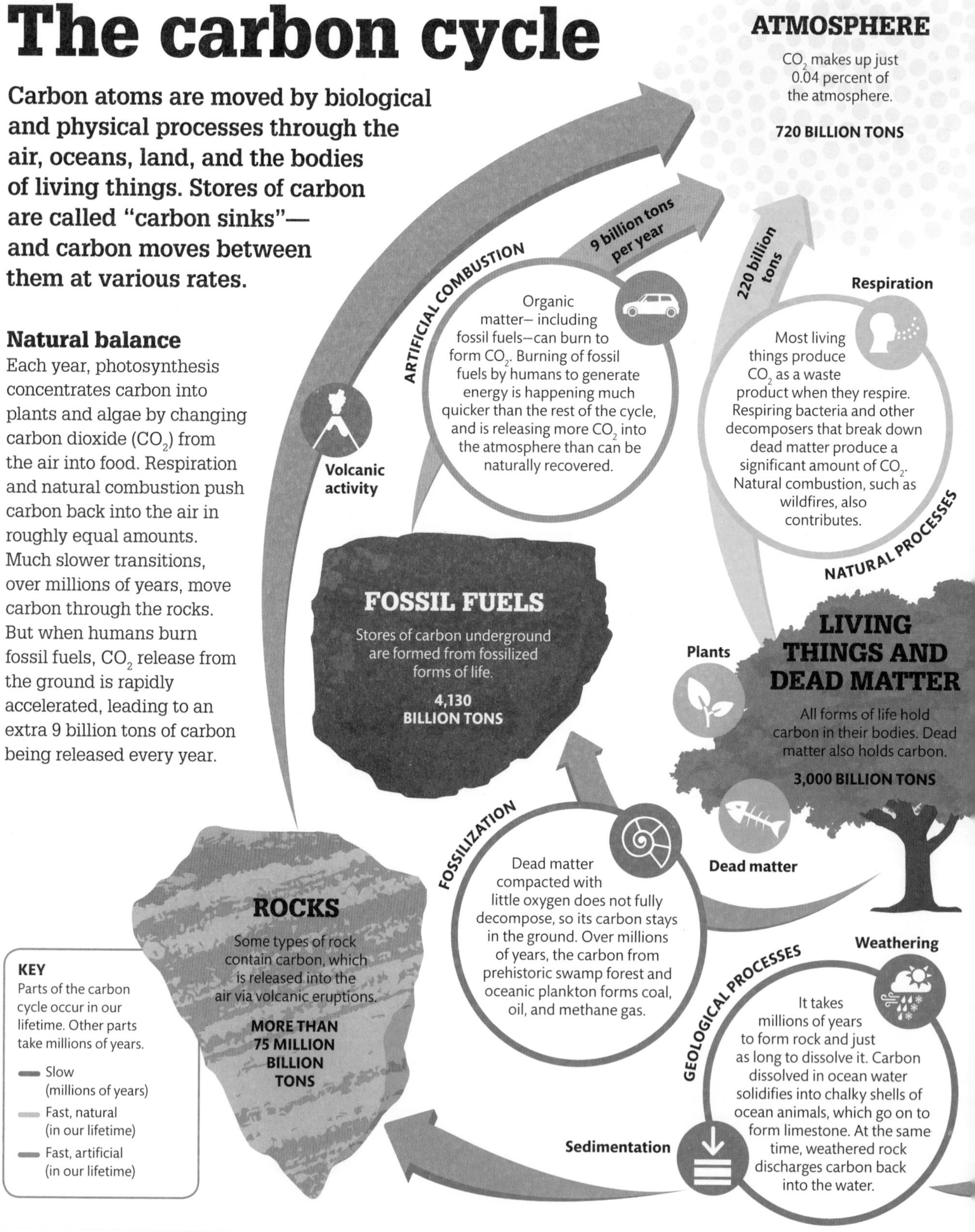

ATMOSPHERE
CO_2 makes up just 0.04 percent of the atmosphere.
720 BILLION TONS
ARTIFICIAL COMBUSTION
9 billion tons per year
Organic matter– including fossil fuels—can burn to form CO_2. Burning of fossil fuels by humans to generate energy is happening much quicker than the rest of the cycle, and is releasing more CO_2 into the atmosphere than can be naturally recovered.
Volcanic activity
220 billion tons
Respiration
Most living things produce CO_2 as a waste product when they respire. Respiring bacteria and other decomposers that break down dead matter produce a significant amount of CO_2. Natural combustion, such as wildfires, also contributes.
NATURAL PROCESSES
FOSSIL FUELS
Stores of carbon underground are formed from fossilized forms of life.
4,130 BILLION TONS
Plants
LIVING THINGS AND DEAD MATTER
All forms of life hold carbon in their bodies. Dead matter also holds carbon.
3,000 BILLION TONS
Dead matter
FOSSILIZATION
Dead matter compacted with little oxygen does not fully decompose, so its carbon stays in the ground. Over millions of years, the carbon from prehistoric swamp forest and oceanic plankton forms coal, oil, and methane gas.
ROCKS
Some types of rock contain carbon, which is released into the air via volcanic eruptions.
MORE THAN 75 MILLION BILLION TONS
GEOLOGICAL PROCESSES
Weathering
It takes millions of years to form rock and just as long to dissolve it. Carbon dissolved in ocean water solidifies into chalky shells of ocean animals, which go on to form limestone. At the same time, weathered rock discharges carbon back into the water.
Sedimentation
KEY
Parts of the carbon cycle occur in our lifetime. Other parts take millions of years.
Slow (millions of years)
Fast, natural (in our lifetime)
Fast, artificial (in our lifetime)

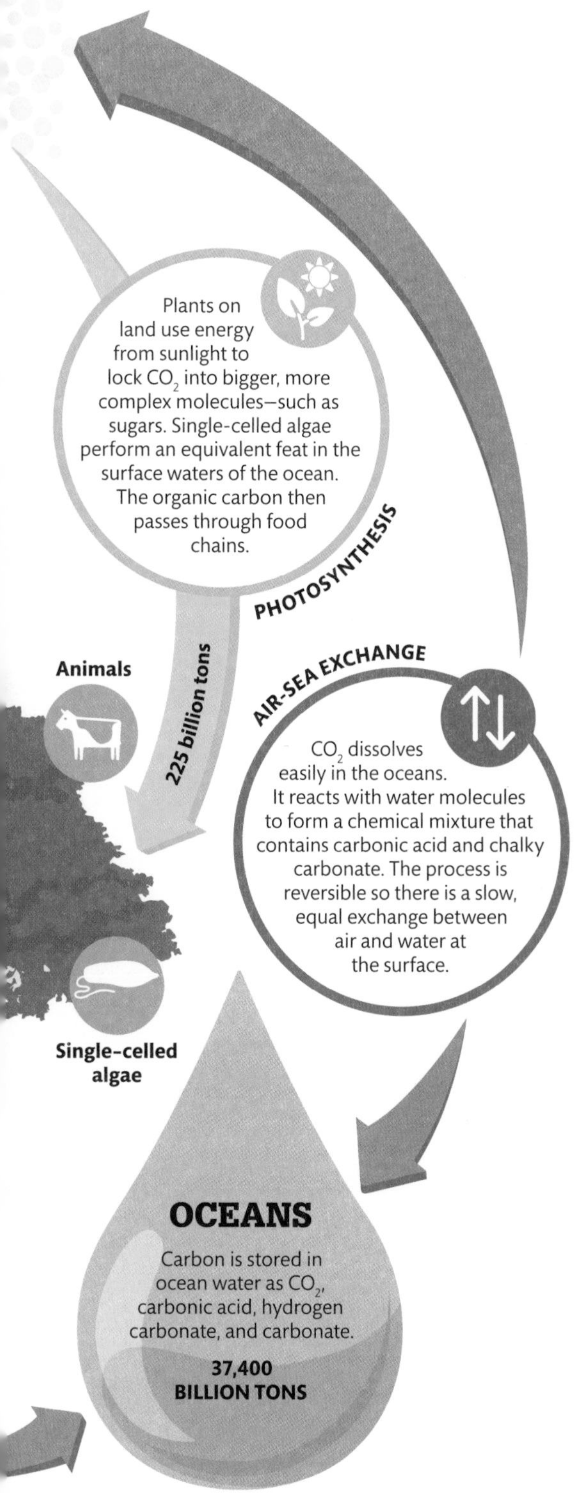
Plants on land use energy from sunlight to lock CO_2 into bigger, more complex molecules—such as sugars. Single-celled algae perform an equivalent feat in the surface waters of the ocean. The organic carbon then passes through food chains.
PHOTOSYNTHESIS
225 billion tons
Animals
AIR-SEA EXCHANGE
CO_2 dissolves easily in the oceans. It reacts with water molecules to form a chemical mixture that contains carbonic acid and chalky carbonate. The process is reversible so there is a slow, equal exchange between air and water at the surface.
Single-celled algae
OCEANS
Carbon is stored in ocean water as CO_2, carbonic acid, hydrogen carbonate, and carbonate.
37,400 BILLION TONS

Carbon capture

Human-influenced combustion and respiration releases 229 billion tons of CO_2 into the atmosphere each year. Photosynthesis absorbs 225 billion tons—so an extra 4 billion tons accumulates. Rises in CO_2—one of many greenhouse gases (see pp.245)—cause global warming (see pp.246–47). Technology offers ways for industrial practices to capture carbon instead of releasing it into the atmosphere.

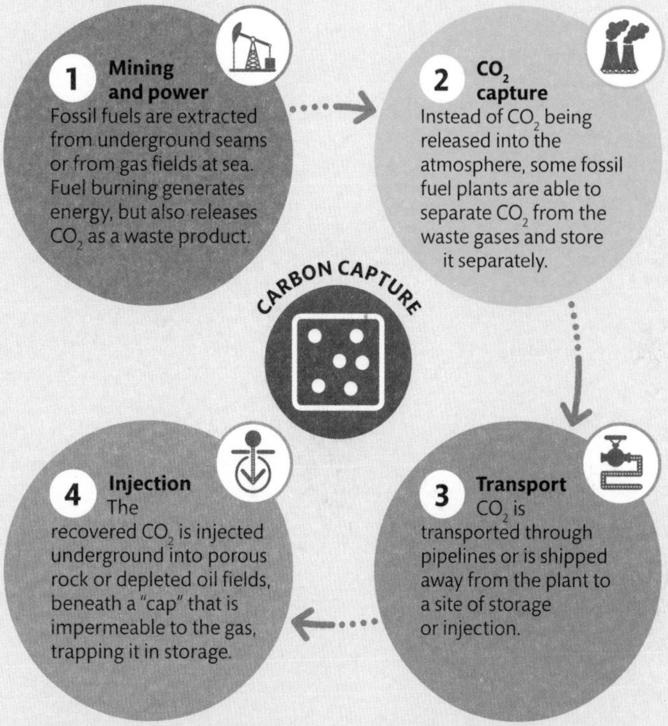
1 Mining and power
Fossil fuels are extracted from underground seams or from gas fields at sea. Fuel burning generates energy, but also releases CO_2 as a waste product.
2 CO_2 capture
Instead of CO_2 being released into the atmosphere, some fossil fuel plants are able to separate CO_2 from the waste gases and store it separately.
CARBON CAPTURE
3 Transport
CO_2 is transported through pipelines or is shipped away from the plant to a site of storage or injection.
4 Injection
The recovered CO_2 is injected underground into porous rock or depleted oil fields, beneath a "cap" that is impermeable to the gas, trapping it in storage.

OCEAN ACIDIFICATION

As levels of CO_2 in the atmosphere rise, more CO_2 enters the ocean to react with water, producing more carbonic acid. A 30 percent rise in ocean acidity since 1750 has had significant consequences for marine life, causing corrosion of animal shells and dieback of rocky corals.

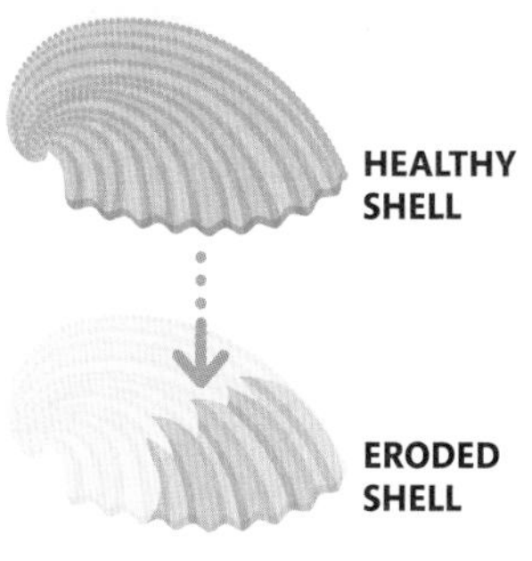
HEALTHY SHELL
ERODED SHELL

Aging

Like anything with many working parts, a living thing shows signs of its age. Living things can check and repair themselves, but over time their bodies will start to malfunction.

What is aging?

A decline in biological function with age can be traced to the waning properties of cells, chromosomes, and genes. Cells of multicelled living things constantly divide to create new ones and typically start to deteriorate after 50 rounds of division, so the production of new cells declines and eventually stops altogether. This is associated with the genetic makeup becoming increasingly unstable, which ultimately leads to cells—and consequently the body—malfunctioning. The effects are linked to many degenerative phenomena, from slowing of repair after injury to dementia.

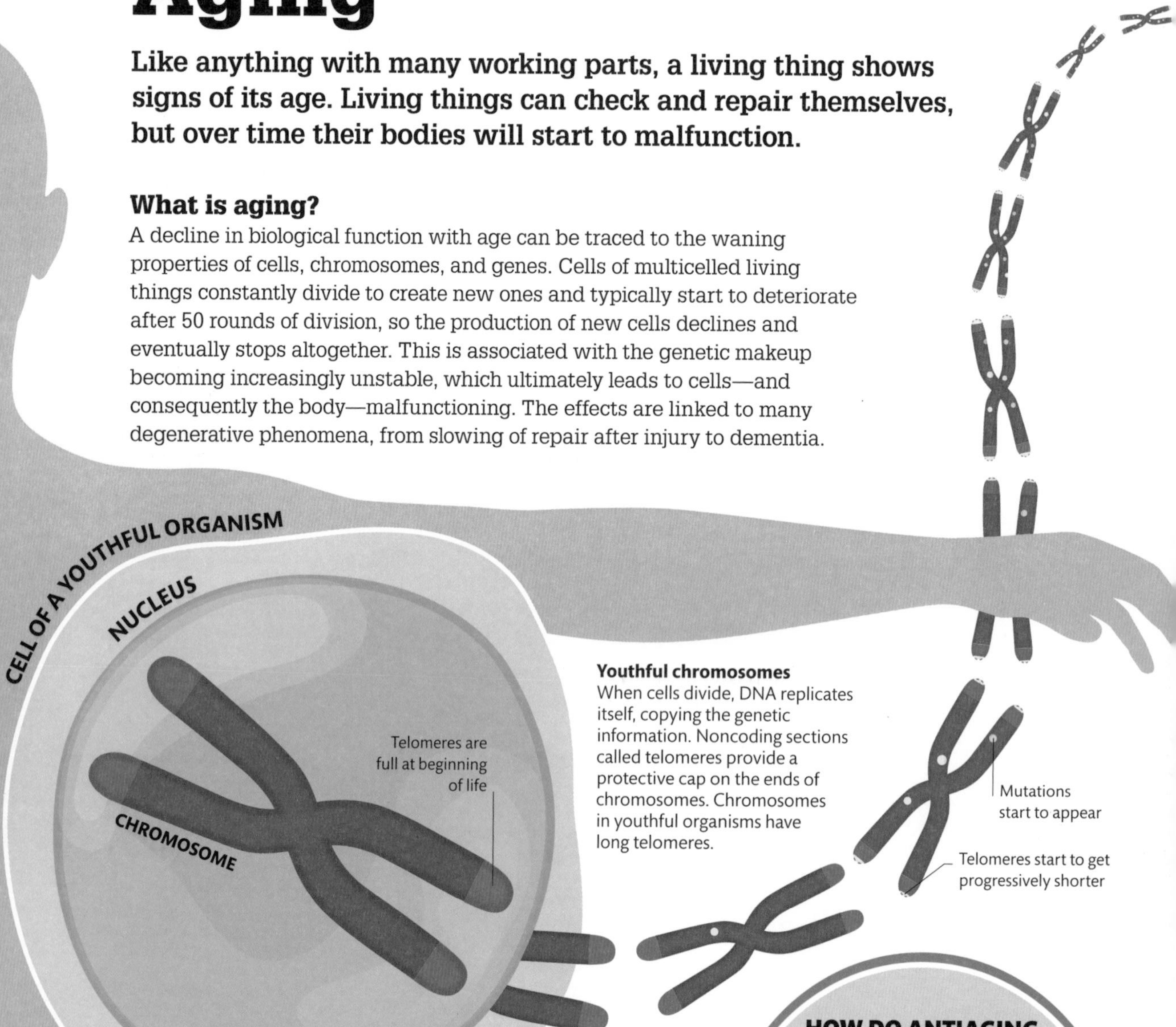

CELL OF A YOUTHFUL ORGANISM
NUCLEUS
CHROMOSOME
Telomeres are full at beginning of life
Mutations start to appear
Telomeres start to get progressively shorter

Youthful chromosomes
When cells divide, DNA replicates itself, copying the genetic information. Noncoding sections called telomeres provide a protective cap on the ends of chromosomes. Chromosomes in youthful organisms have long telomeres.

HOW DO ANTIAGING CREAMS WORK?

Wrinkly skin is caused by a loss of protein fibers. Antiaging creams that contain antioxidants and protein building blocks increase production of these fibers, so firming up the skin.

ONE OF THE **OLDEST LIVING THINGS** IS A **BRISTLECONE PINE TREE** THAT IS ESTIMATED TO BE OVER **5,000 YEARS OLD**

Degrading chromosomes
Mutations (copying errors) accumulate over time, and telomeres shorten each time DNA replicates. Once this erosion reaches the coding section beneath the cap, underlying genes may malfunction.

Cellular breakdown
Genes code for proteins that perform tasks ranging from driving chemical reactions to intercepting signals, so faulty genes inevitably lead to faulty functions. Over time, cells work less efficiently.

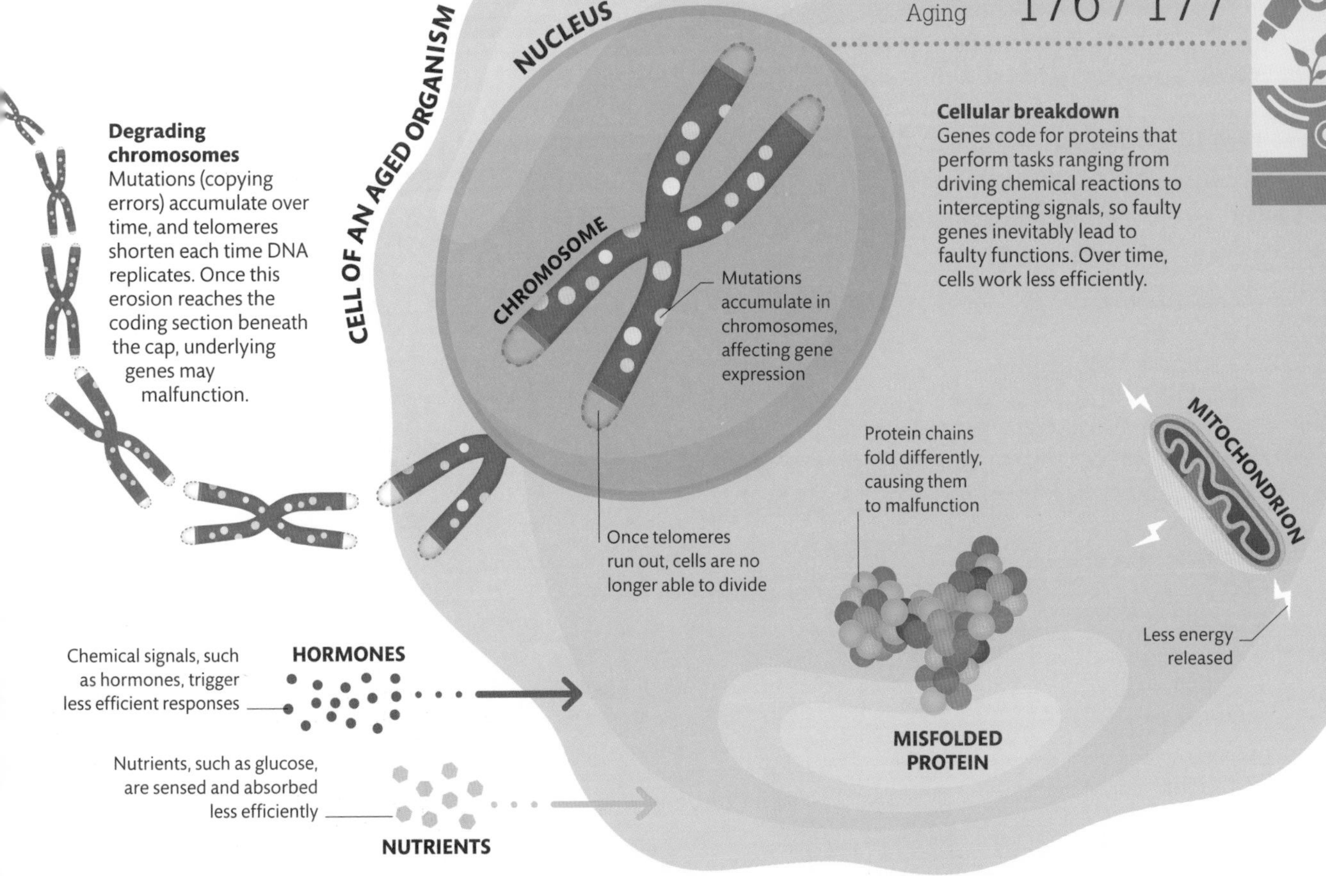
CELL OF AN AGED ORGANISM
NUCLEUS
CHROMOSOME
Mutations accumulate in chromosomes, affecting gene expression
Once telomeres run out, cells are no longer able to divide
Protein chains fold differently, causing them to malfunction
MISFOLDED PROTEIN
MITOCHONDRION
Less energy released
Chemical signals, such as hormones, trigger less efficient responses
HORMONES
Nutrients, such as glucose, are sensed and absorbed less efficiently
NUTRIENTS

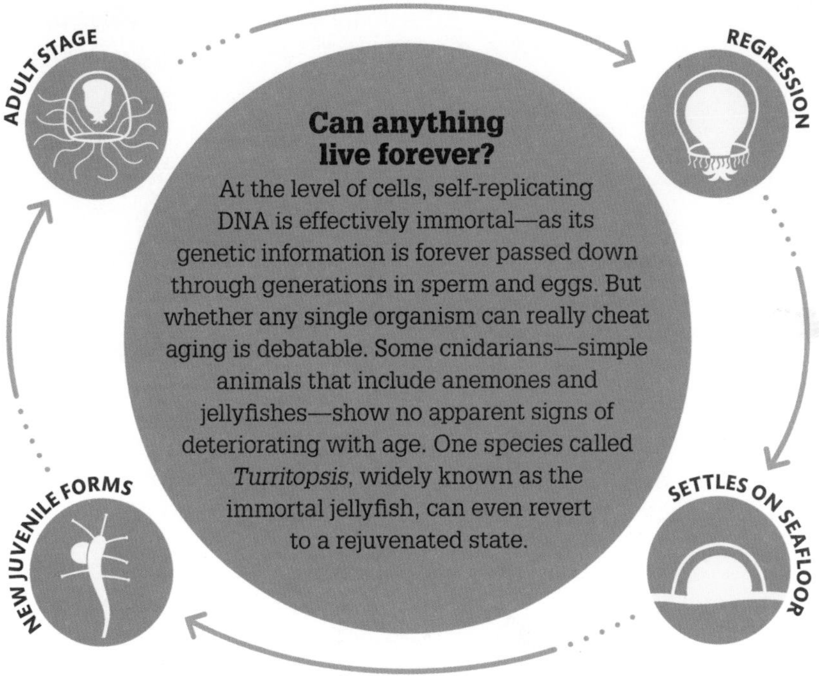
ADULT STAGE
REGRESSION
SETTLES ON SEAFLOOR
NEW JUVENILE FORMS

Can anything live forever?

At the level of cells, self-replicating DNA is effectively immortal—as its genetic information is forever passed down through generations in sperm and eggs. But whether any single organism can really cheat aging is debatable. Some cnidarians—simple animals that include anemones and jellyfishes—show no apparent signs of deteriorating with age. One species called *Turritopsis*, widely known as the immortal jellyfish, can even revert to a rejuvenated state.

DELAYING AGING

Experimental drugs have been found that counteract or repair DNA damage, and, in the future, gene therapy (see pp.182–83) could be used to "reboot" aged cells. However, attempts to slow aging or even reverse it remain unproven and controversial. Lifestyle changes, such as regular exercise and a good diet, remain the best way to reduce the risk of degenerative disorders occurring in old age—and so prolong lifespan.

Drugs **Gene therapy**
Diet **Exercise**

Genomes

The genetic information of a living thing is contained in molecules of DNA, the complete set of which is called a genome. Analyzing the genome in a laboratory can allow us to pinpoint certain genes, understand how they work, and even produce a "DNA fingerprint" that is unique to an individual.

How DNA is organized

DNA contains sections called genes that provide information for making proteins (see pp.158–59). Molecules of DNA inside bacteria are free in the cell's cytoplasm, but organisms with more complex cells—such as plants and animals—have many extremely long strands of DNA, which are packed inside the nucleus. During cell division, they coil more tightly, forming a chromosome, to prevent them from becoming entangled.

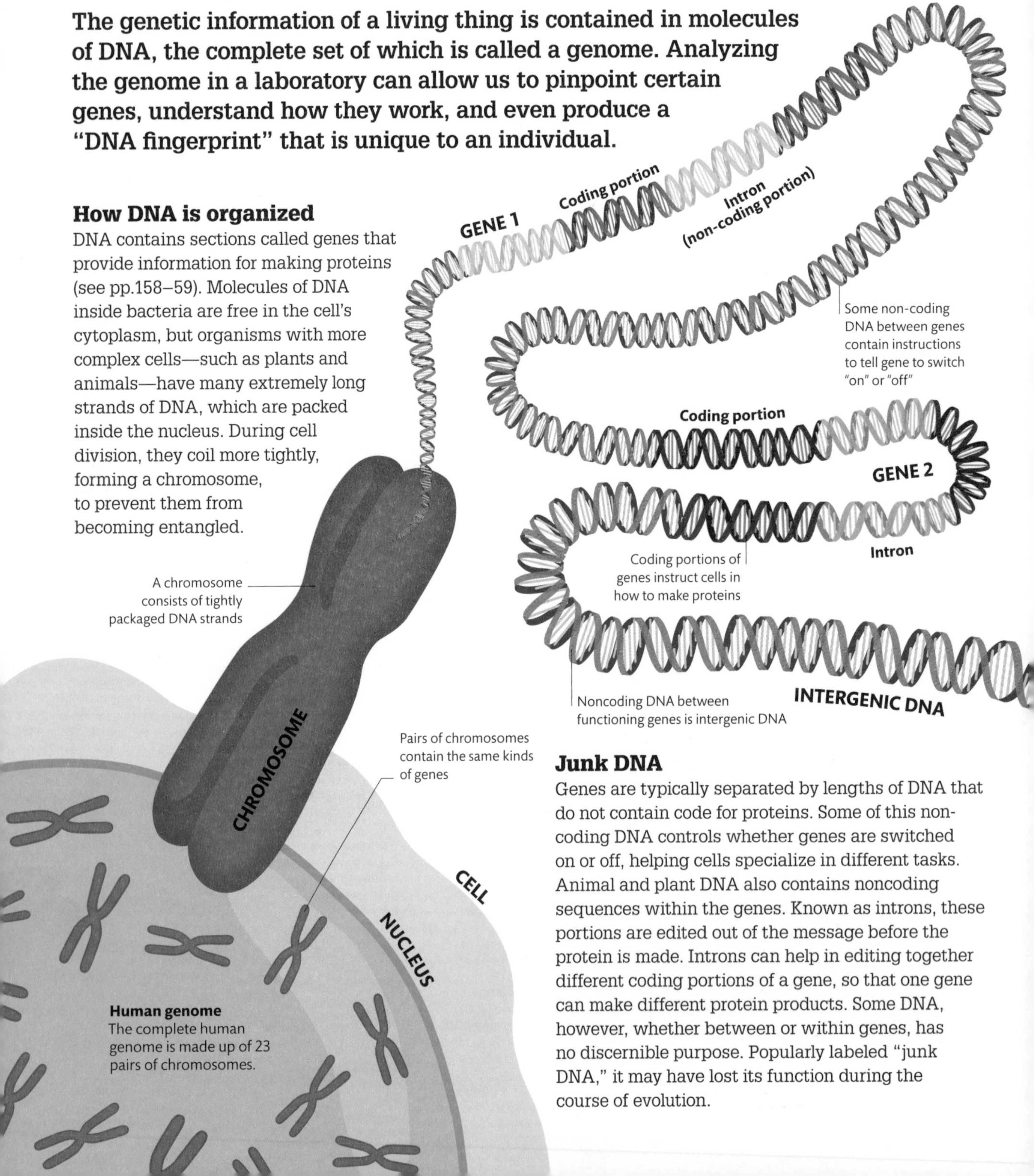

GENE 1
Coding portion
Intron (non-coding portion)
Some non-coding DNA between genes contain instructions to tell gene to switch "on" or "off"
Coding portion
GENE 2
Intron
Coding portions of genes instruct cells in how to make proteins
A chromosome consists of tightly packaged DNA strands
Noncoding DNA between functioning genes is intergenic DNA
INTERGENIC DNA
Pairs of chromosomes contain the same kinds of genes
CHROMOSOME
CELL
NUCLEUS
Human genome
The complete human genome is made up of 23 pairs of chromosomes.

Junk DNA

Genes are typically separated by lengths of DNA that do not contain code for proteins. Some of this non-coding DNA controls whether genes are switched on or off, helping cells specialize in different tasks. Animal and plant DNA also contains noncoding sequences within the genes. Known as introns, these portions are edited out of the message before the protein is made. Introns can help in editing together different coding portions of a gene, so that one gene can make different protein products. Some DNA, however, whether between or within genes, has no discernible purpose. Popularly labeled "junk DNA," it may have lost its function during the course of evolution.

DNA profiling

The sequence of chemical bases in an individual's DNA (see pp.158–59) is unique—except for identical twins. This means DNA is a formidable tool when cross-matching samples of blood, saliva, semen, or other biological material. DNA profiling—or DNA fingerprinting—compares repetitive sections of DNA called short tandem repeats (STRs) that differ in length between individuals.

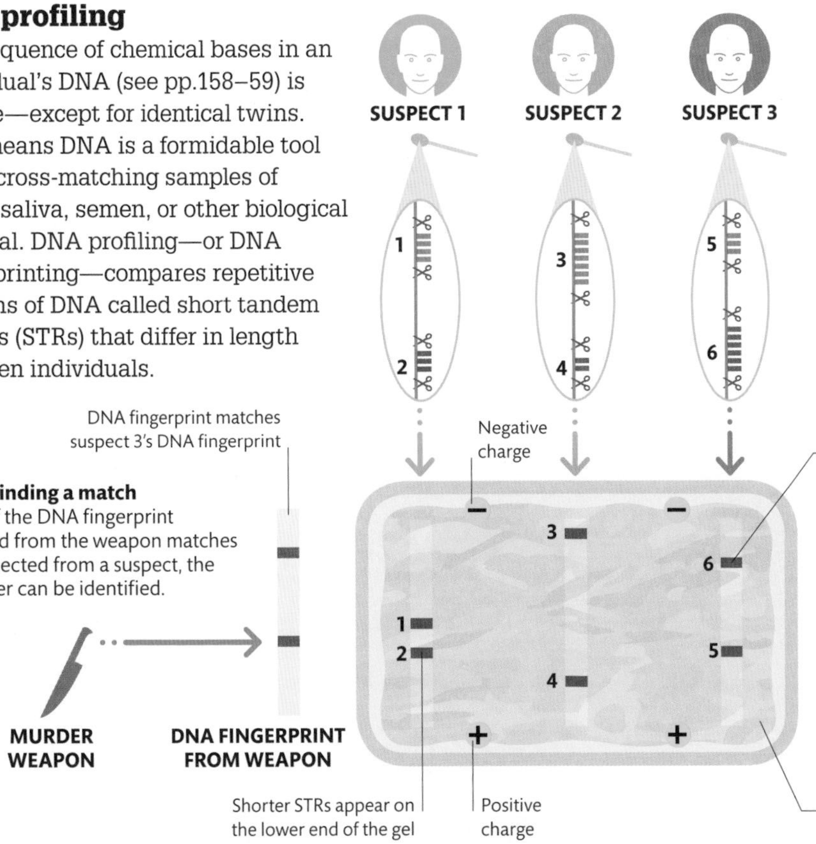
SUSPECT 1
SUSPECT 2
SUSPECT 3
1
2
3
4
5
6
DNA fingerprint matches suspect 3's DNA fingerprint
Negative charge
Longer STRs appear on the upper end of the gel
MURDER WEAPON
DNA FINGERPRINT FROM WEAPON
Shorter STRs appear on the lower end of the gel
Positive charge
Gel provides a medium in which DNA strands can travel

1 Collecting samples DNA samples are collected both from a murder weapon and the suspects of a crime–usually by taking a mouth swab. The DNA is copied over and over to maximize the quantity for analysis.

2 Fragmenting the DNA DNA is cut into fragments that specifically excise the STRs. This gives a mixture of fragments of various sizes, depending upon the lengths of the STRs.

3 Separating the fragments An electrical charge is applied across a block of gel that separates the negatively charged DNA. Small fragments travel faster toward the positive end, so move the greatest distance. Each cluster of fragments is then stained to produce a pattern of bands unique to each individual.

4 Finding a match If the DNA fingerprint collected from the weapon matches that collected from a suspect, the murderer can be identified.

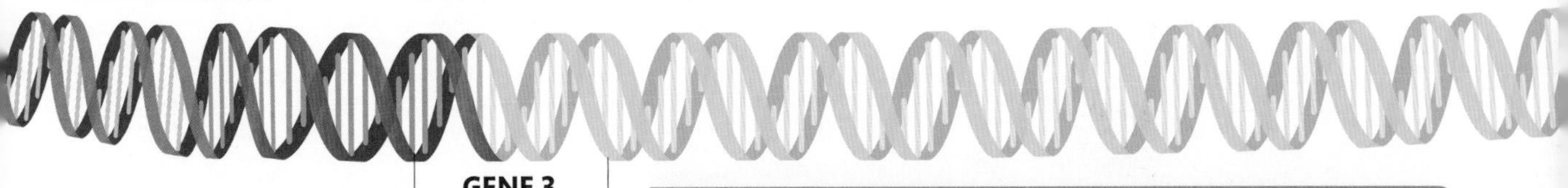
GENE 3
As in the other genes, only a small portion of gene 3 codes for proteins
Intron in the gene might control when the gene is active, or it may contain useless "junk" DNA

IF THE DNA IN A HUMAN CELL WERE UNRAVELED, IT WOULD STRETCH TO OVER 6FT (2M)

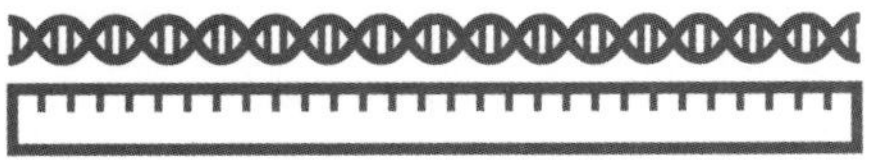

THE HUMAN GENOME PROJECT

In 2003, the Human Genome Project was completed–it was an international collaboration of researchers that started in 1990 and sought to document the sequence of 3 billion human DNA building blocks called bases. Although the specific sequence differs between individuals, the project published an average sequence using several anonymous donors. It paved the way for a greater understanding of human genes in general.

Genetic engineering

Genetic information is so tightly linked to the identity of a living thing that it seems amazing that we can manipulate it at all. But science can alter the information to change characteristics for the benefit of medicine and other fields.

Rewriting the genetic data

Genetic engineering involves changing the genetic makeup of a living thing by adding, removing, or altering genes. Since genes are sections of DNA that code for proteins (see p.158), altering them in precise ways changes their protein-making capabilities, resulting in changed characteristics in an organism. Target genes can be snipped from chromosomes (see p.178) or copied from genetic material called RNA (see pp.158–59). Each step is driven by a specific chemical catalyst called an enzyme.

GENETICALLY MODIFIED **FISH THAT GLOW IN THE DARK** ARE **SOLD AS PETS** IN THE US

Making insulin
The genetic code for producing insulin can be extracted from human cells and inserted into bacteria, which can then be used as living factories to supply insulin for treating diabetic people. The code is obtained from the cell's RNA copies, which are easier to extract than DNA and have also been edited to remove their noncoding portions (see p.179).

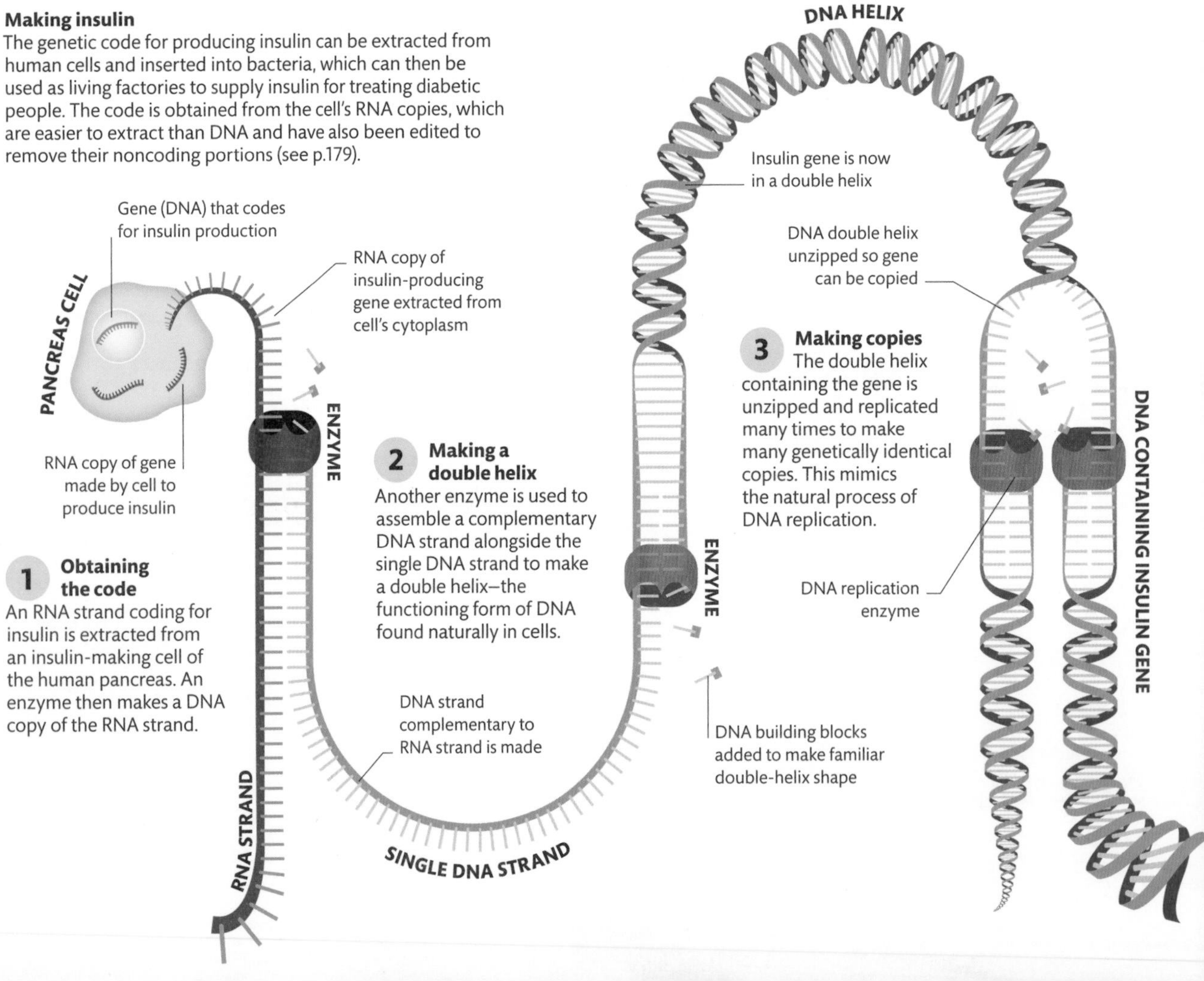
DNA HELIX
Insulin gene is now in a double helix
DNA double helix unzipped so gene can be copied
Gene (DNA) that codes for insulin production
RNA copy of insulin-producing gene extracted from cell's cytoplasm
PANCREAS CELL
RNA copy of gene made by cell to produce insulin
ENZYME
1 Obtaining the code
An RNA strand coding for insulin is extracted from an insulin-making cell of the human pancreas. An enzyme then makes a DNA copy of the RNA strand.
2 Making a double helix
Another enzyme is used to assemble a complementary DNA strand alongside the single DNA strand to make a double helix—the functioning form of DNA found naturally in cells.
3 Making copies
The double helix containing the gene is unzipped and replicated many times to make many genetically identical copies. This mimics the natural process of DNA replication.
DNA replication enzyme
DNA CONTAINING INSULIN GENE
ENZYME
DNA building blocks added to make familiar double-helix shape
DNA strand complementary to RNA strand is made
RNA STRAND
SINGLE DNA STRAND

Why do we want to change genes?

Genetic engineering can be incredibly useful. As well as engineering microbes to mass-produce medically important proteins, plants and animals can acquire agriculturally desirable traits, while gene therapy has the potential to treat genetic disorders.

EXAMPLES OF GENETIC ENGINEERING

Medical products
Unlike proteins sourced from animals, those made by genetically modified microbes can be mass-produced.

Genetically modified (GM) plants and animals
Plants and animals can be modified to improve their nutritional value or increase resistance to drought, diseases, or pests.

Gene therapy (see pp.182–83).
Cells that carry a genetic disorder can temporarily be made to work normally by inserting a functioning gene.

CAN INSERTED GENES SPREAD?

Introducing plants with alien genes into the environment raises concerns that they might spread out of control and become "super weeds" in the wild. GM crop plants might even accidentally pollinate wild plants, which in turn could become damaging agricultural weeds. "Gene flow" between GM and non-GM plants has been documented, but there is no scientific consensus on its potential environmental impact.

4 Preparing to join
Rings of DNA called plasmids (these occur naturally inside bacteria) are cut open with a specific enzyme, which leaves behind overhanging single-stranded sections on the cut ends that have a specific sequence of bases.

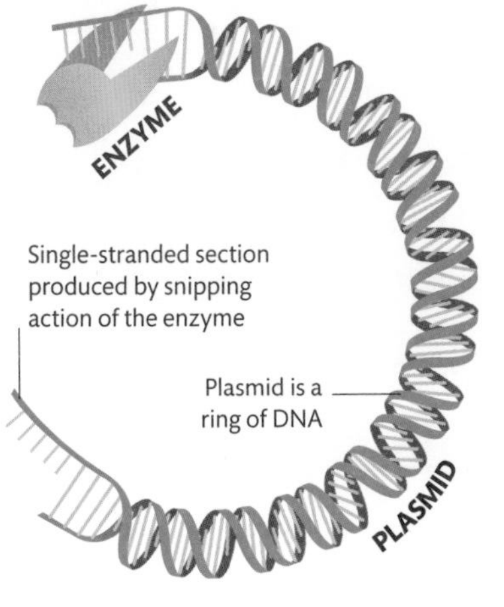
ENZYME
Single-stranded section produced by snipping action of the enzyme
Plasmid is a ring of DNA
PLASMID

5 Inserting the gene
Single-stranded ends must be added to the DNA containing the gene. These overhangs complement those on the plasmid, so the strands readily combine. The connection is sealed with another enzyme to make plasmids that now contain the insulin gene.

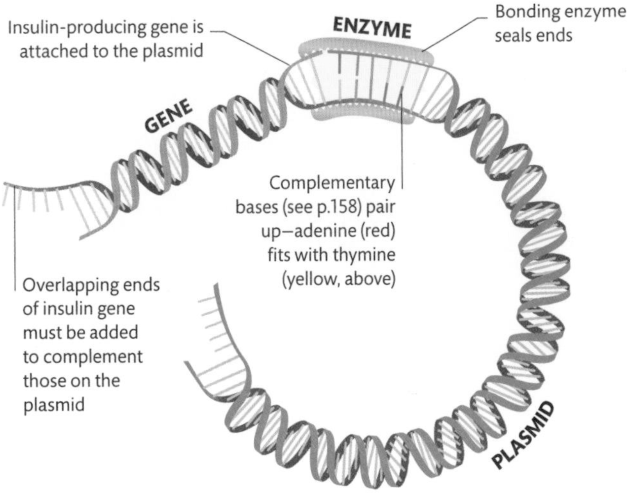
Insulin-producing gene is attached to the plasmid
ENZYME
Bonding enzyme seals ends
GENE
Complementary bases (see p.158) pair up—adenine (red) fits with thymine (yellow, above)
Overlapping ends of insulin gene must be added to complement those on the plasmid
PLASMID

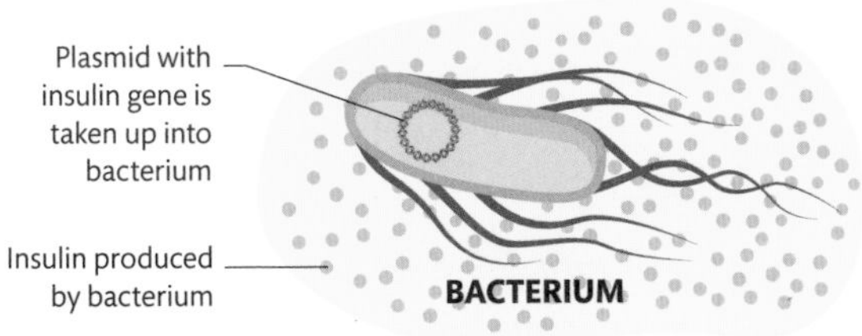
Plasmid with insulin gene is taken up into bacterium
Insulin produced by bacterium
BACTERIUM

6 Producing insulin
Bacteria take up the genetically engineered plasmids that contain the insulin gene. The plasmids replicate as the bacteria reproduce. The bacteria make insulin, which can be separated and purified from the culture.

Gene therapy

Some kinds of diseases call for a particularly sophisticated approach to treatment—one that uses DNA as the medicine. Gene therapy supplies cells with the genetic information that changes their behavior to remedy a disease.

How gene therapy works

Genes are sections of DNA that instruct cells to make specific kinds of proteins. By inserting a gene into a cell, gene therapy can compensate for faulty DNA that fails to produce a working protein or prompt a new task that counteracts a disease. The technique can work for diseases that are caused by a single gene (such as cystic fibrosis) rather than diseases caused by the combined effects of many. The curative gene is carried into the cell inside a particle called a vector, which can be a deactivated virus or an oily droplet called a liposome.

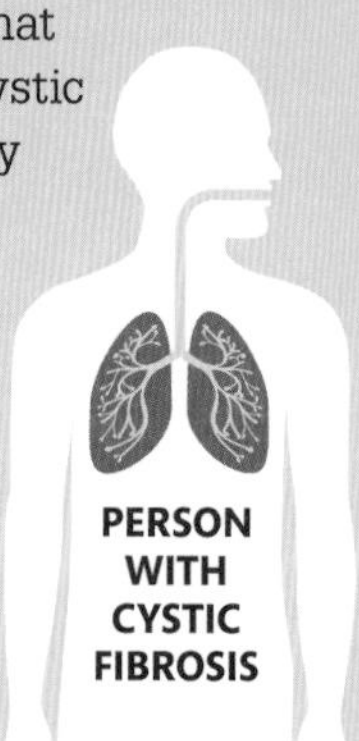
PERSON
WITH
CYSTIC
FIBROSIS

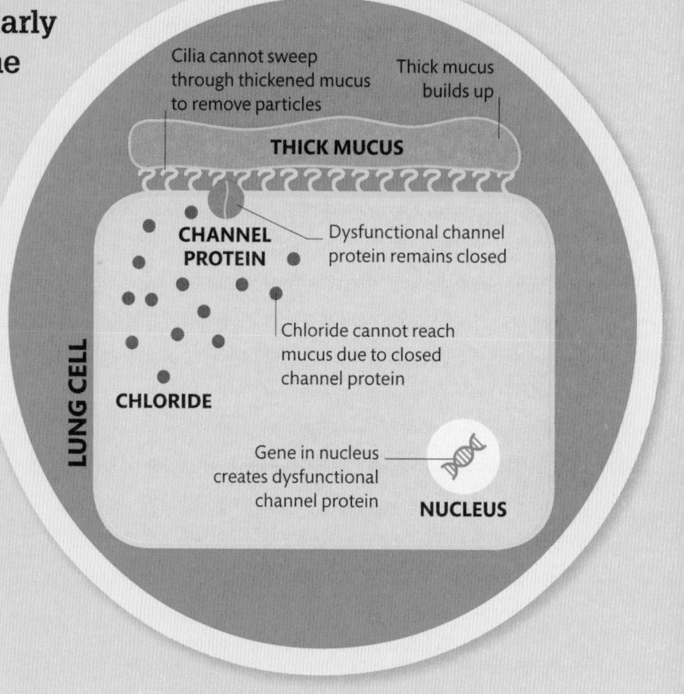
Cilia cannot sweep through thickened mucus to remove particles
Thick mucus builds up
THICK MUCUS
CHANNEL PROTEIN
Dysfunctional channel protein remains closed
Chloride cannot reach mucus due to closed channel protein
LUNG CELL
CHLORIDE
Gene in nucleus creates dysfunctional channel protein
NUCLEUS

1 Cystic fibrosis
People with cystic fibrosis have diseased lung cells with dysfunctional genes that code for closed channel proteins. This means mucus lining the airways is too thick, causing breathing difficulties.

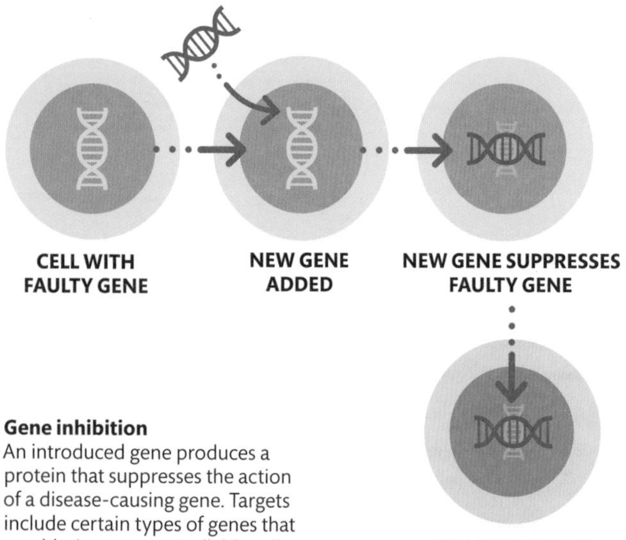
CELL WITH FAULTY GENE
NEW GENE ADDED
NEW GENE SUPPRESSES FAULTY GENE
CELL FUNCTIONS NORMALLY

Gene inhibition
An introduced gene produces a protein that suppresses the action of a disease-causing gene. Targets include certain types of genes that would trigger uncontrollable cell division, causing cancer.

GENE THERAPY RESEARCH IS CURRENTLY TARGETING CERTAIN KINDS OF CANCER

IS GENE THERAPY A PERMANENT CURE?

Treated cells divide but they eventually die and are replaced by diseased cells, so currently therapies are short-lived and multiple treatments are needed.

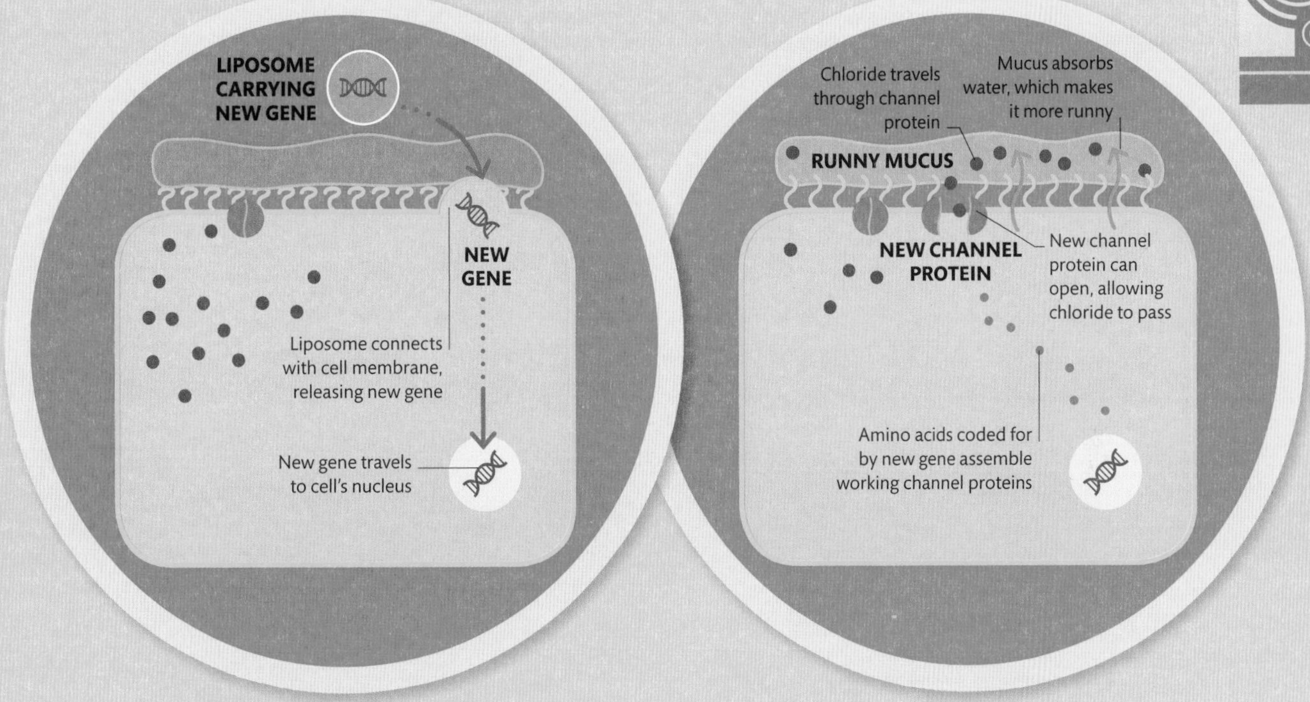
LIPOSOME
CARRYING
NEW GENE
NEW
GENE
Liposome connects
with cell membrane,
releasing new gene
New gene travels
to cell's nucleus
Chloride travels
through channel
protein
Mucus absorbs
water, which makes
it more runny
RUNNY MUCUS
NEW CHANNEL
PROTEIN
New channel
protein can
open, allowing
chloride to pass
Amino acids coded for
by new gene assemble
working channel proteins

2 Gene is added
Brought into the body by an inhaler, liposomes carrying functioning channel-protein genes sweep through the airways and are absorbed into cells of the lining. Here, they combine with other DNA inside the cells' nuclei.

3 Gene restores function
The new genes instruct cells to make functioning channel proteins, letting chloride into the mucus. The saltier mucus absorbs water from the cells and so gets runnier—making breathing easier.

Killing specific cells
Suicide genes that specifically target diseased cells can cause these cells to self-destruct, or to mark them as targets for attack from the immune system.

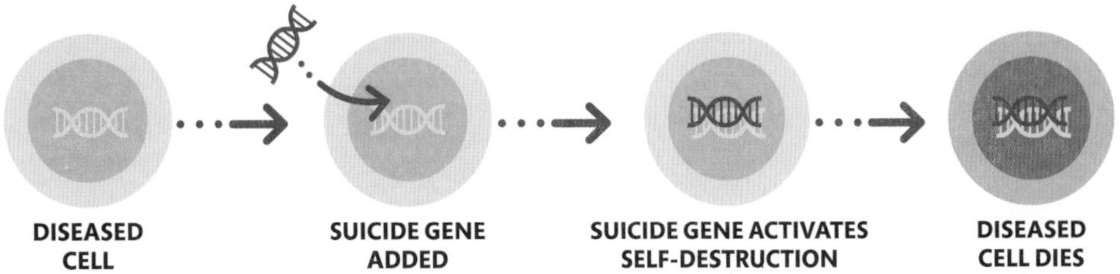
DISEASED
CELL
SUICIDE GENE
ADDED
SUICIDE GENE ACTIVATES
SELF-DESTRUCTION
DISEASED
CELL DIES

CAN NEW GENES BE INHERITED?

Conventional gene therapy—called somatic gene therapy—inserts genes into body cells that are not involved in producing eggs or sperm. When these cells multiply, the replicated genes stay in the diseased tissues and are not passed to offspring. Germline gene therapy—widely rejected as being unethical—would add genes to sperm or eggs so genes could be inherited.

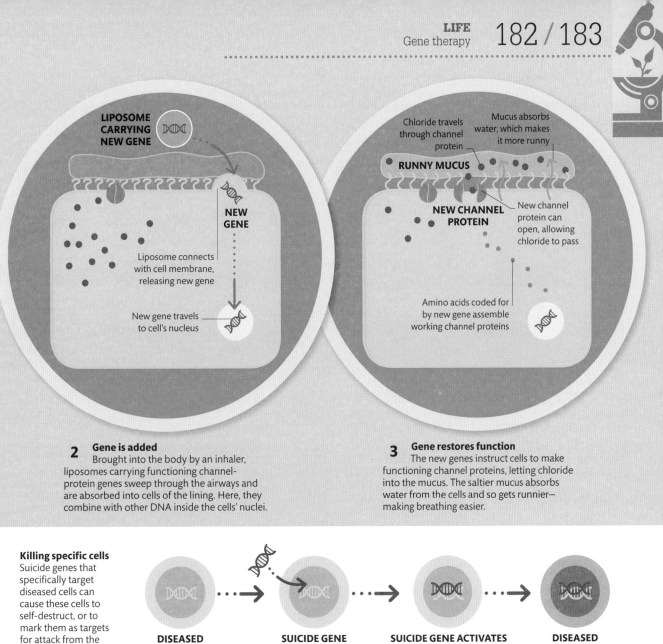

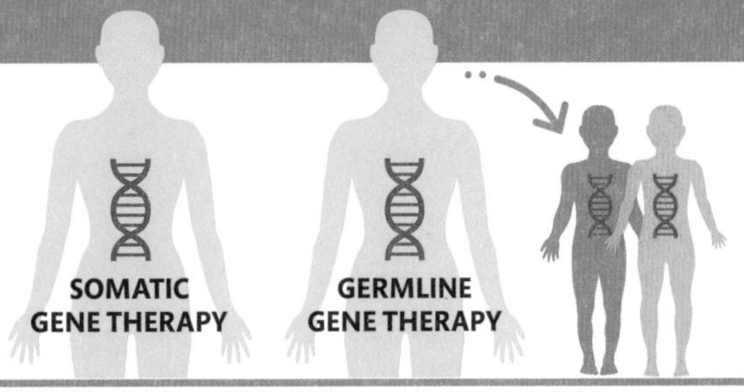
SOMATIC
GENE THERAPY
GERMLINE
GENE THERAPY

Stem cells

Animal bodies are made of cells specialized for tasks such as carrying oxygen or nerve impulses. From embryo to adulthood, only a small bank of unspecialized founding cells, called stem cells, retain the ability to give rise to this diversity—a potential that can be used to cure disease.

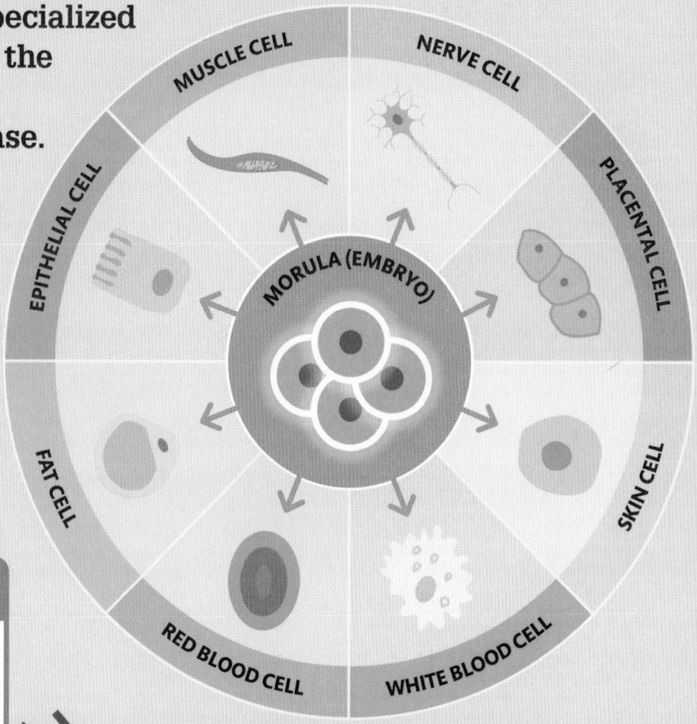
MUSCLE CELL
NERVE CELL
PLACENTAL CELL
SKIN CELL
WHITE BLOOD CELL
RED BLOOD CELL
FAT CELL
EPITHELIAL CELL
MORULA (EMBRYO)

Types of stem cells

Embryos, unsurprisingly, have cells with the greatest potential for forming different tissues. A small ball of embryonic cells must develop into all the parts of the body. But as these parts become distinct, their cells lose their versatility as they become committed to their specialized tasks. Only some parts of the body, such as the bone marrow, retain stem cells, but their ability to diversify is limited.

ETHICS OF HARVESTING STEM CELLS

Embryonic stem cells have the greatest potential for therapeutic use, but many people view using human embryos as ethically unacceptable, and taking stem cells from embryos is illegal in some countries. Adult stem cells—such as from bone marrow or the umbilical cord—bypass these concerns, but they have limited potential and are not as useful in researching treatments for conditions including diabetes and Parkinson's disease.

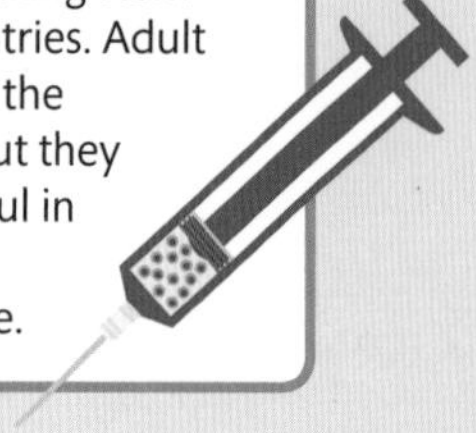

Earliest embryo stem cells
When it is still a solid ball called a morula, the earliest embryo has cells with the maximum developmental potential. Each so-called "totipotent" stem cell has the potential to form any part of the embryo; in most mammals, this includes the membranes that will ultimately form the placenta.

Stem cell therapy

The developmental potential of stem cells can help grow healthy tissues for treating illness. Bone marrow transplants, for example, rely on the blood-cell-forming capacity of adult stem cells to treat blood disorders such as leukemia. Stem cell therapy could also restore insulin-producing cells in diabetic people. Experimental trials, usually on animals, are using stem cells from embryos or adult cells chemically treated to increase their potential.

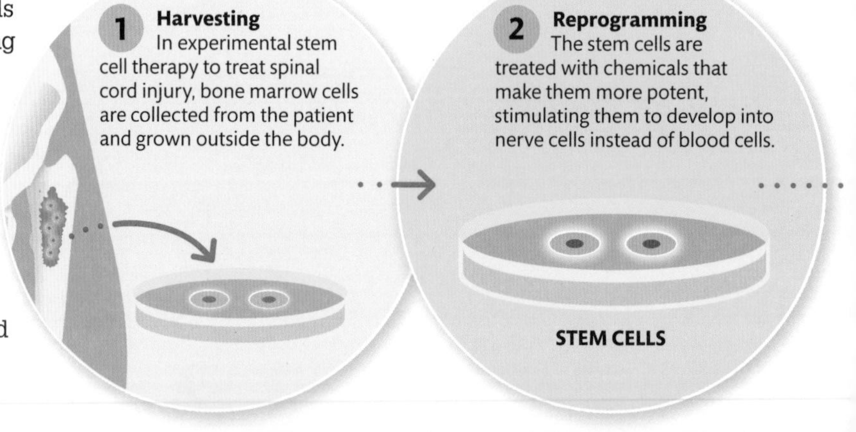
1 Harvesting
In experimental stem cell therapy to treat spinal cord injury, bone marrow cells are collected from the patient and grown outside the body.
2 Reprogramming
The stem cells are treated with chemicals that make them more potent, stimulating them to develop into nerve cells instead of blood cells.
STEM CELLS

IN A TRIAL, **50 PERCENT** OF **STEM CELL THERAPY** PATIENTS WITH **SPINAL CORD INJURY** REGAINED SOME **MOVEMENT**

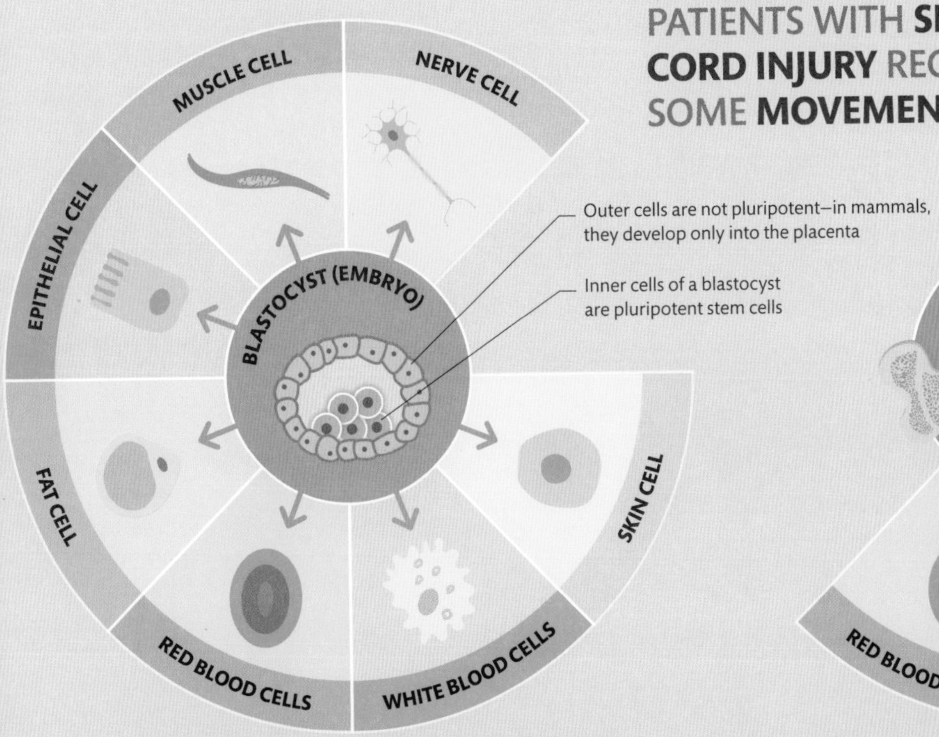
MUSCLE CELL
NERVE CELL
EPITHELIAL CELL
BLASTOCYST (EMBRYO)
FAT CELL
SKIN CELL
RED BLOOD CELLS
WHITE BLOOD CELLS
Outer cells are not pluripotent—in mammals, they develop only into the placenta
Inner cells of a blastocyst are pluripotent stem cells

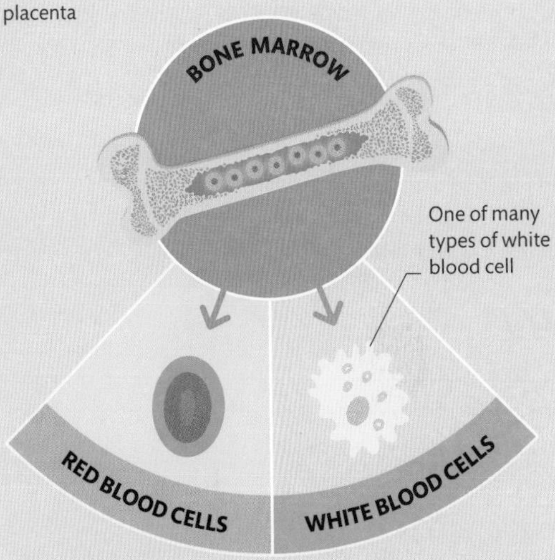
BONE MARROW
One of many types of white blood cell
RED BLOOD CELLS
WHITE BLOOD CELLS

Early embryo stem cell
Once the embryo has reached the next stage—a hollow sphere of cells called a blastocyst—the first step in specialization has been reached. In most mammals, the outer cell layer helps form the placenta. Only the inner cell mass, containing "pluripotent" stem cells, will form parts of the embryo's body.

Adult stem cells
Stem cells persist in parts of the adult's body, but they can develop into only a restricted range of cell types and are described as "multipotent." For example, most bones of the body contain multipotent stem cells in their inner marrow that can differentiate into various kinds of blood cells.

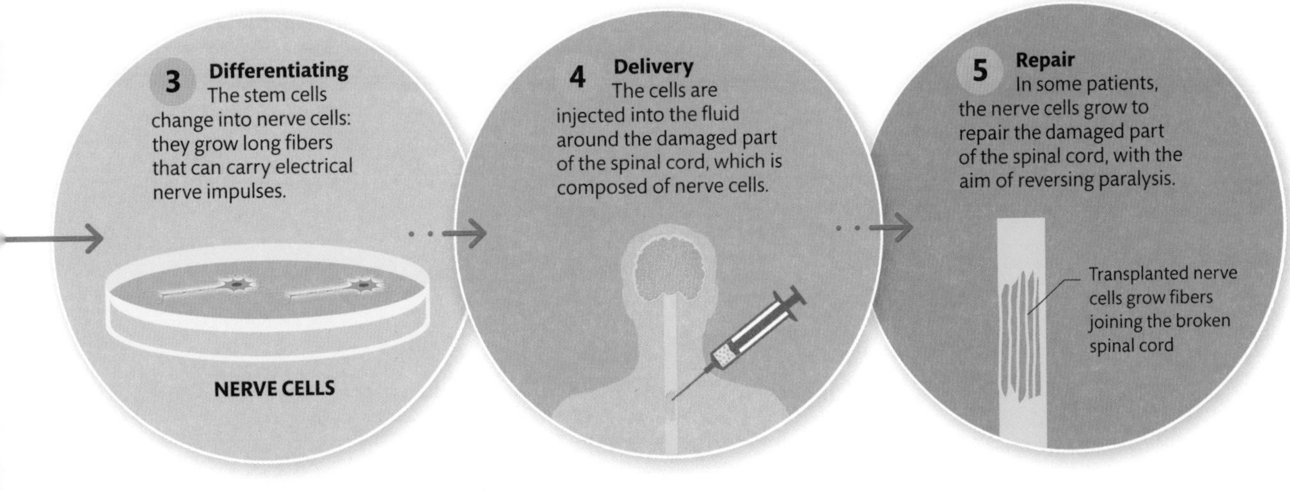
3 **Differentiating**
The stem cells change into nerve cells: they grow long fibers that can carry electrical nerve impulses.
NERVE CELLS
4 **Delivery**
The cells are injected into the fluid around the damaged part of the spinal cord, which is composed of nerve cells.
5 **Repair**
In some patients, the nerve cells grow to repair the damaged part of the spinal cord, with the aim of reversing paralysis.
Transplanted nerve cells grow fibers joining the broken spinal cord

Cloning

Clones are genetically identical living things. Technology can manipulate cloning artificially, with implications for medicine and beyond.

How cloning works

At the heart of cloning is the self-replicating DNA that drives the division of cells and multiplies any living thing that can reproduce asexually. Laboratory techniques go beyond this, manipulating particular kinds of unspecialized cells and tissues to produce clones in ways that do not happen naturally.

ARE TWINS TECHNICALLY CLONES?

Yes—identical twins are clones. These arise when a single fertilized egg splits into two separate cells inside the womb. These then develop into genetically identical embryos.

NATURAL CLONING

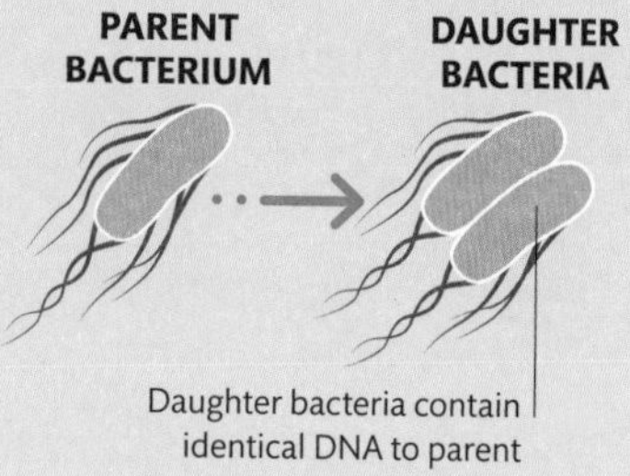
PARENT BACTERIUM
DAUGHTER BACTERIA
Daughter bacteria contain identical DNA to parent

Asexual reproduction in microbes
Microbes, such as bacteria, reproduce asexually by cloning themselves. Just before cell division, the DNA replicates. Identical copies of the DNA will go to each cell.

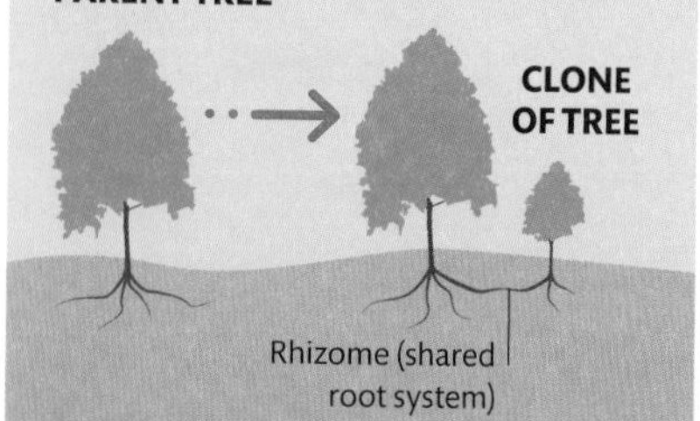
PARENT TREE
CLONE OF TREE
Rhizome (shared root system)

Asexual reproduction in plants
Underground root systems called rhizomes contain the tissue needed to sprout new trees that are genetically identical to parent plants. Aspen trees produce some of the biggest clonal patches on the planet.

ARTIFICIAL CLONING

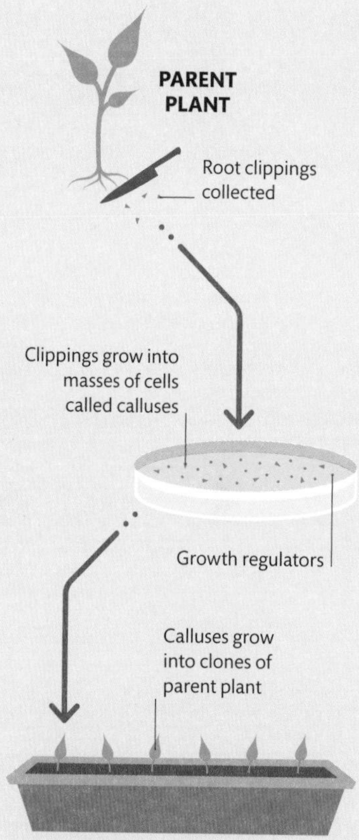
PARENT PLANT
Root clippings collected
Clippings grow into masses of cells called calluses
Growth regulators
Calluses grow into clones of parent plant

Tissue culture
Parts of plants can be encouraged to grow into new plants if they are treated with chemicals called growth regulators. Tiny plants sprout in a sterile nutrient-rich jelly, before being transferred to soil.

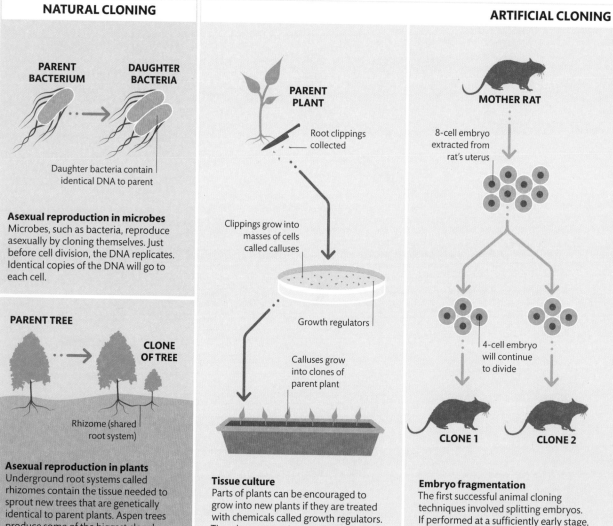

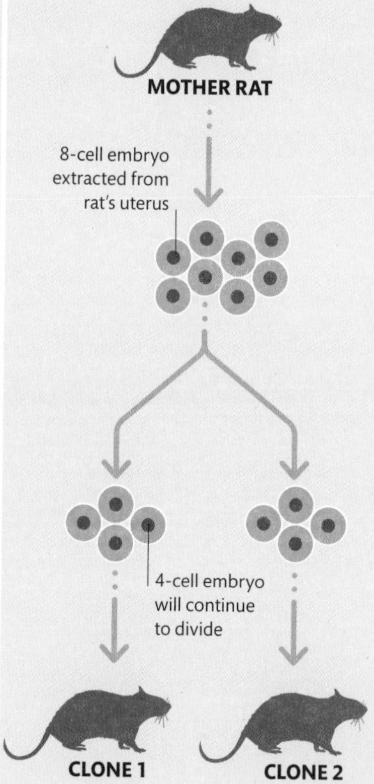
MOTHER RAT
8-cell embryo extracted from rat's uterus
4-cell embryo will continue to divide
CLONE 1
CLONE 2

Embryo fragmentation
The first successful animal cloning techniques involved splitting embryos. If performed at a sufficiently early stage, an embryo's unspecialized cells retain the potential to form all parts of a body.

A PYRENEAN IBEX WAS THE FIRST ANIMAL TO BE REVIVED FROM EXTINCTION, BUT IT DIED AFTER 7 MINUTES

RESURRECTING EXTINCT SPECIES?

Preserved specimens offer the tantalizing prospect of resurrecting extinct species. However, DNA degrades over time—this means that old DNA lacks key instructions vital for making a viable embryo. Scientists have remarkably intact sequences of DNA from frozen mammoth tissues—but it is too damaged and incomplete for cloning to work. Scientists are planning to splice (join) mammoth and Asian elephant (the mammoth's closest living relative) genes to create a hybrid embryo that could be reared in an artificial womb. However, this raises ethical concerns.

WOOLLY MAMMOTH

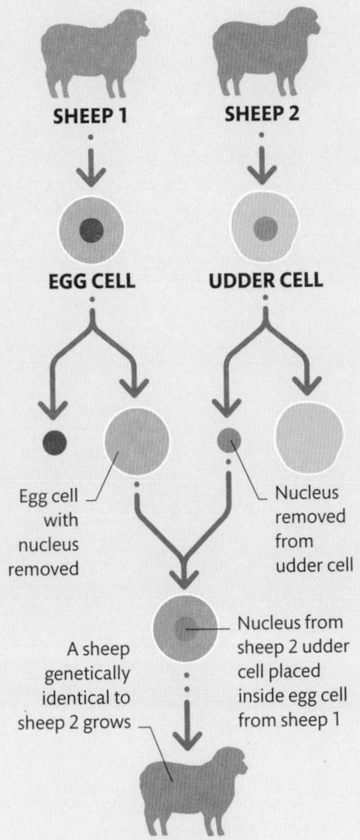
SHEEP 1
SHEEP 2
EGG CELL
UDDER CELL
Egg cell with nucleus removed
Nucleus removed from udder cell
Nucleus from sheep 2 udder cell placed inside egg cell from sheep 1
A sheep genetically identical to sheep 2 grows
CLONE OF SHEEP 2

Somatic cell nuclear transfer
Clones can be produced from body (somatic) tissue. An egg cell, stripped of its own nucleus, reprograms the donor body cell nucleus with the potential to produce a clone. Dolly the sheep was cloned using this technique.

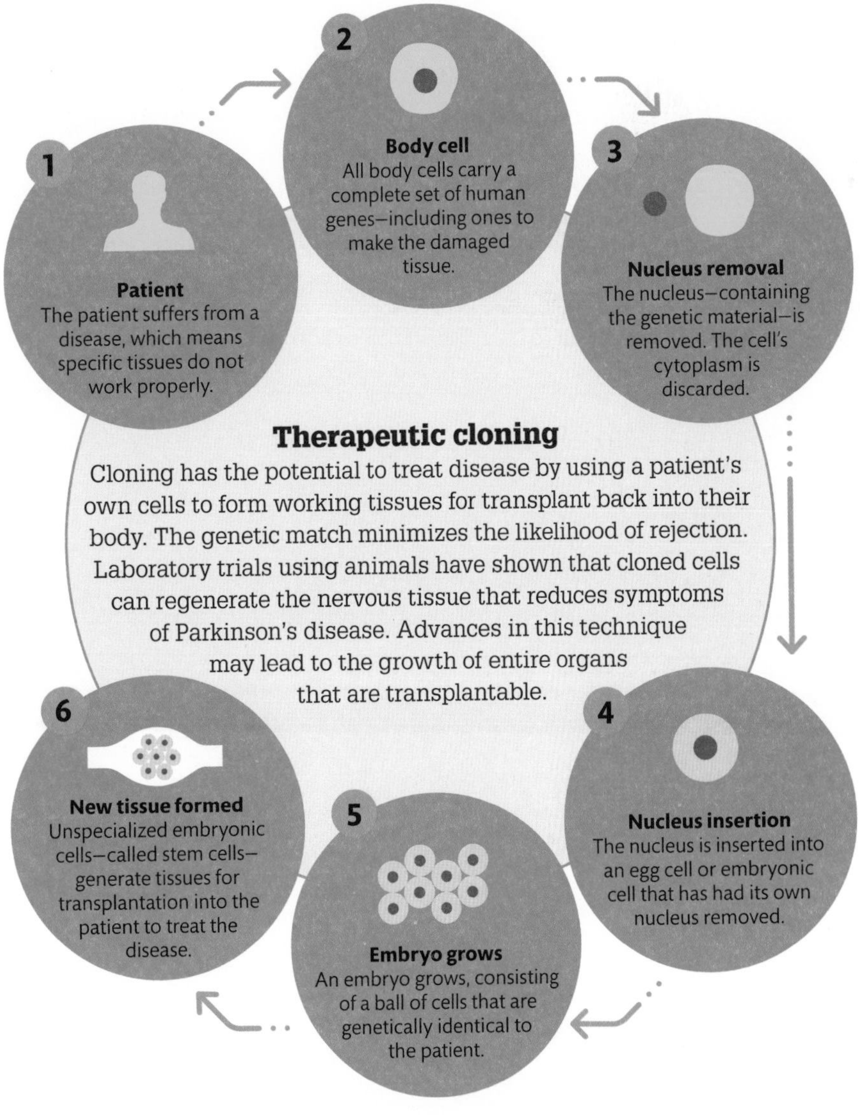
1
Patient
The patient suffers from a disease, which means specific tissues do not work properly.
2
Body cell
All body cells carry a complete set of human genes—including ones to make the damaged tissue.
3
Nucleus removal
The nucleus—containing the genetic material—is removed. The cell's cytoplasm is discarded.
4
Nucleus insertion
The nucleus is inserted into an egg cell or embryonic cell that has had its own nucleus removed.
5
Embryo grows
An embryo grows, consisting of a ball of cells that are genetically identical to the patient.
6
New tissue formed
Unspecialized embryonic cells—called stem cells—generate tissues for transplantation into the patient to treat the disease.

Therapeutic cloning

Cloning has the potential to treat disease by using a patient's own cells to form working tissues for transplant back into their body. The genetic match minimizes the likelihood of rejection. Laboratory trials using animals have shown that cloned cells can regenerate the nervous tissue that reduces symptoms of Parkinson's disease. Advances in this technique may lead to the growth of entire organs that are transplantable.

SPACE

Stars

A star is a huge glowing ball of gas that comes to life when nuclear reactions ignite its core. The largest stars burn brightly but fade more quickly than smaller stars, which burn their fuel slowly. A star's mass also determines the nature of its death.

HOW LONG DO STARS LIVE?

The length of a star's life depends on its size. The most massive stars may last for a few hundred thousand years, while the smallest can burn for trillions of years.

A star is born

Stars arise in freezing clouds of interstellar dust and gas known as nebulae. Clumps of gas break into fragments, and if they become dense enough they collapse under their own gravity, releasing heat. If enough heat is produced for thermonuclear fusion to take place (see p.193), a star is born. This process can take several million years.

Dust and gas (mainly hydrogen)
Core collapses under its own weight
Infalling matter
Stellar winds flow outward

1 Molecular cloud At just above absolute zero, gases become molecular (formed of molecules, not ions) and bind together. Denser cloud fragments break off.

2 Collapsing fragment A dense gas fragment collapses, raising the temperature at its center. Angular momentum turns the fragment into a rotating disc.

3 Protostar forms The dense central region forms a protostar, while the disc may become a planetary system. The protostar's size increases 100 times as matter falls into it.

4 Fusion begins The infall of matter stops when core pressure causes thermonuclear fusion to begin. The protostar burns hydrogen, producing strong stellar winds.

The life and death of a star

Most protostars will go on to become average or "main sequence" stars, which remain stable due to a balance of forces: the outward pressure of expanding hot gases versus the inward pull of gravity. A star's life cycle depends on its mass, and it changes in size, temperature, and color as it ages. Some stars fade away, but others will meet an explosive end in a supernova, providing material for new stars and planets. Since most elements in the Universe were created by nuclear reactions in stars, it may be said that our world is made of stardust.

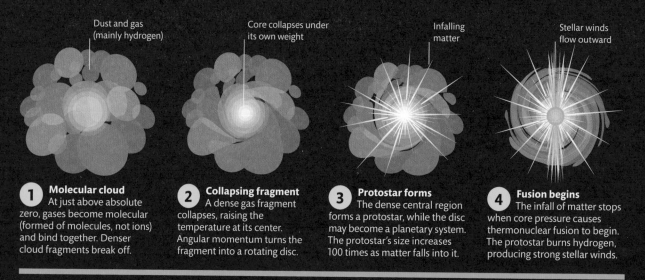

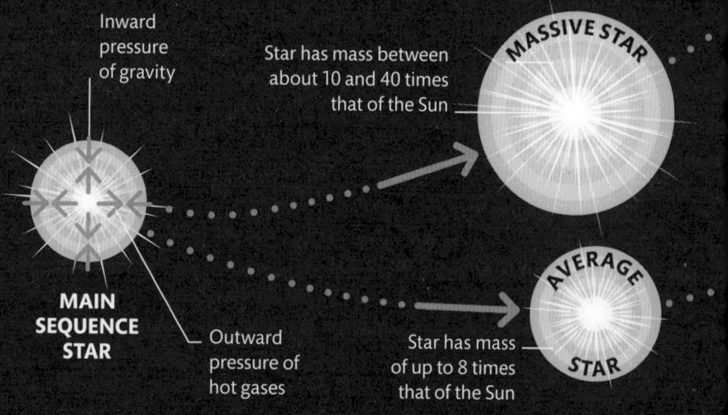
Inward pressure of gravity
MAIN SEQUENCE STAR
Outward pressure of hot gases
Star has mass between about 10 and 40 times that of the Sun
MASSIVE STAR
AVERAGE STAR
Star has mass of up to 8 times that of the Sun

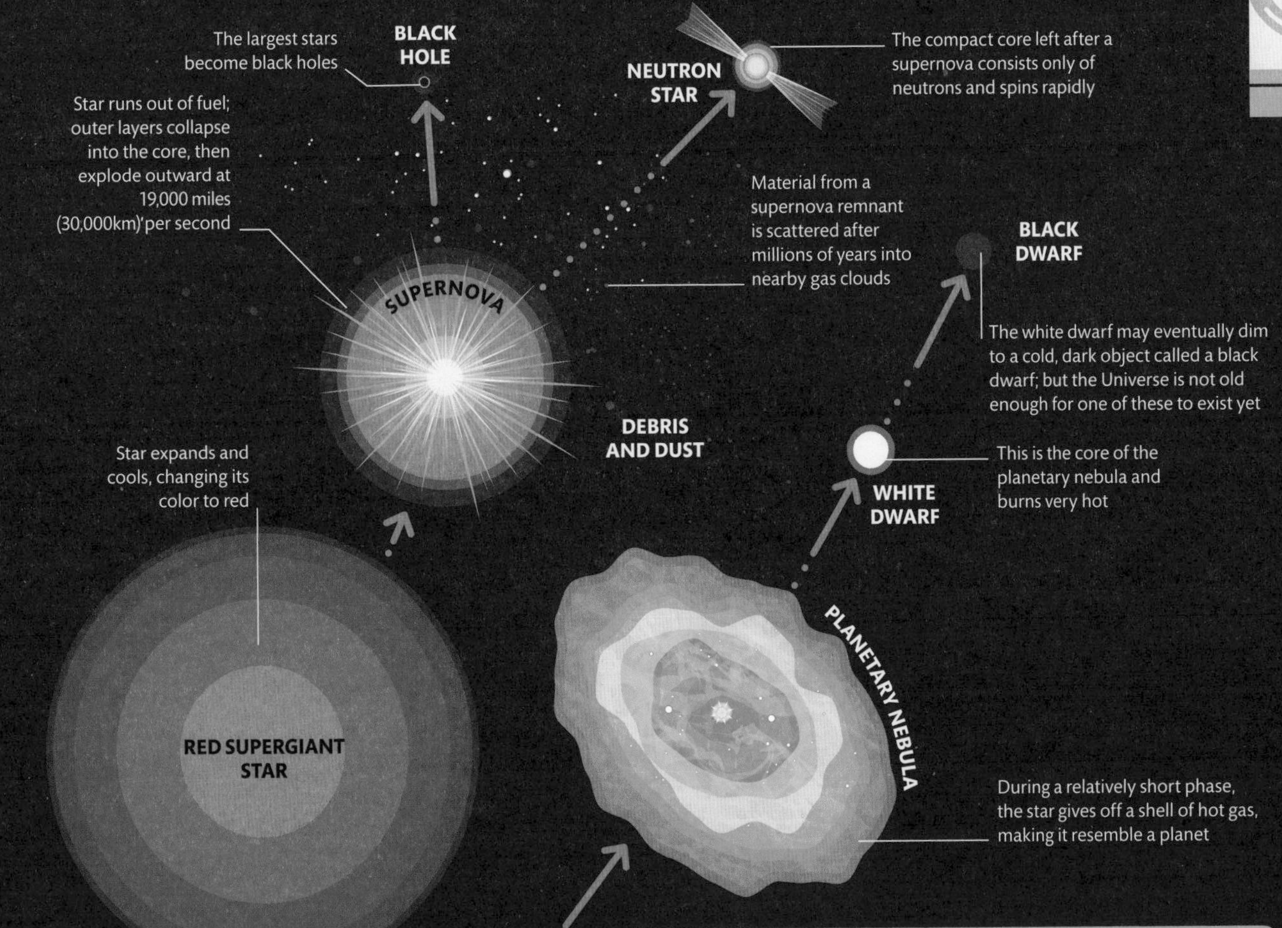
The largest stars become black holes
BLACK HOLE
NEUTRON STAR
The compact core left after a supernova consists only of neutrons and spins rapidly
Star runs out of fuel; outer layers collapse into the core, then explode outward at 19,000 miles (30,000km) per second
Material from a supernova remnant is scattered after millions of years into nearby gas clouds
BLACK DWARF
SUPERNOVA
The white dwarf may eventually dim to a cold, dark object called a black dwarf; but the Universe is not old enough for one of these to exist yet
DEBRIS AND DUST
Star expands and cools, changing its color to red
This is the core of the planetary nebula and burns very hot
WHITE DWARF
PLANETARY NEBULA
RED SUPERGIANT STAR
During a relatively short phase, the star gives off a shell of hot gas, making it resemble a planet

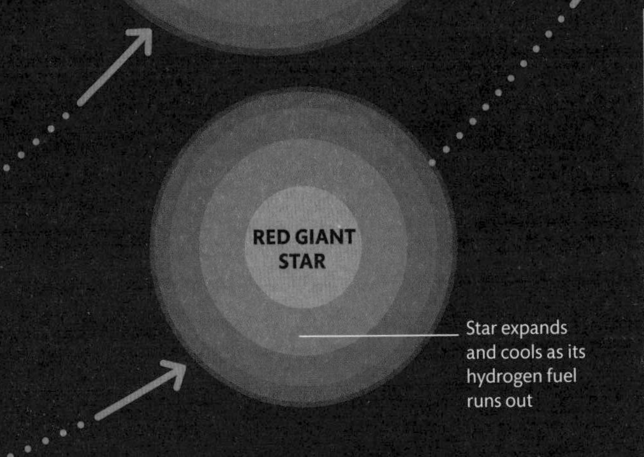
RED GIANT STAR
Star expands and cools as its hydrogen fuel runs out

STELLAR RECYCLING

The Big Bang created only hydrogen, helium, and a little lithium. Nearly all the other heavier elements were forged in stars or during supernovas. The latter release these materials, seeding new stars and planets.

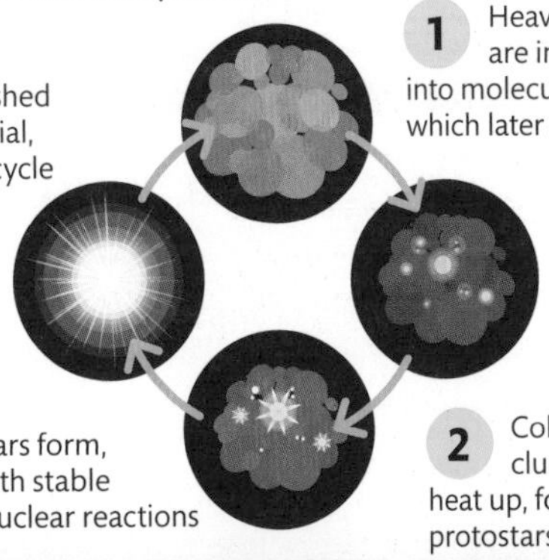
1 Heavy elements are introduced into molecular clouds, which later collapse
2 Collapsing clumps of gases heat up, forming protostars
3 Stars form, with stable thermonuclear reactions
4 Stars shed material, starting the cycle again

A TEASPOON OF NEUTRON STAR MATERIAL WEIGHS OVER 5.5 BILLION TONS

The Sun

The Sun is our closest star. It is a yellow dwarf—a star of average size—and generates its energy through nuclear fusion. The Sun is estimated to be halfway through its life span and will probably remain stable for another 5 billion years.

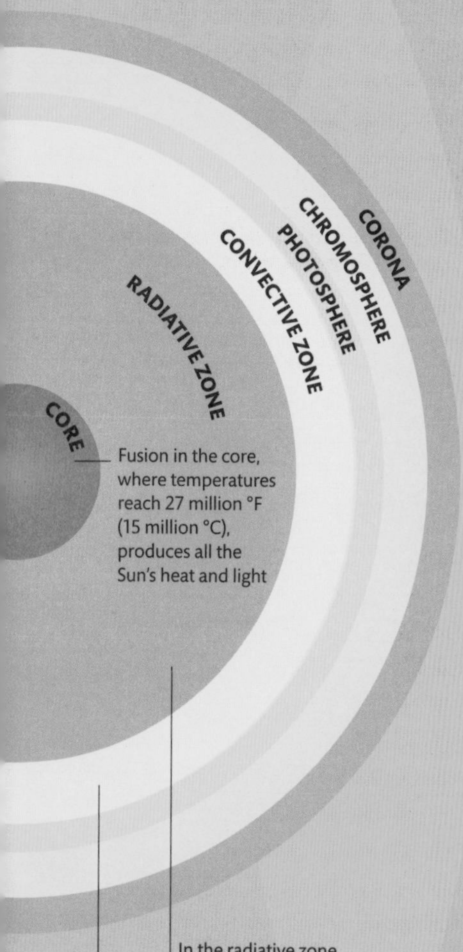
CORONA
CHROMOSPHERE
PHOTOSPHERE
CONVECTIVE ZONE
RADIATIVE ZONE
CORE
Fusion in the core, where temperatures reach 27 million °F (15 million °C), produces all the Sun's heat and light
In the radiative zone, photons skip from particle to particle before eventually escaping outward
The temperature in the convective zone, where bubbles of hot plasma move upward, drops to 2.7 million °F (1.5 million °C)

Inside and outside the Sun

The Sun consists mainly of hydrogen and helium gas in a plasma state, the gas being so hot that its atoms have lost their electrons and become ionized (see pp.20–21). The Sun has six regions: inside is a central core, where nuclear fusion takes place, enclosed by the radiative and convective zones; on the outside, the visible surface, or photosphere, is surrounded by the chromosphere and the outermost region, the corona.

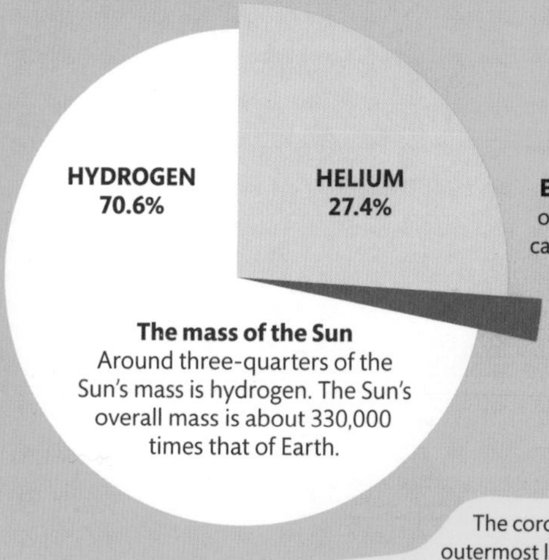
HYDROGEN
70.6%
HELIUM
27.4%
HEAVY ELEMENTS 2%
oxygen, nitrogen, carbon, neon, iron, and others
The mass of the Sun
Around three-quarters of the Sun's mass is hydrogen. The Sun's overall mass is about 330,000 times that of Earth.

The corona, the Sun's outermost layer, extends far out into space

Sunspots are relatively cool, dark areas of the photosphere caused by concentrations of the Sun's magnetic field, which inhibit outward heat transfer

THE SUN IS THE CLOSEST OBJECT IN THE SOLAR SYSTEM TO A PERFECT SPHERE

Solar activity and Earth

Changes in activity on the Sun's surface can be felt on Earth. On reaching Earth, the particles in a coronal mass ejection can penetrate the walls of spacecraft (posing a hazard to astronauts), disable satellites, and cause high currents in power grids on the surface. Sunspot activity also affects Earth's climate. Peak sunspot activity leads to a small increase in solar radiation. Times when sunspots are absent have also been linked to cold periods in Earth's history.

THE SUN'S ENERGY SOURCE

The Sun's huge mass creates immense pressure and temperature at the core, where nuclear fusion occurs. Nuclei of hydrogen atoms, each consisting of a single proton, fuse with other hydrogen nuclei to form a helium nucleus. In the process, other subatomic particles and radiation are released—as well as a huge amount of energy.

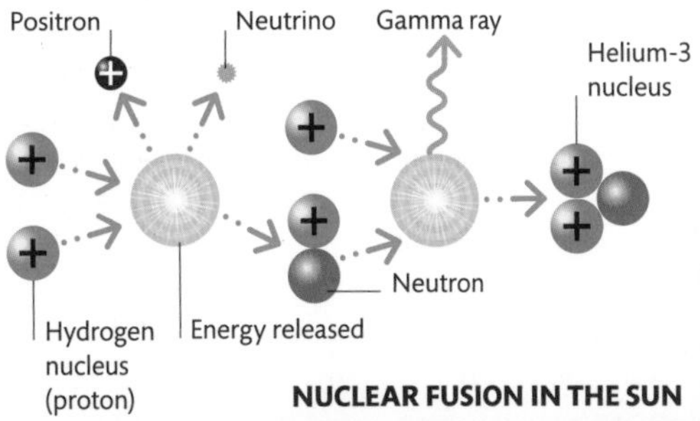

NUCLEAR FUSION IN THE SUN

A solar flare is an intense burst of radiation resulting from the release of magnetic energy associated with sunspots

A prominence is a loop of plasma that extends out into space, although it remains attached to the photosphere

A coronal mass ejection is an unusually large release of plasma from the corona

Coronal holes are areas where the plasma is less dense and relatively cold and dark

The chromosphere is a thin layer of the Sun's atmosphere—we see it during a total solar eclipse as a red rim around the edge of the Sun

Radiation escaping from the photosphere, where the temperature is 9,900°F (5,500°C), appears to us as sunlight

HOW LONG DOES SUNLIGHT TAKE TO REACH EARTH?

It can take several hundred thousand years for a photon to travel from the Sun's core to its surface. But after that, it takes just eight minutes for the photon to reach Earth.

The Solar System

The Solar System consists of the Sun—our local star—at its center, and its eight orbiting planets. It also includes more than 170 moons, several dwarf planets, asteroids, comets, and other space bodies.

How it formed

Our Solar System arose when a cloud of freezing gas and dust called a nebula condensed and started to spin (see p.190). The Sun formed at the hot center of the disc, while matter farther away became the planets and moons. Only rocky material withstood the heat near the Sun, forming the inner planets, while freezing gaseous matter settled in the disc's outer regions to form the outer planets.

Our place in the Milky Way
Our Solar System is located in an inner arm of the Milky Way galaxy. The Sun is one of 100–400 billion stars.

HOW OLD IS THE SOLAR SYSTEM?

The Solar System is about 4.6 billion years old. This age has been estimated by measuring the radioactive decay of the material in meteorites that have fallen to Earth.

SATURN'S DENSITY IS SO LOW THAT THE PLANET WOULD **FLOAT IN WATER**

SUN

36 million miles (58 million km) average distance from Sun

Mercury
The smallest planet, Mercury orbits at 29 miles (47km) per second.

Diameter 3,032 miles (4,879km)

67 million miles (108 million km) from Sun

Venus
The hottest planet, Venus rotates so slowly that its day is longer than its year.

Diameter 7,521 miles (12,104km)

93 million miles (150 million km) from Sun

Earth
Earth is the densest planet. Water covers 70 percent of its surface.

Diameter 7,926 miles (12,756km)

142 million miles (228 million km) average distance from Sun

Mars
Gravity on the frozen Red Planet is about one-third of Earth's.

Diameter 4,220 miles (6,792km)

Asteroid Belt
The Asteroid Belt is located between the orbits of Mars and Jupiter. It is home to the dwarf planet Ceres.

484 million miles (779 million km) from Sun

Jupiter
The largest planet, Jupiter has a giant red spot that is a 300-year-old storm.

Jupiter's moons
Jupiter has 69 moons, the largest of which, Ganymede, is bigger than Mercury. The moon Europa is thought to have liquid water beneath its icy surface.

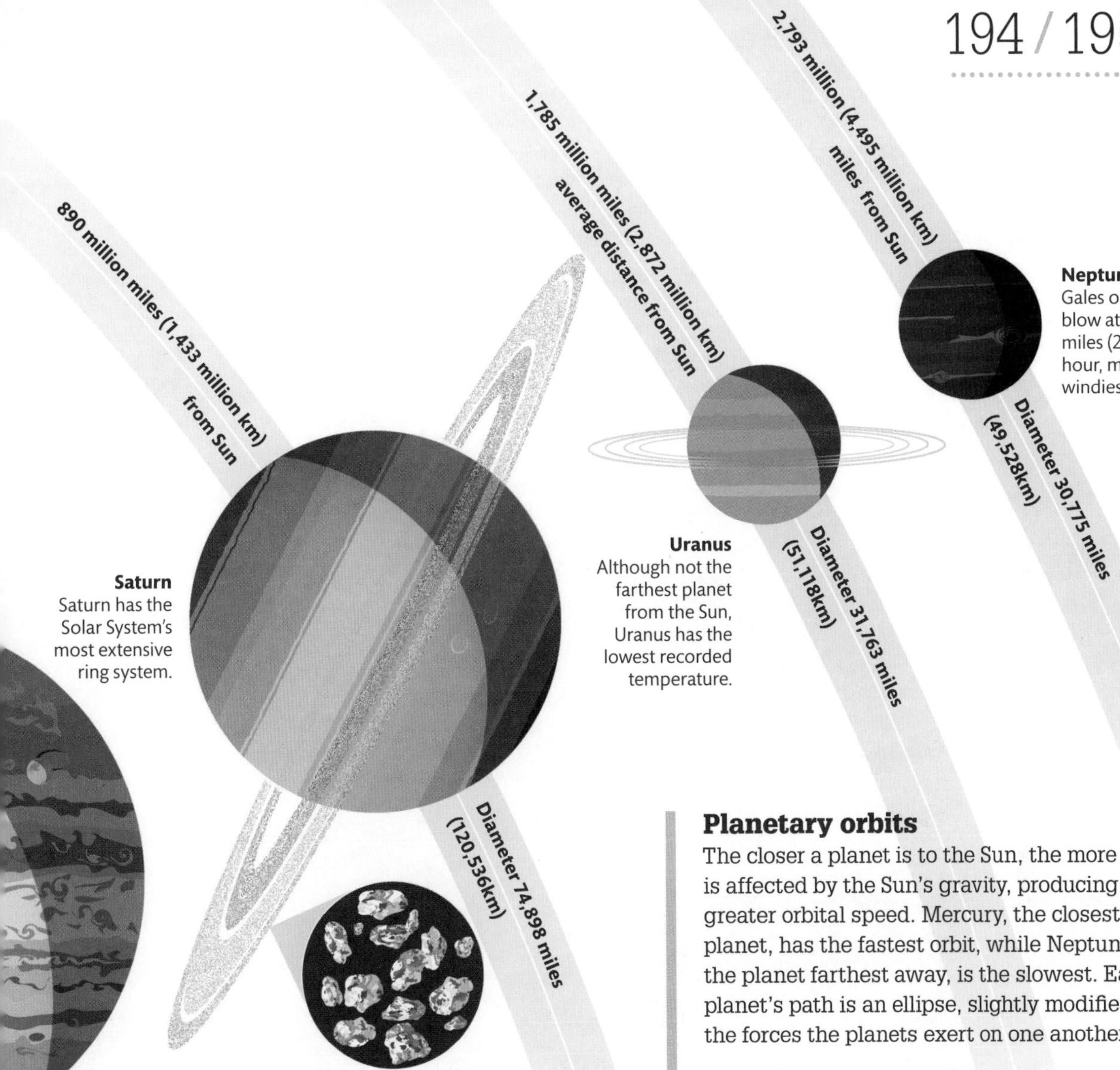

Saturn
Saturn has the Solar System's most extensive ring system.

Uranus
Although not the farthest planet from the Sun, Uranus has the lowest recorded temperature.

Neptune
Gales on Neptune blow at up to 1,200 miles (2,000km) per hour, making it the windiest planet.

Saturn's rings
The rings are made mainly of highly reflective water ice, with traces of rocky materials. They are thought to be the remains of one or more moons that collided with asteroids or comets.

Planetary orbits

The closer a planet is to the Sun, the more it is affected by the Sun's gravity, producing a greater orbital speed. Mercury, the closest planet, has the fastest orbit, while Neptune, the planet farthest away, is the slowest. Each planet's path is an ellipse, slightly modified by the forces the planets exert on one another.

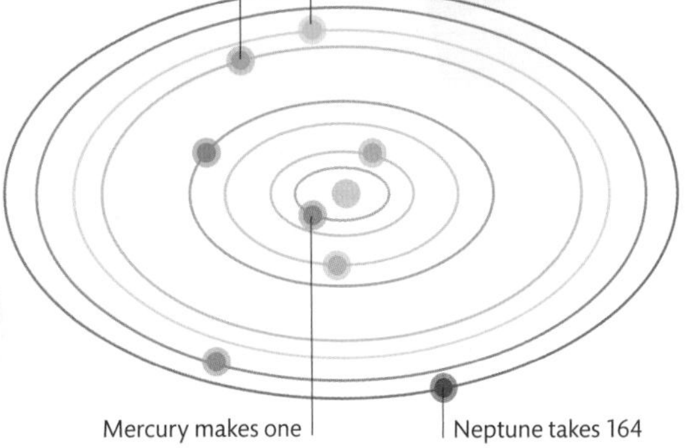

DWARF PLANETS

Dwarf planets, such as Pluto, have sufficient gravity and mass to form a spherical body and directly orbit the Sun. However, unlike the other planets, they have not cleared their orbital path, and still share it with asteroids and comets.

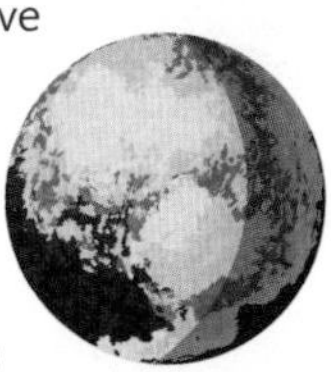
PLUTO

Space flotsam

As the Solar System formed, pieces of rock and ice created bodies of different sizes, with the largest becoming planets. Some pieces remain as meteoroids, asteroids, and comets, which sometimes fall to Earth.

CAN WE STOP A DEADLY IMPACT?

Dusting a comet or asteroid with chalk or charcoal could change the way it is heated by sunlight and modify its orbit. Detonating explosives near an object could change its orbit more quickly.

Meteoroids

Meteoroids are particles of asteroids or comets. These small rocky or metallic bodies are usually around the size of a sand grain or pebble but can reach over a meter across. Meteoroids that fall through a planet's atmosphere, becoming incandescent as they fall, are called meteors. Pieces that survive entry to reach the ground are known as meteorites. Around 90 to 95 percent of meteors completely burn up traveling through Earth's atmosphere. Their brightness in the sky has more to do with their speed of entry than their size.

The ISS sometimes changes course to avoid space debris. A potential collision is judged dangerous if the probability of impact is 0.001 percent or more

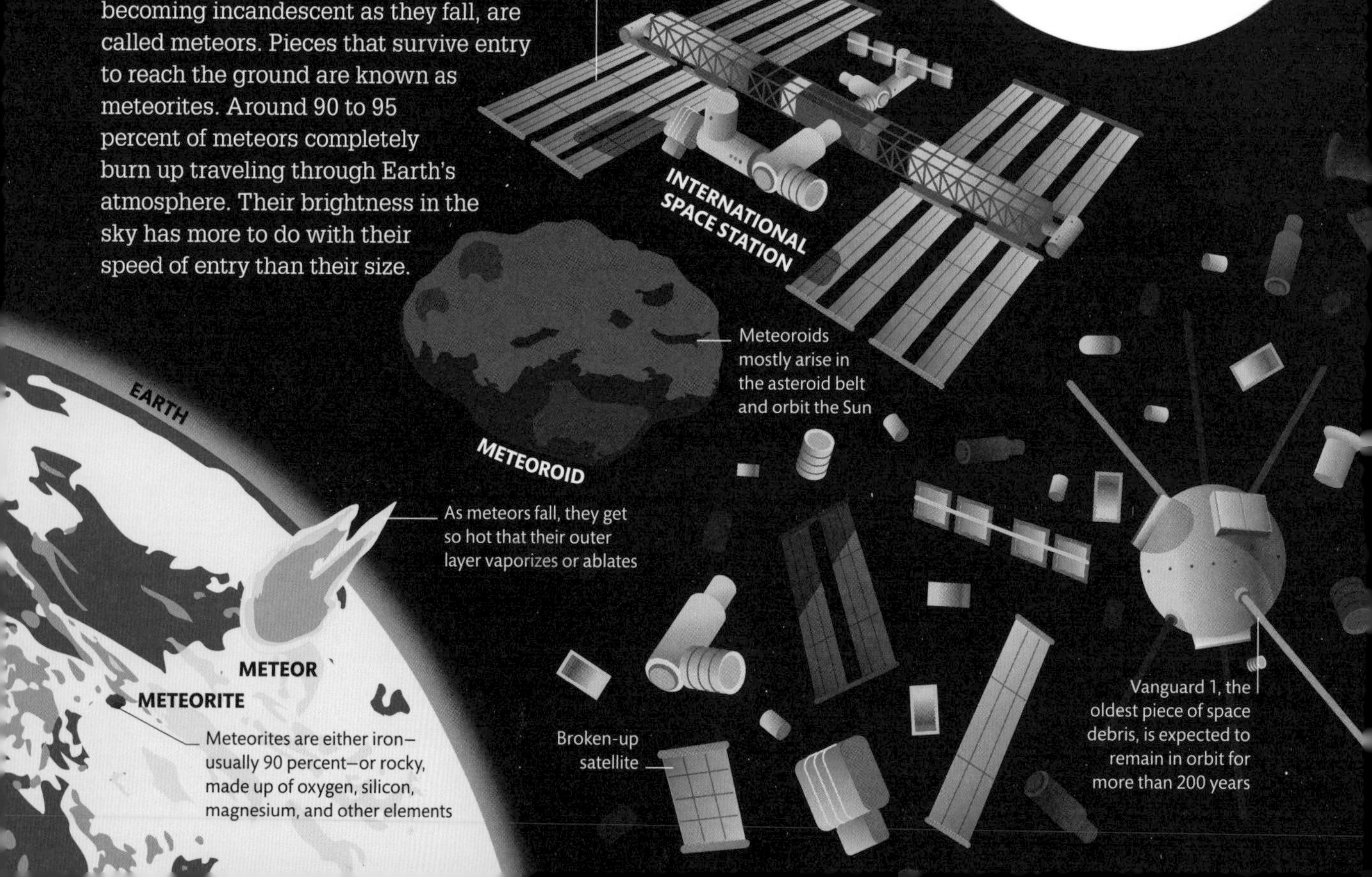

INTERNATIONAL SPACE STATION
Meteoroids mostly arise in the asteroid belt and orbit the Sun
METEOROID
EARTH
As meteors fall, they get so hot that their outer layer vaporizes or ablates
METEOR
METEORITE
Meteorites are either iron—usually 90 percent—or rocky, made up of oxygen, silicon, magnesium, and other elements
Broken-up satellite
Vanguard 1, the oldest piece of space debris, is expected to remain in orbit for more than 200 years

Asteroids

Asteroids are rocky or metallic objects circling the Sun, mainly between the orbits of Mars and Jupiter in what is known as the asteroid belt. Most have diameters of up to ⅔ mile (1km), but some—like the largest dwarf planet, Ceres—are more than 60 miles (100km) across and exert a significant gravitational pull. Jupiter's gravity prevents the asteroids from joining together to form planets.

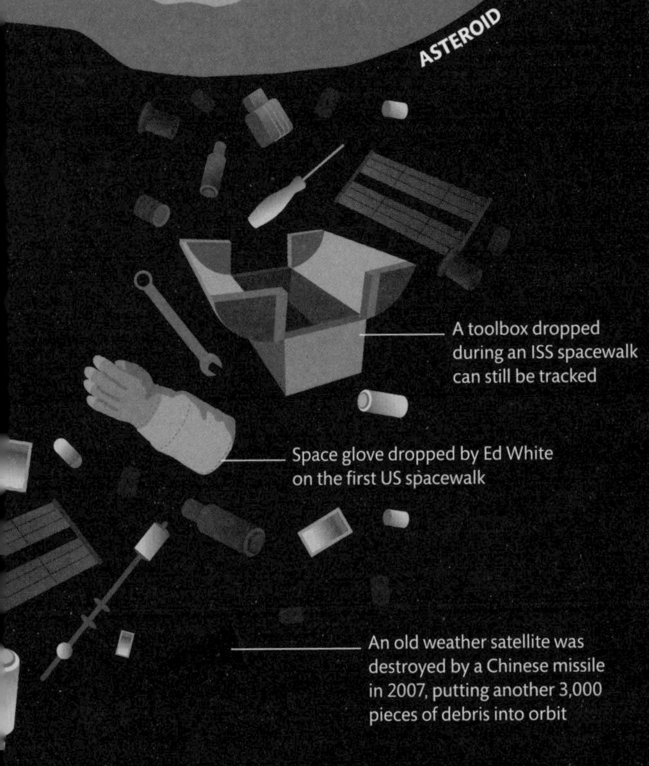
ASTEROID
A toolbox dropped during an ISS spacewalk can still be tracked
Space glove dropped by Ed White on the first US spacewalk
An old weather satellite was destroyed by a Chinese missile in 2007, putting another 3,000 pieces of debris into orbit

Space junk

Millions of man-made objects, ranging from tiny flakes of paint to truck-sized lumps of metal, are floating around the Solar System, most of it orbiting Earth. Moving at high speeds and increasing in number, space junk poses a growing threat to manned spacecraft like the International Space Station. There are also derelict spacecraft on the surfaces of Venus, Mars, and the Moon.

The Kuiper Belt and Oort Cloud

Icy bodies in the Kuiper Belt, a disc-shaped band of objects beyond Neptune's orbit, are tugged inward by planets to become comets. Those in the Oort Cloud, a huge spherical cloud of icy debris in the outer Solar System, are affected by the gravity of passing stars.

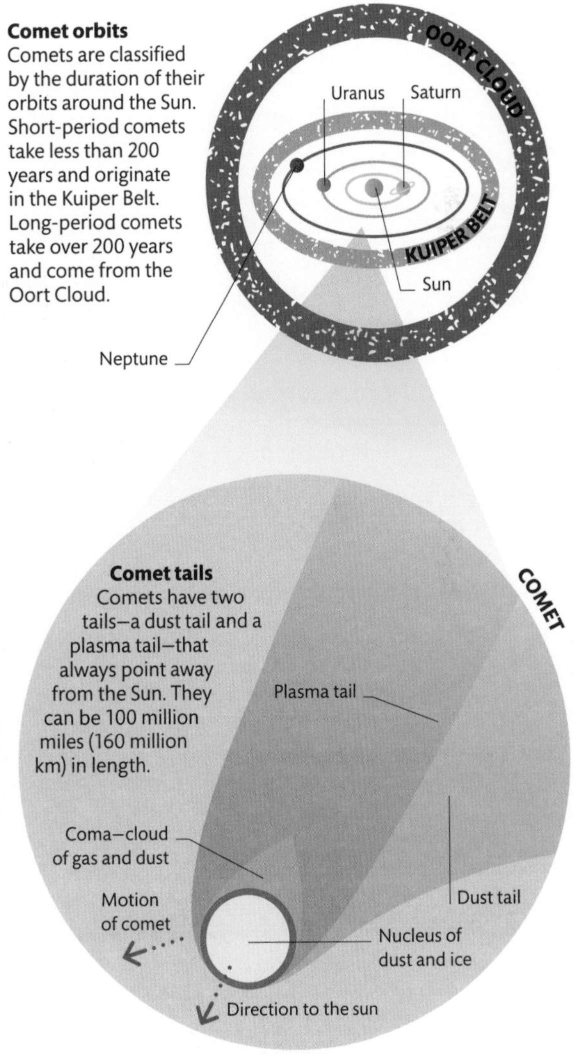
Comet orbits
Comets are classified by the duration of their orbits around the Sun. Short-period comets take less than 200 years and originate in the Kuiper Belt. Long-period comets take over 200 years and come from the Oort Cloud.
OORT CLOUD
Uranus
Saturn
KUIPER BELT
Sun
Neptune
COMET
Comet tails
Comets have two tails—a dust tail and a plasma tail—that always point away from the Sun. They can be 100 million miles (160 million km) in length.
Plasma tail
Coma—cloud of gas and dust
Dust tail
Motion of comet
Nucleus of dust and ice
Direction to the sun

MOVING AT 22,000MPH (36,000 KPH) A 4IN (10CM) OBJECT CAN CAUSE DAMAGE EQUAL TO 25 STICKS OF DYNAMITE

Black holes

A black hole is a region of space where matter is squashed into an infinitesimally small point of infinite density. It is so dense that its gravitational pull does not let anything escape. Even light is dragged in, making a black hole invisible—the only way to detect it is by observing the effect it has on its surroundings.

THE NEAREST SUSPECTED BLACK HOLE IS ABOUT 3,000 LIGHT-YEARS AWAY

Complete collapse

Most black holes form as a result of the death of a massive star (with about 10 or more times the mass of the Sun). Matter pulled toward a black hole by gravity often forms a spinning disc, which throws off X-rays and other kinds of radiation and is detectable by astronomers.

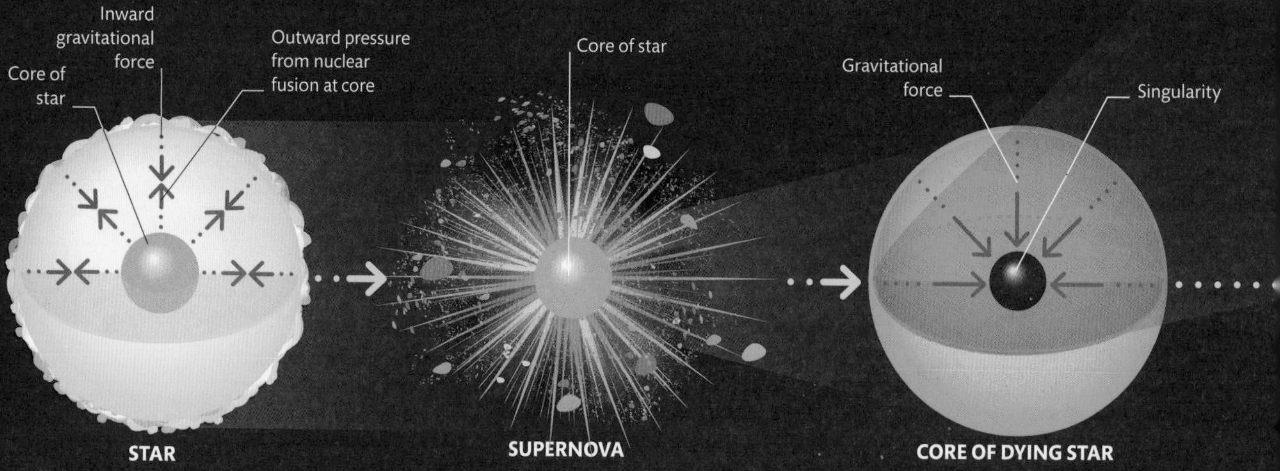

1 A stable star
Nuclear reactions in the core of a star create energy and outward pressure. When these are in balance with the force of gravity pulling inward, the star remains stable. But when the fuel runs out, gravity takes over.

2 Explosive death
As the nuclear reactions stop, the star dies. Unable to resist the crushing force of its own gravity, it collapses. This also causes a supernova explosion, blasting the star's outer parts into space.

3 Core collapse
If the core that remains after the supernova is still massive (more than three times the mass of the Sun), it keeps shrinking and collapses under its own weight into a point of infinite density called a singularity.

TYPES OF BLACK HOLE

There are two main types of black hole: stellar and supermassive. A stellar black hole forms when an enormous star goes supernova at the end of its life (see above). Supermassive black holes are bigger and are found at the centers of galaxies, often surrounded by whirlpools of intensely hot, glowing matter. Black holes of a third type, called primordial black holes, may have formed in the Big Bang. If indeed they ever existed, most were probably tiny and quickly evaporated. To have survived to the present, one would have needed to start with at least the mass of a large mountain.

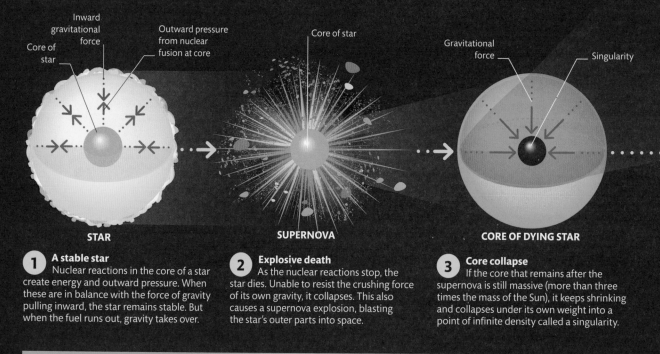

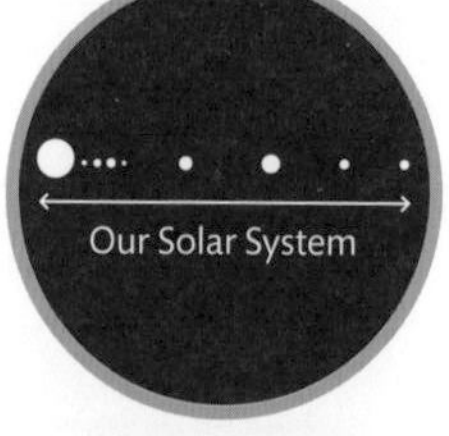

SUPERMASSIVE
Event horizon diameter: up to Solar System size
Mass: up to billions of suns

STELLAR
Event horizon diameter: 20–200 miles (30–300km)
Mass: 5–50 suns

PRIMORDIAL
Event horizon diameter: width of a small atomic nucleus or greater
Mass: more than that of a mountain

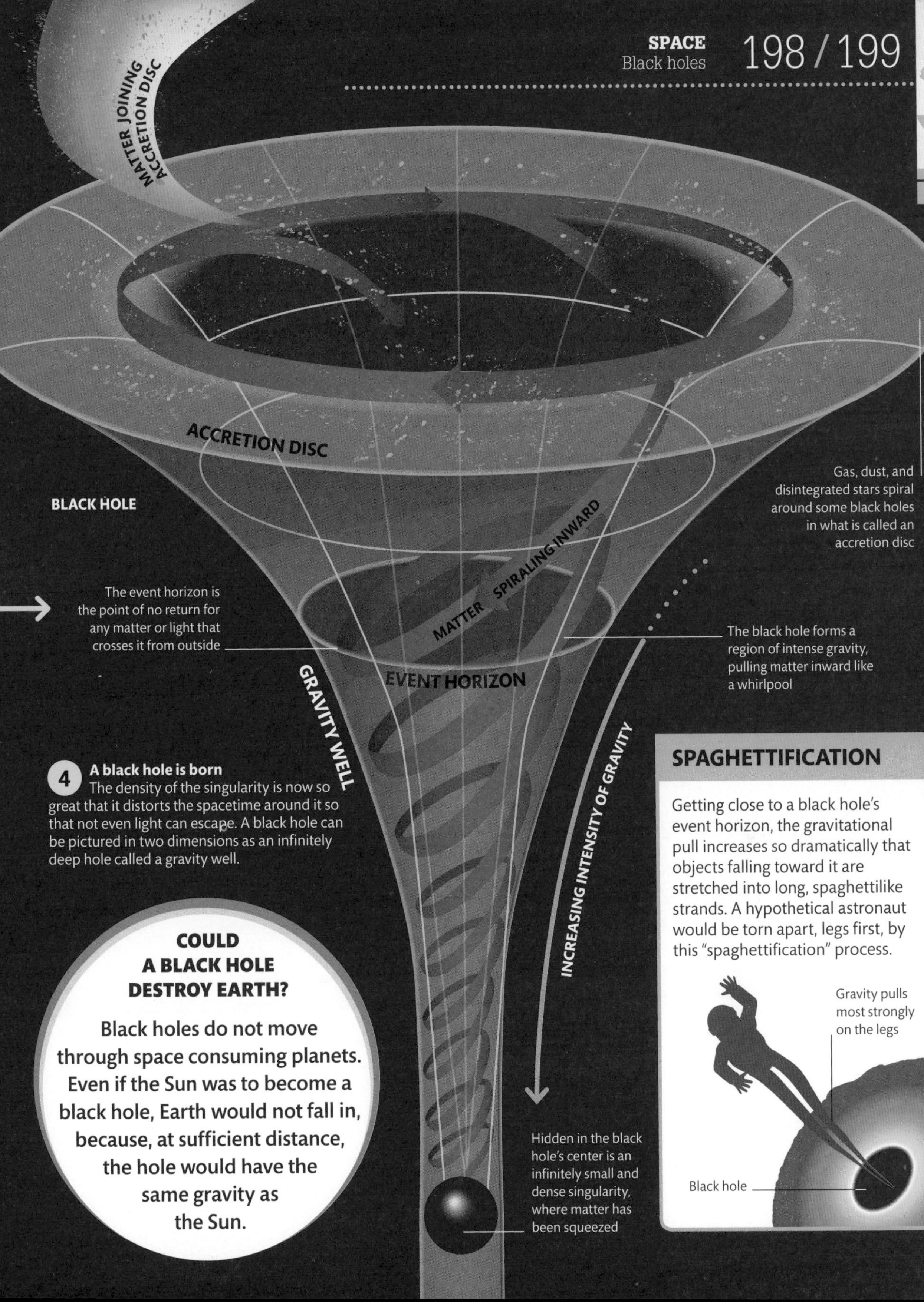
MATTER JOINING ACCRETION DISC
ACCRETION DISC
BLACK HOLE
Gas, dust, and disintegrated stars spiral around some black holes in what is called an accretion disc
MATTER SPIRALING INWARD
The event horizon is the point of no return for any matter or light that crosses it from outside
The black hole forms a region of intense gravity, pulling matter inward like a whirlpool
EVENT HORIZON
GRAVITY WELL
INCREASING INTENSITY OF GRAVITY
Hidden in the black hole's center is an infinitely small and dense singularity, where matter has been squeezed
Gravity pulls most strongly on the legs
Black hole

4 A black hole is born
The density of the singularity is now so great that it distorts the spacetime around it so that not even light can escape. A black hole can be pictured in two dimensions as an infinitely deep hole called a gravity well.

COULD A BLACK HOLE DESTROY EARTH?

Black holes do not move through space consuming planets. Even if the Sun was to become a black hole, Earth would not fall in, because, at sufficient distance, the hole would have the same gravity as the Sun.

SPAGHETTIFICATION

Getting close to a black hole's event horizon, the gravitational pull increases so dramatically that objects falling toward it are stretched into long, spaghettilike strands. A hypothetical astronaut would be torn apart, legs first, by this "spaghettification" process.

Galaxies

Galaxies are massive systems containing millions to billions of stars, clouds of gas and dust called nebulas, and an unknown quantity of dark matter (see pp.206–07). They are held together by gravitational attraction. Our galaxy is known as the Milky Way.

HOW BIG IS THE MILKY WAY?

It is about 100,000 light-years across, and the disk is some 1,000 light-years thick. Our Solar System takes about 230 million years to revolve around its central black hole.

The Milky Way
Our Solar System is located on the Orion Arm of a large, barred spiral galaxy, which contains around 100–400 billion stars revolving around a supermassive black hole. Viewed side-on, our galaxy appears flattened, with a bright bulge at the center, and a halo region that contains star clusters.

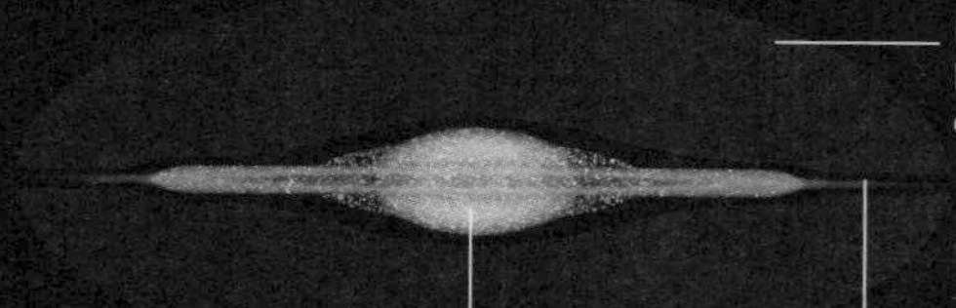

SIDE VIEW OF MILKY WAY

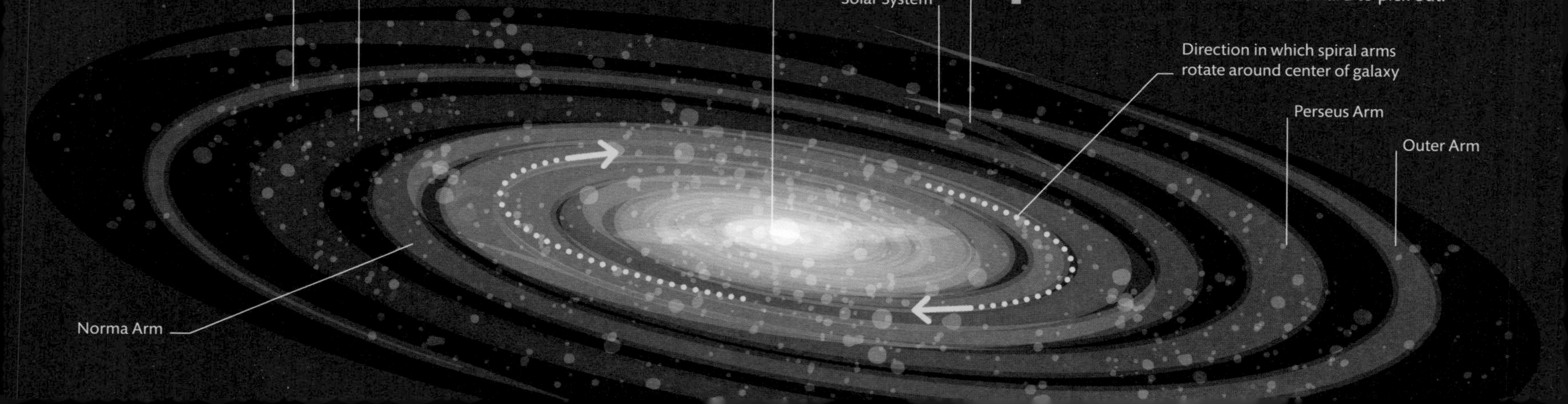

Types of galaxies

There are around 2 trillion galaxies in the observable Universe, although more may yet be revealed (see pp.204–05).
The three main galaxy types are elliptical, spiral, and irregular. Some galaxies are combinations of these types, such as lenticular galaxies. Part ellipse, part spiral, they are flattened but lack clear spiral arms.

Spiral galaxies
Spirals are flat, rotating disks with arm structures, a nuclear bulge, and a surrounding halo. In barred spirals, the arms emanate from a central bar rather than from the nucleus.

Elliptical galaxies
Elliptical galaxies vary from nearly spherical to football shaped, and are classified by how circular or flattened they are. Unlike spirals, they have no single axis of rotation.

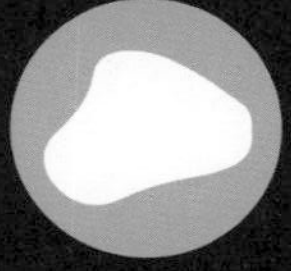

Irregular galaxies
These galaxies have no symmetrical structure and little or no nucleus. Some contain new, hot stars. Others have large quantities of dust, making individual stars hard to pick out.

Colliding galaxies

Collisions between galaxies are common—the Milky Way is currently colliding with the Sagittarius Dwarf galaxy. However, the distance between stars is so vast, they almost never collide. Galaxies that narrowly miss each other can still distort each other's shapes, and interactions can compress the clouds of gas in each galaxy, kick-starting new star formation.

Galactic crash
These two spiral galaxies are colliding, attracting each other's main spiral arms. Over millions of years, they might eventually come together to form an elliptical galaxy.

ACTIVE GALAXIES

Unlike normal galaxies, active galaxies emit much more energy than their stars can produce due to the accretion (accumulation) of material by the supermassive black hole at each galaxy's center. Some active galaxies send out jets of energetic particles.

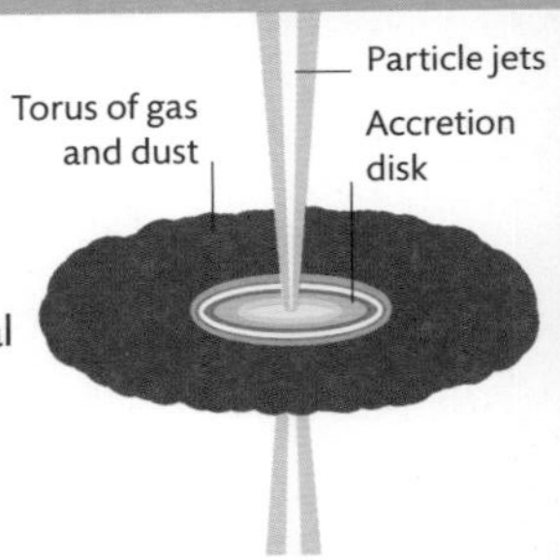

NUCLEUS AND TORUS

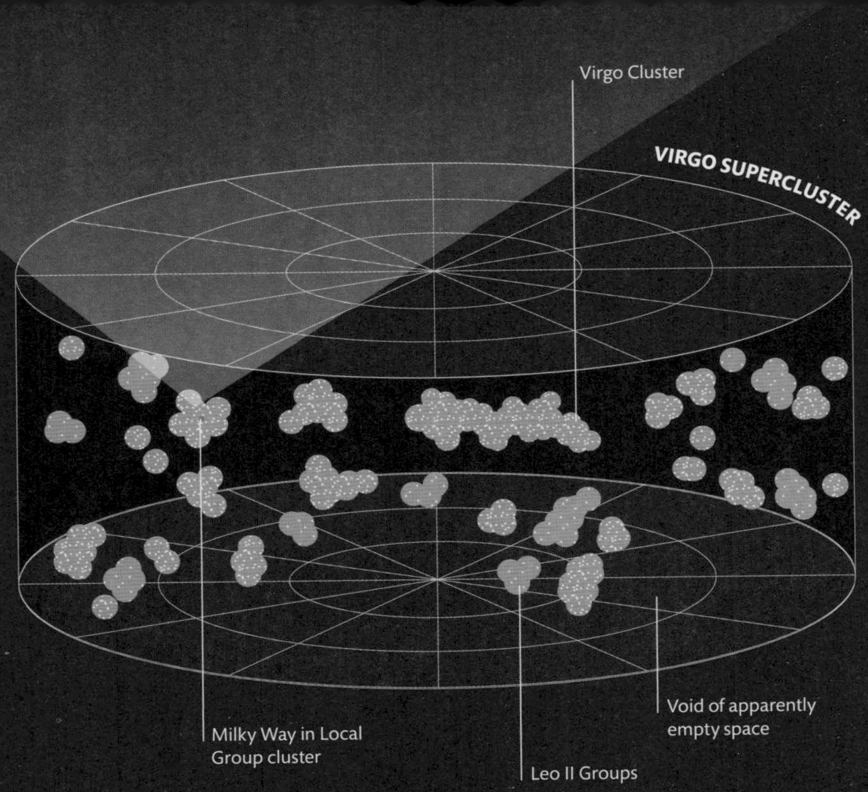

Virgo Supercluster
Our galaxy is part of a cluster called the Local Group, which is part of the Virgo Supercluster. The supercluster is dominated by the Virgo Cluster, which contains up to 2,000 galaxies.

Clusters and superclusters

Three-quarters of galaxies are not randomly distributed but clumped together. They are connected by a cosmic web of ordinary and dark matter filaments, and clusters of galaxies form at the points where these filaments intersect. Where galaxy clusters collide with one another, superclusters form. There are around 10 million of these. The largest, the Sloan Great Wall, is 1.4 billion light-years across. Dark energy is expected to eventually tear these superclusters apart.

The Big Bang

Most astronomers think that the Universe had a definite beginning 13.8 billion years ago in an event called the Big Bang. Starting from an infinitesimally small, dense, hot point, all matter, energy, space, and time were formed. Ever since the Big Bang, the Universe has been getting bigger and cooler.

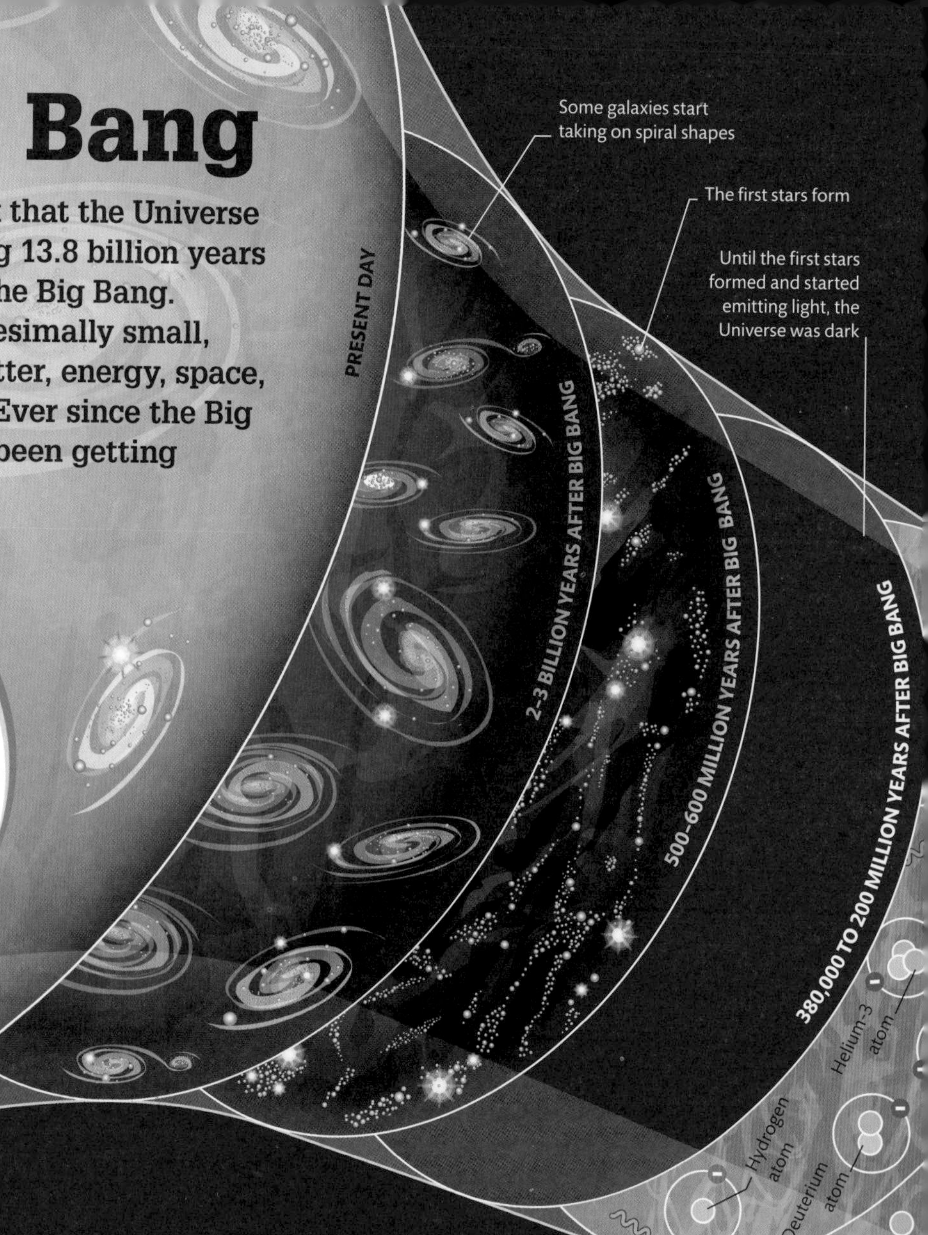

WHAT CAME BEFORE THE BIG BANG?

If time began with the Big Bang, then ... nothing. Or perhaps our universe is material from a parent universe.

Expanding space

Scientists have observed that the Universe is expanding, suggesting it was once much smaller. During a tiny fraction of its first second, part of the Universe grew faster than the speed of light in an episode called inflation. The rate of expansion soon slowed, but the Universe is still getting bigger. At large scales, all objects are moving away from each other—and the farther away they are, the faster they are receding. This

Redshift

As an object recedes from an observer at high speed, light waves from it appear stretched. This causes lines in the object's spectrum (see p.211) to shift toward the red end. An object's distance from Earth can be calculated from the

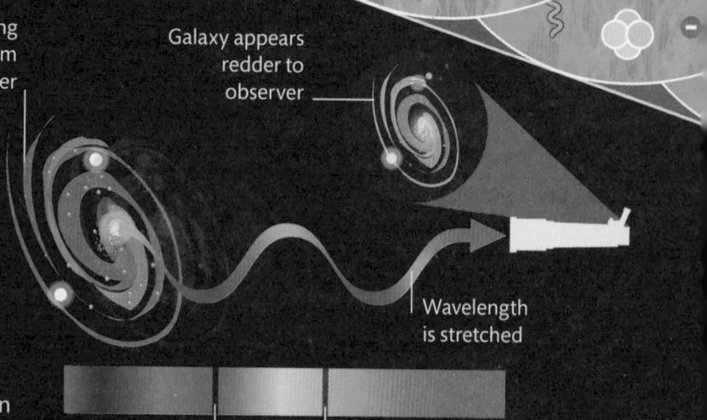

In the beginning

The Universe was initially pure energy. As it cooled, energy and matter were in an interchangeable state called mass-energy. After the end of inflation, the first subatomic particles began to emerge. Many of these are no longer present, but the remnants make up all the matter in the Universe today. By the time about 400,000 years had passed, the first atoms had formed.

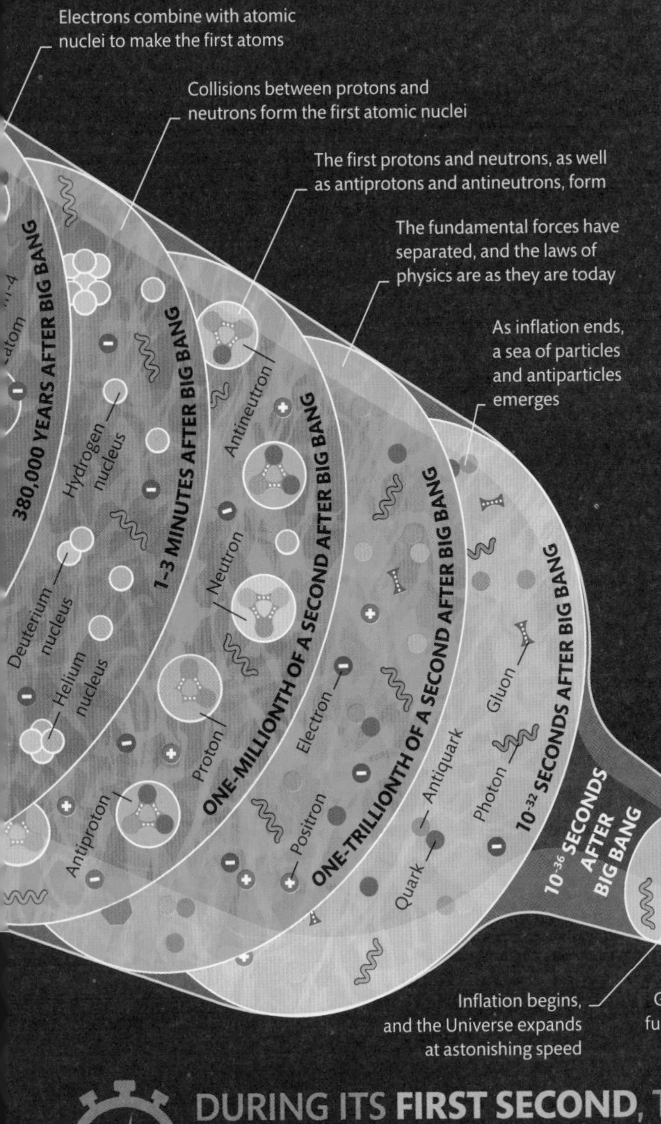

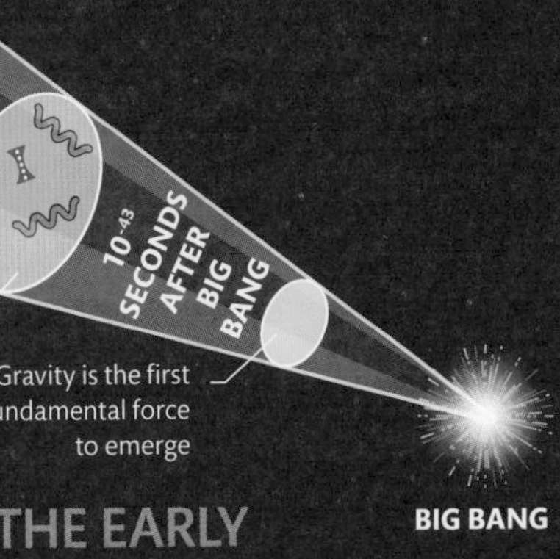

EVIDENCE FOR THE BIG BANG

The scientists who proposed the Big Bang theory predicted that it would leave behind a faint heat radiation coming from all directions in the sky. In 1964, this radiation, called the cosmic microwave background, was found by two US astronomers using a large horn-shaped radio antenna in New Jersey.

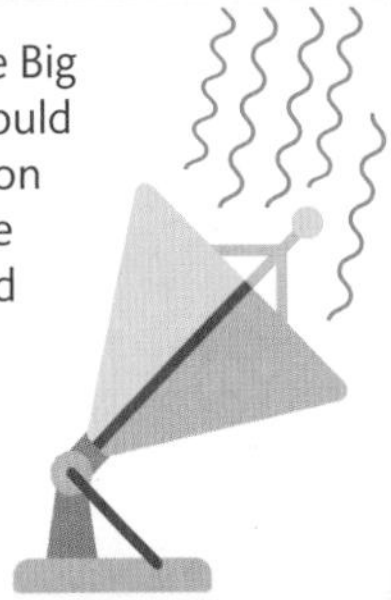

The laws of physics

The four basic forces that govern the interaction between particles (see pp.26–27) did not initially exist but were established soon after the Universe's birth. Just after the Big Bang, a time known as the Planck Era, when matter and energy were not yet separate, there was a single unified force or superforce. A trillionth of a second after the Big Bang, this had separated into electromagnetism, the strong nuclear force, the weak nuclear force, and gravity.

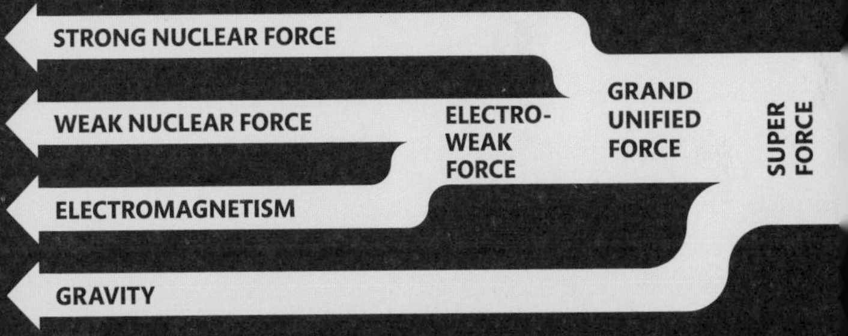

Big Bang
During the first second of time, the fundamental forces and subatomic particles formed. It would take several hundred thousand more years for atoms to emerge and millions of years for stars and then galaxies to develop.

DURING ITS **FIRST SECOND**, THE EARLY UNIVERSE GREW FROM NOTHING TO **BILLIONS OF MILES** ACROSS

How big is the Universe?

Is space infinite? And what shape is the Universe? Although astronomers haven't answered these questions, they can estimate the size of the part of the Universe that we can see. By studying the density of mass and energy, they can also draw conclusions about the geometry of space.

WHAT IS A LIGHT-YEAR?

A unit of distance (not time), a light-year is the distance light travels in a year. Light moves at 190,000 miles (300,000km) per second, so a light-year is about 6 trillion miles (9.5 trillion km).

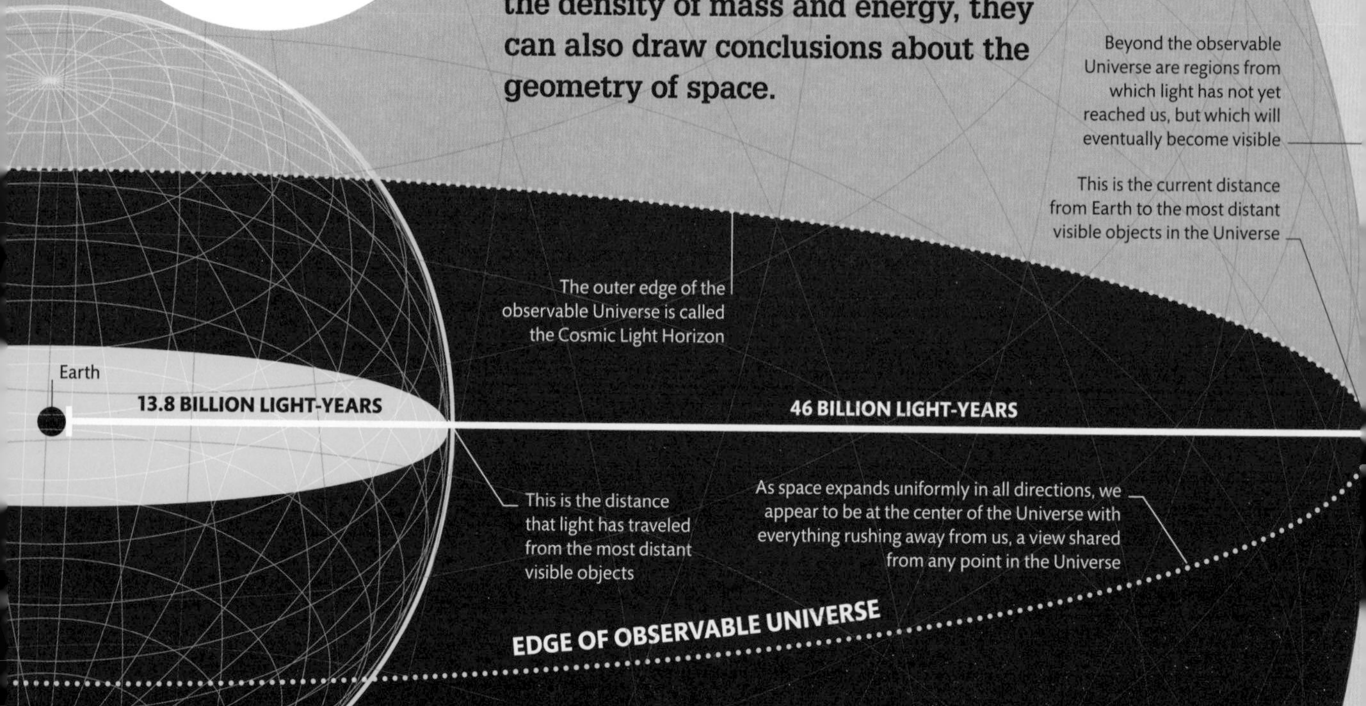
Beyond the observable Universe are regions from which light has not yet reached us, but which will eventually become visible
This is the current distance from Earth to the most distant visible objects in the Universe
The outer edge of the observable Universe is called the Cosmic Light Horizon
Earth
13.8 BILLION LIGHT-YEARS
46 BILLION LIGHT-YEARS
This is the distance that light has traveled from the most distant visible objects
As space expands uniformly in all directions, we appear to be at the center of the Universe with everything rushing away from us, a view shared from any point in the Universe
EDGE OF OBSERVABLE UNIVERSE

The observable Universe

The part of space we can see and study is called the observable Universe. A spherical region centered on Earth, it is the volume of space from which light has had time to reach us since the Big Bang. As an object recedes from us, the light it emits shifts toward the red end of the spectrum as it crosses space toward us (see p.202). The most redshifted light detectable has come from 13.8 billion light-years away. This tells us how big the Universe would be if it was static. It also reveals that it must be about 13.8 billion years old. But we know that, ever since it began, the Universe has been expanding.

THE MOST DISTANT GALAXIES APPEAR TEN BILLION TIMES FAINTER THAN THE DIMMEST OBJECTS VISIBLE TO THE NAKED EYE

Measuring distance in expanding space

As space is expanding, the true distance to an object in space, called its co-moving distance, is greater than the distance light from the object has traveled to reach us, known as the lookback distance. Taking the expansion of space into account, the edge of the observable Universe is about 46.5 billion light-years away.

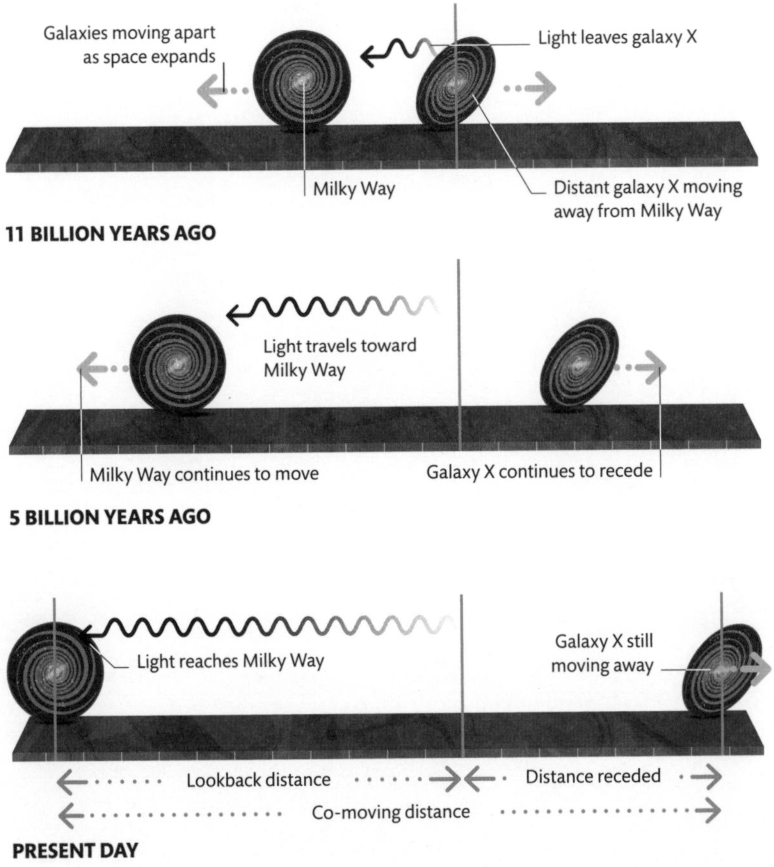

HOW FAST IS SPACE EXPANDING?

At relatively small scales, such as within galaxies, objects in space are held at fixed distances from each other by gravity. But at bigger scales, the expansion of space means that objects are moving away from each other, just like points on the surface of an expanding balloon. And the farther away two objects are, the faster they are moving apart. The latest measurements suggest that two objects separated by a megaparsec (about 3 million light-years) are moving apart at about 46 miles (74km) per second.

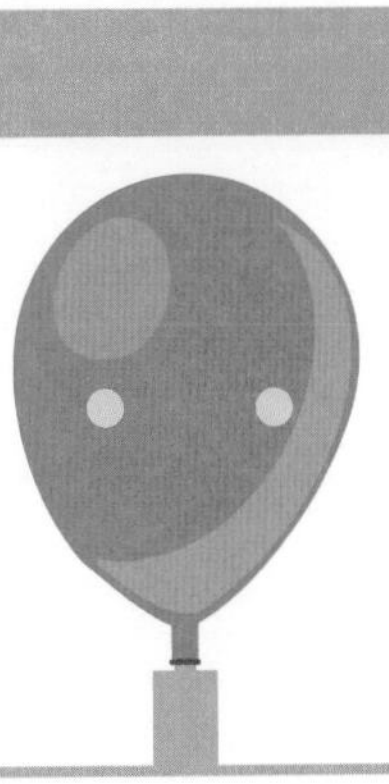

Universe shapes

The Universe has three possible geometries. Each has a different spacetime curvature. This is not the kind of curvature we are used to, but it can be represented by a 2-D shape. Our Universe is thought to be flat or nearly flat. Several theories about the Universe's fate are based on these geometries (pp.208–09).

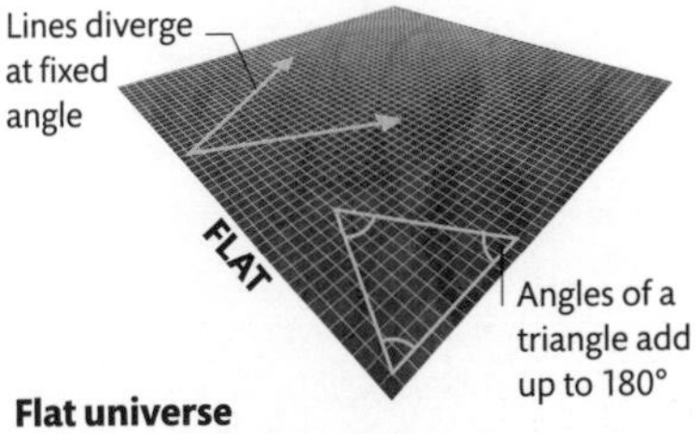

Flat universe
The 2-D analogy for a flat universe is a plane where the familiar rules of geometry apply. For example, parallel lines never meet.

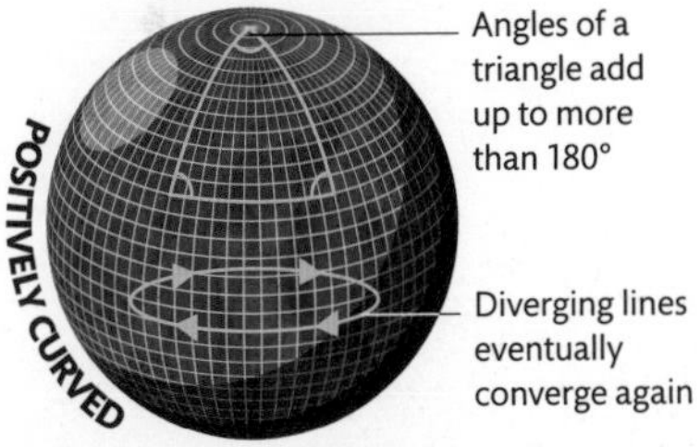

Positively curved universe
A universe in which spacetime is positively curved is "closed" and finite in mass and extent. Parallel lines converge on a spherical surface in this 2-D analogy.

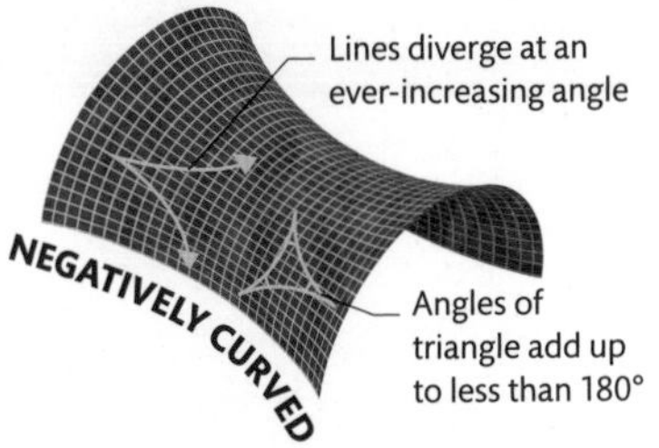

Negatively curved universe
In this scenario, the universe is "open" and infinite. The 2-D analogy is a saddle-shaped space in which diverging lines gradually get farther apart.

Dark matter and dark energy

Most of the Universe consists of what astronomers call dark matter and dark energy. We can't observe these types of matter and energy directly, but we know they exist because of how they interact with ordinary matter and light waves.

Missing mass and energy

Mass and energy are two forms of a single phenomenon called mass–energy (see p.141). When astronomers try to track down all the Universe's mass–energy, they find that most of it can't be seen. But there must be more mass than we can see because without it galaxy clusters would fly apart. And there has to be more energy because something is opposing gravity and causing the expansion of space to speed up.

THE WORLD'S MOST SENSITIVE DARK MATTER DETECTOR IS 1 MILE (1.5KM) UNDERGROUND

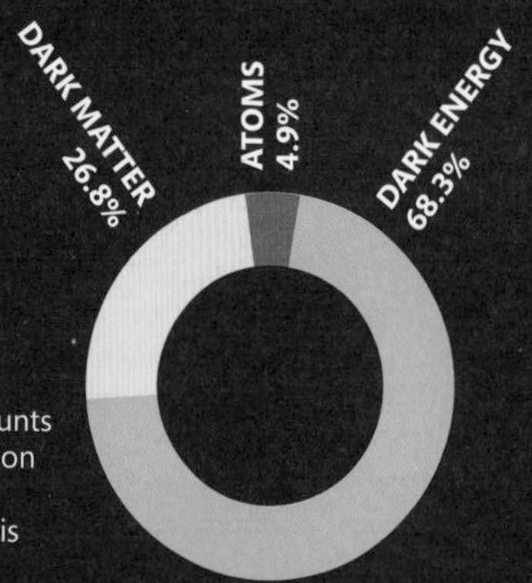

How much is missing?
Ordinary visible matter, made up of atoms, accounts for only a small proportion of the Universe's mass-energy. Most of the rest is dark energy.

Dark matter

Dark matter forms in haloes around ordinary, or "baryonic," matter but to a large extent doesn't interact with it, doesn't reflect or absorb light, and can't be detected by electromagnetic radiation. However, its gravitational effects on galaxies and stars and its distorting effects on the paths of light waves can be observed. Dark matter's nature is unknown, but two of the forms astronomers think it could take are called MACHOs and WIMPs.

Gravitational lensing
A large mass can act like a lens, distorting gravitational fields, which alters the paths of light waves and changes the appearance of galaxies. Weak lensing effects elongate galaxy shapes, while a strong effect alters their positions or even duplicates them.

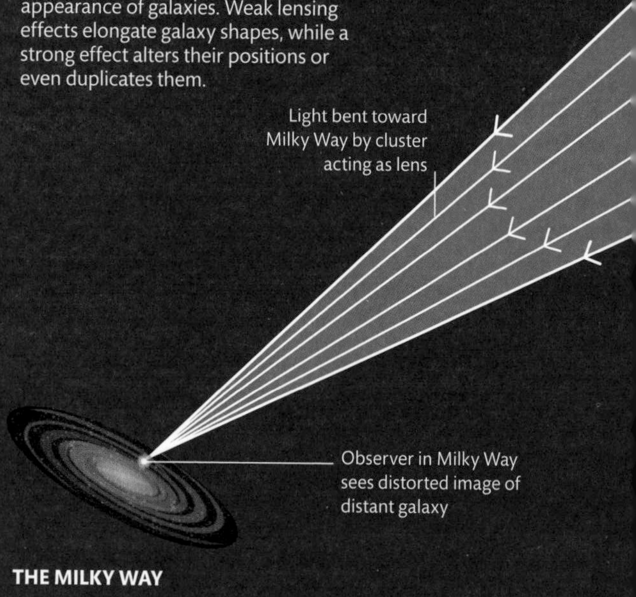

<table>
<tr><th rowspan="1">MACHOs</th><th colspan="2">WIMPs</th></tr>
<tr><td rowspan="3">Some dark matter might consist of dense objects such as black holes and brown dwarfs—collectively called MACHOs (MAssive Compact Halo Objects)—which emit so little light they can only be detected by gravitational lensing (see above). However, MACHOs cannot account for all of dark matter's mass.</td><td colspan="2">Other hypothetical candidates are Weakly Interacting Massive Particles (WIMPs), strange particles created in the early Universe, which interact via the weak force (see p.27) and gravity.</td></tr>
<tr><th>Hot</th><th>Cold</th></tr>
<tr><td>This theoretical form of dark matter consists of particles traveling close to the speed of light.</td><td>Most dark matter, such as WIMPS, is thought to be cold—a form of matter moving slowly.</td></tr>
</table>

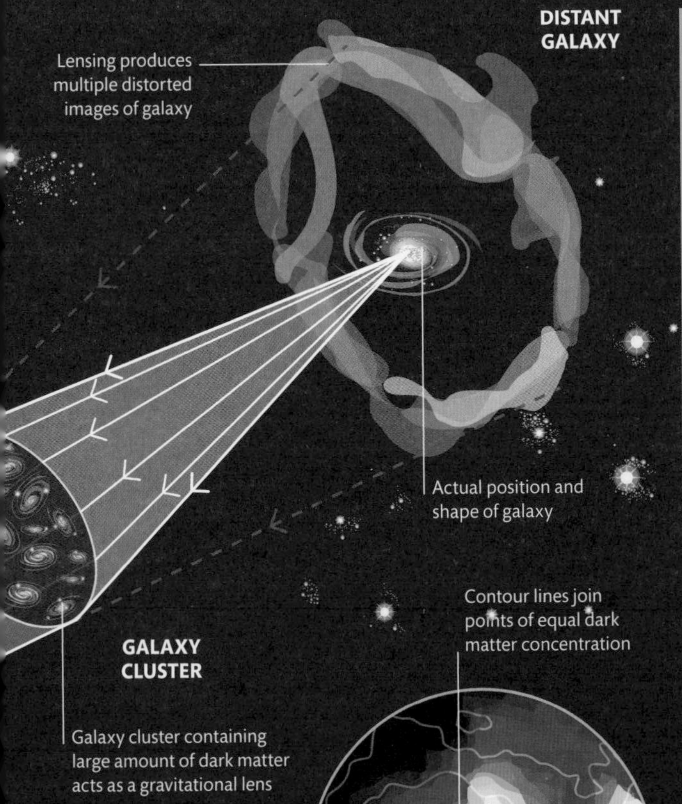

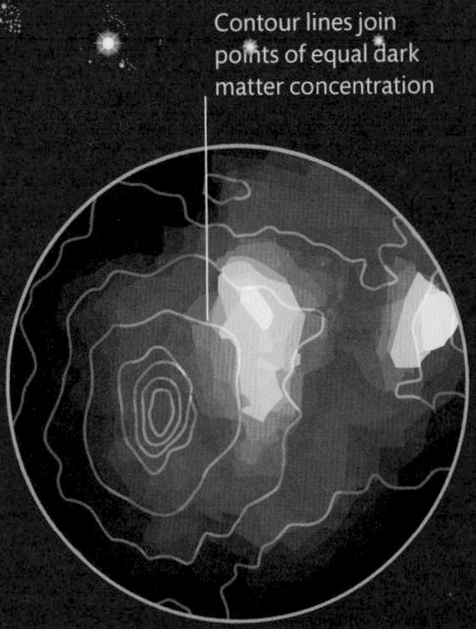

Mapping dark matter
By observing a body of dark matter's lensing effects, astronomers can map its shape, a bit like inferring the shape of a ripple on water from the apparent distortions to a pebble seen on a riverbed.

IS THERE ANY DARK MATTER ON EARTH?

Yes—probably. According to some estimates, billions of dark matter particles pass through our bodies every second.

Dark energy

Measurements of the distances to remote supernovas have shown that the Universe's expansion is speeding up. This finding led to the theory of dark energy, a force that opposes gravity and explains both the flatness of our Universe and its accelerating expansion. Dark matter dominated the early Universe, but now dark energy has overtaken it, and its effects are increasing as the Universe gets larger.

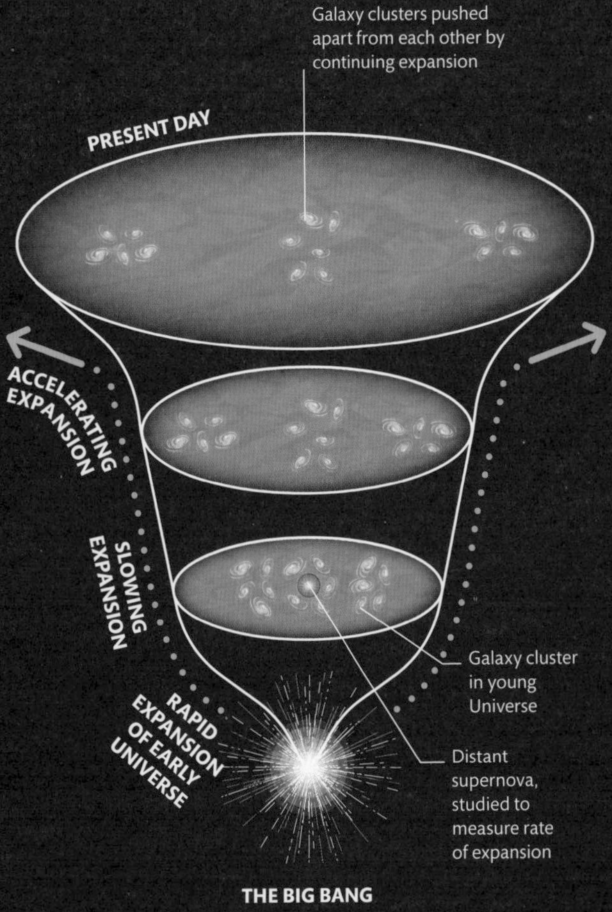

Accelerating expansion
After the Big Bang, an initial rapid expansion was followed by a slowdown. But from around 7.5 billion years ago, as shown by the rapidly widening curve, objects flew apart at a faster rate due to the force of dark energy.

How it ends

The ultimate fate of the Universe remains uncertain. Whether it will collapse and terminate with another Big Bang, come to a cold and silent close, meet with a violent and permanent end, or expand infinitely remains a subject of scientific speculation.

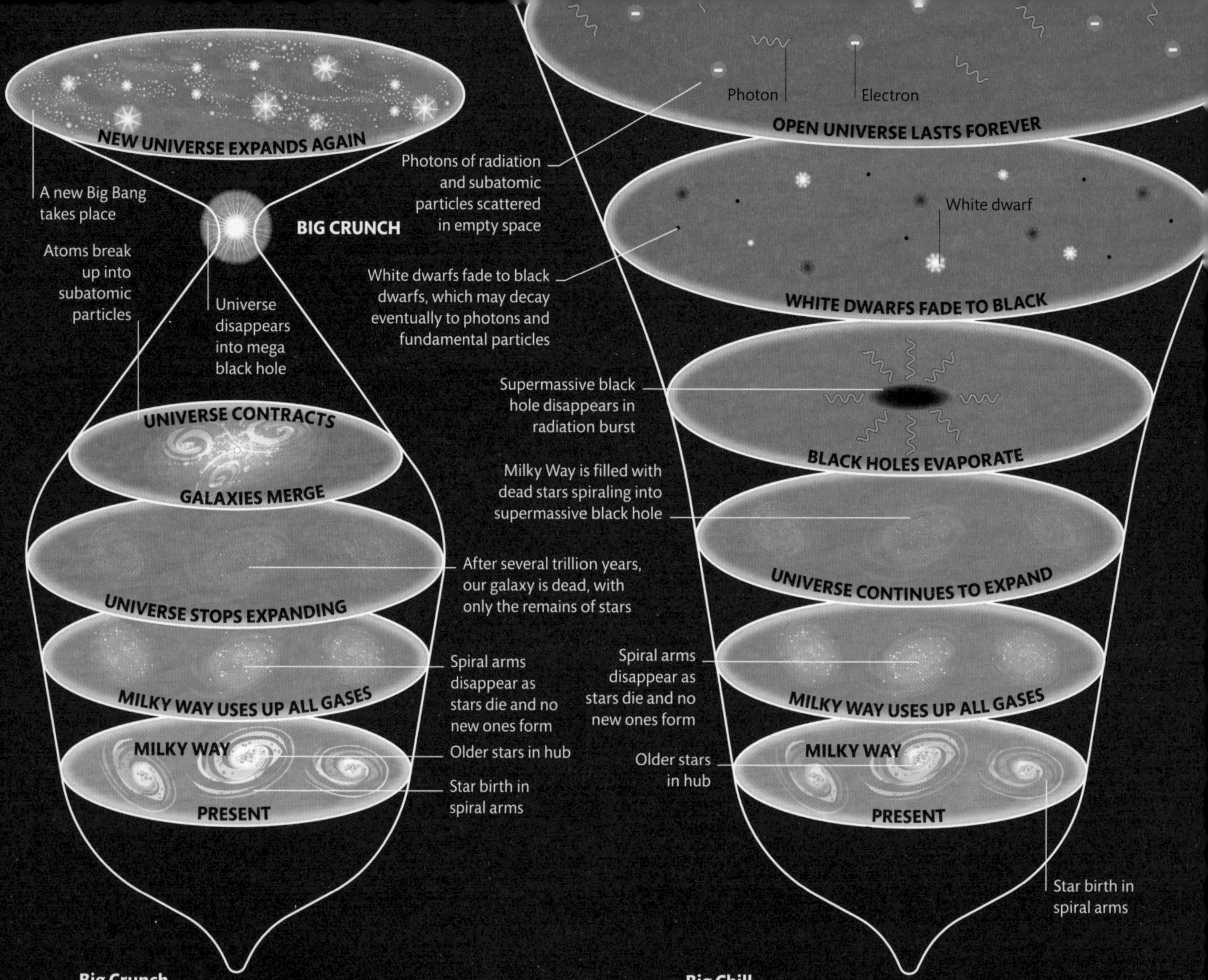
NEW UNIVERSE EXPANDS AGAIN
A new Big Bang takes place
BIG CRUNCH
Atoms break up into subatomic particles
Universe disappears into mega black hole
UNIVERSE CONTRACTS
GALAXIES MERGE
After several trillion years, our galaxy is dead, with only the remains of stars
UNIVERSE STOPS EXPANDING
Spiral arms disappear as stars die and no new ones form
MILKY WAY USES UP ALL GASES
MILKY WAY
Older stars in hub
Star birth in spiral arms
PRESENT
Photon
Electron
OPEN UNIVERSE LASTS FOREVER
Photons of radiation and subatomic particles scattered in empty space
White dwarf
White dwarfs fade to black dwarfs, which may decay eventually to photons and fundamental particles
WHITE DWARFS FADE TO BLACK
Supermassive black hole disappears in radiation burst
BLACK HOLES EVAPORATE
Milky Way is filled with dead stars spiraling into supermassive black hole
UNIVERSE CONTINUES TO EXPAND
Spiral arms disappear as stars die and no new ones form
MILKY WAY USES UP ALL GASES
MILKY WAY
Older stars in hub
PRESENT
Star birth in spiral arms

Big Crunch
Some cosmologists believe that dark energy will weaken over time, allowing gravity to win the battle and cause the Universe to stop expanding and contract. Over trillions of years, galaxies would collide and the Universe's temperature would rise, even incinerating stars. Atoms would tear apart and a giant black hole would devour everything, including itself. Some theorize that, as particles smash into one another, a second Big Bang would take place—the Big Bounce.

Big Chill
The Big Chill theory suggests that the Universe will continue to expand until energy and matter are evenly spread throughout the Universe. As a result, there would not be enough concentrated energy to create new stars. Temperatures would fall to absolute zero, stars would die, and the Universe would go dark.

WHEN MIGHT THE UNIVERSE END?

In most possible scenarios, the end of the Universe will not occur for billions of years. However, theoretically, the Big Change could happen at any time.

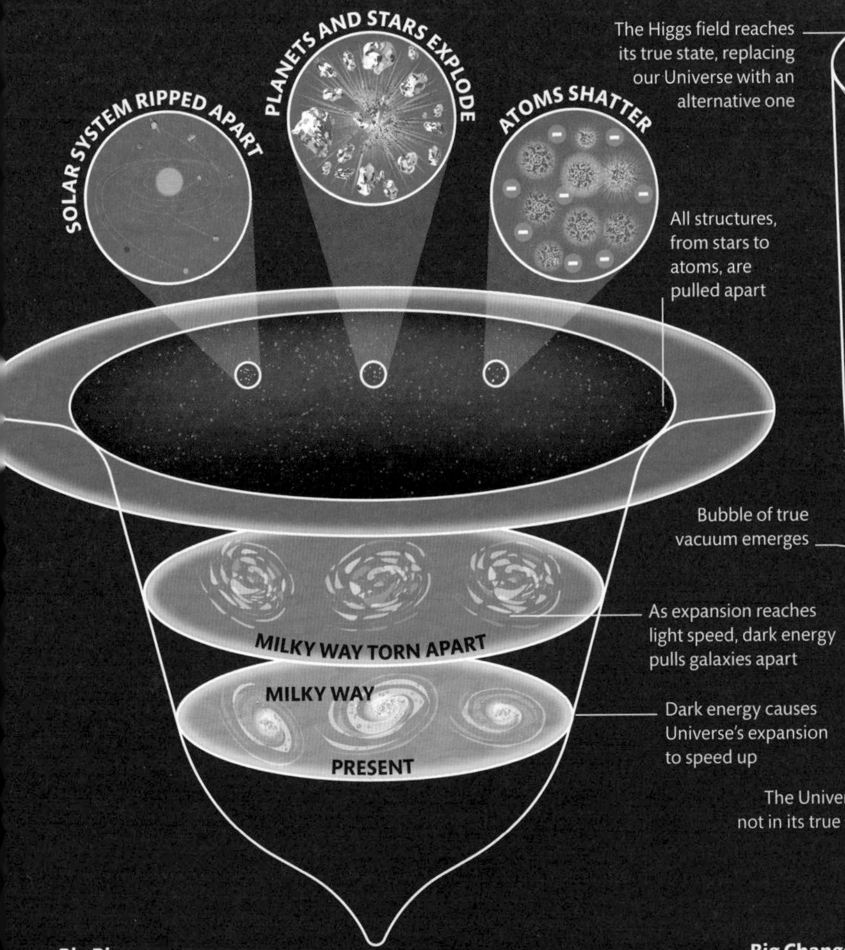
SOLAR SYSTEM RIPPED APART
PLANETS AND STARS EXPLODE
ATOMS SHATTER
All structures, from stars to atoms, are pulled apart
MILKY WAY TORN APART
As expansion reaches light speed, dark energy pulls galaxies apart
MILKY WAY
Dark energy causes Universe's expansion to speed up
PRESENT

Big Rip
In a scenario known as the Big Rip, the Universe will ultimately tear itself apart. If the space in between the galaxies is filled with dark energy, which counters the effects of gravity, the Universe would continue to expand at a faster and faster rate, eventually reaching the speed of light. Unable to be contained by gravity anymore, all the matter in the Universe, including galaxies and black holes, and even spacetime itself, would rip apart.

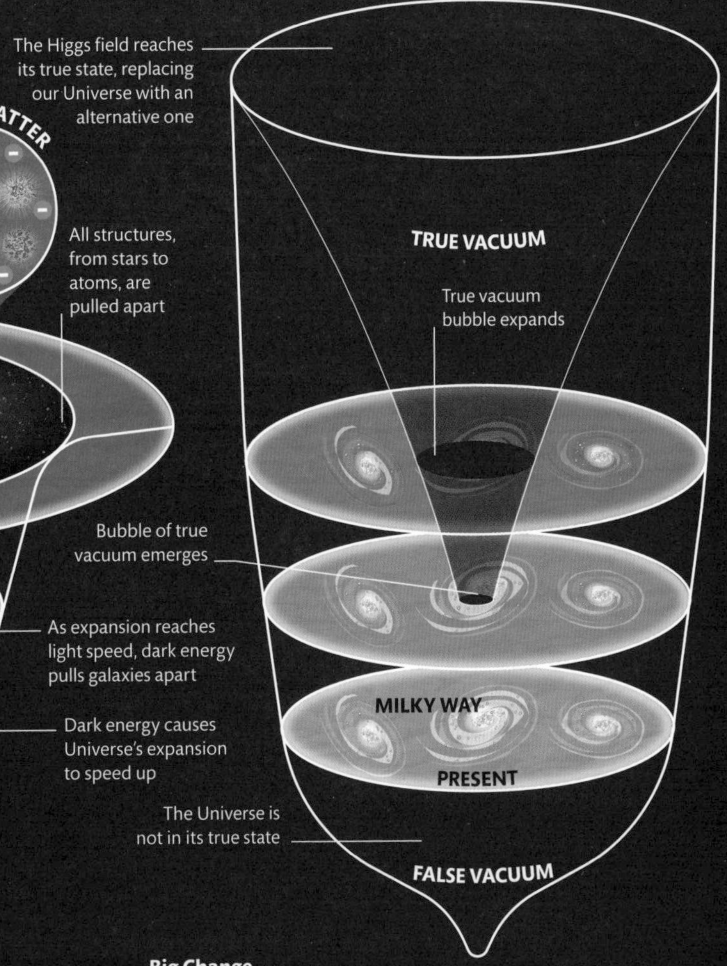
The Higgs field reaches its true state, replacing our Universe with an alternative one
TRUE VACUUM
True vacuum bubble expands
Bubble of true vacuum emerges
MILKY WAY
PRESENT
The Universe is not in its true state
FALSE VACUUM

Big Change
The Big Change theory involves the Higgs Boson particle and the Higgs field—a bit like an omnipresent electromagnetic field—which is thought to have not yet reached its lowest energy or "vacuum" state. Should it reach its true vacuum state, the Higgs field could fundamentally transform matter, energy, and spacetime to create an alternative universe spreading out like a bubble at the speed of light. Everything in the Universe, in its present form, would end.

Our current Universe

The Universe has been constantly expanding since it formed almost 14 billion years ago. Galaxies continue to move farther away from each other, and observations of distant supernovas suggest the expansion is accelerating. This implies the presence of a force with negative pressure, known as dark energy (see pp.206–07), which counters gravity. If this force plays a significant role, infinite expansion is our Universe's most probable fate.

THE HIGGS BOSON IS AROUND **130 TIMES** THE MASS OF THE PROTON, MAKING IT **HIGHLY UNSTABLE**

Seeing the Universe

Astronomers have looked into space since the earliest times, initially with the naked eye and, more recently, using sophisticated equipment that can detect light waves from the farthest reaches of space.

SPIRAL GALAXY

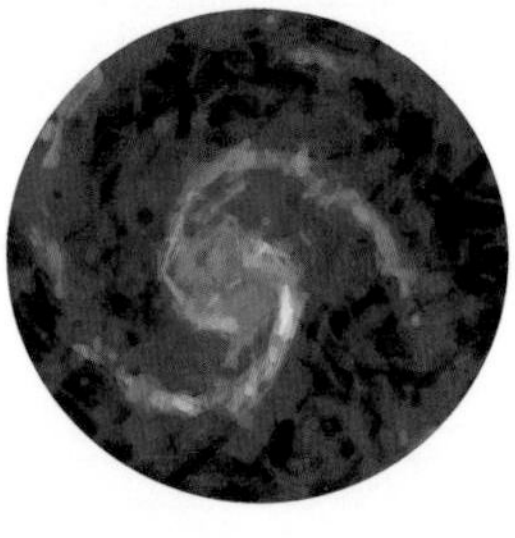

Radio waves
The longest light waves, radio waves are emitted by many objects, including the Sun, planets, many galaxies, and nebulas. Most penetrate Earth's atmosphere to reach its surface.

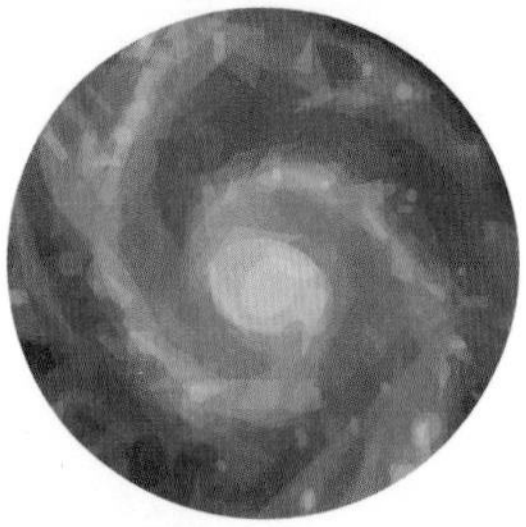

Infrared light
Infrared light is heat energy, like the Sun's warmth. Everything in the Universe radiates some of its energy as infrared. Most is absorbed by Earth's atmosphere.

Visible light
Astronomers can see objects emitting visible light using telescopes on Earth, but much clearer views are obtained without light pollution and atmospheric interference.

Ultraviolet light
The Sun and stars emit ultraviolet (UV) light, which is mostly blocked by Earth's ozone layer. Studying UV light can tell us about the structure and evolution of galaxies.

Across the spectrum
A complex object, like a spiral galaxy, emits radiation across the spectrum. To learn as much as possible, astronomers study it with a range of instruments.

The WMAP probe measured microwave radiation, revealing the composition of the early Universe

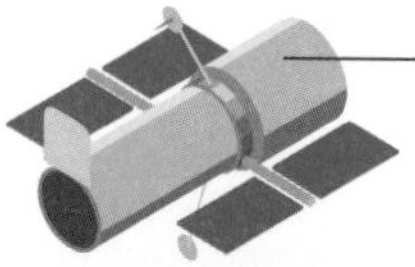

The Hubble telescope has taken famous views of distant stars, nebulas, and galaxies, capturing infrared, visible, and ultraviolet light

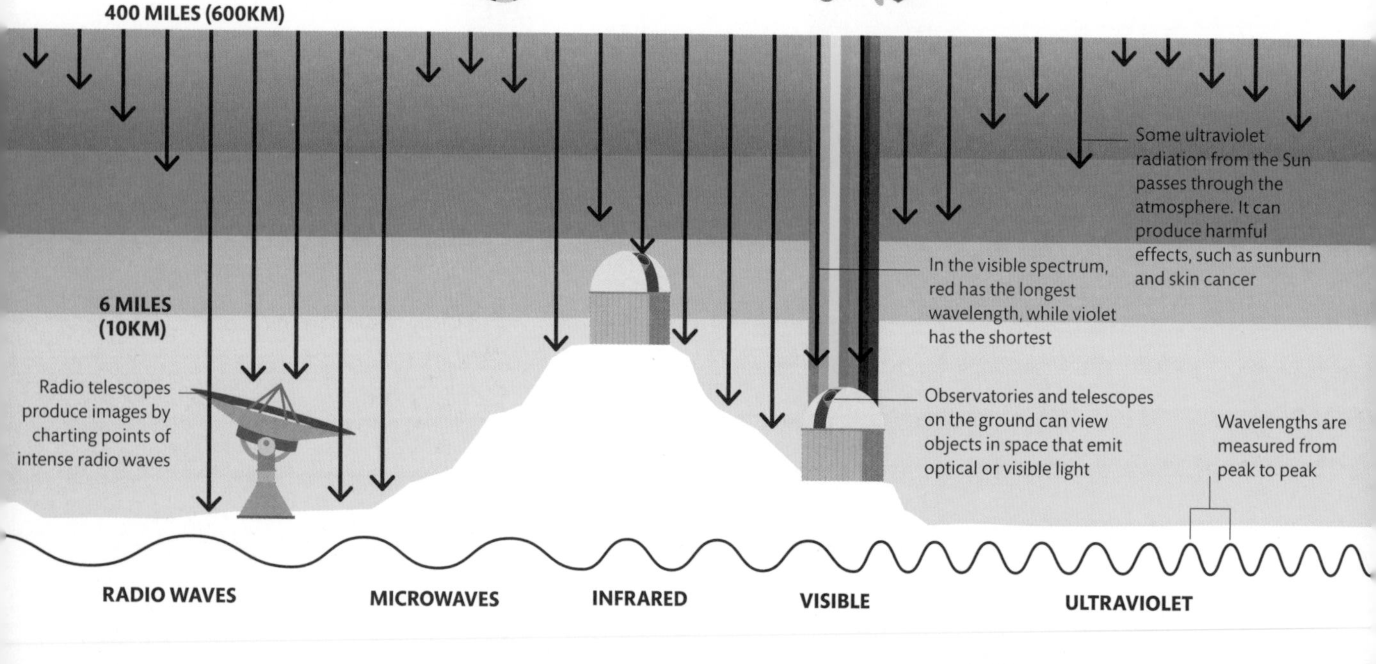
400 MILES (600KM)
6 MILES (10KM)
Radio telescopes produce images by charting points of intense radio waves
Some ultraviolet radiation from the Sun passes through the atmosphere. It can produce harmful effects, such as sunburn and skin cancer
In the visible spectrum, red has the longest wavelength, while violet has the shortest
Observatories and telescopes on the ground can view objects in space that emit optical or visible light
Wavelengths are measured from peak to peak
RADIO WAVES
MICROWAVES
INFRARED
VISIBLE
ULTRAVIOLET

Seeing the light

The electromagnetic spectrum is a continuous range of radiation types, of different wavelengths, that can all be described as forms of light. It includes visible light, which is seen as colors depending on its wavelength, but also various forms that are invisible to the human eye, such as radio waves and X-rays. Each travels through space at the speed of light.

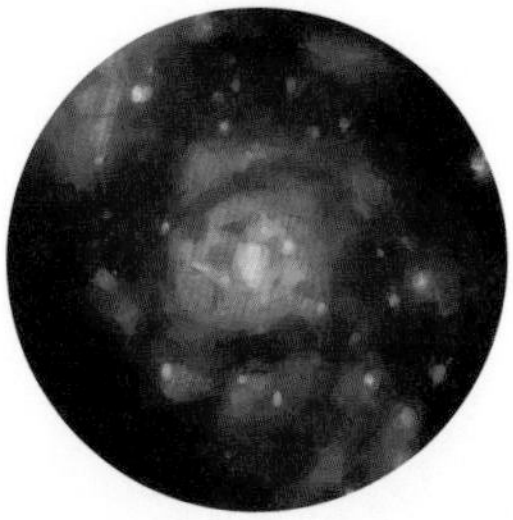

X-rays
X-rays are emitted by black holes, neutron stars, binary star systems, supernova remnants, the Sun and other stars, and some comets. Most are blocked by Earth's atmosphere.

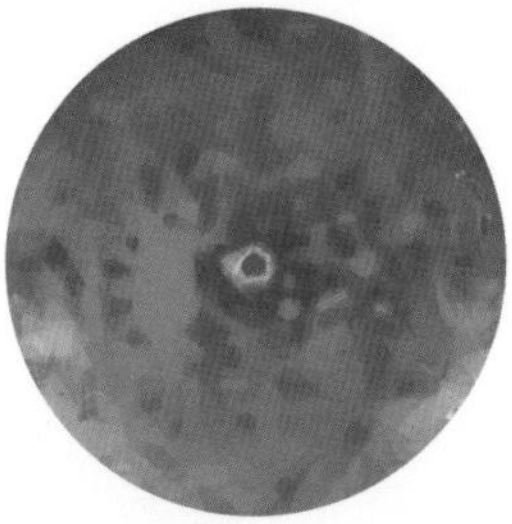

Gamma rays
The smallest and most energetic waves, gamma rays are produced by neutron stars, pulsars, supernova explosions, and regions around black holes.

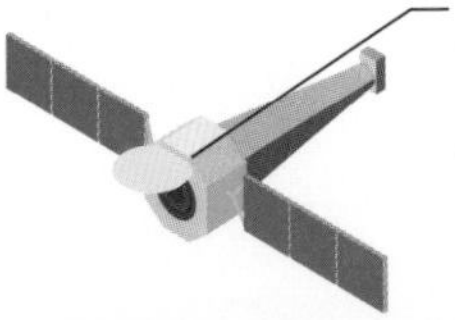

The Chandra X-ray Observatory's eight mirrors focus incoming X-rays to a point where other instruments capture sharp images

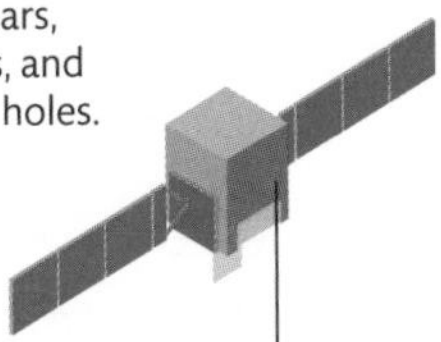

The Fermi telescope has towers of metal and silicon sheets to detect gamma rays

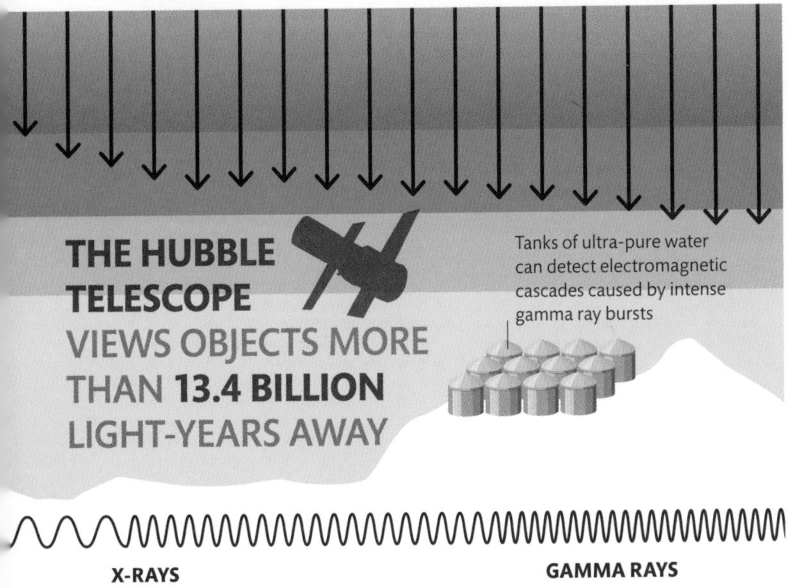
THE HUBBLE TELESCOPE VIEWS OBJECTS MORE THAN 13.4 BILLION LIGHT-YEARS AWAY
Tanks of ultra-pure water can detect electromagnetic cascades caused by intense gamma ray bursts
X-RAYS
GAMMA RAYS

Spectroscopy

The atoms of an element emit light of specific wavelengths when heated. In a technique called spectroscopy, the light from an object is split using a prism, then the pattern of wavelengths, called a spectrum, is studied to see what types of atoms are in the object. This is how scientists can tell what remote objects are made of.

Lines correspond to emission by neon atoms at different wavelengths

EMISSION SPECTRUM FOR NEON

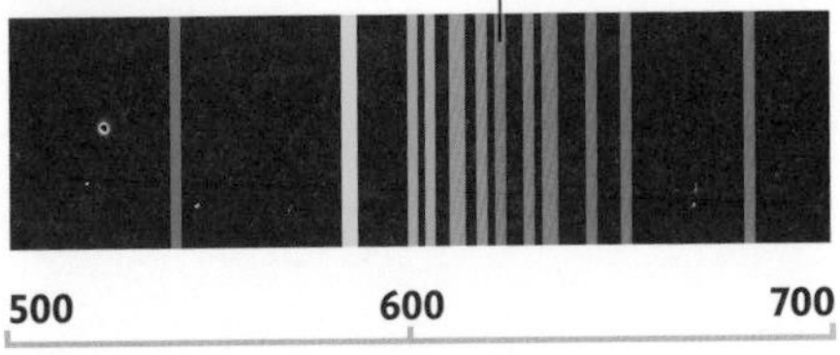
500
600
700
WAVELENGTH (NANOMETERS)

FALSE-COLOR IMAGING

Our eyes can only detect light from a narrow part of the spectrum. To make images using radiation gathered from beyond that range, astronomers use the colors that we can see to represent varying levels of radiation intensity. This is known as false-color imaging.

Low-energy UV
High-energy UV

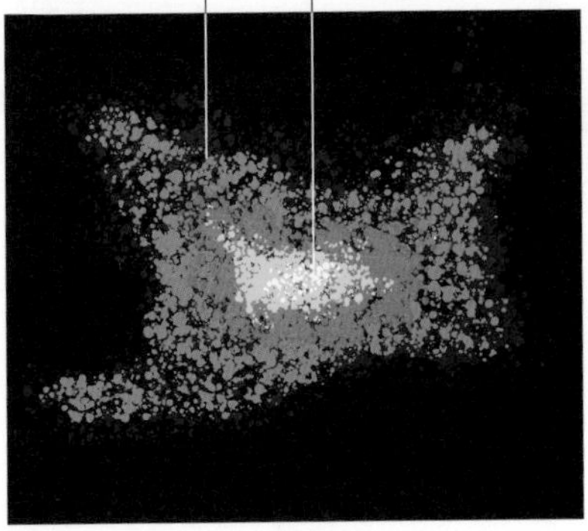

NEBULA IN ULTRAVIOLET

Are we alone?

We have found thousands of extrasolar planets, or exoplanets—planets outside our Solar System. We can also calculate that there must be tens of billions of potentially habitable planets in our galaxy. Could we find life on other worlds?

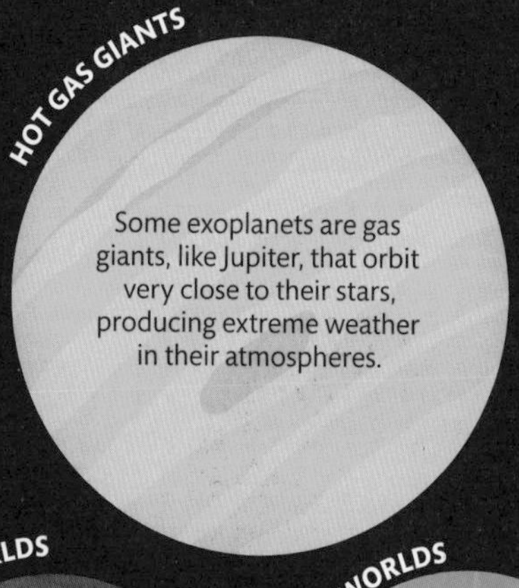
HOT GAS GIANTS
Some exoplanets are gas giants, like Jupiter, that orbit very close to their stars, producing extreme weather in their atmospheres.

Finding another Earth

One way astronomers try to detect exoplanets involves searching for the minute effects they can have on their stars. If a planet similar to Earth in size and distance from its star is found, we can analyze its atmosphere to tell if the elements needed for life are present. Many of the exoplanets discovered are nothing like Earth.

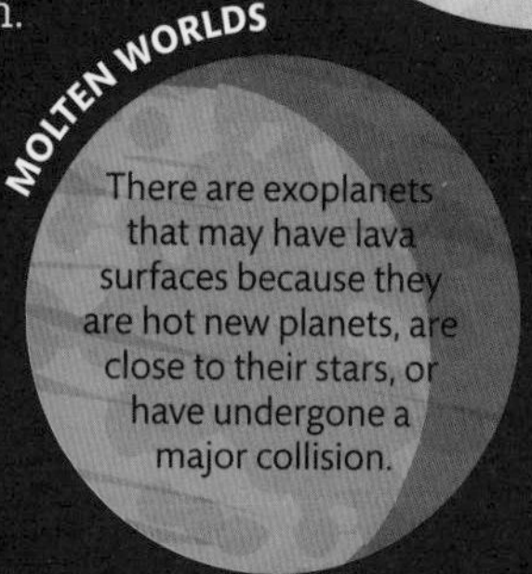
MOLTEN WORLDS
There are exoplanets that may have lava surfaces because they are hot new planets, are close to their stars, or have undergone a major collision.

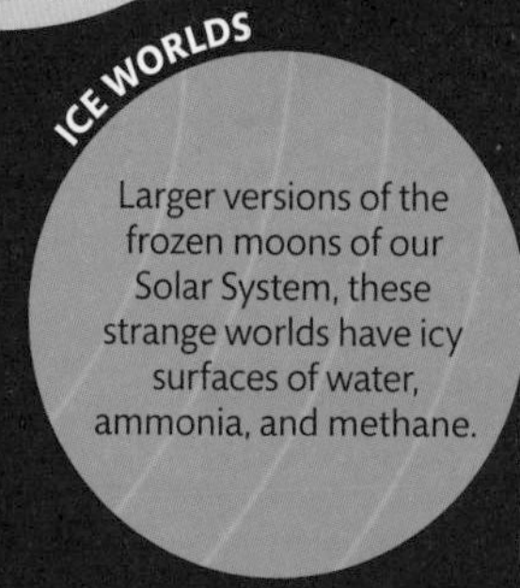
ICE WORLDS
Larger versions of the frozen moons of our Solar System, these strange worlds have icy surfaces of water, ammonia, and methane.

The Goldilocks zone

The habitable zone is also called the Goldilocks zone—a reference to the fairytale in which Goldilocks prefers the bowl of porridge that is neither too hot nor too cold, but "just right." A Goldilocks planet will have the right temperature to maintain liquid surface water, although for life to evolve, other criteria must also be met (see below). However, it is now thought that large amounts of liquid surface water can exist outside these zones.

Habitable zones

The habitable zone near a star would be a point that is not too close or far but ideal for potential life. By locating appropriate stars and this suitable range, astronomers can begin their search for Earth-type planets.

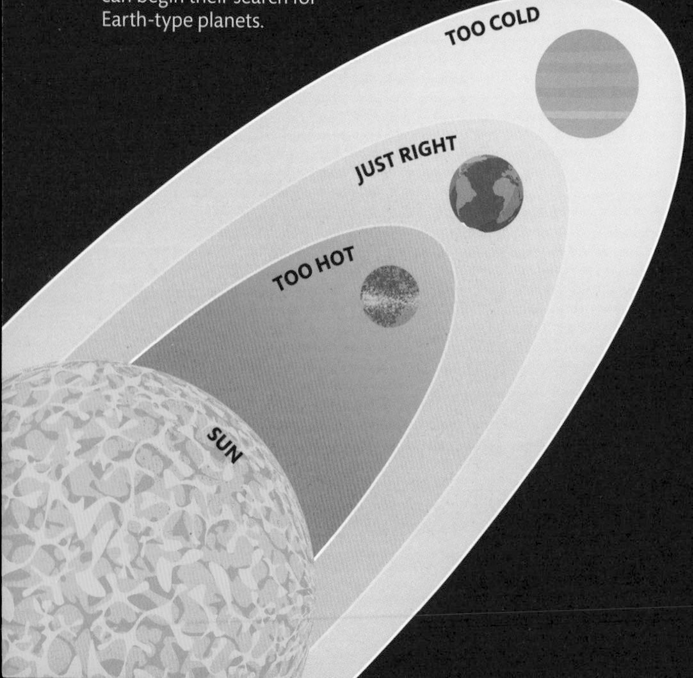
TOO COLD
JUST RIGHT
TOO HOT
SUN

What makes a planet habitable?

There are several criteria for a planet to be suitable for the development of life. Temperature and water are key.

Right temperature
A moderate surface temperature is needed. Too near to a star and the planet boils; too far and it freezes.

Surface water
There must be liquid surface water or humidity (or another liquid that can perform a similar function).

Reliable Sun
The nearest star must remain stable and shine for long enough for life to evolve on a rocky planet.

Elements
The building blocks of life, such as carbon, nitrogen, oxygen, hydrogen, and sulfur, must be present.

Spin and tilt
A spinning planet, with a tilted spin axis, has days, nights, and seasons, which prevents regional temperature extremes.

Atmosphere
A dense atmosphere will protect against radiation, prevent the escape of gases, and keep in warmth.

Molten core
A planet with a molten core can generate a magnetic field to protect potential life against space radiation.

Sufficient mass
A planet with a high-enough mass exerts the gravity required to retain its atmosphere.

Searching for intelligent life

One way of detecting intelligent life is to listen for it. SETI (Search for Extraterrestrial Intelligence) is an organization that searches for radio or optical signals that signify highly evolved alien life. Radio telescopes look for narrow-band radio signals that suggest an artificial source. Scientists also look for very brief flashes of light lasting only nanoseconds. So far, no verifiable signs have been detected.

SETI
SETI's Allen Telescope Array in California targets particular areas in the sky, based on data gathered by the Kepler Space Telescope, which hunts for exoplanets.

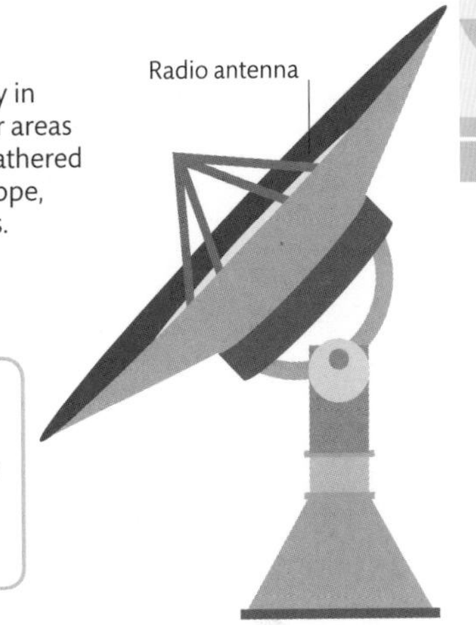
Radio antenna

KEY

- Drake's 1961 estimates
- Recent estimates

Drake's equation
Proposed by astronomer Frank Drake in 1961, this equation is used to estimate the number of communicating civilizations that might exist in our galaxy.

$$N = R \times f_p \times n_e \times f_1 \times f_i \times f_c \times L$$

Term	Meaning	Drake's 1961 estimates	Recent estimates
N	Number of alien civilizations sending out signals	500	2,100
R	Yearly rate of star formation in the galaxy	10	7
f_p	Fraction of stars with planetary systems	0.5	1
n_e	Average number of life-supporting worlds per planetary system	1	3
f_1	Fraction of those worlds that give rise to life	0.1	0.1
f_i	Fraction of such worlds on which intelligent life appears	0.1	0.1
f_c	Fraction of civilizations with communication technology	1.0	1.0
L	Average lifetime (in years) of communicating civilizations	10,000	10,000

Where is everyone?

There are billions of planets potentially suitable for life, and enough time has passed since the Milky Way's formation for an advanced civilization to colonize it. So why haven't we been contacted yet? It may be that life is, in fact, so rare that we are truly alone in the Universe.

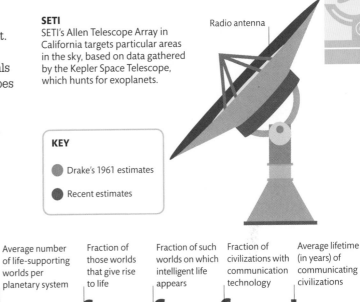

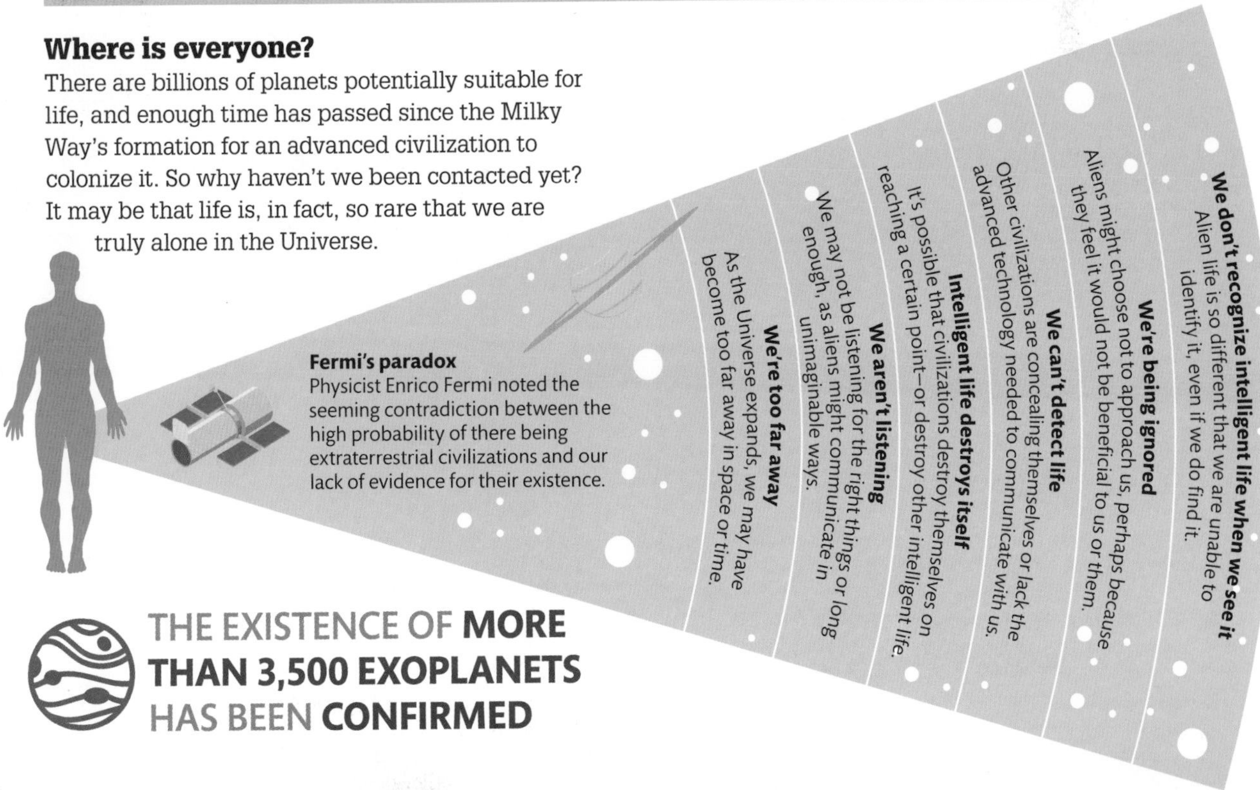
We're too far away As the Universe expands, we may have become too far away in space or time.
We aren't listening We may not be listening for the right things or long enough, as aliens might communicate in unimaginable ways.
Intelligent life destroys itself It's possible that civilizations destroy themselves on reaching a certain point—or destroy other intelligent life.
We can't detect life Other civilizations are concealing themselves or lack the advanced technology needed to communicate with us.
We're being ignored Aliens might choose not to approach us, perhaps because they feel it would not be beneficial to us or them.
We don't recognize intelligent life when we see it Alien life is so different that we are unable to identify it, even if we do find it.

Fermi's paradox
Physicist Enrico Fermi noted the seeming contradiction between the high probability of there being extraterrestrial civilizations and our lack of evidence for their existence.

THE EXISTENCE OF **MORE THAN 3,500 EXOPLANETS** HAS BEEN **CONFIRMED**

Spaceflight

All spacecraft are projectiles on ballistic courses resulting from an initial burst of power. They are in free fall, at the mercy of gravity from large celestial bodies, although some can adjust their course a little with small steering rockets.

Free-falling through space

After being launched from Earth, a spacecraft is not really flying, but falling. Astronauts in space are still under the influence of gravity—that of Earth or the Sun—but experience weightlessness as they fall around these bodies. An orbiting spacecraft falls around Earth but never collides with it because its forward velocity, combined with gravity, produces a curved path, or trajectory, that follows Earth's curvature.

Destination Mars

Counterintuitively, it is more efficient to travel to Mars when it is farthest away, or in "opposition" with the Sun between Mars and Earth. This is because it is easiest to travel along an ellipse that follows the curve of Earth's orbit at one end and Mars's orbit at the other.

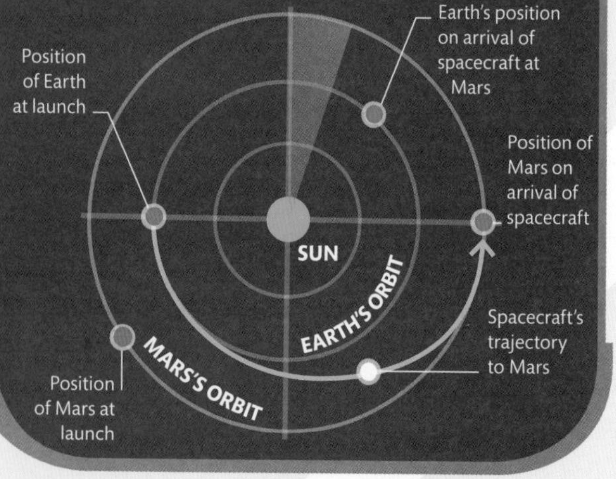

VOYAGER 2 USED NEPTUNE'S GRAVITY TO SLOW DOWN TO CAPTURE IMAGES OF ITS MOON, TRITON

Escape velocity

An object fired with enough velocity can escape Earth's gravity and follow an open curve into space to fall around another celestial body. A spacecraft's initial launch trajectory and velocity are key. For example, if a craft is launched to the Moon with too much velocity, it might not be able to decelerate on arrival as the Moon's weak gravity would not stop it flying past.

1 Failing to leave Earth

A spacecraft that does not reach sufficient velocity on takeoff is unable to achieve a trajectory that will take it into orbit. It fails to escape the planet's gravity and falls back down to Earth.

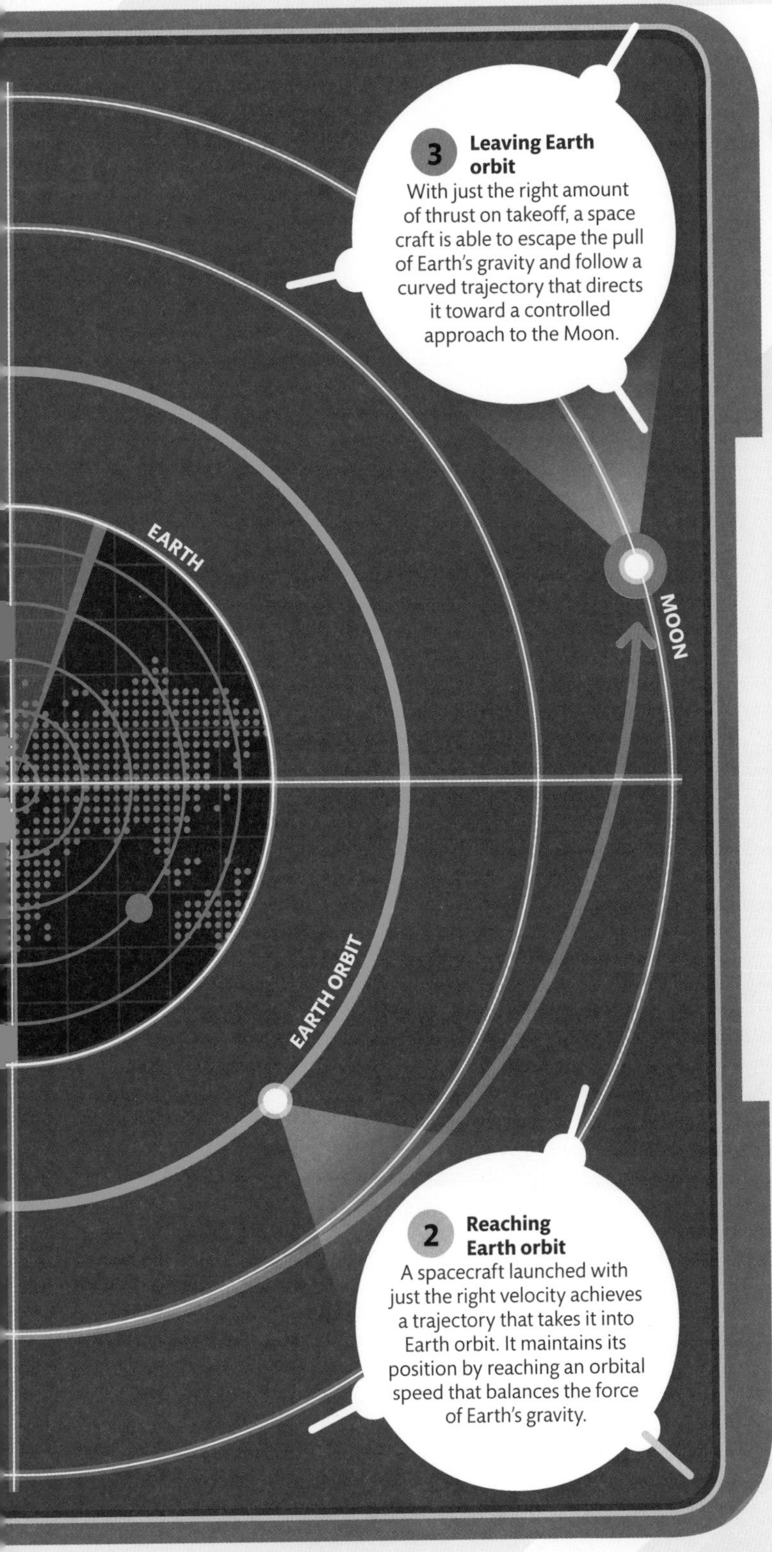

3 Leaving Earth orbit
With just the right amount of thrust on takeoff, a space craft is able to escape the pull of Earth's gravity and follow a curved trajectory that directs it toward a controlled approach to the Moon.

2 Reaching Earth orbit
A spacecraft launched with just the right velocity achieves a trajectory that takes it into Earth orbit. It maintains its position by reaching an orbital speed that balances the force of Earth's gravity.

Slingshots

A craft traveling through space can save time and fuel by using an orbit around a planet to change direction, speed up, or slow down. The gravity of the planet pulls on the spacecraft, and the closer it gets to the planet's surface, the greater the velocity it picks up. This kind of maneuver is known as a slingshot or gravity assist.

Multiple assists
The interplanetary probe Voyager 2 used slingshots from Jupiter, Saturn, Uranus, and Neptune, allowing it to eventually reach the outer Solar System.

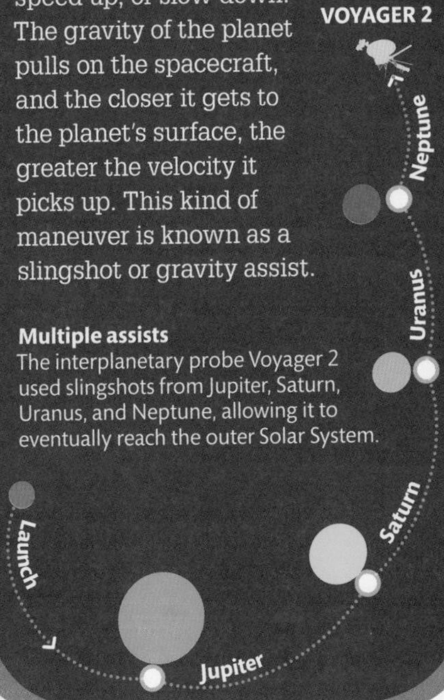

PARKING SPACES

The five Lagrange points (L1–L5) are points in space at which a small object can keep a stable position relative to two large bodies using the joint gravitational forces of the two large bodies. For example, an object at L1 is pulled by equal force toward both the Sun and Earth. These positions can be good places to "park" satellites in space.

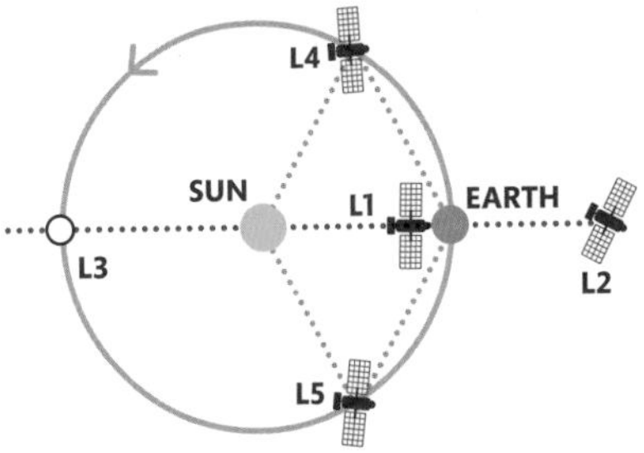

Living in space

Space is a hostile and strange environment. Traveling in a vacuum with no protective atmosphere to reduce radiation, astronauts must also contend with apparent weightlessness due to free fall. Even supposed constants, such as time, cannot be taken for granted.

Fire
As hot air does not rise in space, flames burn in an orb shape. In the event of fire, astronauts must quickly adjust the ventilation and use fire extinguishers.

World of weightlessness

Astronauts and everything in their spacecraft are in a constant state of free fall—either in orbit "falling around" Earth or in a larger orbit free falling around the Sun. While weightless, the human body undergoes numerous stresses (see pp.218–19) and materials behave very differently—for example, water does not flow and hot air does not rise. Therefore, keeping astronauts aboard a spacecraft safe and healthy involves careful preparation and considerable adaptation of their regular environment and behavior.

Microgravity
Astronauts propel themselves through the spacecraft by gently pushing off surfaces. The International Space Station (ISS) is equipped with footholds and bungees for astronauts to steady themselves.

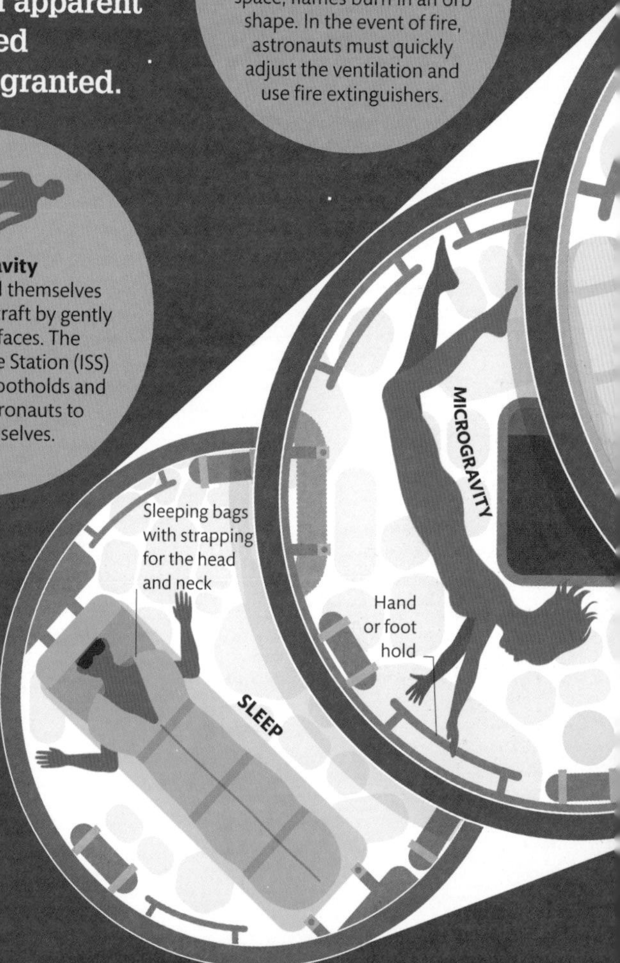

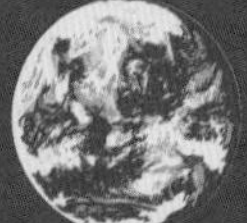

Life in space
Aboard a spacecraft, everyday activities can become complex, but astronauts aim to maintain daily routines that match those on Earth to keep both physically and mentally fit.

Space toilets
Toilets in space use suction cups and recycle urine as drinking water. Feces is stored, rather than dumped, so as not to become a projectile in space.

THE TWIN PARADOX

In this puzzle, one twin leaves Earth and, after traveling close to light speed or near a strong gravitational field, returns home to find his or her twin has aged faster. Special relativity (see pp.140–41) explains how the space traveler's experience of time has slowed relative to the twin's.

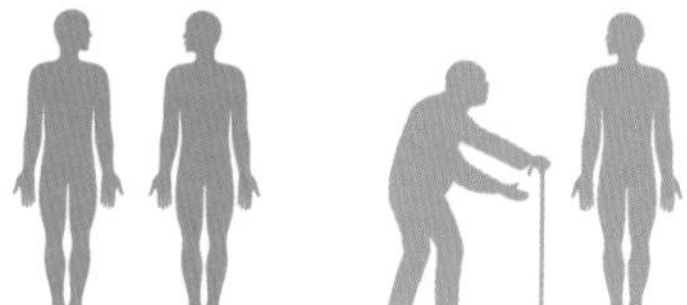

Sleeping in space
Without gravity, there is no sense of lying down. Astronauts strap themselves into sleeping bags and secure their arms. An astronaut's head can also be strapped down to relieve strain on his or her neck.

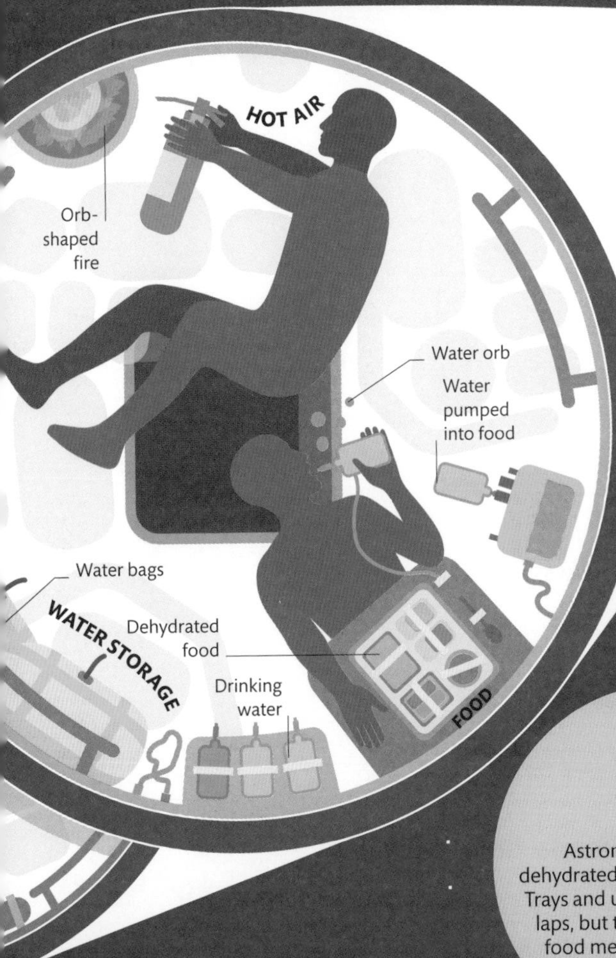

Unmoving air
Without ventilation, air does not circulate, allowing carbon dioxide to gather around the head and hot air to stay around the body. Sweat does not evaporate.

Water
Water does not flow but coalesces in orbs due to surface tension. Astronauts have to dry shower and use face cloths for washing. They drink using straws or specially designed cups.

HOW LONG CAN WE LIVE IN SPACE?

We are still finding our limits. The record holder, Russian cosmonaut Valeri Polyakov, stayed on the Mir space station for 437 days in 1994–1995.

Food
Astronauts add liquid to dehydrated food to make it edible. Trays and utensils are strapped to laps, but the surface tension of food means it sticks to plates and does not float around.

AN ASTRONAUT CAN GROW IN HEIGHT **UP TO 3 PERCENT** WHILE LIVING IN SPACE

Radiation in space

Radiation consists of charged particles and electromagnetic (EM) waves traveling through space. On Earth, we are protected from most of these by the atmosphere, but when astronauts travel beyond a low-Earth orbit, radiation starts to pose a serious hazard. Radiation can be ionizing or nonionizing. Ionizing radiation can strip atoms of electrons, which in turn causes cells to die or lose the ability to reproduce or mutations to occur.

Earth's trapped radiation
This ionizing form of radiation is caused by charged particles trapped within Earth's magnetic field. The areas of trapped radiation above a low-Earth orbit are called the Van Allen Radiation Belts.

Solar particle radiation
This ionizing radiation is caused by the Sun's release of energetic particles from its surface. This type of radiation can be protected against using shielding materials for astronaut suits and their equipment.

Ultraviolet radiation
Ultraviolet (UV) radiation is nonionizing: although particles impart energy to atoms, they do not strip off electrons. UV radiation is easily deflected by wearing reflecting visors and opaque suits outside the craft.

Galactic cosmic radiation
This ionizing radiation includes cosmic rays—high-energy charged particles thought to come from supernovas and high-energy EM radiation such as X-rays from objects like neutron stars. Thick shielding is needed to protect against these.

Traveling to other worlds

Space travel has a significant impact on the human body and mind, with astronauts enduring various physical discomforts and potential health risks. Those traveling to colonize a new planet will need to be well prepared and take active measures to minimize the risks.

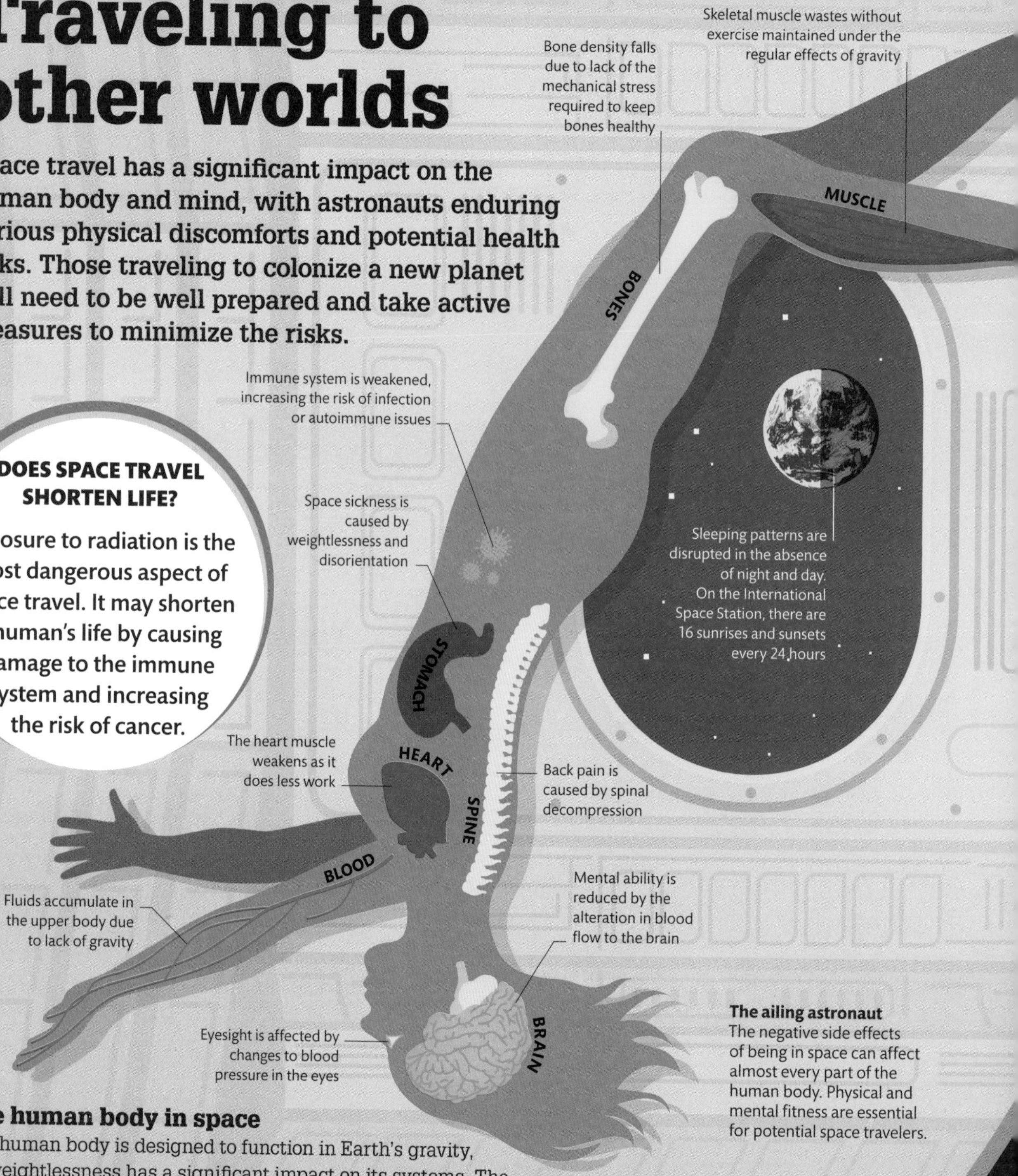
Skeletal muscle wastes without exercise maintained under the regular effects of gravity
Bone density falls due to lack of the mechanical stress required to keep bones healthy
MUSCLE
BONES
Immune system is weakened, increasing the risk of infection or autoimmune issues
Space sickness is caused by weightlessness and disorientation
Sleeping patterns are disrupted in the absence of night and day. On the International Space Station, there are 16 sunrises and sunsets every 24 hours
STOMACH
The heart muscle weakens as it does less work
HEART
SPINE
Back pain is caused by spinal decompression
BLOOD
Fluids accumulate in the upper body due to lack of gravity
Mental ability is reduced by the alteration in blood flow to the brain
Eyesight is affected by changes to blood pressure in the eyes
BRAIN

The ailing astronaut
The negative side effects of being in space can affect almost every part of the human body. Physical and mental fitness are essential for potential space travelers.

DOES SPACE TRAVEL SHORTEN LIFE?

Exposure to radiation is the most dangerous aspect of space travel. It may shorten a human's life by causing damage to the immune system and increasing the risk of cancer.

The human body in space

The human body is designed to function in Earth's gravity, so weightlessness has a significant impact on its systems. The lack of physical stress and exercise results in the rapid loss of bone and muscle mass and a drop in cardiovascular performance. Without gravity, body fluids are redistributed into the upper body, which can cause eye problems and affect blood pressure.

MINIMIZING NEGATIVE EFFECTS

Exercise is vital in maintaining bone density and muscle mass, so astronauts work out for up to 2 hours a day in space. They do resistive training using elastic bands and tether themselves to cycling and running machines for cardiovascular sessions. Astronauts mostly exercise the lower body, as this deteriorates most quickly in low-gravity conditions.

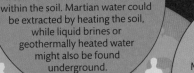

Activity stimulates the heart and exercises the muscles of the lower body

How could we colonize Mars?

Mars is within reach: we can travel there directly in relatively small spacecraft using the technology that took us to the Moon. Although complete self-sufficiency on Mars is unlikely for a long time, early colonizers could live off the land to a large degree, later even manufacturing items to trade with Earth.

Mining for water
Water on Mars is plentiful but frozen in ice fields and contained within the soil. Martian water could be extracted by heating the soil, while liquid brines or geothermally heated water might also be found underground.

Growing food
Martian soil is very fertile. Plants grown in domes could be provided with water and carbon dioxide. The plants would produce oxygen, while inedible plant matter could be used as fertilizer.

Built of bricks
The first housing might comprise interlinked metal and plastic pods transported to Mars by spacecraft. Later, buildings could be made using bricks, as Martian soil is ideal for making bricks and mortar.

Terraforming Mars
Mars is cold and dry but has the elements needed to support life. Its atmosphere could initially be built up by raising carbon dioxide levels, creating a greenhouse effect that would also raise the temperature.

Getting there
The most direct flight path to Mars would take a spacecraft there in 180 days. The crew would stay on Mars for a year and half before the launch window opened for the return trip. The craft would land in an area likely to have water.

Preparing the way
An unmanned craft sent to Mars could deploy a nuclear reactor to react carbon dioxide in Martian air with hydrogen from Earth, to make methane for fuel. The water by-product could be stored, or split into hydrogen and oxygen.

IT WOULD TAKE **900 YEARS** TO MAKE THE MARTIAN ATMOSPHERE **FULLY BREATHABLE**

EARTH

Inside Earth

Earth is one of four small rocky planets orbiting close to the Sun. Formed by the force of gravity, it developed into a dynamic, multilayered world with a searingly hot interior, a cool rocky crust, broad oceans of liquid water, and an airy atmosphere.

HOW CAN HOT ROCK STAY SOLID?

The rock inside Earth is much hotter than molten volcanic lava. But it is under intense pressure, which keeps most of it solid. If the pressure eases, it melts.

How did Earth form?

When the Sun formed some 4.6 billion years ago, it was surrounded by a disc-shaped cloud of orbiting rocky and icy debris. Attracted to each other by gravity, the drifting fragments clumped together—a process called accretion—to form larger masses; these eventually grew into Earth and the other planets of the Solar System. The intense heat generated by the process created Earth's layered structure.

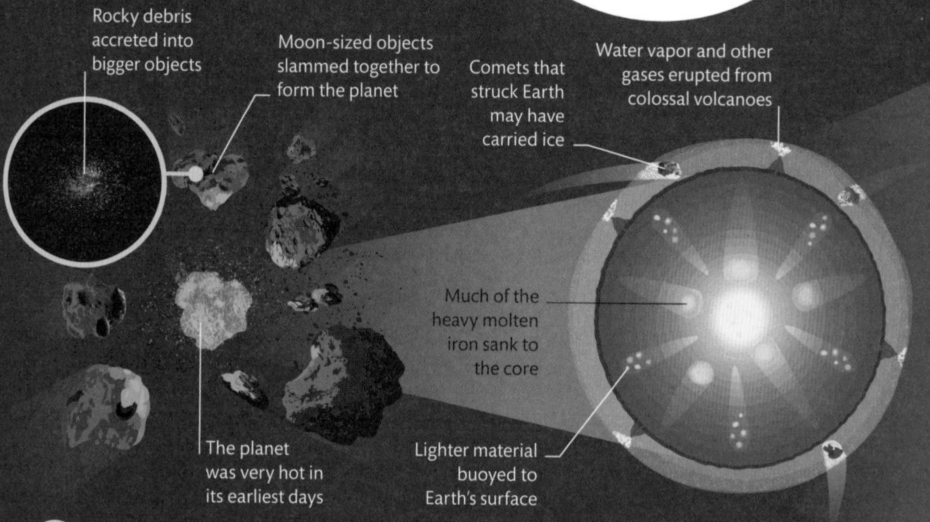
Rocky debris accreted into bigger objects
Moon-sized objects slammed together to form the planet
Comets that struck Earth may have carried ice
Water vapor and other gases erupted from colossal volcanoes
Much of the heavy molten iron sank to the core
The planet was very hot in its earliest days
Lighter material buoyed to Earth's surface

1 Growing planet Every physical object has gravity that attracts other objects. The large objects that formed Earth were attracted with such force that their impact energy turned to heat, partly melting and welding them together.

2 Meltdown and layering As Earth grew by accretion, impact energy generated enough heat to melt the entire planet. The heaviest material sank to the center to form a metallic core, surrounded by deep layers of lighter rock.

Oceans and continents

The crust beneath the oceans (oceanic crust) consists mainly of basalt and gabbro—fairly dense, iron-rich rocks similar to the even denser rock of the underlying mantle. But over time, volcanoes and other geological processes have built up thick layers of silica-rich rock such as granite, forming the continents. This thick continental crust is a lot less dense than mantle rock, so it floats on it like icebergs on polar oceans. This is why the continents rise high above the ocean floors.

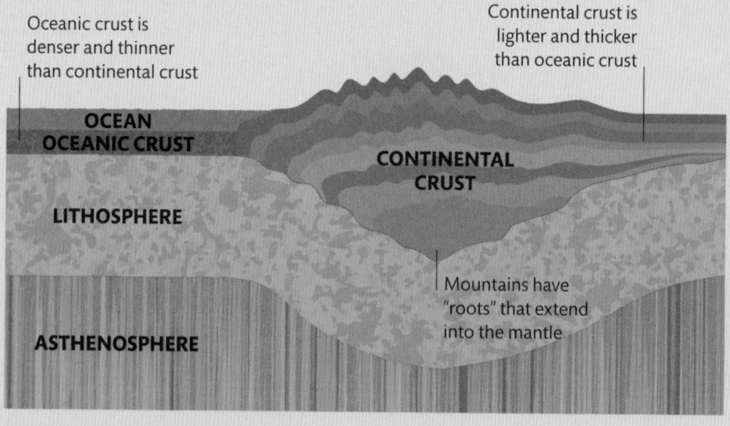
Oceanic crust is denser and thinner than continental crust
Continental crust is lighter and thicker than oceanic crust
OCEAN
OCEANIC CRUST
CONTINENTAL CRUST
LITHOSPHERE
ASTHENOSPHERE
Mountains have "roots" that extend into the mantle

3 Earth today
After the meltdown early in its history, the layered planet cooled enough to support oceans of liquid water. Most of the rock solidified, but the outer core has remained molten.

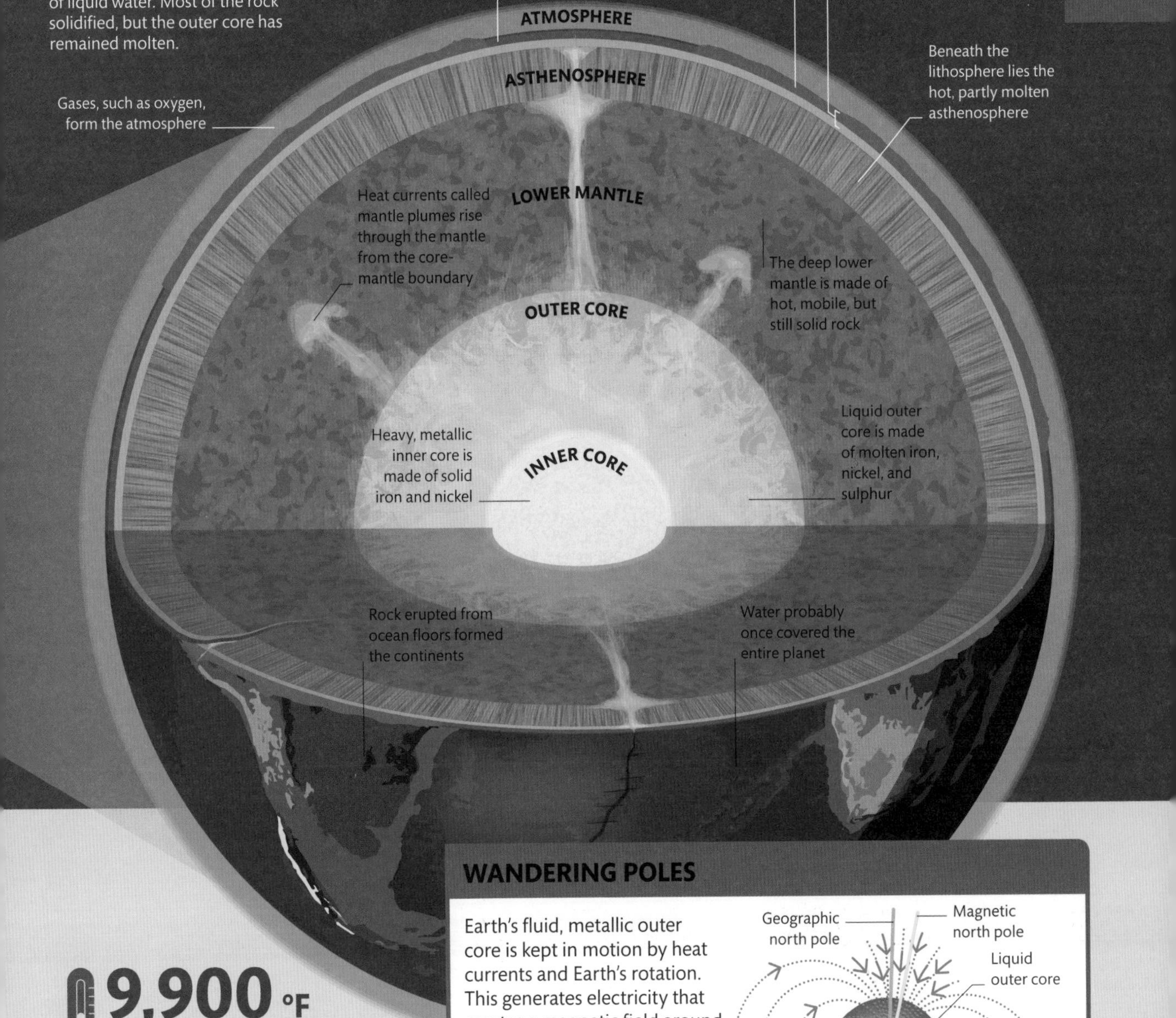
Thin oceanic crust made of dense, iron-rich rock underlies the ocean floors
Thick continental crust consists of relatively light, silica-rich rock
Cool crust and uppermost mantle form the rocky lithosphere
ATMOSPHERE
ASTHENOSPHERE
Beneath the lithosphere lies the hot, partly molten asthenosphere
Gases, such as oxygen, form the atmosphere
Heat currents called mantle plumes rise through the mantle from the core-mantle boundary
LOWER MANTLE
The deep lower mantle is made of hot, mobile, but still solid rock
OUTER CORE
Liquid outer core is made of molten iron, nickel, and sulphur
Heavy, metallic inner core is made of solid iron and nickel
INNER CORE
Rock erupted from ocean floors formed the continents
Water probably once covered the entire planet

9,900 °F (5,500°C) IS THE TEMPERATURE OF EARTH'S INNER CORE—AND THE SURFACE OF THE SUN

WANDERING POLES

Earth's fluid, metallic outer core is kept in motion by heat currents and Earth's rotation. This generates electricity that creates a magnetic field around the planet. The field is roughly aligned with Earth's axis, so magnetic north is close to true north. But its position is always moving, at up to 30miles (50km) per year.

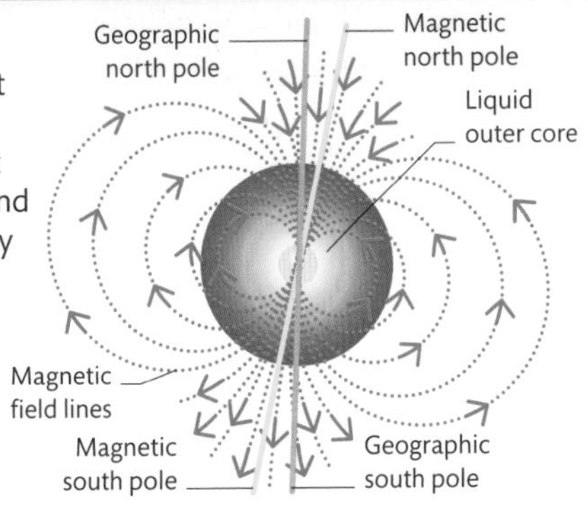
Geographic north pole
Magnetic north pole
Liquid outer core
Magnetic field lines
Magnetic south pole
Geographic south pole

Plate tectonics

Earth's lithosphere (its brittle crust and top layer of the mantle) is split into sections called tectonic plates. Heat rising from Earth's core keeps these plates constantly on the move, either dragging them apart or pushing them together to move continents, build mountains, and fuel spectacular volcanoes.

HOW QUICKLY DO PLATES MOVE?

On average, plates move at the speed your fingernails grow. The fastest-spreading rift, the East Pacific Rise, is spreading at less than 6½in (16cm) per year.

Trenches, rifts, and mountains

Deep within the planet, radioactive elements generate heat (see pp.36–37) that, together with heat escaping from the core, causes the mantle to circulate in very slow convection currents. The movement pulls the plates apart in some places to form long rifts. In other places, it drags plates together, creating subduction zones where the edge of one plate sinks into the mantle. Most of the rifting and subduction occurs on ocean floors. Plate tectonics makes some oceans expand while others are shrinking, and has even made continents collide.

THE MID-ATLANTIC RIDGE IS 10,000 MILES (16,000KM) LONG

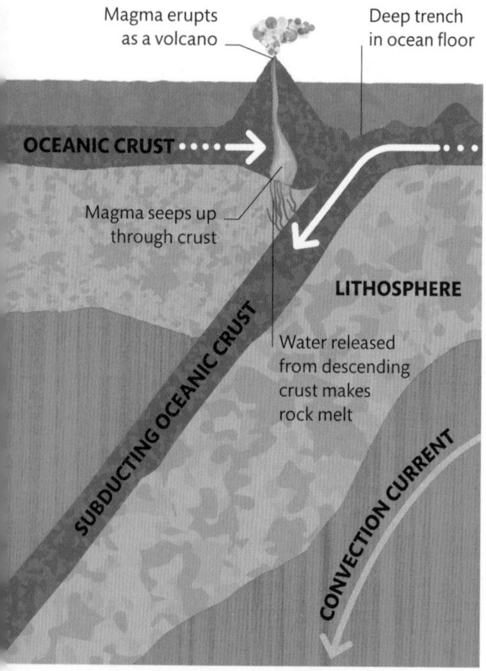

Magma erupts as a volcano
Deep trench in ocean floor
OCEANIC CRUST
Magma seeps up through crust
LITHOSPHERE
SUBDUCTING OCEANIC CRUST
Water released from descending crust makes rock melt
CONVECTION CURRENT

Oceanic subduction zone
Where plates carrying oceanic crust are driven together, the heavier plate moves under the other and melts in the mantle. A deep trench forms in the ocean—such as the Mariana Trench in the Pacific.

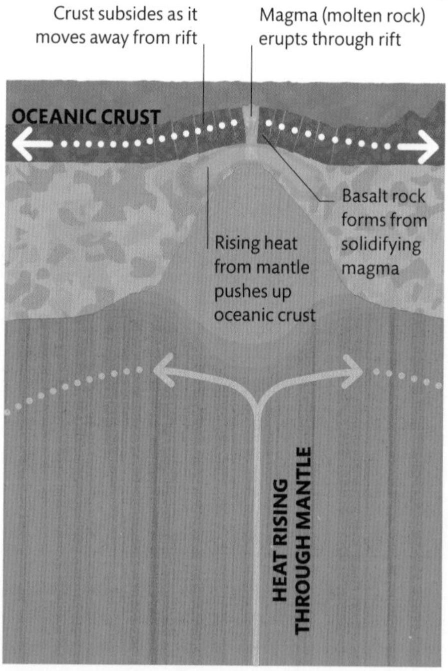

Crust subsides as it moves away from rift
Magma (molten rock) erupts through rift
OCEANIC CRUST
Basalt rock forms from solidifying magma
Rising heat from mantle pushes up oceanic crust
HEAT RISING THROUGH MANTLE

Mid-ocean ridge zone
Long rifts in the ocean floor form where plates are pulled apart. This eases the pressure on hot rock below, allowing it to melt, erupt, and form new oceanic crust, such as at the Mid-Atlantic Ridge.

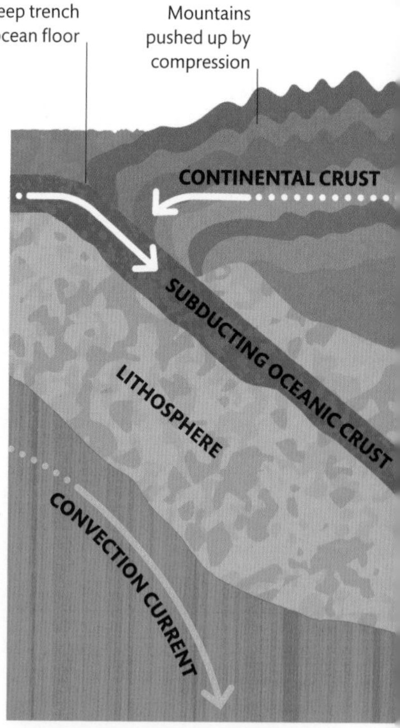

Deep trench in ocean floor
Mountains pushed up by compression
CONTINENTAL CRUST
SUBDUCTING OCEANIC CRUST
LITHOSPHERE
CONVECTION CURRENT

Ocean-continental subduction
Where plates carrying oceanic and continental crust move together, the heavier oceanic crust is dragged down. The continental crust is compressed, forming mountains such as the Andes.

DRIFTING CONTINENTS

Since the continents are rooted in the mobile tectonic plates, the relentless movement of these plates carries them around the globe. This means that the continents are constantly being split up and pushed together in different ways. At one point in time, there was a supercontinent known as Pangaea. It was created about 300 million years ago and broke up about 130 million years later. The continents will continue to move and reform.

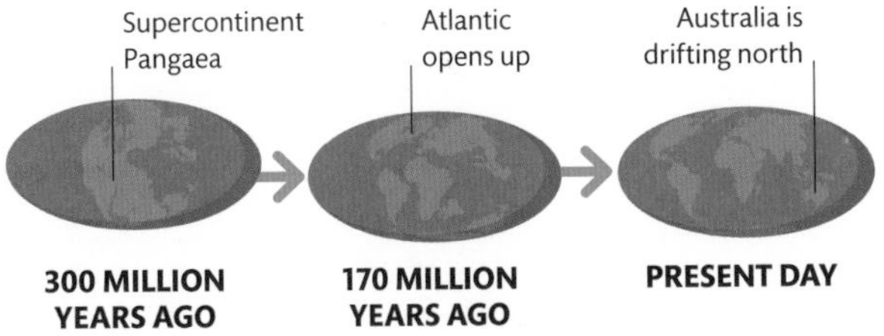
Supercontinent Pangaea
Atlantic opens up
Australia is drifting north
300 MILLION YEARS AGO
170 MILLION YEARS AGO
PRESENT DAY

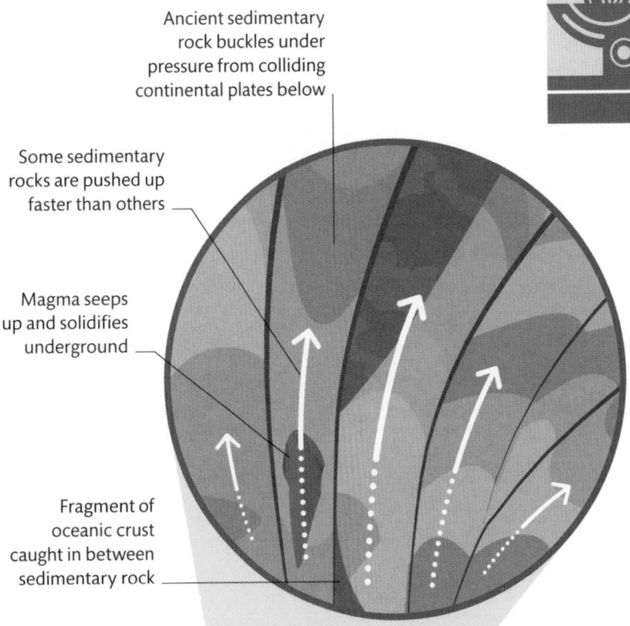
Ancient sedimentary rock buckles under pressure from colliding continental plates below
Some sedimentary rocks are pushed up faster than others
Magma seeps up and solidifies underground
Fragment of oceanic crust caught in between sedimentary rock

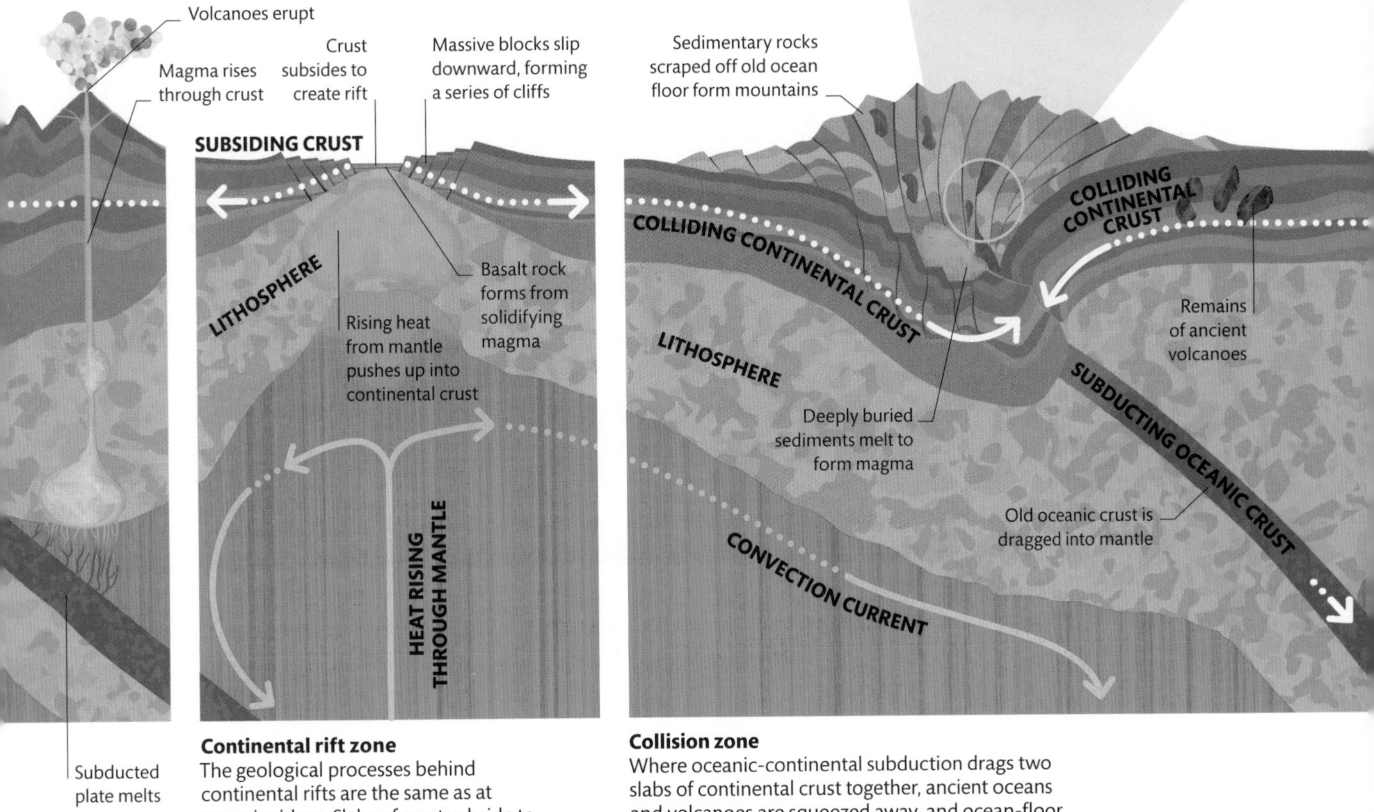
Volcanoes erupt
Magma rises through crust
Crust subsides to create rift
Massive blocks slip downward, forming a series of cliffs
SUBSIDING CRUST
LITHOSPHERE
Rising heat from mantle pushes up into continental crust
Basalt rock forms from solidifying magma
HEAT RISING THROUGH MANTLE
Sedimentary rocks scraped off old ocean floor form mountains
COLLIDING CONTINENTAL CRUST
COLLIDING CONTINENTAL CRUST
LITHOSPHERE
Remains of ancient volcanoes
Deeply buried sediments melt to form magma
SUBDUCTING OCEANIC CRUST
Old oceanic crust is dragged into mantle
CONVECTION CURRENT
Subducted plate melts

Continental rift zone
The geological processes behind continental rifts are the same as at oceanic ridges. Slabs of crust subside to create long rift valleys lined by steep cliffs (such as the Rift Valley in East Africa).

Collision zone
Where oceanic-continental subduction drags two slabs of continental crust together, ancient oceans and volcanoes are squeezed away, and ocean-floor sediments are compressed up into fold mountains. The Himalayas sit on this type of boundary.

Earthquakes

Tectonic plates are constantly moving. But the ragged edges of the plates sometimes lock together until enough strain builds up to rip them apart, generating the shock waves that cause earthquakes.

What is an earthquake?

Where plates are pushing together or past each other, strain builds up at the fault forming the plate boundary. This distorts the edge of each plate until the rocks give way and spring back. If this happens frequently, the recoil is relatively small and causes only minor earth tremors. But if the fault stays locked for a century or more, the rocks could shift several meters within a few seconds—triggering a catastrophic earthquake.

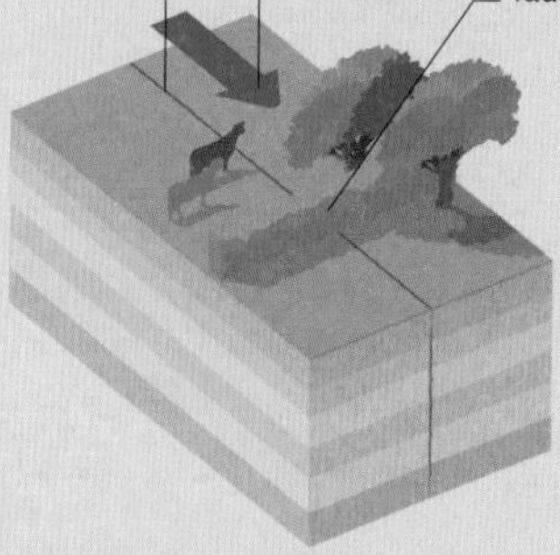

1 At the fault line
This transform fault marks the boundary between two plates that are slipping past each other. Each plate is moving at just 1in (2.5cm) per year.

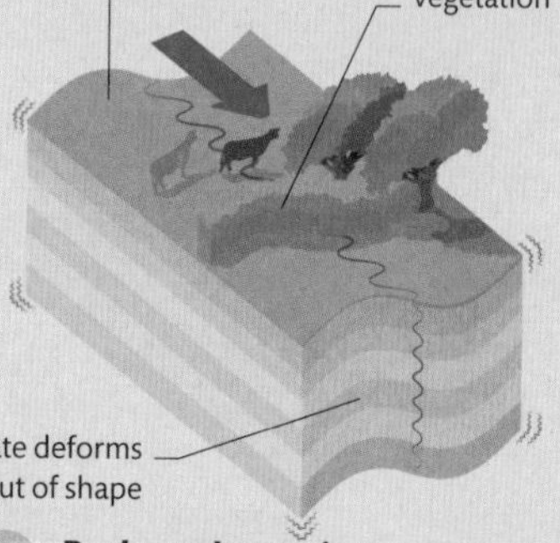

2 Rocks under strain
Many decades later, the plates are still moving past each other, but the fault has stayed locked. This has distorted the plates, building up tension.

WHAT WAS THE STRONGEST RECORDED EARTHQUAKE?

The strongest earthquake recorded to date occurred on May 22, 1960 in Chile. It measured 9.5 on the Richter scale and the subsequent tsunami reached Hawaii, Japan, and the Philippines.

Triggering a tsunami

Where one tectonic plate is grinding beneath another on the ocean floor, this distorts the overlying plate, dragging its edge downward. When the rocks give way, the distorted plate suddenly straightens out, pushing up a big wave that surges rapidly across the ocean. At sea, the wave is long and low, but when it rolls into shallow water, it can build up into a devastating tsunami.

1 Locked fault
A deep ocean trench near land marks a subduction zone where ocean floor is slipping beneath a continent, but the fault between the plates has become locked.

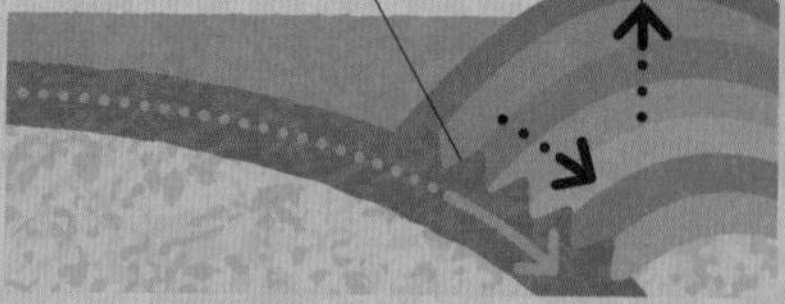

2 Distorted plate
Gripped by the locked fault, the submerged edge of the continental plate is dragged downward. This distorts the plate so the coastal region bulges upward.

3 Rupture and recoil
After a century, the fault gives way under the strain. Within minutes, both plates can spring back by up to 8¼ft (2.5m), generating shock waves that radiate from points below the ground (the focus) and at the surface (the epicenter).

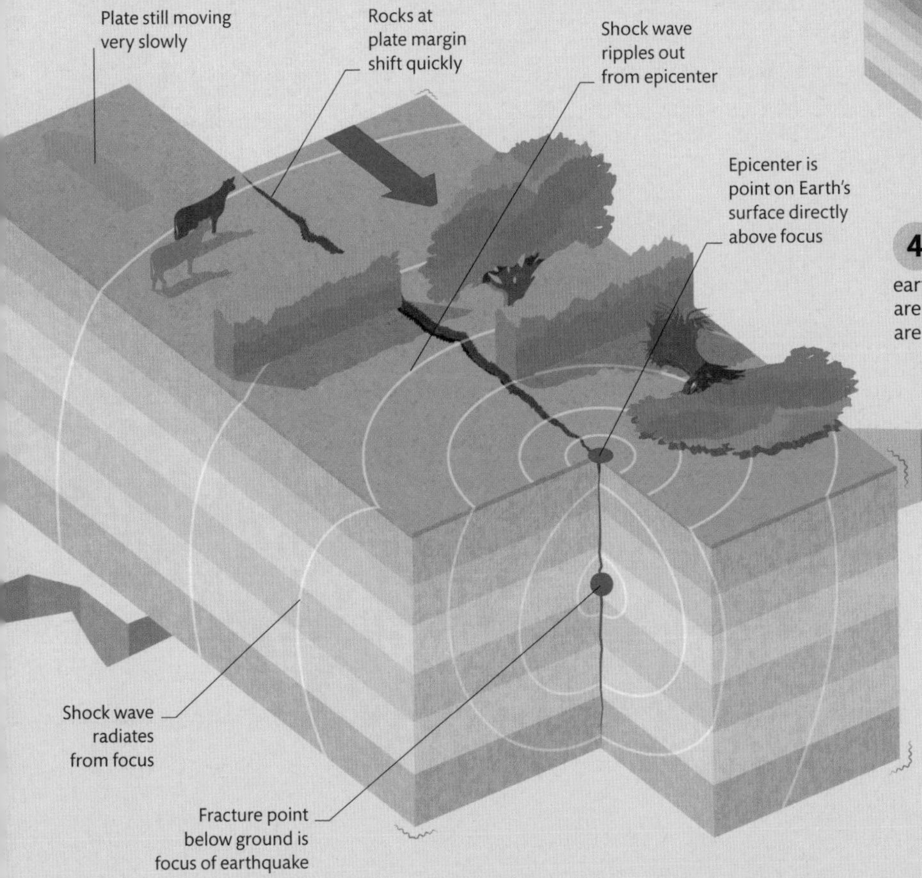
Plate still moving very slowly
Rocks at plate margin shift quickly
Shock wave ripples out from epicenter
Epicenter is point on Earth's surface directly above focus
Shock wave radiates from focus
Fracture point below ground is focus of earthquake

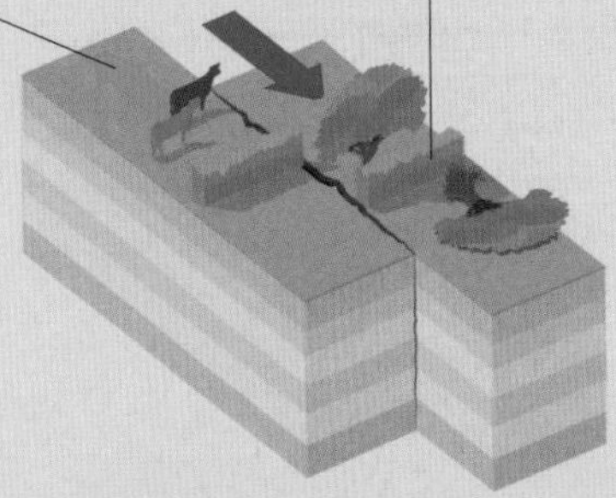
Each plate continues to move as before
Vegetation is offset on line of fault

4 After the earthquake
When the dust settles after the main earthquake and any aftershocks, the rocks are no longer under strain. But the plates are still moving, so the cycle begins again.

500,000 EARTHQUAKES ARE ESTIMATED TO HAPPEN **PER YEAR,** BUT FEWER THAN **100 OF THEM CAUSE DAMAGE**

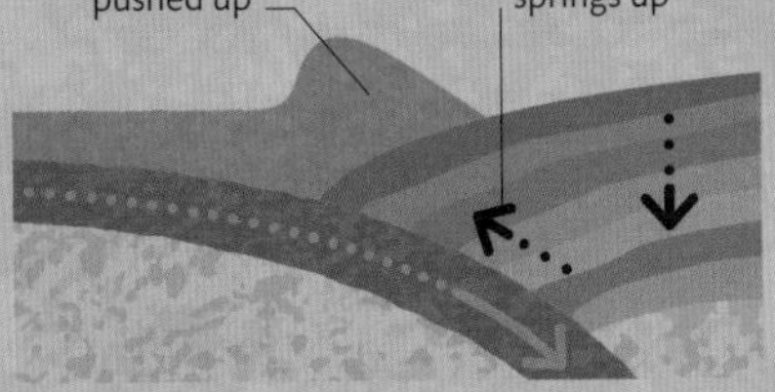
Tsunami wave pushed up
Edge of plate springs up

3 Release and surge
When the fault ruptures, the edge of the continental plate springs up, triggering a tsunami. This surges ashore over a coastline lowered by the straightening of the plate.

MEASURING EARTHQUAKES

Destructive earthquakes are measured using the Moment Magnitude Scale. This has superseded the older Richter Scale because the measurements involved give scientists a more accurate picture of the energy released by the most powerful events. The data is gathered using instruments called seismometers, which produce seismograms that show the degree of plate movement.

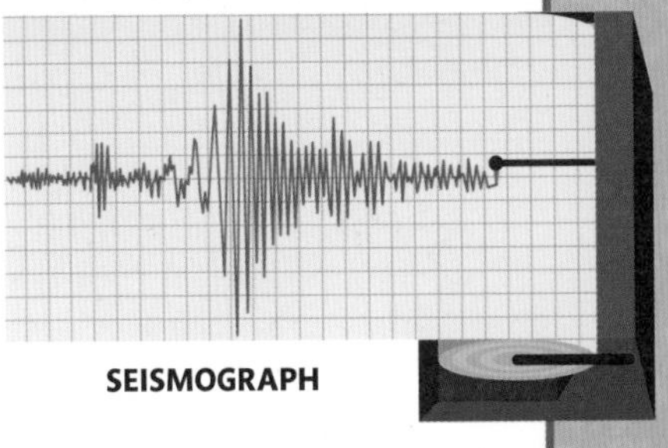

SEISMOGRAPH

Volcanoes

Molten rock and gas erupt from ruptures in Earth's surface called volcanic vents, each usually enclosed in a bowl-shaped crater. Most occur near plate boundaries, created by forces that rip plates apart or push them together.

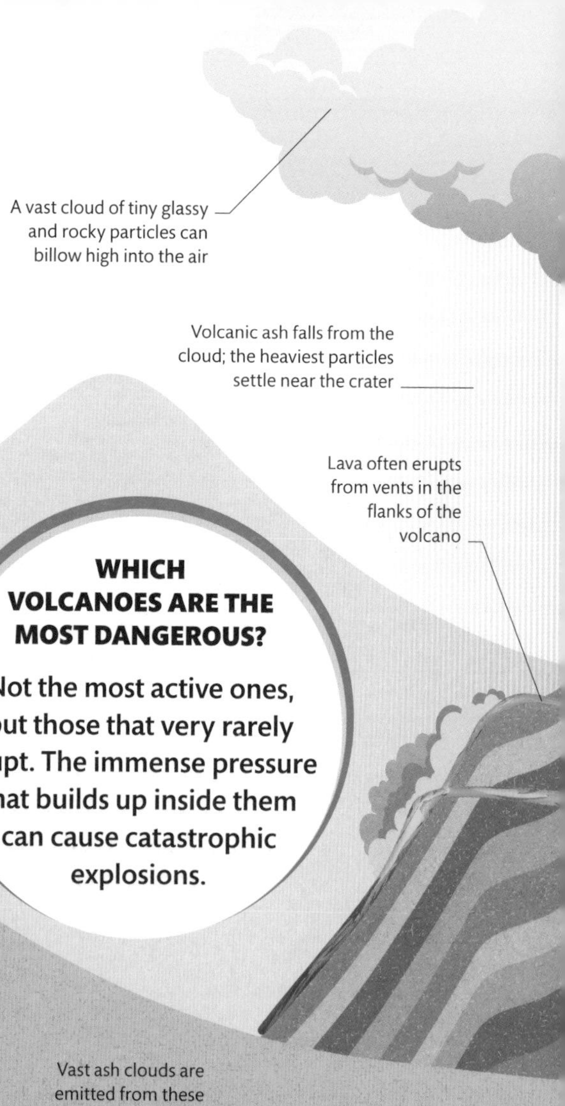
A vast cloud of tiny glassy and rocky particles can billow high into the air
Volcanic ash falls from the cloud; the heaviest particles settle near the crater
Lava often erupts from vents in the flanks of the volcano

Why do volcanoes form?

There are three main types of volcano. Some erupt from rifts between diverging continental or oceanic plates. Others, with different types of lava, erupt above subduction zones where one plate grinds beneath another. A third type is caused by hotspots in the mantle that cause local melting of the rock just below the crust—usually far away from plate boundaries.

WHICH VOLCANOES ARE THE MOST DANGEROUS?

Not the most active ones, but those that very rarely erupt. The immense pressure that builds up inside them can cause catastrophic explosions.

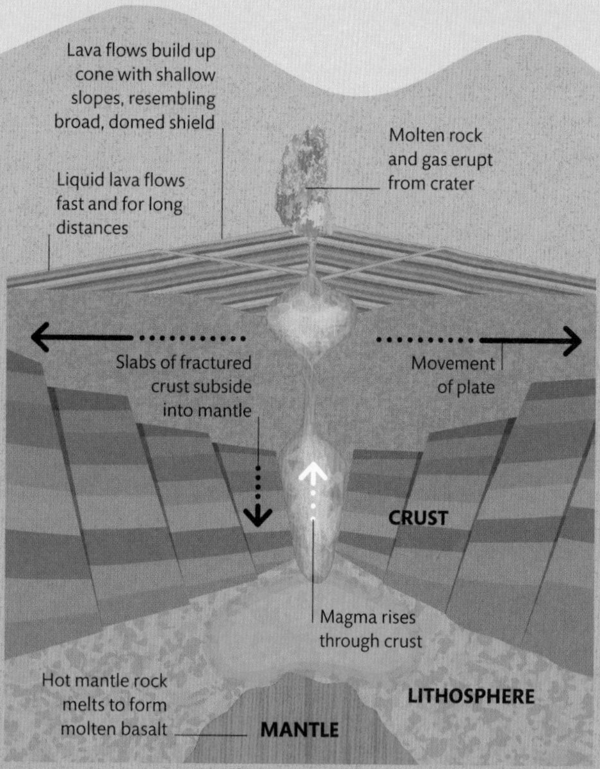
Lava flows build up cone with shallow slopes, resembling broad, domed shield
Liquid lava flows fast and for long distances
Molten rock and gas erupt from crater
Slabs of fractured crust subside into mantle
Movement of plate
CRUST
Magma rises through crust
Hot mantle rock melts to form molten basalt
LITHOSPHERE
MANTLE

Rift volcano
Where plates pull apart, this eases the pressure on the mantle below, allowing some of the hot rock to melt. It erupts as fluid basalt lava, which spreads out to form broad shield volcanoes.

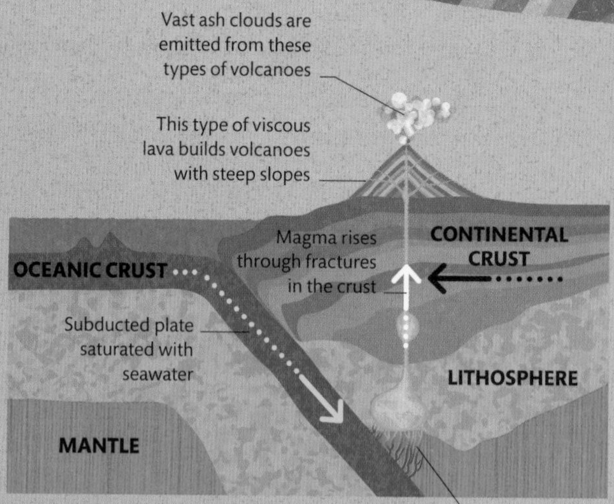
Vast ash clouds are emitted from these types of volcanoes
This type of viscous lava builds volcanoes with steep slopes
CONTINENTAL CRUST
OCEANIC CRUST
Magma rises through fractures in the crust
Subducted plate saturated with seawater
LITHOSPHERE
MANTLE
Water boils up into rock above and makes it melt

Subduction zone volcano
Oceanic crust dragged down into subduction zones carries water that alters the nature of hot rock, making it melt. At these volcanoes, the molten rock erupts as viscous, thick lava.

What's inside a volcano?

A subduction zone volcano has a steep-sided cone called a stratovolcano, built up from layers of lava and volcanic ash. This is because it erupts sticky lava that often blocks the crater, leading to explosive eruptions—blasting rock and ash into the air that fall on the volcano's slopes.

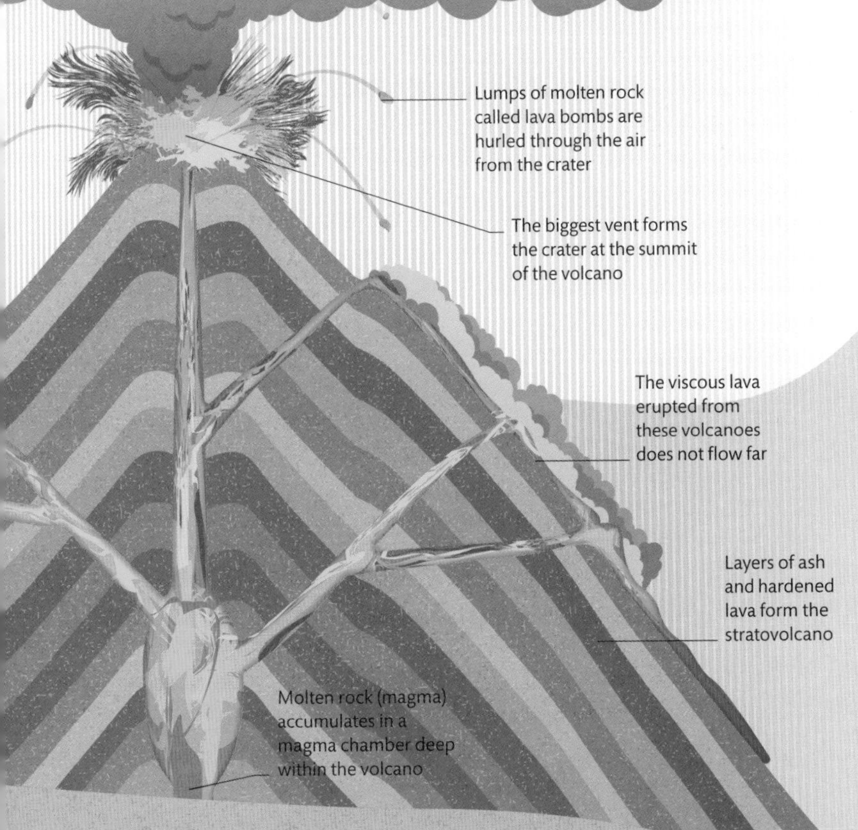
Lumps of molten rock called lava bombs are hurled through the air from the crater
The biggest vent forms the crater at the summit of the volcano
The viscous lava erupted from these volcanoes does not flow far
Layers of ash and hardened lava form the stratovolcano
Molten rock (magma) accumulates in a magma chamber deep within the volcano

Types of eruption

Volcanoes erupt in different ways, depending on the nature of their lava. The fluid lava of rift and hotspot volcanoes causes relatively quiet fissure and Hawaiian-type eruptions. Stickier lava is more explosive, causing Strombolian, Vulcanian, Peléan, and Plinian-type eruptions. The stickier the lava, the more explosive the eruption.

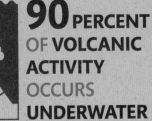

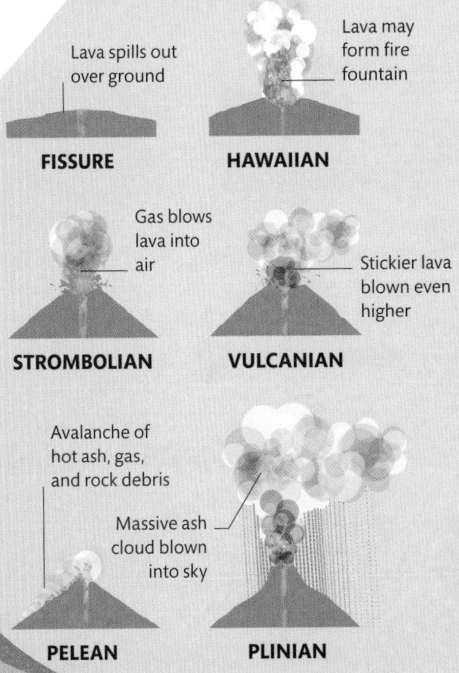
Lava spills out over ground
FISSURE
Lava may form fire fountain
HAWAIIAN
Gas blows lava into air
STROMBOLIAN
Stickier lava blown even higher
VULCANIAN
Avalanche of hot ash, gas, and rock debris
PELEAN
Massive ash cloud blown into sky
PLINIAN

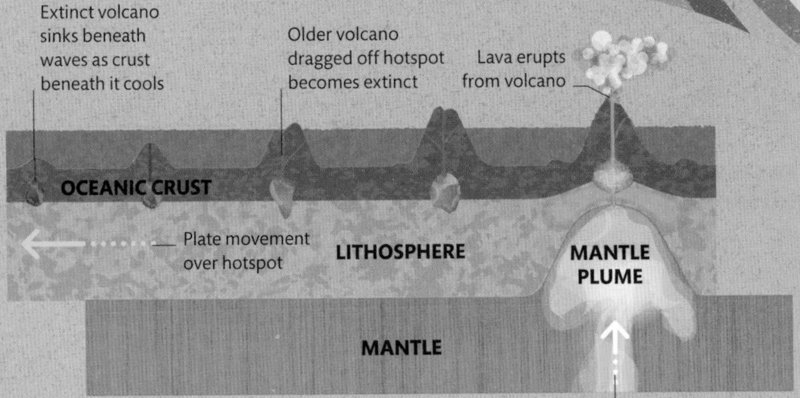
Extinct volcano sinks beneath waves as crust beneath it cools
Older volcano dragged off hotspot becomes extinct
Lava erupts from volcano
OCEANIC CRUST
Plate movement over hotspot
LITHOSPHERE
MANTLE PLUME
MANTLE
Heat rising through mantle forms hotspot under ocean floor

Hotspot volcanoes
These types of volcanoes are fueled by isolated heat currents, called mantle plumes, rising beneath the crust. Plate movement over the hotspot can create chains of volcanoes, such as those of Hawaii and the Galápagos Islands.

90 PERCENT OF VOLCANIC ACTIVITY OCCURS UNDERWATER

DATING ROCKS

Some igneous rocks can be dated by analyzing crystals containing heavy radioactive elements. Since these elements decay (change) to lighter elements at a known rate (see p.37), their ratio gives a measure of time elapsed since the crystal formed. But if the crystal is part of a sedimentary rock, this technique dates the crystal, not the rock. Luckily such rocks can be dated by the age of any fossils they contain.

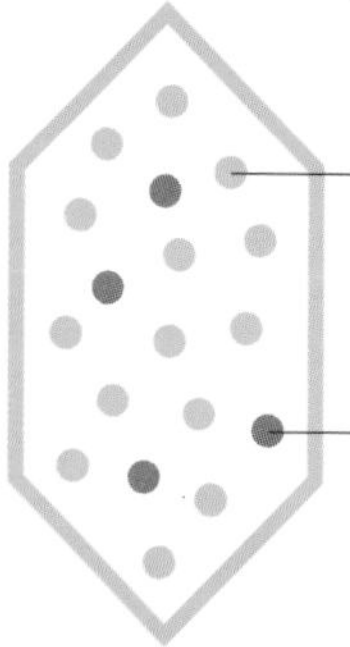
Uranium atoms inside a crystal decay to lead at a steady rate
Ratio of lead atoms to uranium atoms gives the crystal's age

ZIRCON CRYSTAL

The rock cycle

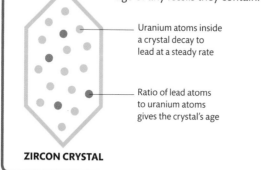

Rocks are made up of mixtures of minerals, such as quartz or calcite. Some rocks are very hard, others much softer, but over time they are all eroded and reworked into different types of rock in a process called the rock cycle.

Constant transformation

When molten rock cools, the minerals it contains crystallize (solidify) to form various types of solid, hard igneous rock. Over time, weathering breaks these down into soft sediments that can form layered sedimentary rocks. They may be transformed by heat and pressure into harder metamorphic rocks. If these are buried deeply, they may melt, eventually cooling to form more igneous rocks.

WHAT IS THE OLDEST ROCK ON EARTH?

Zircon crystals found in the Jack Hills region of western Australia have been dated to be 4.4 billion years old—which is close to Earth's age (4.5 billion years old)!

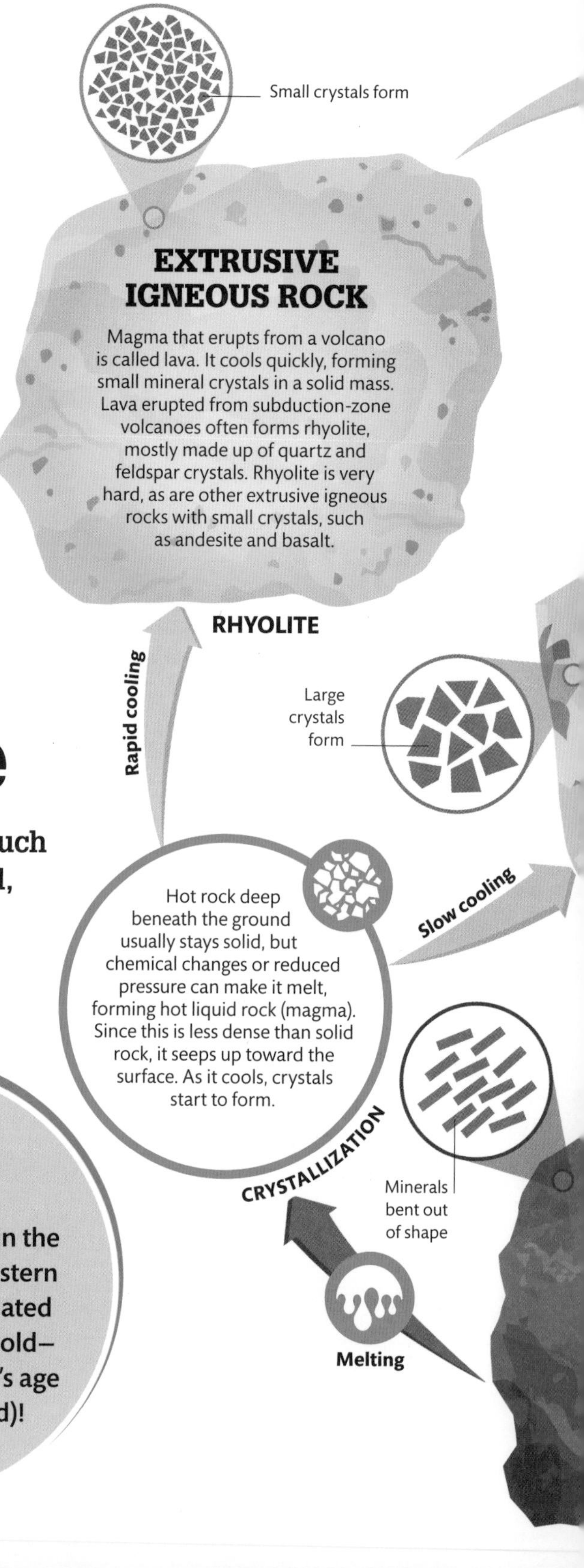
Small crystals form
EXTRUSIVE IGNEOUS ROCK
Magma that erupts from a volcano is called lava. It cools quickly, forming small mineral crystals in a solid mass. Lava erupted from subduction-zone volcanoes often forms rhyolite, mostly made up of quartz and feldspar crystals. Rhyolite is very hard, as are other extrusive igneous rocks with small crystals, such as andesite and basalt.
RHYOLITE
Rapid cooling
Large crystals form
Hot rock deep beneath the ground usually stays solid, but chemical changes or reduced pressure can make it melt, forming hot liquid rock (magma). Since this is less dense than solid rock, it seeps up toward the surface. As it cools, crystals start to form.
Slow cooling
CRYSTALLIZATION
Minerals bent out of shape
Melting

UPTHRUST

The forces of plate tectonics that build mountains can fracture and fold rocks upward so they are exposed to the air, becoming vulnerable to weathering (breakdown into smaller particles) and erosion (removal by rivers, glaciers, or wind).

Pressure

WEATHERING AND EROSION

Freeze-thaw

Rain

Wind

Glaciers

Rivers

Water in rock crevices expands if it freezes, splitting the rock. Rain dissolves carbon dioxide from the air to form weak carbonic acid, which attacks many minerals. Soft rock can be scoured by winds. Smaller bits of rock are carried away by rivers or glaciers.

LITHIFICATION

Deposition

Compaction

Cementation

Sediment (rock particles produced by weathering) carried in rivers or glaciers, or by wind, is deposited and buried. The particles are compacted by the weight of more sediment and form layers. Minerals dissolved in water crystallize, cementing them together. This process is called lithification.

INTRUSIVE IGNEOUS ROCK

Magma that does not erupt at the surface cools slowly underground and forms big mineral crystals—this can take millions of years. Huge masses of intrusive igneous rock, such as granite, can form this way. Granite has the same mineral ingredients as rhyolite but has much bigger crystals.

GRANITE

Pressure

Pressure

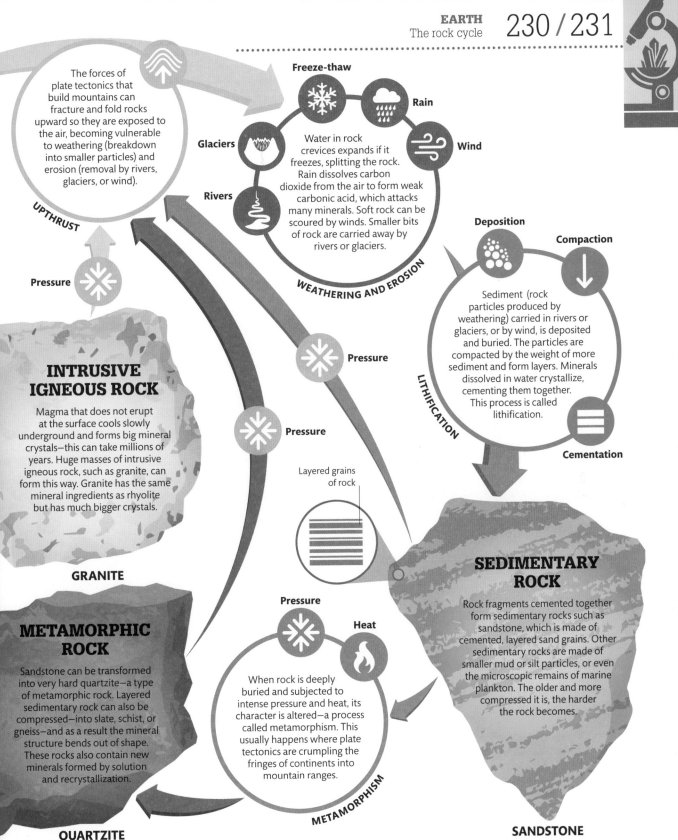

SEDIMENTARY ROCK

Rock fragments cemented together form sedimentary rocks such as sandstone, which is made of cemented, layered sand grains. Other sedimentary rocks are made of smaller mud or silt particles, or even the microscopic remains of marine plankton. The older and more compressed it is, the harder the rock becomes.

SANDSTONE

METAMORPHISM

Pressure

Heat

When rock is deeply buried and subjected to intense pressure and heat, its character is altered—a process called metamorphism. This usually happens where plate tectonics are crumpling the fringes of continents into mountain ranges.

METAMORPHIC ROCK

Sandstone can be transformed into very hard quartzite—a type of metamorphic rock. Layered sedimentary rock can also be compressed—into slate, schist, or gneiss—and as a result the mineral structure bends out of shape. These rocks also contain new minerals formed by solution and recrystallization.

QUARTZITE

Oceans

Earth is a predominantly blue planet—most of its surface is covered by oceans. There are five named oceans—the Pacific, Atlantic, Indian, Arctic, and Southern—but water circulates slowly through them all.

THE **MARIANA TRENCH** IN THE PACIFIC OCEAN CAN **FIT MOUNT EVEREST IN IT** WITH **6,600FT (2,000M) TO SPARE**

WHY IS SEAWATER SALTY?

Rainwater draining off the land over millions of years has carried salty minerals into the sea. These give seawater its salty taste.

OPEN OCEAN

What is an ocean?

Oceans are not just giant puddles of water—they are created by the forces of plate tectonics (see pp.224–25). Where the plates of Earth's crust pull apart, new crust forms. Oceanic crust lies much lower than thicker, lighter continental crust (see p.222), forming the ocean floors. Where plates meet underwater, one subducts beneath another, creating deep ocean trenches. The fringes of continents also lie underwater, having been cut back by coastal erosion. Shelf seas, the coastal seas over continental shelves, are far shallower than true oceans.

The true ocean floor, the abyssal plain, lies 10,000–20,000ft (3,000–6,000m) below the waves

Rocky debris and particles swept off continents build up at foot of continental crust and along abyssal plain

ABYSSAL PLAIN

OCEANS IN MOTION

Winds drive powerful surface currents that swirl around the oceans, carrying cold water into the tropics and warm water toward the poles. They are linked to deepwater currents driven by cool, salty water sinking toward the ocean floor. Together, these currents carry ocean water all around the world in a network often called the global conveyor.

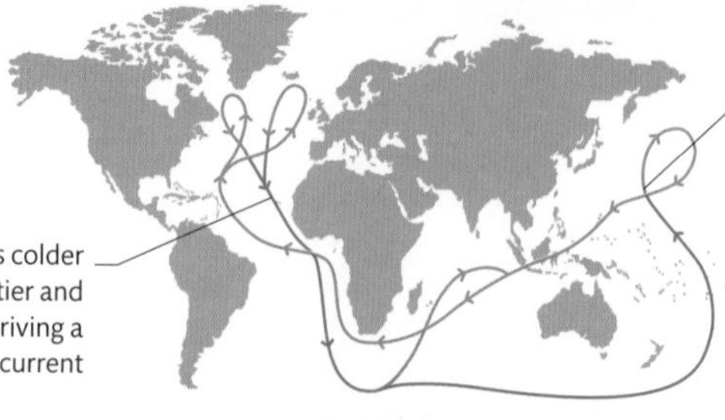
Water gets colder and saltier and sinks, driving a deepwater current
Displaced deep, cold water is forced to the surface, joining warmer surface currents

OCEAN CURRENTS

WHY DO TIDES RISE AND FALL?

The Moon's gravity drags ocean water into an oval with two tidal bulges. As Earth spins, sea shores pass in and out of these bulges and experience daily high and low tides. When the Moon is in line with the Sun at full and new Moon, their combined gravity causes larger tides. At half Moon, when the Moon's gravitational pull is at right angles to the Sun, the tides are weaker.

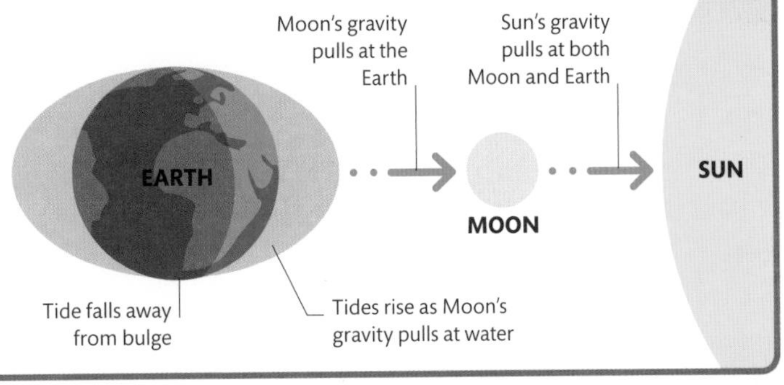
Moon's gravity pulls at the Earth
Sun's gravity pulls at both Moon and Earth
EARTH
MOON
SUN
Tide falls away from bulge
Tides rise as Moon's gravity pulls at water

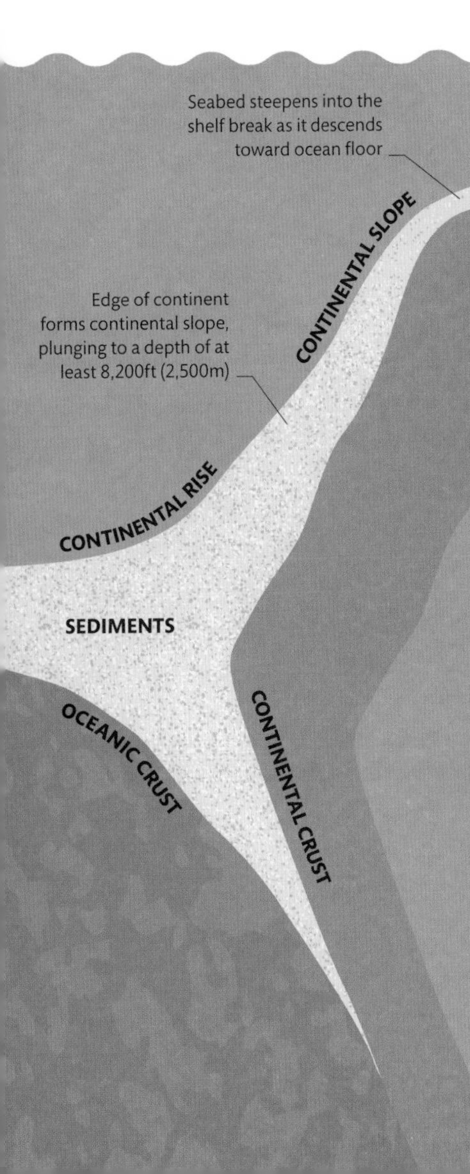
SHELF SEA
SHORELINE
Bed of shelf sea, the continental shelf, usually less than 490ft (150m) below surface
CONTINENTAL SHELF
Seabed steepens into the shelf break as it descends toward ocean floor
CONTINENTAL SLOPE
Edge of continent forms continental slope, plunging to a depth of at least 8,200ft (2,500m)
CONTINENTAL RISE
SEDIMENTS
OCEANIC CRUST
CONTINENTAL CRUST

Making waves

As a wind blows over an ocean, it whips up waves on the surface. The stronger the wind is, and the longer it blows, the bigger the waves get. Waves also get bigger the farther they travel. Water molecules move in a circular path, which is why we are carried up and forward as we are caught by waves, and then fall down and back as the wave passes.

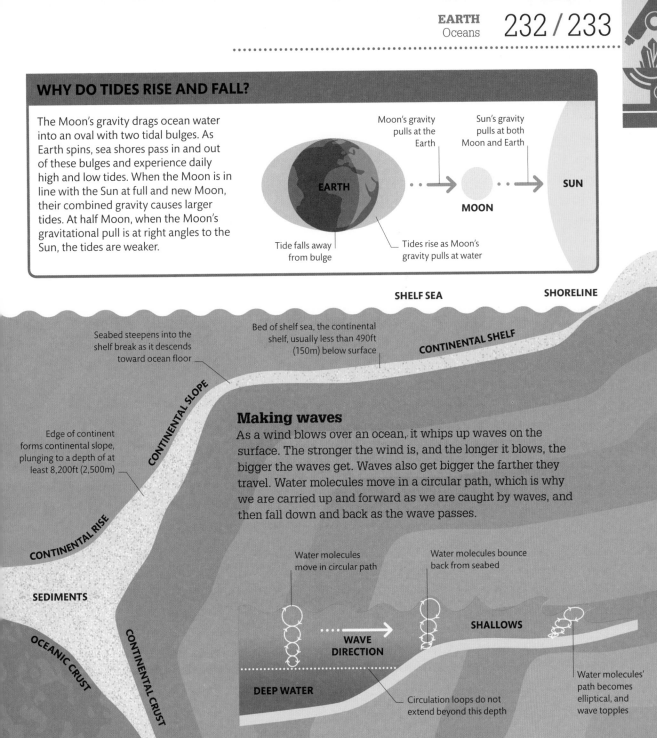

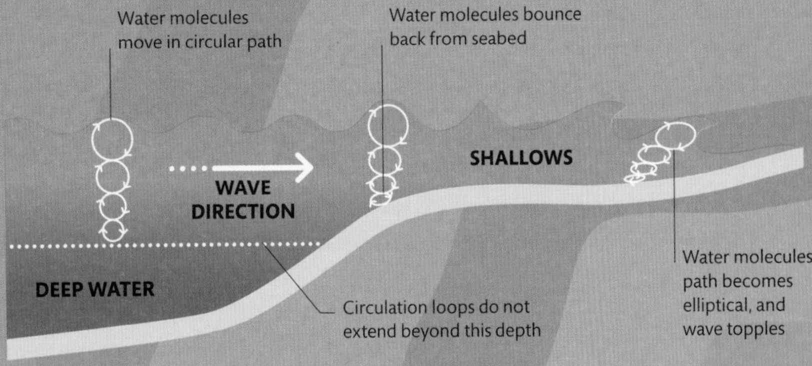
Water molecules move in circular path
Water molecules bounce back from seabed
SHALLOWS
WAVE DIRECTION
DEEP WATER
Circulation loops do not extend beyond this depth
Water molecules' path becomes elliptical, and wave topples

1 Open water
At sea, waves make the water roll up and forward, then down and back. Water molecules move in a circular path.

2 Waves get higher
Water molecules bounce back from the seabed, causing the wave to get shorter and steeper as it approaches the shore.

3 Waves break
As the seabed becomes shallower, the paths get more elliptical, causing the wave crest to grow so tall that it topples and breaks.

Earth's atmosphere

Earth is encircled with gases that shield the planet's surface from damaging solar radiation and retain heat at night, making life possible. Circulation of air in the lower atmosphere causes the phenomena we call weather.

What is the atmosphere?

The atmosphere consists of gases – mainly nitrogen, oxygen, argon, and carbon dioxide. It is split into layers that are defined by their temperature: some layers get colder with altitude, while others get warmer due to the ability of some gases to absorb rays from the Sun. Most of the air is concentrated in the lowest layer, the troposphere, but its density decreases with altitude. This means that just 10 km (6 miles) above sea level there is not enough air to keep people alive.

WHY DOESN'T OUR ATMOSPHERE FLOAT AWAY INTO SPACE?

Gas particles are held near Earth's surface by gravity. The much smaller Moon has far less gravity and so cannot retain an atmosphere.

Atmosphere forms relatively shallow layer around Earth

LAYERS OF THE ATMOSPHERE

TEMPERATURE

400–6,000 MILES (600–10,000KM)

Exosphere
The outermost layer of the atmosphere fades into space, with no clear outer boundary. The air particles are so dispersed that they do not interact at all.

Many artificial satellites orbit the planet in the exosphere

50–370 MILES (80–600KM)

Thermosphere
Above the mesosphere, the thermosphere extends a vast distance, its temperature increasing with altitude to as much as 3,630°F (2,000°C)–this is because gases in this layer absorb x-rays and ultraviolet light from the Sun.

Molecules absorb x-rays and ultraviolet light, then radiate heat

Oxygen and nitrogen atoms energized by solar radiation glow to cause auroras

THE THERMOSPHERE CAN GET AS HOT AS 3,630°F (2,000°C)

Spin and swerve

In the troposphere, warm air rises, flows sideways, cools, and then sinks—these circulation cells distribute heat around the globe (see pp.240–41). Earth's spin makes circulating air swerve off course. North of the equator the airflow swerves to the right, while south of the equator it swerves left. This is called the Coriolis effect and it results in the air in each circulation cell spiraling around the globe.

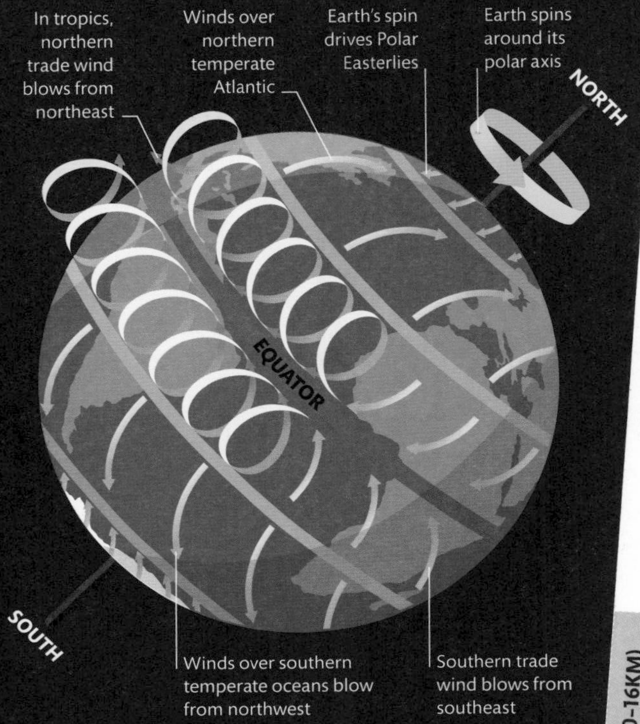
In tropics, northern trade wind blows from northeast
Winds over northern temperate Atlantic
Earth's spin drives Polar Easterlies
Earth spins around its polar axis
NORTH
EQUATOR
SOUTH
Winds over southern temperate oceans blow from northwest
Southern trade wind blows from southeast

Global spirals
Prevailing winds are driven by the spiraling air circulation cells and blow close to Earth's surface. These winds blow most steadily over the oceans.

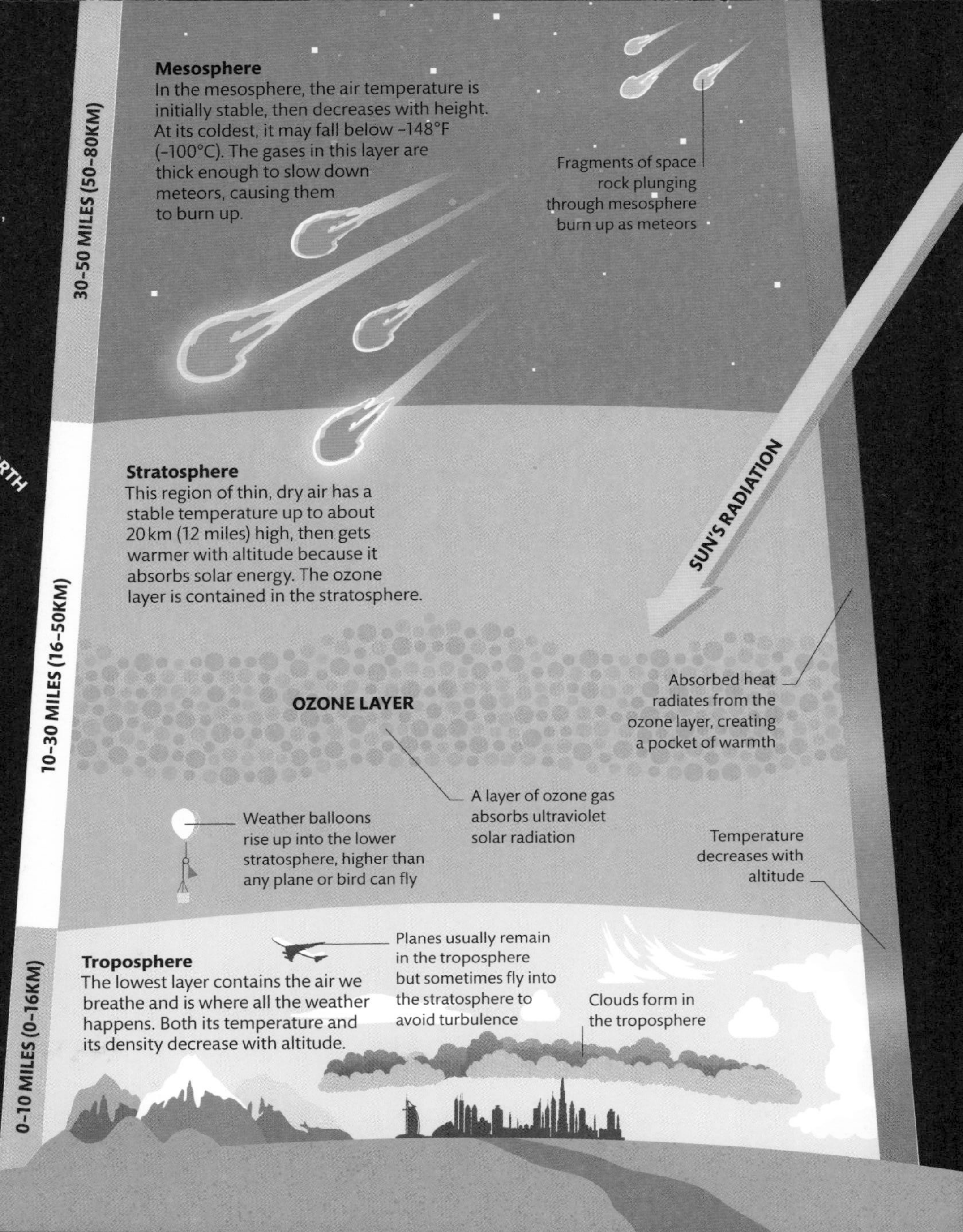
30–50 MILES (50–80KM)
10–30 MILES (16–50KM)
0–10 MILES (0–16KM)
Fragments of space rock plunging through mesosphere burn up as meteors
SUN'S RADIATION
OZONE LAYER
Absorbed heat radiates from the ozone layer, creating a pocket of warmth
A layer of ozone gas absorbs ultraviolet solar radiation
Weather balloons rise up into the lower stratosphere, higher than any plane or bird can fly
Temperature decreases with altitude
Planes usually remain in the troposphere but sometimes fly into the stratosphere to avoid turbulence
Clouds form in the troposphere

Mesosphere
In the mesosphere, the air temperature is initially stable, then decreases with height. At its coldest, it may fall below -148°F (-100°C). The gases in this layer are thick enough to slow down meteors, causing them to burn up.

Stratosphere
This region of thin, dry air has a stable temperature up to about 20 km (12 miles) high, then gets warmer with altitude because it absorbs solar energy. The ozone layer is contained in the stratosphere.

Troposphere
The lowest layer contains the air we breathe and is where all the weather happens. Both its temperature and its density decrease with altitude.

How weather works

Weather is the state of the atmosphere at a particular place and time. It changes constantly as the Sun evaporates moisture into warm air, which rises to form clouds. This process drives the swirling low-pressure systems, or cyclones, that bring wind and rain. These are balanced by the calm of anticyclones.

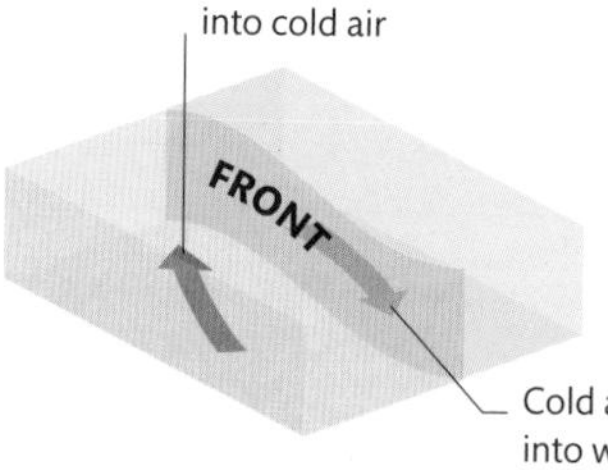

Warm air blows into cold air
FRONT
Cold air blows into warm air

1 Cold meets warm
Cyclones often form over temperate oceans where warm, moist, tropical air masses push into cold, polar air masses. A front is the region where the two air masses meet.

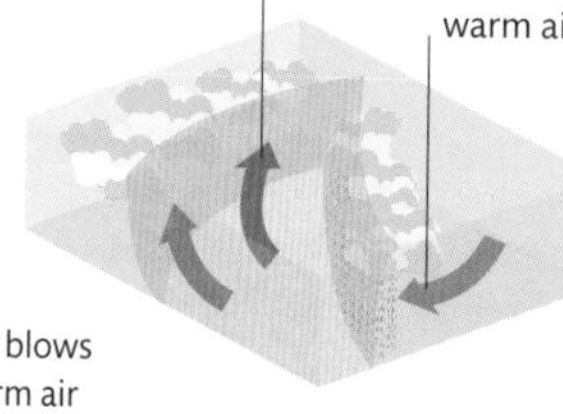

Curving warm air drives a wedge into the cold air
Cold air moves faster and wraps around the back of the wedge of warm air

2 Rotation begins
As they move, both air masses follow curved paths due to Earth's spin—a phenomenon called the Coriolis effect. The curved paths become a rotational pattern, and the air masses begin to spiral.

Birth of a cyclone

When warm, moist air rises, it creates a zone of low air pressure, drawing in surrounding air in a spiral flow called a cyclone, or depression. The warm, moist air is forced upward, where it rides over cooler, denser air, making its moisture condense into clouds and rain. The airflow—which we feel as wind—is strongest when the warm air is rising with the most energy. In the tropics, this generates the powerful storms known as tropical cyclones, hurricanes, or typhoons.

HOW CAN HAIL FALL IN KENYA?

Clouds in the tropics are so tall that moisture reaches the cold upper atmosphere and freezes, eventually falling as hailstones (see pp.238–39).

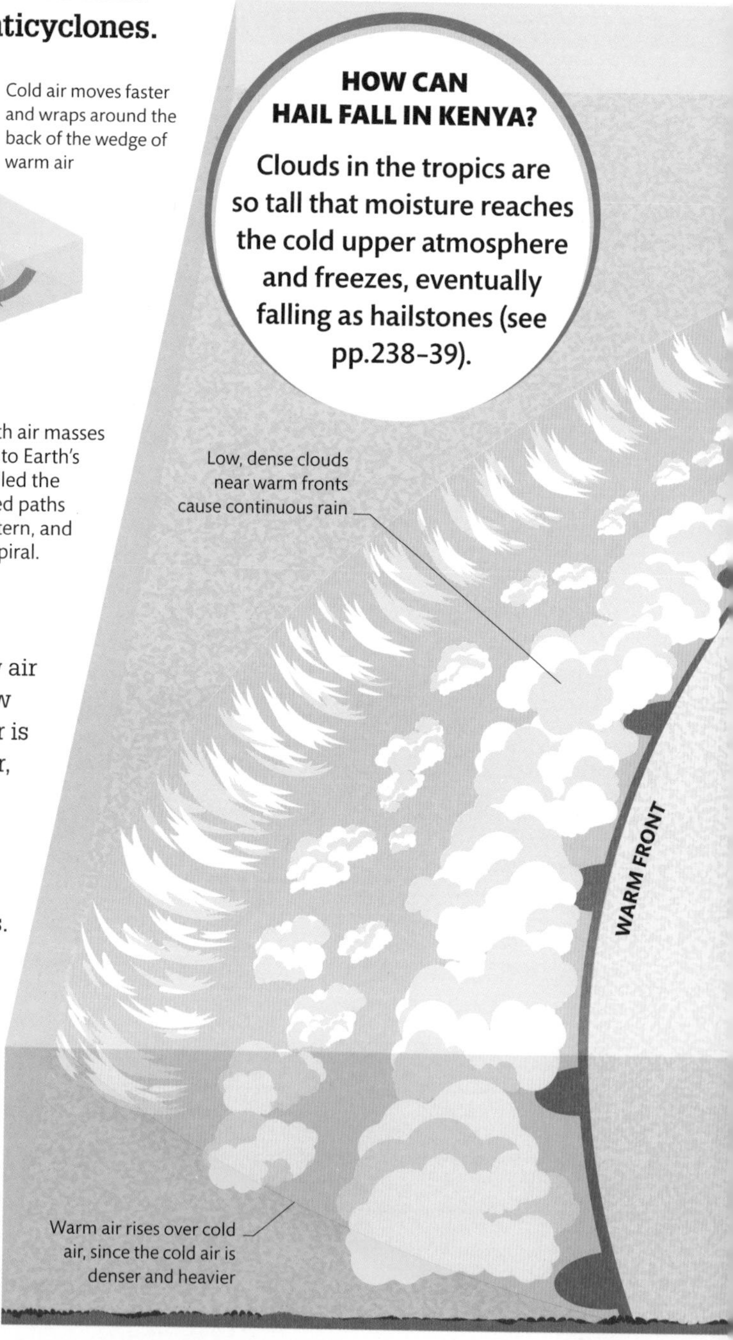

Low, dense clouds near warm fronts cause continuous rain
WARM FRONT
Warm air rises over cold air, since the cold air is denser and heavier

SNOW

If cloud droplets rise high enough, they form microscopic, six-sided ice crystals. Water freezing onto the crystals makes them stick together in six-pointed snowflakes. These clump together in larger fluffy masses that fall as snow.

MOST RAINFALL OUTSIDE THE TROPICS STARTS OUT AS SNOW AND MELTS AS IT FALLS

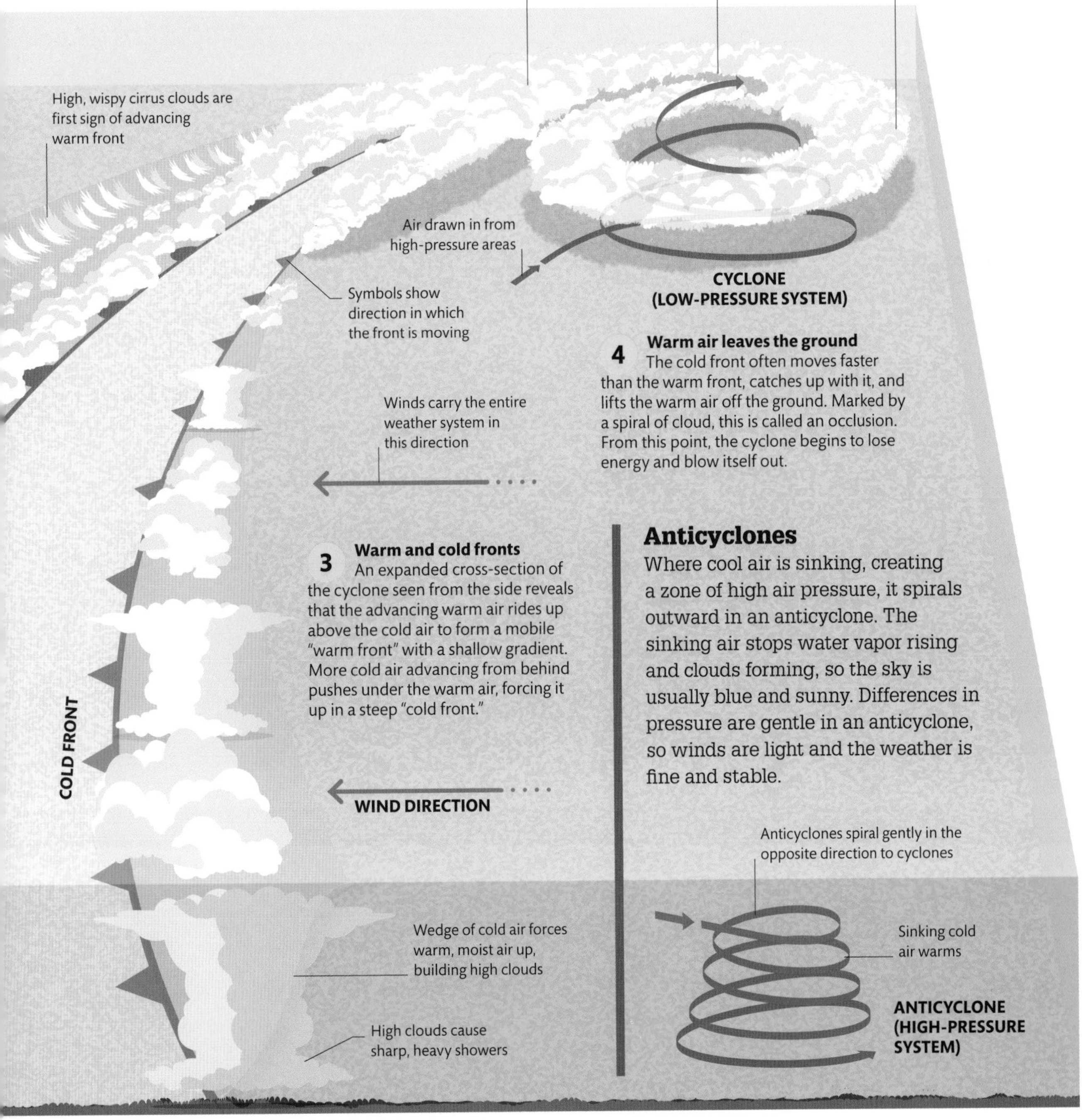
Where fronts meet, they combine into a single occluded front, and the wedge of warm air is lifted clear of the ground
Cyclones rotate clockwise in the southern hemisphere (and counterclockwise in the north)
Air rises in a spiral
High, wispy cirrus clouds are first sign of advancing warm front
Air drawn in from high-pressure areas
Symbols show direction in which the front is moving
CYCLONE (LOW-PRESSURE SYSTEM)
Winds carry the entire weather system in this direction
COLD FRONT
WIND DIRECTION
Anticyclones spiral gently in the opposite direction to cyclones
Sinking cold air warms
ANTICYCLONE (HIGH-PRESSURE SYSTEM)
Wedge of cold air forces warm, moist air up, building high clouds
High clouds cause sharp, heavy showers

3 Warm and cold fronts
An expanded cross-section of the cyclone seen from the side reveals that the advancing warm air rides up above the cold air to form a mobile "warm front" with a shallow gradient. More cold air advancing from behind pushes under the warm air, forcing it up in a steep "cold front."

4 Warm air leaves the ground
The cold front often moves faster than the warm front, catches up with it, and lifts the warm air off the ground. Marked by a spiral of cloud, this is called an occlusion. From this point, the cyclone begins to lose energy and blow itself out.

Anticyclones

Where cool air is sinking, creating a zone of high air pressure, it spirals outward in an anticyclone. The sinking air stops water vapor rising and clouds forming, so the sky is usually blue and sunny. Differences in pressure are gentle in an anticyclone, so winds are light and the weather is fine and stable.

Extreme weather

Most extreme weather events are caused by the build-up of airborne moisture in towering cumulonimbus storm clouds. Powerful air currents within these clouds trigger lightning, hail, and even tornadoes.

Superclouds

Cumulonimbus clouds are much bigger than other clouds, rising from near the ground all the way to the top of the troposphere (see p.235). They are fueled by intense evaporation of moisture from the ground or ocean surface. As the vapor rises and cools, it condenses into water droplets to form giant clouds, releasing energy as heat (see p.117). The heat warms the air, which rises farther, carrying more water vapor that condenses and releases even more energy—and the cycle continues. Eventually, the cloud may grow to more than 6 miles (10km) high.

Vigorous updrafts can make the cloud's core billow up into the stratosphere

Most of the cloud stops rising and spreads sideways, driven by wind

Electricity discharges from cloud and arcs through air as lightning

Heat generated by lightning makes air expand explosively, causing shockwaves we hear as thunder

1 Charged up
Powerful updrafts within the cloud, flanked by sinking cold air, toss water droplets and ice crystals up and down, generating static electricity (see pp.78–79) that charges the cloud like a giant battery.

SINKING COLD AIR

Currents of rising warm air can catch falling ice crystals and lift them back into the air

Extra moisture refreezes at higher altitude

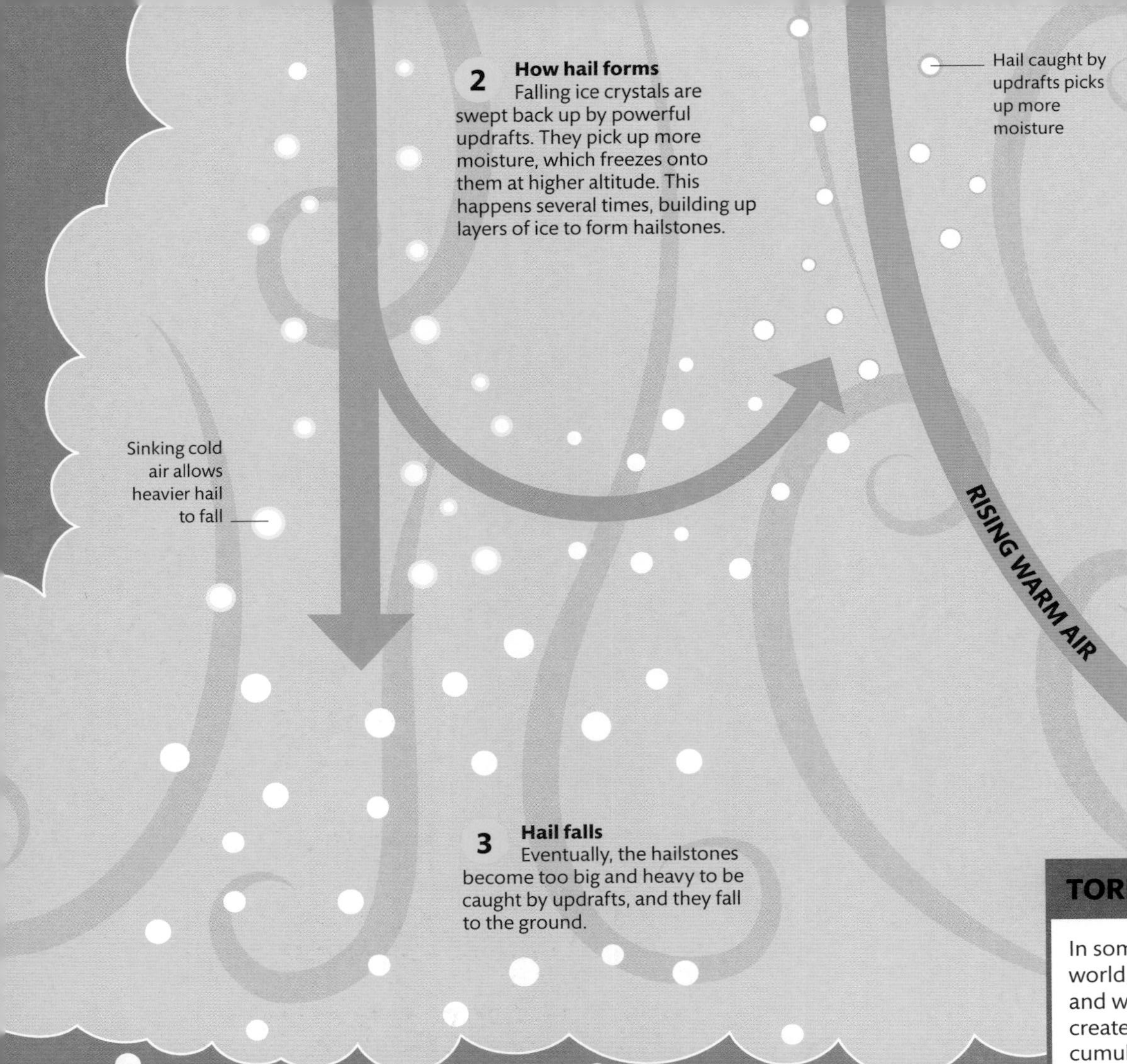
Sinking cold air allows heavier hail to fall
Hail caught by updrafts picks up more moisture
RISING WARM AIR

2 How hail forms
Falling ice crystals are swept back up by powerful updrafts. They pick up more moisture, which freezes onto them at higher altitude. This happens several times, building up layers of ice to form hailstones.

3 Hail falls
Eventually, the hailstones become too big and heavy to be caught by updrafts, and they fall to the ground.

WHAT IS A HURRICANE?

Intense evaporation from tropical oceans forms colossal cloud systems around intense low-pressure zones (see p.236). Air swirls into these at high speeds, causing the strong winds of a hurricane.

TORNADOES

In some parts of the world, swirling cold and warm air masses create huge, rotating cumulonimbus clouds called supercells. The spinning, rapidly rising air can be concentrated into a tight vortex called a tornado, powerful enough to rip a house apart.

HAILSTONES CAN BE THE **SIZE** OF **HUMAN FISTS**

Climate and the seasons

Sunlight and heat are concentrated in the tropics and dispersed near the poles. The heat drives air currents in the atmosphere that create the world's climate zones.

Circulating cells

At the tropics, intense heat evaporates water from the oceans. When the warm, moist air rises—creating a low pressure band called the Intertropical Convergence Zone (ITCZ)—it cools. Water vapor condenses into gigantic clouds, causing heavy rain. The now dry, cool air flows to the subtropical zones and sinks, causing high pressure that inhibits rain. This is the Hadley circulation cell. Two other cells—the Ferrel and Polar cells—have similar effects in cooler regions.

LOCATOR

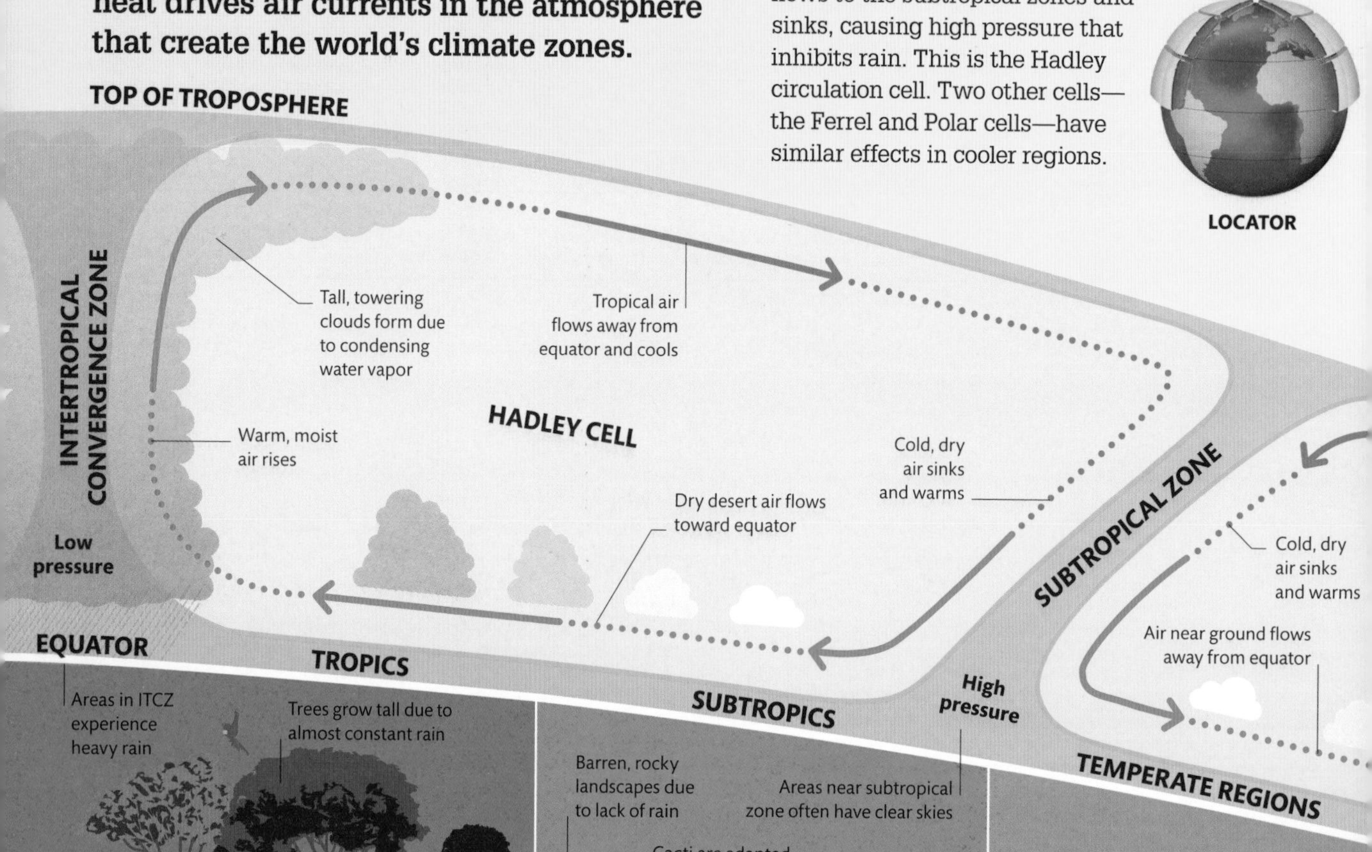
TOP OF TROPOSPHERE
INTERTROPICAL CONVERGENCE ZONE
Tall, towering clouds form due to condensing water vapor
Tropical air flows away from equator and cools
HADLEY CELL
Warm, moist air rises
Cold, dry air sinks and warms
Dry desert air flows toward equator
SUBTROPICAL ZONE
Cold, dry air sinks and warms
Low pressure
Air near ground flows away from equator
EQUATOR
TROPICS
SUBTROPICS
High pressure
TEMPERATE REGIONS

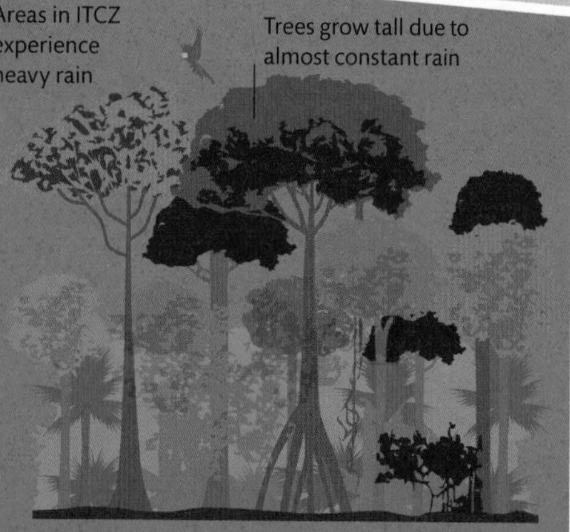
Areas in ITCZ experience heavy rain
Trees grow tall due to almost constant rain

Tropics
Rising moist air near the equator builds huge storm clouds that cause daily heavy rain, fueling the growth of tropical rainforests. The trees produce water vapor, so to some extent they make their own climate.

Barren, rocky landscapes due to lack of rain
Areas near subtropical zone often have clear skies
Cacti are adapted to arid climates

Subtropics
When rising equatorial air reaches the top of the troposphere, it flows horizontally until it cools and sinks over the subtropics. The sinking air stops clouds forming, so there is little rain, creating deserts such as the Sahara.

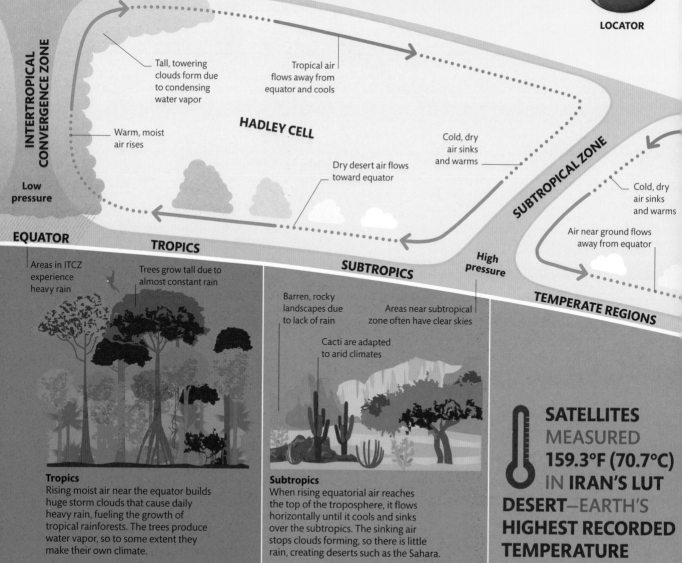

SATELLITES MEASURED 159.3°F (70.7°C) IN IRAN'S LUT DESERT—EARTH'S HIGHEST RECORDED TEMPERATURE

Seasonal cycles

As Earth orbits the Sun, its tilted spin axis always points to Polaris, the north star. This means polar and temperate latitudes move toward and then away from the Sun, causing summer and winter. Seasons are most extreme near the poles. The ITCZ also moves north and south, causing tropical wet and dry seasons. Monsoon seasons are caused by a shift in wind direction that brings moist air from the oceans and, with it, heavy rains.

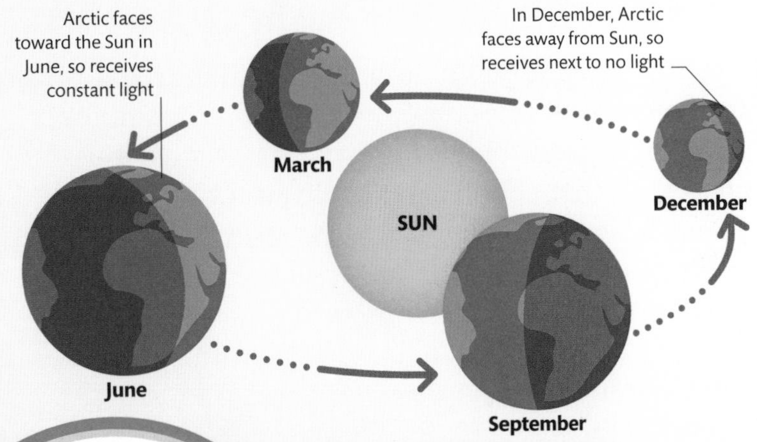
Arctic faces toward the Sun in June, so receives constant light
In December, Arctic faces away from Sun, so receives next to no light
March
SUN
December
June
September

WHERE IS THE DRIEST PLACE ON EARTH?

The McMurdo Dry Valleys in Antarctica have had no rain or snow for nearly 2 million years. The landscape is mostly bare rock and gravel.

Polar regions
Cold, dry air sinks over polar regions, forming cold deserts. It flows away from the poles at low level, warms up, and gathers moisture. In temperate regions, it is dragged up by rising subtropical air and flows back toward the poles at higher levels.

Areas near the polar front are often cloudy
Cold, dry air flows toward equator
FERREL CELL
Warm, moist air rises
POLAR FRONT
Warm, moist air rises
Low pressure
POLAR CELL
Cold air sinks and flows away from pole
High pressure
POLAR CIRCLE

Temperate regions
In temperate regions, warm air flowing from the subtropics at low level meets colder polar air. This makes it rise, forming clouds and rain, especially over and near oceans. The rain creates forests and grasslands.

THE HIGHEST RAINFALL IN 1 DAY WAS ON RÉUNION ISLAND—RECORDED AT 73IN (1,870MM) IN 1952

The water cycle

Water is the planet's life force. Life cannot exist without water because it is essential to all the biochemical processes that enable living things to thrive and multiply. Without the water cycle supplying the land with water, the continents would be lifeless deserts. Water also shapes the planet by eroding its land surface.

Earth's circulation system

Our Sun drives the water cycle—heating up the oceans so that evaporation occurs constantly from their surfaces. Water is carried over land by clouds borne on the wind. The clouds spill rain, which soaks into the ground. Some of this water is absorbed by plants, which pump it back into the air to form more clouds. Most of the rest flows off the land in rivers and finds its way back to the sea, and the cycle continues.

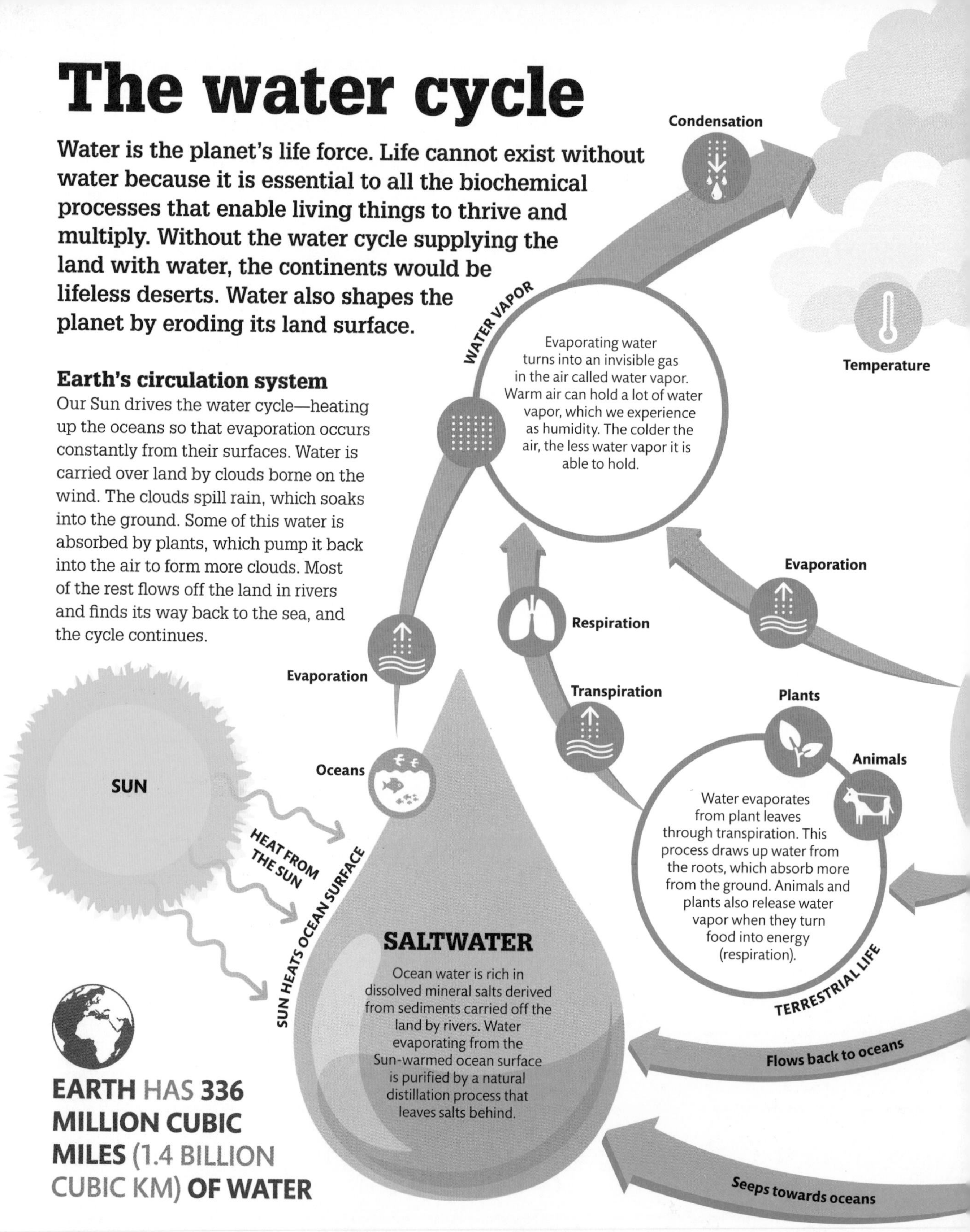
Condensation
Temperature
WATER VAPOR
Evaporating water turns into an invisible gas in the air called water vapor. Warm air can hold a lot of water vapor, which we experience as humidity. The colder the air, the less water vapor it is able to hold.
Evaporation
Respiration
Evaporation
Transpiration
Plants
Animals
Oceans
SUN
HEAT FROM THE SUN
SUN HEATS OCEAN SURFACE
SALTWATER
Ocean water is rich in dissolved mineral salts derived from sediments carried off the land by rivers. Water evaporating from the Sun-warmed ocean surface is purified by a natural distillation process that leaves salts behind.
Water evaporates from plant leaves through transpiration. This process draws up water from the roots, which absorb more from the ground. Animals and plants also release water vapor when they turn food into energy (respiration).
TERRESTRIAL LIFE
Flows back to oceans
Seeps towards oceans

EARTH HAS 336 MILLION CUBIC MILES (1.4 BILLION CUBIC KM) OF WATER

CLOUDS

Rising warm air that carries water vapor cools at higher altitudes—this causes the vapor to condense into microscopic water droplets and ice crystals. These droplets and crystals are visible as clouds that can be carried vast distances on the wind.

Altitude

Winds

PARTS OF THE **ANTARCTIC ICE SHEET** ARE MORE THAN **2.5 MILLION YEARS OLD**

PRECIPITATION

Snow

If a cloud cools, its water droplets and ice crystals grow and combine, eventually forming bigger raindrops or snowflakes that become so heavy they fall from the cloud. Snowflakes usually clump together in larger, fluffy masses.

Rain

ICE

Snow falling in cold climates does not melt. It builds up and is compressed by the weight of more snow—turning it to ice. On mountain slopes, the ice flows slowly downhill as glaciers and eventually melts, but polar ice sheets may never melt. Over the course of thousands of years, glaciers carve deep valleys.

Glaciers

Rivers

Lakes

Surface runoff

Melting

Surface runoff

Seeps underground

FRESHWATER

Rainfall that stays on the ground surface, and melting snow, is called surface runoff. This collects as rivers and lakes that eventually flow back to the oceans. Rain reacts with carbon dioxide gas in the air to form carbonic acid that weathers rocks, breaking down minerals that dissolve in the water.

WHERE IS ALL THE WATER?

Two-thirds of the planet is covered by oceans, which contain 97.5 percent of the world's water. Just 2.5 percent is freshwater. Most of this is locked up as ice in the polar regions and on high mountains, or hidden deep below ground. Only a tiny fraction forms rivers and lakes.

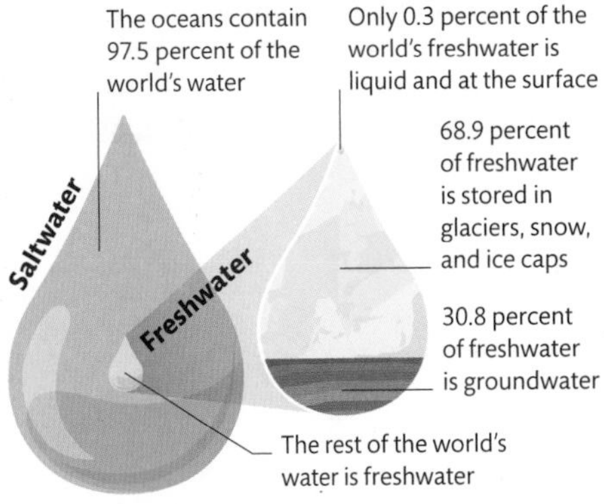
The oceans contain 97.5 percent of the world's water
Only 0.3 percent of the world's freshwater is liquid and at the surface
68.9 percent of freshwater is stored in glaciers, snow, and ice caps
30.8 percent of freshwater is groundwater
Saltwater
Freshwater
The rest of the world's water is freshwater

ALL OF EARTH'S WATER

GROUNDWATER

Rain and melting snow seep below the surface to become groundwater. At lower levels, the rock is saturated, forming underground reservoirs called aquifers. Limestone can be dissolved to create caves. Groundwater eventually seeps back into the oceans.

Caves

The greenhouse effect

Life on Earth depends on the greenhouse effect, which is the way some gases in our atmosphere—notably carbon dioxide and methane—absorb some of the infrared radiation emitted from Earth's surface. Like glass in a greenhouse, these gases trap heat.

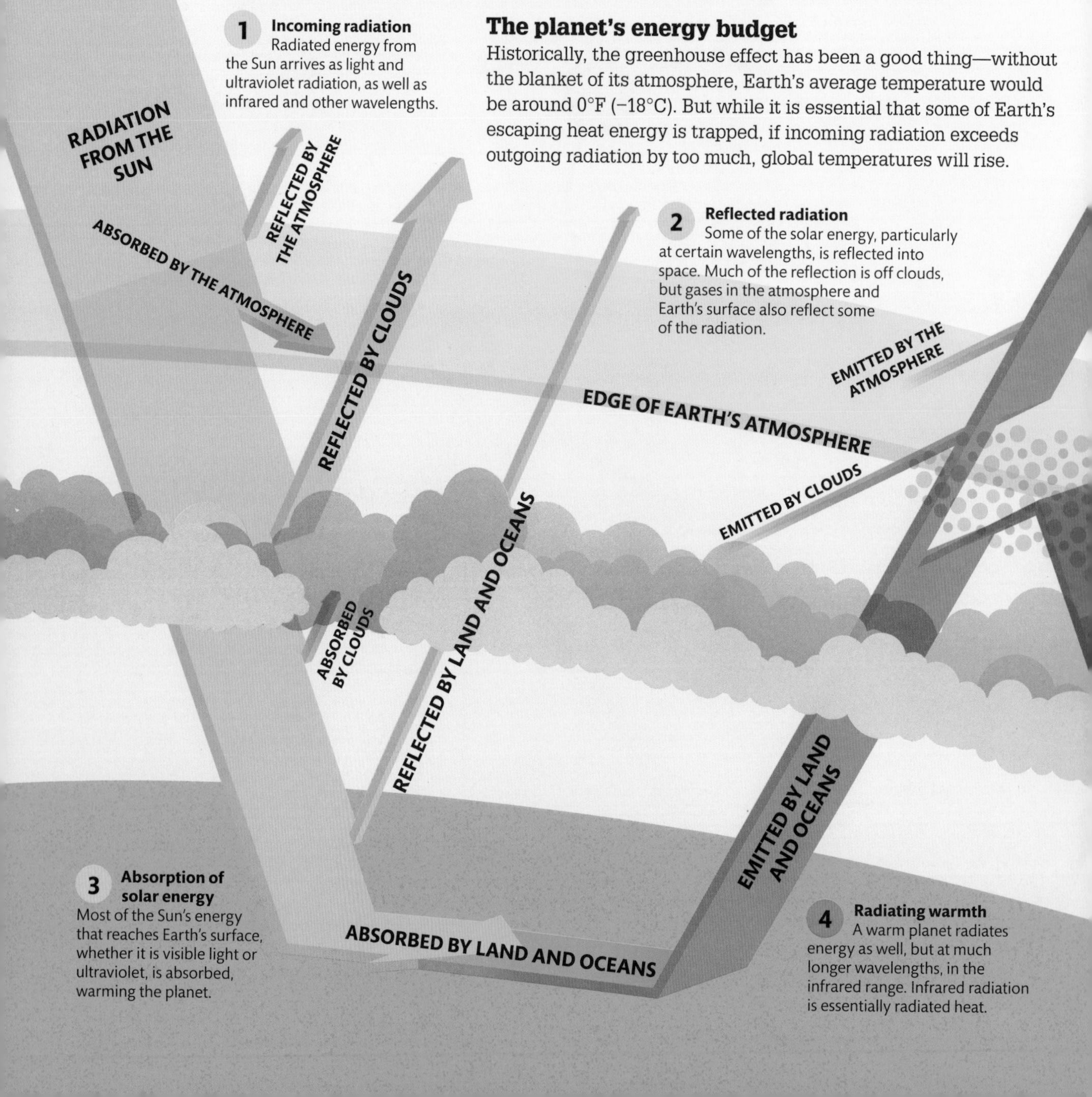

1 Incoming radiation Radiated energy from the Sun arrives as light and ultraviolet radiation, as well as infrared and other wavelengths.

The planet's energy budget

Historically, the greenhouse effect has been a good thing—without the blanket of its atmosphere, Earth's average temperature would be around 0°F (−18°C). But while it is essential that some of Earth's escaping heat energy is trapped, if incoming radiation exceeds outgoing radiation by too much, global temperatures will rise.

2 Reflected radiation Some of the solar energy, particularly at certain wavelengths, is reflected into space. Much of the reflection is off clouds, but gases in the atmosphere and Earth's surface also reflect some of the radiation.

3 Absorption of solar energy Most of the Sun's energy that reaches Earth's surface, whether it is visible light or ultraviolet, is absorbed, warming the planet.

4 Radiating warmth A warm planet radiates energy as well, but at much longer wavelengths, in the infrared range. Infrared radiation is essentially radiated heat.

RADIATION EMITTED INTO SPACE

5 Escaping radiation
Much of the radiation absorbed and re-emitted by Earth's atmosphere, clouds, and surface escapes into space.

GREENHOUSE EFFECTS ON OTHER WORLDS

Venus has a far stronger greenhouse effect than Earth. Its thick carbon dioxide atmosphere retains nearly all the solar energy that reaches its surface, creating temperatures hot enough to melt lead. By contrast, Titan, Saturn's largest moon, has an antigreenhouse effect created by a thick orange haze, which blocks 90 percent of sunlight. A similar but much weaker antigreenhouse effect can be caused on Earth by gas and dust erupted from volcanoes.

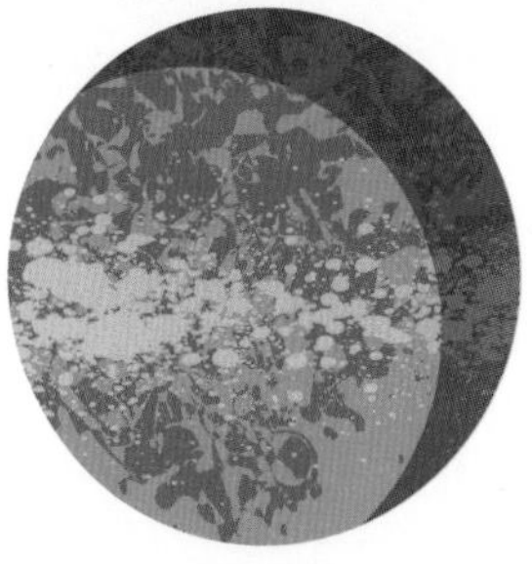

VENUS

HAS EARTH EVER BEEN WARMER THAN IT IS TODAY?

Near the end of the Mesozoic Era (the time of the dinosaurs), Earth was so warm that there was no ice at the poles in summer and sea level was 560ft (170m) higher than today.

GREENHOUSE GASES

RE-EMITTED BACK BY WARM GREENHOUSE GASES

6 Downward re-emission
Some of the infrared energy re-emitted by Earth is trapped by greenhouse gases in the atmosphere. The gases warm up and radiate heat back toward Earth's surface, raising global temperatures.

GREENHOUSE GASES IN THE ATMOSPHERE IN 2013 (MEASURED IN PARTS PER BILLION—PPB)

What are the culprits?

The main greenhouse gases in Earth's atmosphere are water vapor, carbon dioxide, methane, nitrous oxide, and ozone. The molecular structure of these gases allows them to absorb energy from infrared radiation, heat up, and then re-emit radiation to keep the planet warm. Some gases absorb heat better than others because of the way their molecules interact with heat radiation. This means that some gases, even though there are less of them in the atmosphere, have a more potent greenhouse effect than others.

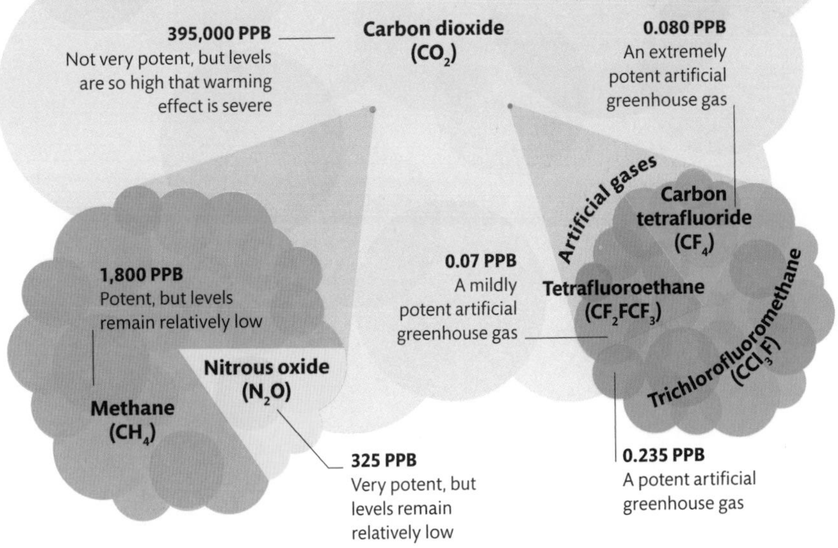

Climate change

Climates are always changing for natural reasons. These changes take place slowly, over thousands or millions of years. But we are now living in a period of rapid climate change caused by pollution of the atmosphere with gases that increase the greenhouse effect (see pp.244–45).

HOW MUCH COULD SEA LEVELS RISE?

If the melting polar ice sheets start to collapse, sea levels could rise by up to 80ft (25m), swamping coastal cities such as London, New York, Tokyo, and Shanghai.

What's happening?

The world is getting warmer. Temperatures have been rising since at least 1910, and 16 of the 17 warmest years on record have occurred since 2001. Meanwhile, analysis of the atmosphere since 1958 shows a steady rise in carbon dioxide (CO_2), the most important of the greenhouse gases that keep the planet warm. The extra CO_2 has been generated by our modern, energy-hungry way of life.

On the rise

Average global air temperatures have been recorded since the late 19th century. There have been ups and downs, but the trend is upward. It is closely matched by the increase in atmospheric CO_2.

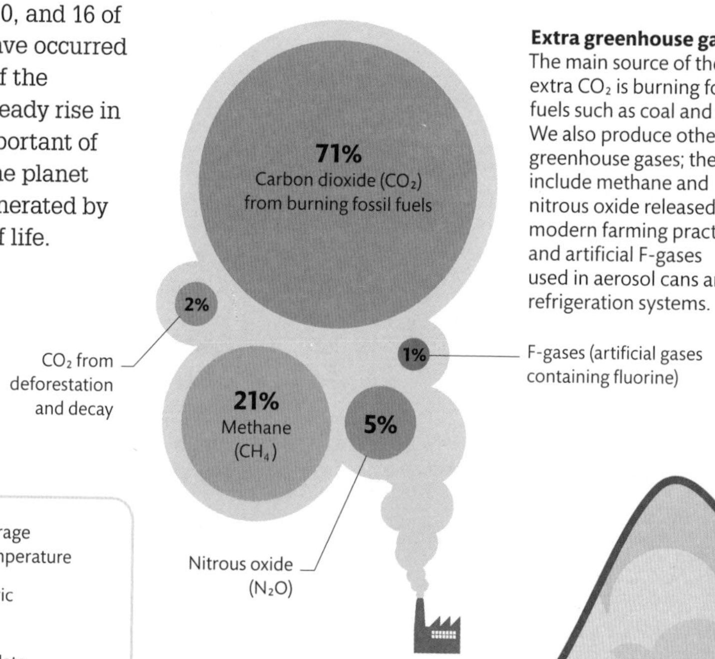

Extra greenhouse gases

The main source of the extra CO_2 is burning fossil fuels such as coal and oil. We also produce other greenhouse gases; they include methane and nitrous oxide released by modern farming practices, and artificial F-gases used in aerosol cans and refrigeration systems.

KEY

Average temperature has been recorded since 1880. Historic CO_2 levels are measured by analyzing tree rings and ice cores.

- Global average surface temperature
- Atmospheric CO_2 levels
- Projected data

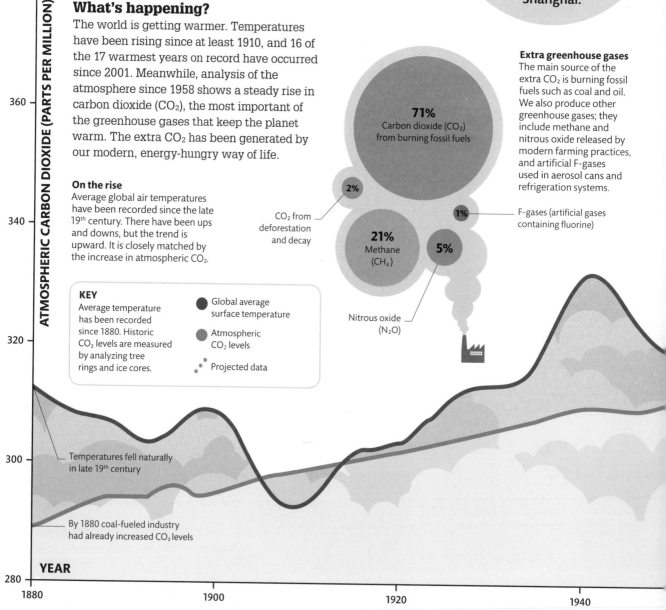

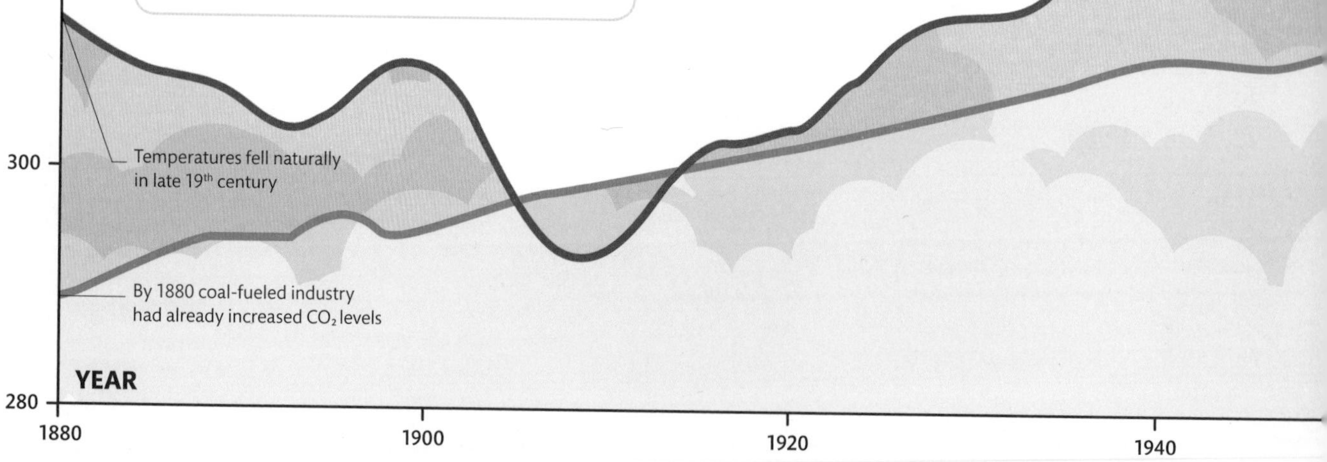

Vicious circles

If temperatures continue to rise, this could trigger feedback effects that will make the problem worse. For example, deforestation of tropical rainforests means there are fewer trees to remove CO_2 from the atmosphere. Higher atmospheric CO_2 levels add to global warming and alter atmospheric circulation systems, leading to prolonged drought and yet more tropical rainforest dieback. Other feedback effects involve the release of seabed methane and melting of Arctic sea ice.

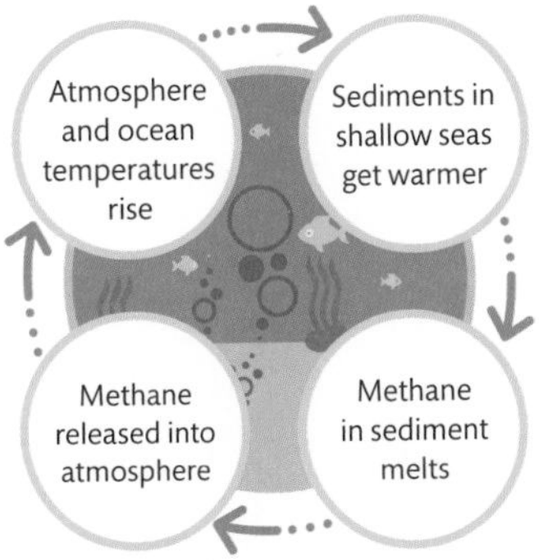
Atmosphere and ocean temperatures rise
Sediments in shallow seas get warmer
Methane in sediment melts
Methane released into atmosphere

SEABED METHANE RELEASE

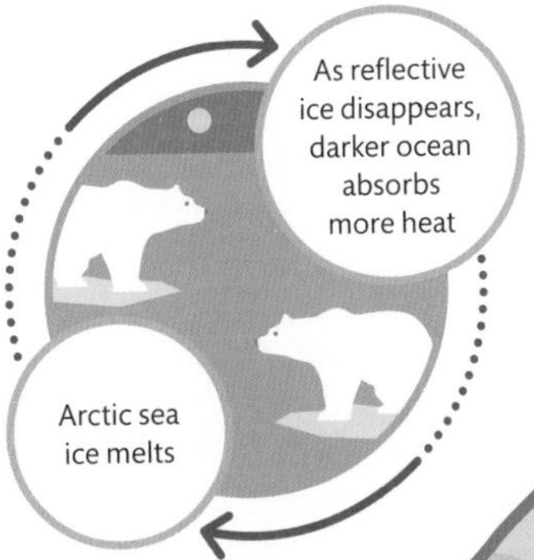
As reflective ice disappears, darker ocean absorbs more heat
Arctic sea ice melts

ARCTIC SEA ICE MELT

2016 WAS THE WARMEST YEAR ON RECORD

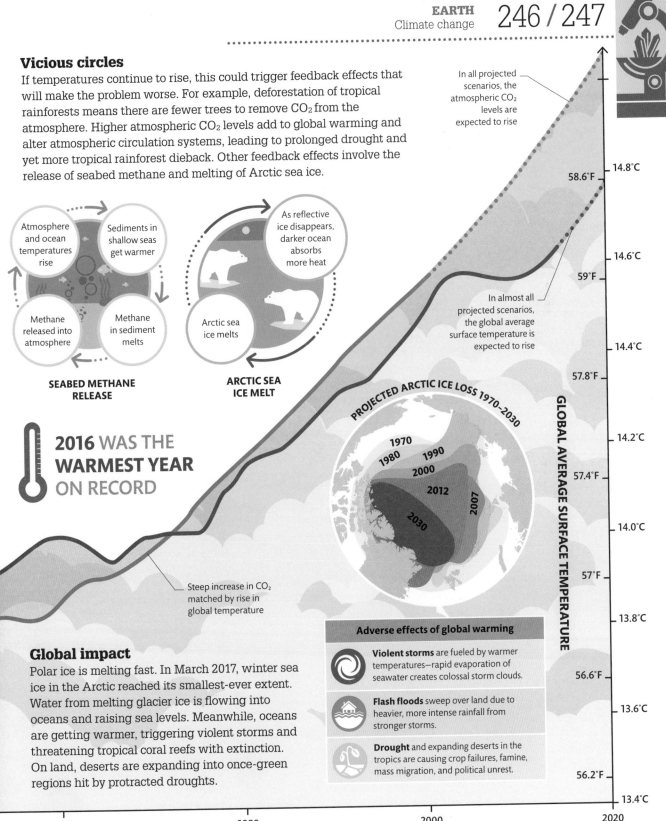

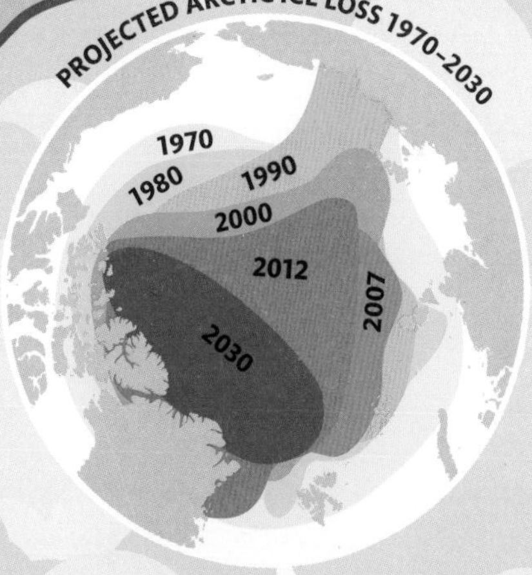
PROJECTED ARCTIC ICE LOSS 1970–2030
1970
1980
1990
2000
2012
2007
2030

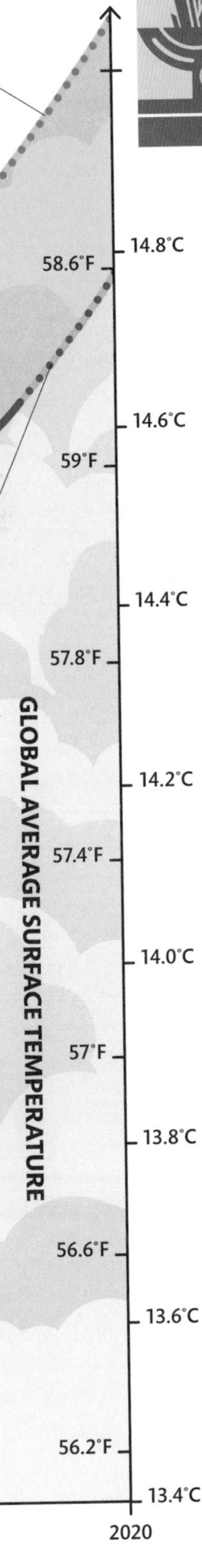
In all projected scenarios, the atmospheric CO_2 levels are expected to rise
In almost all projected scenarios, the global average surface temperature is expected to rise
Steep increase in CO_2 matched by rise in global temperature
GLOBAL AVERAGE SURFACE TEMPERATURE
14.8°C
58.6°F
14.6°C
59°F
14.4°C
57.8°F
14.2°C
57.4°F
14.0°C
57°F
13.8°C
56.6°F
13.6°C
56.2°F
13.4°C
1960
1980
2000
2020

Global impact

Polar ice is melting fast. In March 2017, winter sea ice in the Arctic reached its smallest-ever extent. Water from melting glacier ice is flowing into oceans and raising sea levels. Meanwhile, oceans are getting warmer, triggering violent storms and threatening tropical coral reefs with extinction. On land, deserts are expanding into once-green regions hit by protracted droughts.

Adverse effects of global warming
Violent storms are fueled by warmer temperatures—rapid evaporation of seawater creates colossal storm clouds.
Flash floods sweep over land due to heavier, more intense rainfall from stronger storms.
Drought and expanding deserts in the tropics are causing crop failures, famine, mass migration, and political unrest.

Index

Page numbers in **bold** refer to main entries.

Q

R

S

T

Acknowledgments

DK would like to thank the following people for help in preparing this book: Michael Parkin for illustrations; Suhel Ahmed and David Summers for editorial help; Briony Corbett for design assistance; Helen Peters for indexing; and Katie John for proofreading.